# Springer Series in Materials Science

Volume 303

The Springer Series in Materials Science covers the complete spectrum of materials research and technology, including fundamental principles, physical properties, materials theory and design. Recognizing the increasing importance of materials science in future device technologies, the book titles in this series reflect the state-of-the-art in understanding and controlling the structure and properties of all important classes of materials.

More information about this series at http://www.springer.com/series/856

Ye Zhou · Yan Wang
Editors

# Perovskite Quantum Dots

## Synthesis, Properties and Applications

Springer

*Editors*
Ye Zhou
Institute for Advanced Study
Shenzhen University
Shenzhen, Guangdong, China

Yan Wang
Institute of Microscale Optoelectronics
Shenzhen University
Shenzhen, Guangdong, China

ISSN 0933-033X ISSN 2196-2812 (electronic)
Springer Series in Materials Science
ISBN 978-981-15-6639-4 ISBN 978-981-15-6637-0 (eBook)
https://doi.org/10.1007/978-981-15-6637-0



This Springer imprint is published by the registered company Springer Nature Singapore Pte Ltd.
The registered company address is: 152 Beach Road, #21-01/04 Gateway East, Singapore 189721, Singapore

# Preface

Perovskite quantum dots based materials have elicited intense research in recent years thanks to their inherent narrow emission, tunable wavelength and high photoluminescence quantum efficiency. Various types of perovskite quantum dots featuring several outstanding performances have been synthesized and utilized for implementing optoelectronic devices, and their further emerging commercial applications comprising lighting and backlight displays can be developed on account of their manufacturability challenges in stability. Within this book, recent achievements of perovskite quantum dots involving materials manufacturing, surface treating, characterization, on-chip design, the stability improving strategy and promising applications will be highlighted. The prospect towards the existed manufacturability challenges for researching perovskite quantum dot-based materials in optoelectronics will also be discussed. It is expected that our book are well-timed and instrumental for the rapid progress of academic sector of perovskite quantum dots in energy conversion, lighting, display and memory application as well as neuromorphic processing systems.

We would like to acknowledge all the authors who have contributed in this book. We also want to express our gratitude to Dr. Mengchu Huang, Praveen Anand and Nobuko Kamikaw at Springer Nature for all the help during the book editorial process and for the excellent experience of working with them. We also want to thank all the readers for their interest in our book. Through this book, we have summarized the recent progress in the research on synthesis and properties of perovskite quantum dots. The challenges and possible strategy for improving their stabilities are also pointed out, and prospects for commercial applications are evaluated. Our aim is to give a comprehensive, critical and up-to-date book here.

We hope that this book can be useful as a reference guide for researchers who work in the field, as well as an introduction for students entering the field of perovskite quantum dots and related optoelectronic devices.

Shenzhen, China

Ye Zhou
Yan Wang

# Preface

# Contents

# Contributors

**Jianhua Chang** School of Electronic and Information Engineering, Nanjing University of Information Science and Technology, Nanjing, China

**Xu Chen** State Key Laboratory of Modern Optical Instrumentation, College of Optical Science and Engineering, Zhejiang University, Hangzhou, Zhejiang, China

**Chia-Jung Cho** Institute of Organic and Polymeric Materials, Research and Development Center of Smart Textile Technology, National Taipei University of Technology, Taipei, Taiwan

**Zhaohui Dong** Shanghai Synchrotron Radiation Facility (SSRF), Shanghai Advanced Research Institute, Chinese Academy of Science, Shanghai, People's Republic of China

**Jing Feng** State Key Laboratory of Integrated Optoelectronics, College of Electronic Science and Engineering, Jilin University, Changchun, China

**Kai Gu** College of Science, Minzu University of China, Beijing, China

**He Huang** Department of Physics and Materials Science, and Centre for Functional Photonics (CFP), City University of Hong Kong, Kowloon Tong, Hong Kong, People's Republic of China;
Chair for Photonics and Optoelectronics, Nano-Institute Munich, Department of Physics, Ludwig-Maximilians-Universität (LMU), Munich, Germany

**Siyuan Huang** State Key Laboratory of Modern Optical Instrumentation, College of Optical Science and Engineering, Zhejiang University, Hangzhou, Zhejiang, China

**Tingming Jiang** State Key Laboratory of Modern Optical Instrumentation, College of Optical Science and Engineering, Zhejiang University, Hangzhou, Zhejiang, China

**Qiang Jing** Department of Electrical and Electronic Engineering, Southern University of Science and Technology, Shenzhen, China

**Chi-Ching Kuo** Institute of Organic and Polymeric Materials, Research and Development Center of Smart Textile Technology, National Taipei University of Technology, Taipei, Taiwan

**Yun-Fei Li** Tianjin Key Laboratory of Electronic Materials and Devices, School of Electronics and Information Engineering, Hebei University of Technology, Beichen District, Tianjin, People's Republic of China

**Fang-Cheng Liang** Institute of Organic and Polymeric Materials, Research and Development Center of Smart Textile Technology, National Taipei University of Technology, Taipei, Taiwan

**Lijia Liu** Department of Chemistry, University of Western Ontario, London, ON, Canada

**Xiang Liu** School of Electronic and Information Engineering, Nanjing University of Information Science and Technology, Nanjing, China

**David Parobek** Department of Chemistry, Texas A&M University, Texas, USA

**Hongshang Peng** College of Science, Minzu University of China, Beijing, China

**Daniel Rossi** Center for Nanomedicine, Institute for Basic Science (IBS), and Graduate Program of Nano Biomedical Engineering, Advanced Science Institute, Yonsei University, Seoul, Republic of Korea

**Dong Hee Son** Department of Chemistry, Texas A&M University, Texas, USA

**Hong-Bo Sun** State Key Laboratory of Integrated Optoelectronics, College of Electronic Science and Engineering, Jilin University, Changchun, China;
State Key Lab of Precision Measurement Technology and Instruments, Department of Precision Instrument, Tsinghua University, Haidian, Beijing, China

**Yuan Tao** School of Electronic and Information Engineering, Nanjing University of Information Science and Technology, Nanjing, China

**Zhi Tao** School of Electronic and Information Engineering, Nanjing University of Information Science and Technology, Nanjing, China

**Yue Tian** State Key Laboratory of Modern Optical Instrumentation, College of Optical Science and Engineering, Zhejiang University, Hangzhou, Zhejiang, China

**Loganathan Veeramuthu** Institute of Organic and Polymeric Materials, Research and Development Center of Smart Textile Technology, National Taipei University of Technology, Taipei, Taiwan

**Manikandan Venkatesan** Institute of Organic and Polymeric Materials, Research and Development Center of Smart Textile Technology, National Taipei University of Technology, Taipei, Taiwan

**Kai Wang** Department of Electrical and Electronic Engineering, Southern University of Science and Technology, Shenzhen, China

**Yue Wang** College of Materials Science and Engineering, Nanjing University of Science and Technology, Nanjing, China

**Siyang Xia** College of Materials Science and Engineering, Nanjing University of Science and Technology, Nanjing, China

**Xiangxing Xu** School of Chemistry and Materials Science, Nanjing Normal University, Nanjing, China

**Dongdong Yan** Key Laboratory of Optoelectronic Technology & Systems (Ministry of Education), Chongqing University, Chongqing, China

**Mu Yang** College of Science, Minzu University of China, Beijing, China

**Yang (Michael) Yang** State Key Laboratory of Modern Optical Instrumentation, College of Optical Science and Engineering, Zhejiang University, Hangzhou, Zhejiang, China

**Linwei Yu** School of Electronic Science and Engineering, Nanjing University, Nanjing, China

**Zhigang Zang** Key Laboratory of Optoelectronic Technology & Systems (Ministry of Education), Chongqing University, Chongqing, China

**You Zhang** School of Electronic and Information Engineering, Nanjing University of Information Science and Technology, Nanjing, China

**Haiguang Zhao** College of Physics & State Key Laboratory of Bio-Fibers and Eco-Textiles, Qingdao University, Qingdao, People's Republic of China

# Chapter 1
# Synthesis of Perovskite Nanocrystals

**He Huang**

**Abstract** This chapter gives an overview of the synthesis of perovskite nanocrystals. The most commonly used methods are the precipitation method and the injection. Other methods are also mentioned briefly with a few examples.

## 1.1 Introduction

Metal halide perovskite materials have attracted great scientific and technological interest in recent years, due to their attractive optical and electronic properties [1–4] combined with low cost and solution processability, which greatly contribute to their high potential for photovoltaic and light-emitting devices [1, 5–11]. It is worth mentioning that the earliest scientific work on perovskites was conducted in the late 70s by Weber, who characterized these materials in detail [12–14]. Significant advances were made in the 90s and 2000s by Mitzi et al. leading to the first implementation of perovskites in devices [15–17]. Interest seemed to subside for a while until the first reports on the utilization of organic–inorganic lead halide perovskites as sensitizers in solar cells in 2009 [5] and the demonstration of their light emission in 2012 [18]. Since then many studies have contributed towards the fundamental understanding of this fascinating material as well as to its various technological applications.

Halide perovskites have a general formula of $ABX_3$, where A and B are monovalent and divalent cations, respectively, and X is a monovalent halide (Cl, Br, I) anion. The three-dimensional crystal structure of lead halide perovskites is shown in Fig. 1.1a, where the B cation, commonly Pb, Sn, or (Ag with Bi for double perovskite), is coordinated to six halide ions in an octahedral configuration. The octahedra are corner-sharing, with the A cation located in between those octahedra [19]. Lead

H. Huang (✉)
Department of Physics and Materials Science, and Centre for Functional Photonics (CFP), City University of Hong Kong, Kowloon Tong, Hong Kong, People's Republic of China
e-mail: hh@live.cn

Chair for Photonics and Optoelectronics, Nano-Institute Munich, Department of Physics, Ludwig-Maximilians-Universität (LMU), Königinstr. 10, 80539 Munich, Germany

 
Y. Zhou and Y. Wang (eds.), *Perovskite Quantum Dots*, Springer Series in Materials Science 303, https://doi.org/10.1007/978-981-15-6637-0_1

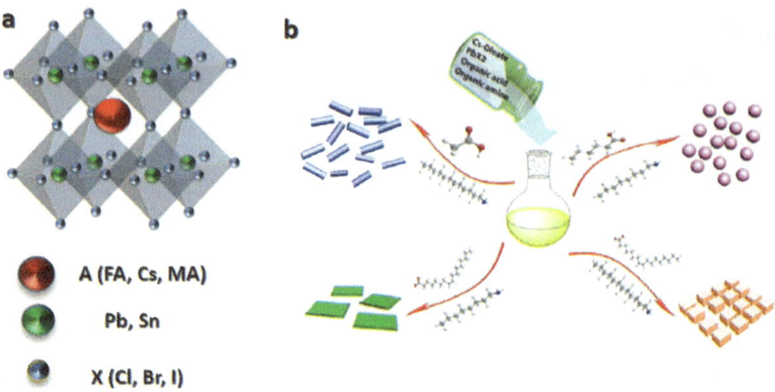


**Fig. 1.1** **a** Illustration of the perovskite crystal structure. Adapted from Ref [27]. **b** Tuning of the dimensionality of inorganic perovskites by varying the organic ligands used. Reproduced from Ref. [28]

halide perovskites are further classified into either organic–inorganic (hybrid) or all-inorganic, depending on whether the A cation is an organic molecule, methylammonium ($CH_3NH_3^+$) and formamidinium ($FA^+$), or an inorganic cation (commonly $Cs^+$), respectively. The optical and electronic properties of perovskites are tunable by varying the composition of constituted halide ions and to a smaller degree of the cations [20, 21]. In addition, the size and dimensionality of perovskites can also be used to tune their optical properties, similar to conventional metal chalcogenide semiconductors [22, 23]. Perovskites can be obtained with dimensionality ranging from the 3D to 0D by varying the synthetic conditions used (Fig. 1.1b). Perovskite nanocrystals (NCs) can exhibit very high photoluminescence quantum yields (PLQY) partly due to their defect tolerance [21–26].

## 1.2 Early Works

Methylammonium lead halide perovskite ($CH_3NH_3PbX_3$) NCs were first explored as components of mesoporous titania-based dye-sensitized solar cells in 2009 [5]. By spin-coating a precursor solution containing methylammonium halide and $PbX_2$ onto a $TiO_2$ thin film, perovskite NCs of 2–3 nm in size were formed. The first report on PL of $CH_3NH_3PbBr_3$ NCs was obtained from mesoporous $Al_2O_3$ by Kojima et al. [18]. As shown in Fig. 1.2a, taking advantage of quick vaporization with the organic solvent, perovskite NCs formed into mesoporous $Al_2O_3$ structure. The SEM image (Fig. 1.2b) shows the perovskite NCs are separately distributed on $Al_2O_3$ surface. The formation of perovskite on $Al_2O_3$ film can be observed by strong green PL during the spin-coating process in Fig. 1.2c. The two methods were similar, with a precursor solution spin-coated onto a porous $Al_2O_3$ film and a rapid crystallization of

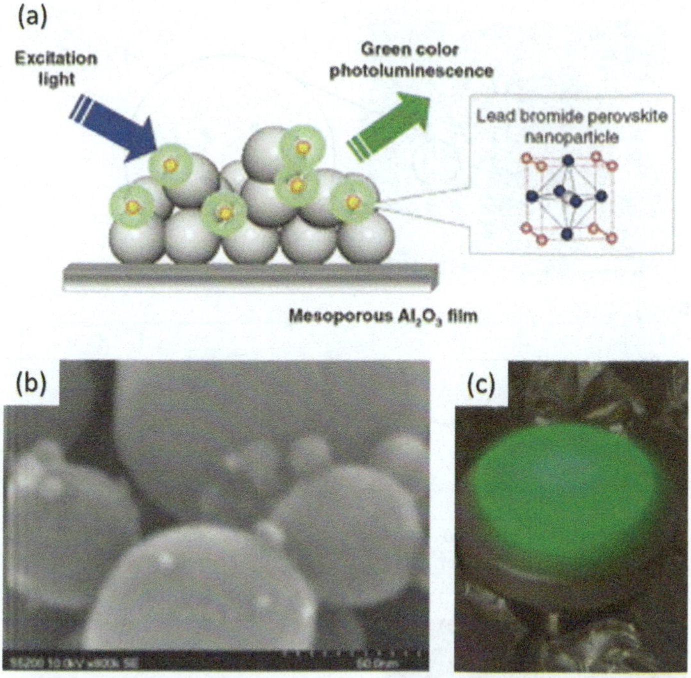


**Fig. 1.2** **a** Illustration for the synthesis process of perovskite NCs on a mesoporous $Al_2O_3$ film. **b** SEM image for $CH_3NH_3PbBr_3$ NCs in $Al_2O_3$ surfaces. **c** PL of $CH_3NH_3PbBr_3$ NCs after synthesis. Adapted from Ref. [18]

the perovskite upon evaporation of the solvent in the latter case. This also constituted the first report on the preparation of luminescent perovskite NCs.

## 1.3 Precipitation Method

The first solution-based synthesis of colloidal $CH_3NH_3PbBr_3$ NCs was carried out by the Pérez-Prieto and Galian group, who employed medium length alkyl chain organic ammonium cations, octylammonium bromide and octadecylammonium bromide, as capping ligands to obtain luminescent NCs via the solvent-induced reprecipitation method [26]. The role of these ligands was to provide a self-termination of the crystallization, leading to the formation of discrete nanoparticles in solution. The NC dispersions were described quite stable with surviving for at least three months. The absorption and PL peaks of these highly crystalline (Fig. 1.3a) $CH_3NH_3PbBr_3$ NCs were at 527 nm and 530 nm, respectively (Fig. 1.3b), with a PL QY of approximately 20%. The same group further enhanced the PL QY of these NCs to 83% by optimizing the molar ratio of octylammonium bromide: methylammonium bromide:$PbBr_2$ (8:12:5) in a typical reprecipitation method while maintaining the 1-octadecene:$PbBr_2$ molar

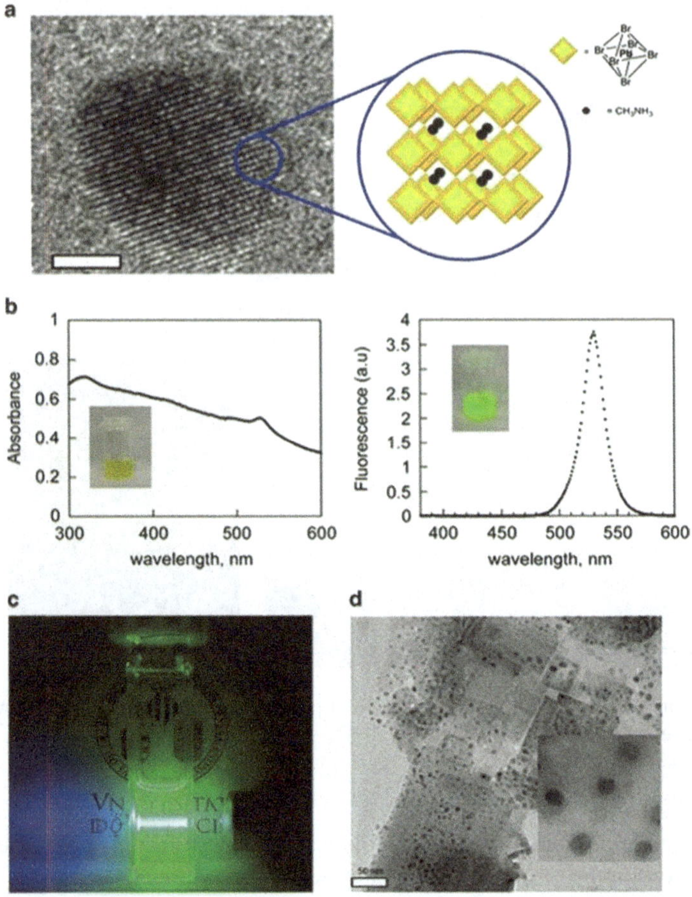


**Fig. 1.3** **a** HRTEM image of a single, highly crystalline $CH_3NH_3PbBr_3$ NC (scale bar 2 nm), shown along with a schematic representation of the perovskite unit cell. **b** UV— visible absorption (left) and PL spectra (right) of $CH_3NH_3PbBr_3$ NCs in toluene. Insets show photographs of the respective solution under ambient light and 365 nm UV lamp. Reproduced from Ref. [26]. **c** Image of the toluene dispersion of $MAPbBr_3$NCs under UV-laser pointer excitation. **d** TEM images of perovskite NCs. Adapted from Ref. [24]

ratio of 62.6:1.0 [24]. They focused on improving the organic capping of the bromide perovskite NCs, taking into account the fact that surface states would be highly accessible for passivation treatment. This resulted in easily dispersible, strongly luminescent $CH_3NH_3PbBr_3$ NCs (Fig. 1.3c). Some rectangular features were visible in electron microscopy images along with spherical nanoparticles, but especially also behind them (Fig. 1.3d), hinting at a more complex morphology of the obtained NCs.

This synthetic approach was subsequently adopted by several groups, leading for example to the demonstration of efficient perovskite-based chemical sensors [29]. Luo et al. used either octylammonium bromide or octadecylammonium bromide to produce perovskite NCs of 3.9 or 6.5 nm size, respectively, by means of the reprecipitation method [30]. Tyagi et al. employed a similar approach, synthesizing colloidal nanoplatelets of $CH_3NH_3PbBr_3$ [31]. They claimed that the excitonic-like features previously attributed to quantum confinement in spherical $CH_3NH_3PbBr_3$ nanoparticles were in fact a property of the nanoplatelets with decremental thickness, down to a single unit cell. By systematically varying the ratio of the organic cations used (octylammonium and methylammonium). Sichert et al. showed how the perovskite nanocrystals gradually decrease in thickness from large cubic crystals to ultrathin nanoplatelets down to a thickness of only a single perovskite unit cell [23]. The octylammonium molecule is too large to fit into the perovskite crystal structure and so leads to termination of the crystal growth upon binding to the perovskite, however predominantly in one direction.

Zhang et al. developed a ligand-assisted re-precipitation (LARP) technique to produce brightly luminescent colloidal $CH_3NH_3PbBr_3$NC with a PL QY up to 70% [21]. Unlike the Pérez-Prieto group, [26] they choose a reverse way of the reagent mixing: The nanoparticles were formed with n-octylamine and oleic acid as ligands through addition of the perovskite precursors in a good solventsdimethylformamide (DMF) or gamma-butyrolactone (GBL), into a bad solvent, *e.g.* toluene or acetone, as shown in Fig. 1.4a. It was assumed that n-octylamine controls the kinetics of crystallization and mainly contributes to the size control of NCs formed, while oleic acid suppresses aggregation effects and ensures their colloidal stability. From TEM studies, $CH_3NH_3PbBr_3$ NCs were quasi-spherical and had an average diameter of $3.3 \pm 0.7$ nm. The authors reasoned that the increased PL QY of the NCs with respect to micrometer-sized bulk perovskite particles resulted from higher excitonic binding energies of up to nearly 400 meV, when the size of the NCs approached the Bohr radius of the respective perovskite, on the order of 13–30 Å, depending on the material [23, 32, 33]. By replacing the bromide anion with chloride or iodide, or by using mixtures of the three anions, the authors were able to obtain perovskite NC dispersions emitting throughout the full range of the visible spectrum (Fig. 1.4b). The average PL lifetimes of these NCs were considerably shorter than for the respective bulk films.

Signs of strong 3D quantum confinement in 0D perovskite NCs were observed in the work by Huang et al. (Fig. 1.4c) [25]. Herein, the re-precipitation technique was used to prepare $CH_3NH_3PbBr_3$ NCs with tunable emission by altering the temperature of the bad solvent. The PL emission wavelength of the resulting NCs ranged from 475–520 nm, with a high PL QY of 74–93%. Sichert et al. showed that the perovskite crystal structure can be progressively reduced in a lateral dimension, yielding quasi-2D and ultimately 2-D nanoplatelets with a thickness of only a single unit cell [23]. By varying the ratio of two organic cations, the commonly used methylammonium and the significantly longer octylammonium, the thickness of platelets could be controlled, resulting in pronounced quantum size effects in $CH_3NH_3PbBr_3$ nanoplatelets [23] with an integer amount of layers varying from 1 to 7. Similar effects have been also reported by other groups [31, 35–37]. For instance, Snaith and co-workers used the same approach to demonstrate tunable PL from $MAPbI_3$ NCs

[38]. The increase of OA content in the reaction medium resulted similarly in the formation of nanoplatelets.

Besides of the commonly used hot-injection technique, room temperature reprecipitation methods have also been used for the synthesis of $CsPbX_3$ NCs of different morphologies. For instance, Li et al. showed that $CsPbX_3$ NCs can be synthesized at room temperature, similar to the LARP method used for the $MAPbX_3$ system (Fig. 1.4d) [34]. Manna and co-workers have demonstrated the synthesis of quantum confined $CsPbBr_3$ nanoplatelets at room temperature [39]. In addition, it was shown by Sun et al. [28] that the shape of the NCs could be controlled using different

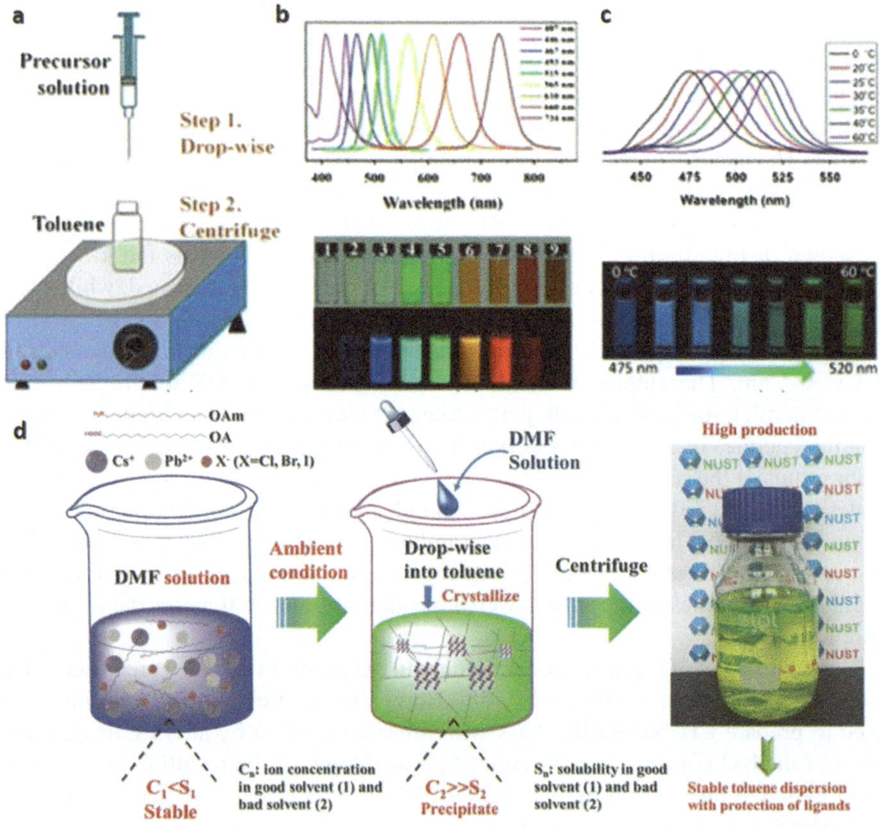


**Fig. 1.4** **a** Schematics of LARP technique used to produce $CH_3NH_3PbX_3$ NCs. **b** PL spectra and optical images of color-tunable $CH_3NH_3PbX_3$NCs produced by LARP technique, under room light and UV lamp. Reproduced from Ref. [21]. **c** PL spectra and photograph of size-tunable emission of strongly emitting $CH_3NH_3PbX_3$ NCs, where the size was controlled by the reaction temperature. Reproduced from Ref. [29]. **d** Perovskite precursor solution in DMF (or DMSO) with high concentration. The dropping of a precursor solution triggered the reaction process to obtain CsPbBr3 NC dispersion. Reproduced with permission. Reproduced from Ref. [34]

ligand molecules. In this regard, $CsPbX_3$ NCs of different shapes, such as spherical dots, nanocubes, nanorods, and nanoplatelets were prepared by using different organic acids and amine ligands. The room-temperature solution-phase synthesis of nanoplatelets can be easily scaled up as shown by Yu and co-workers [40]. Driven by the wish to replace the toxic lead, Jellicoe et al. synthesized $CsSnX_3$ perovskite NCs, with optical properties tunable both by quantum confinement and halide composition [41]. Qiao's group used primary amines (L-type ligands) to stabilize a $PbBr_x$-rich surface with a room-temperature reprecipitation method using L-type ligands (OAm, n-octylamine, or undecylamine) as the sole capping ligand has been developed to synthesize high-quality $CsPbBr_3$ NCs [42].

Methods with similar concept such as emulsion synthesis [43] and electrospray technique coupled antisolvent-solvent extraction [44] have been used for the synthesis of perovskite NCs of different morphologies, mostly yielding rather polydisperse nanoparticles. It has been shown that nearly monodisperse blue-emitting nanoplatelets and green-emitting nanorods can be prepared in 1-octadecene at high temperature [36]. Micro- and nanocrystals comprising layered quasi-2D perovskites in the form of $A_2PbX_4$, $A_4PbX_6$ and $A(A')_2Pb_2X_7$ were also reported to be luminescent [45–47]. Zhu et al. reported different morphologies for $CH_3NH_3PbX_3$ NCs with shape-correlated PL by varying the solvents and using different capping ligands [48].

## 1.4 Injection Method

Perovskite NCs with all-inorganic cations have been receiving significant attention, partly due to their higher stability compared to the hybrid organic–inorganic perovskites. Kovalenko's group was first to report on cesium lead halide perovskite NCs ($CsPbX_3$, X = Cl, Br, and I), with a high degree of compositional bandgap engineering [22]. They were produced as monodisperse nanocubes with edge lengths of 4–15 nm using the arrested precipitation of Cs-oleate and Pb(II)-halide precursors in octadecene. It was found that the size of the NCs decreases by decreasing the reaction temperature. Mixed-halide perovskite NCs could be conveniently produced by varying the ratios of precursor salts ($PbCl_2/PbBr_2$ or $PbBr_2/PbI_2$). Initially, it is believed that $CsPbX_3$ NCs crystallized in the cubic phase (Fig. 1.5a, b), however, with more detailed characterization, the actual crystallization is the orthorhombic phases [3]. The bandgap energies and emission spectra of these NCs were tunable over the entire visible spectral region of 410–700 nm by varying the halide composition as well as the size of the NCs (Fig. 1.5c, d). The PL of $CsPbX_3$ NCs was characterized by narrow emission linewidths of 12–s42 nm, a wide color gamut covering up to 140% of the National Television System Committee (NTSC) color standard, high PL QYs of up to 90%, and PL lifetimes in the range of 1–29 ns (Fig. 1.5e). The similar method was used to prepare $CH_3NH_3PbX_3$ and $FAPbX_3$ NCs [36, 49]. Later on, by means of a droplet-based microfluidic platform, allowing for online absorption/PL measurements, the same group explored the formation mechanism of these perovskite NCs [50]. Koolyk et al. investigated the growth process of cesium lead halide perovskite NCs by time-dependent TEM characterization [51]. They found that the growth kinetics are different for $CsPbBr_3$ and $CsPbI_3$, providing a model for each growth process. From the Full Width at Half Maximum (FWHM) of the PL spectra it was observed that $CsPbI_3$ NCs exhibited size focusing during the first 20s of growth, followed by a subsequent size de-focusing (the FWHM increased due to the broadening size distribution). On the contrary, $CsPbBr_3$ NCs exhibited a de-focusing of the size distribution from the beginning of the reaction and no size focusing occurred within the full reaction time of 40s. Yassitepe et al. reported hot-injection method with only oleic acid and quaternary alkylammonium halides as precursors to enhances the colloidal stability [52]. Imran et al. showed that benzoyl halides as halide sources can be injected in a solution of metal cations for the synthesis of perovskite NCs with improved properties due to alkylammonium ions replace Cs + on the surface [53].

Different morphology are also obtained by hot-injection method. A prolonged reaction at higher temperatures leads to the formation of light-emitting single crystalline $CsPbX_3$ NWs that crystallize in the orthorhombic phase as reported by Yang and co-workers (Fig. 1.5f–h) [54]. The initially formed cubic NCs transformed into NWs with a few percent of nanosheets as a side product for the prolonged reaction time. The same group future obtained single crystal ultrathin CsPbBr3 NWs by step-wise purifications [55]. Imran et al. used mixture of alkyl amines and a short alkyl carboxylic acid as growth medium to get tunable width of NWs [56]. Song et al. reported monolayer and few-layer of $CsPbX_3$ nanosheets [57] while Bekenstein et al. reported quasi-two-dimensional nanoplates and their self-assemble [58]. Doping of

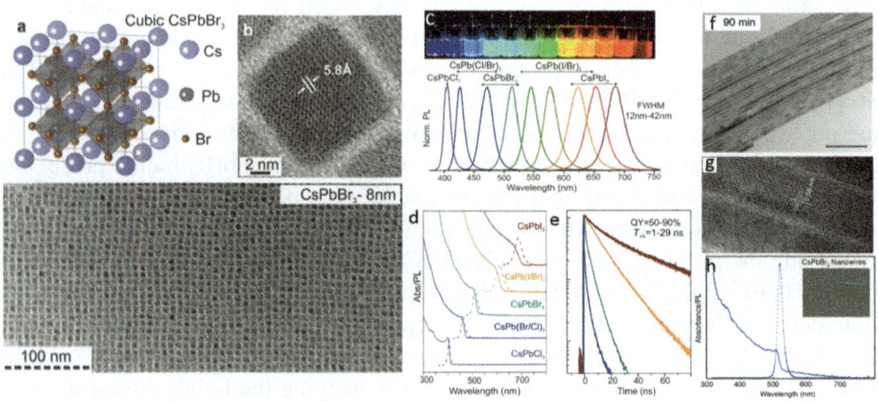


**Fig. 1.5** **a** Schematics of the cubic $CsPbBr_3$ perovskite lattice. **b** HRTEM and TEM images of $CsPbBr_3$ NCs. **c** Photograph of $CsPbX_3$NC solutions taken under UV lamp, with size- and composition-tunable bandgap and corresponding PL spectra. **d** Absorption and PL spectra of $CsPbX_3$NCs with different X anions employed. **e** Time-resolved PL decays of $CsPbBr_3$ NCs. Reproduced from Ref. [22]. **f** TEM and **g** HRTEM image of $CsPbBr_3$ NWs. **h** Absorption and PL spectra of $CsPbBr_3$ NWs; inset shows optical image of a single NWs excited under a laser beam. Reproduced from Ref. [54]

perovskites are also reported by different group. For example, Mn doping are shown by Liu et al. [59] and Parobek et al. with $CsPbX_3$ [60]. Pan et al. reported doping of Lanthanide with $CsPbX_3$[61]. In order to reduce or replace lead, several attempts are shown with hot-injection method. Jellicoe et al. reported CsSnX3 perovskite NCs [41]. By choosing precursor and reaction temperature, Wang et al. obtained $CsSnBr_3$ hollow cubic nanocages [62]. Pal et al. and Pradhan et al. replace lead with antimony to get $Cs_3Sb_2X_9$ NCs [63, 64]. Zhou et al. and Creutz et al. reported lead free double perovskite NCs in order to reduce the toxic of lead [65, 66].

Other methods with similar concept are also reported. Imran et al. described a ligand exchange strategy involving the displacement of both cationic and anionic ligands on native model systems of CsPbBr3 NCs, which are exclusively coated with Cs-oleate [67]. Huang et al. reported on the spontaneous crystallization of perovskite NCs in nonpolar organic media at ambient conditions by simple mixing of precursor–ligand complexes. The shape of the NCs can be controlled from nanocubes, nanoplatelets and nanorods with different compositions [68, 69].

## 1.5 Methods Beyond Precipitation and Hot-Injection

### *1.5.1 Sonication*

Feldmann et al. reported the large-scale tip-unltrasonication method for the synthesis of $CsPbX_3$ perovskite NCs and NPLs [70]. The setup was shown as Fig. 1.6a. They can get different haide compostion fo $CsPbX_3$ (Fig. 1.6b). The same group further used this method to obtain nanwires [71], cutting nanowire [72] for nanorods. Huang et al. also reported tip-unltrasonication method for the synthesis of $CH_3NH_3PbX_3$ perovskite NCs, employing a mixture of ligands oleic acid and oleylamine as coordinating solvents (Fig. 1.6c) [73]. This approach avoids the use of any polar solvents, skips multiple reaction steps by employing a simple ultrasonic treatment of the perovskite precursors, and yields rather monodisperse blue-, green-, and red-emitting methylammonium lead halide nanocrystals with a high photoluminescence quantum yield (up to 72% for the green-emitting nanocrystals) and remarkably improved stability.

### *1.5.2 Solvothermal*

Qiao et al. proposed a solvothermal approach for the first time. The authors can get $CsPbX_3$ nanocrystals with tunable and bright photoluminescent (PL) properties, controllable composition, and morphology (Fig. 1.7a) [74]. The ultrathin $CsPbX_3$ (X = Cl/Br, Br, and Br/I) nanowires (with a diameter as small as $\approx$2.6 nm) can be also

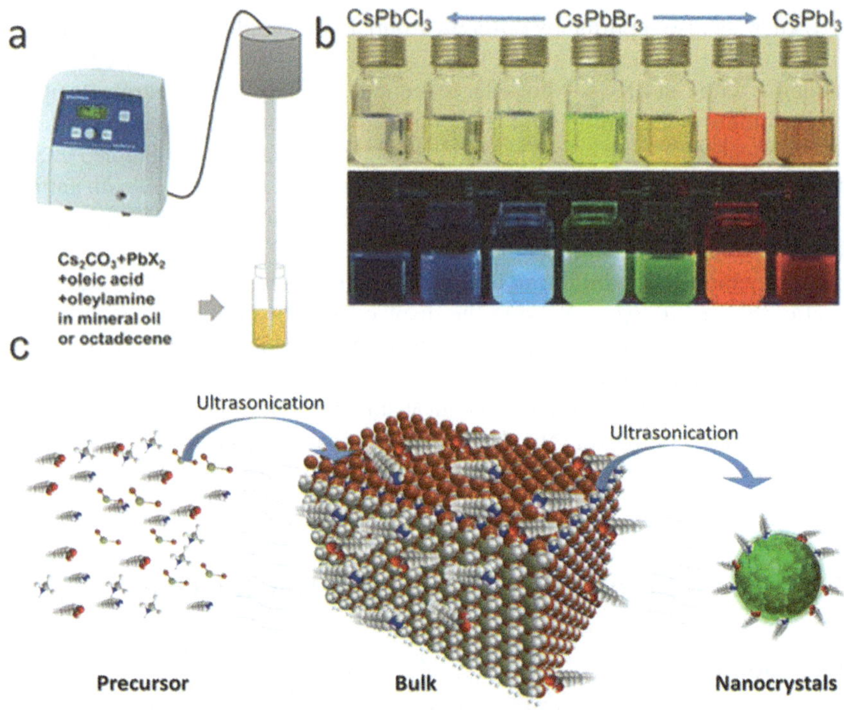


**Fig. 1.6** **a** Synthesis of CsPbX3 NCs through single-step tip sonication. **b** Photograph of colloidal dispersions of CsPbX3 NCs with different halide (X = Cl, Br, and I) compositions in hexane under room light (top) and UV light (bottom l = 367 nm). Reproduced from Ref. [70] **c** The top-down synthetic method towards methylammonium lead halide perovskite NCs. Two precursors (such as $CH_3NH_3Br$ and $PbBr_2$) are mixed under ultrasonication with oleic acid and oleylamine to quickly form bulk $CH_3NH_3PbBr_3$, which is then broken into perovskite NCs under continuous ultrasonication with an assistance of these two ligands acting as coordinating solvents. Reproduced from Ref. [73]

prepared in a very high morphological yield. Tang et al. also reported the controllable synthesis of brightly emitting cesium lead halide $CsPbX_3$ (X = Cl, Br) NWs via a solvothermal method and their assembly into high-performance photodetector nanodevices. Li et al. reported Mn-doped $CsPbCl_3$ nanocrystals by solvothermal method (Fig. 1.7b) [75]. They claimed that by the solvothermal method have better stability than those prepared by traditional hot injection. The authors also used Cl-to-Br anion exchange to fabricate Mn-doped $CsPb(Cl/Br)_3$ NCs with tunable multi-color luminescence.

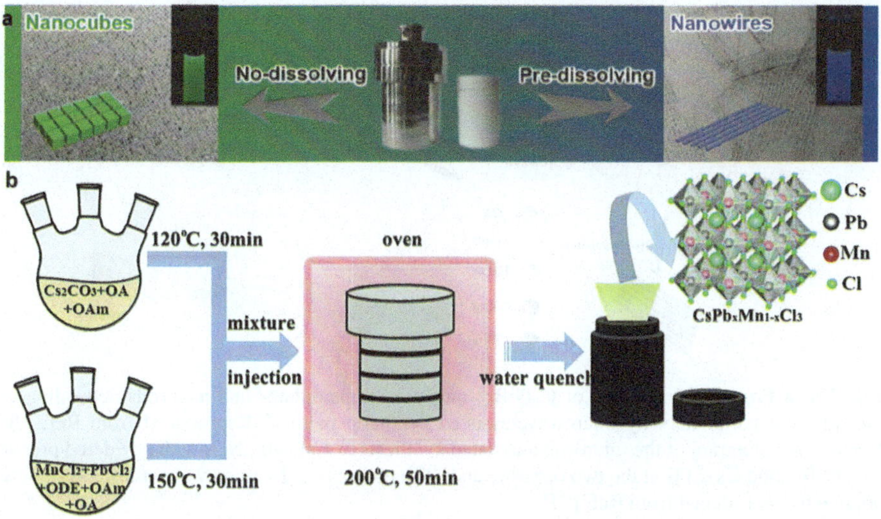


**Fig. 1.7** **a** Synthesis of $CsPbX_3$ through solvothermal approach. TEM images of $CsPbBr_3$ nanocubes on the left and nanowires on the right. Reproduced from Ref. [74] **b** Synthesis of Mn-doped $CsPbCl_3$ NCs by solvothermal method. Reproduced from Ref. [75]

### *1.5.3 Microwave*

Qiao et al. reported microwave-assisted synthesis of $CsPbX_3$ NCs with different morphologies (nanocube, nanoplate and nanorod) for the first time [76]. Without pre-dissolving precursors, low reaction temperature lead to nanoplatelets and high one lead to nanocube. The nanorods can be formed with pre-dissolved precursors (Fig. 1.8a). Later, Rogach et al. further employed microwave for a slowed-down synthesis of $CsPbBr_3$ [77]. This retards the reaction and allows authors to gather useful insights into the formation mechanism of these nanoparticles, by examining the intermediate stages of their growth. The schematic diagram of the synthetic process $CsPbBr_3$ NCs is shown in Fig. 1.8b. At the two opposite ends of a quartz boat, $PbBr_2$ and $Cs_2CO_3$ are placed. A mixture of 1-octadecene, oleic acid, and oleylamine is placed on top of powdered precursors without any disturbance. The boat was treated in a microwave oven with different times. The reactants were not mixed directly during the reaction. Yang et al. also reported microwave-assisted method for $CsPbX_3$ NCs. The authors optimized the preparation conditions to get tunable reaction rate and enhanced stability (Fig. 1.8).

### *1.5.4 Balling Milling*

Kovalenko et al. reported wet ball milling method. The colloidal NCs are obtained from bulk $APbBr_3$ mixed with solvents and oleylammonium bromide ligands.

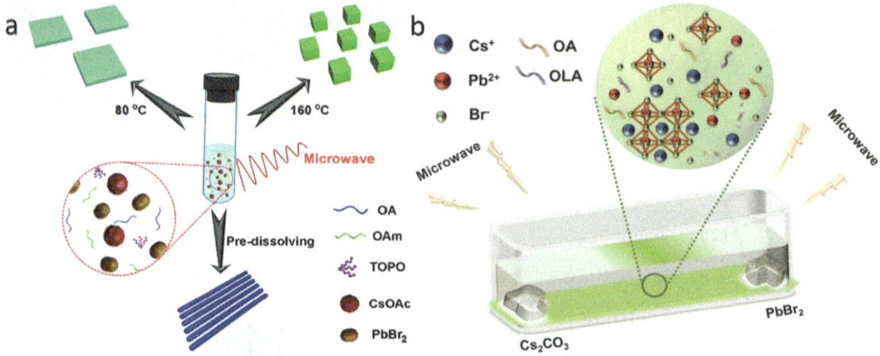


**Fig. 1.8** **a** Proposed formation of $CsPbBr_3$ nanocubes, nanoplates and nanorods depending on the different parameters by microwave-assisted synthetic method. Reproduced from Ref. [76] **b** Schematic diagram of the microwave-assisted synthesis of $CsPbBr_3$ NCs with powdered precursors $PbBr_2$ and $Cs_2CO_3$ at the two opposite ends of a quartz boat by microwave-assisted synthesis approach. Reproduced from Ref. [77]

Figure 1.9 shows the working principle of ball milling method and the quality is similar to the NCs obtained by hot-injection method [78]. Bolink et al. revealed the process mechanisms, kinetics, and possible side-effects of the dry ball milling method [79]. The authors investigated in detail the mechanochemical synthesis as a function of time. They found that dry, and fast (5 min) process leads to excellent phase-purity of fully-inorganic lead halide perovskites. Choi et al. reported ligand-assisted ball milling for MA or $FAPbBr_3$. The formation of perovskite nanocrystals as well as exfoliated nanoplatelets was preferentially developed in the presence of ligands (n-octylamine or octadecylamine).

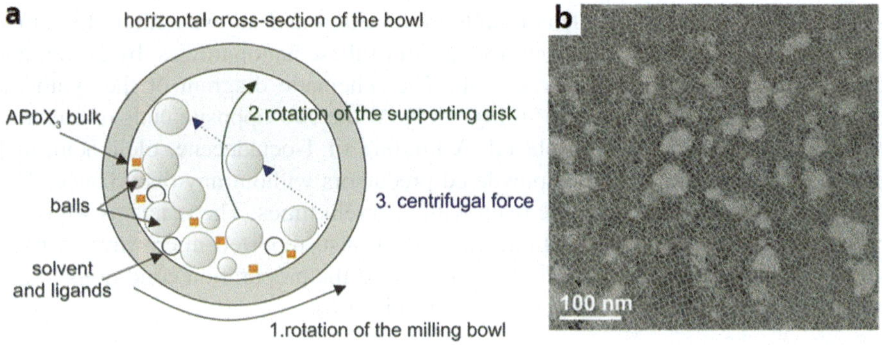


**Fig. 1.9** **a** Schematic of the working principle of the planetary ball milling. **b** TEM image of the resulting $CsPbBr_3$ NCs. Reproduced from Ref. [78]

## 1.6 Summary and Outlook

During the last few years, perovskite materials have received great attention, first in the form of thin films and afterwards as colloidal NCs. This chapter gives an overview of the common synthetic strategies for the fabrication and control over the size and morphology of both hybrid organic/inorganic and all-inorganic perovskite NCs.

In spite of the impressive progress in all the areas related to perovskite NCs which have been discussed in this chapter, there are still several challenges to be addressed in the synthesis of perovskite NCs. As for their thin film counterparts, the stability of perovskite NCs is an important issue as they tend to degrade rapidly. Degradation can happen due to several external factors, *e.g.* moisture, oxygen, high temperature, and UV-light. In terms of degradation due to moisture, this is inherent to the perovskite, which is easily dissolved in polar solvents, in particularly water. UV-light possibly enhances ion migration in the perovskite or leads to the formation of free radicals. In addition, perovskite NCs are extremely sensitive and tend to degrade or transform into other compounds (such as metallic lead) under exposure to electron beams [23]. On the other hand, inorganic metal halide perovskites seem to be more stable under low-energy electron illumination, enabling high-resolution transmission electron microscopy characterization. Nevertheless, they can still degrade under a high energy electron beam, similarly leading to the formation of small dots on the NC surface.

Apart from the stability issues, there are a number of other challenges related to perovskite NCs. Foremost is the desire to eliminate toxic lead from both hybrid and all-inorganic perovskites, either produced in nanoparticle form or as thin films. This is especially necessary for upscaling the perovskite fabrication for wide-spread photovoltaic use. There have been some attempts to use tin instead of Pb, but this element is also not nontoxic and very unstable, as Sn(II) can easily be oxidized, transforming to the Sn(IV) state. $CsSnX_3$ NCs have be prepared using synthesis methods developed for their lead-containing counterparts [41], however the particles degrade within a few hours after preparation. Finding methods for enhancing the viability of lead-free perovskites deserves more attention in the future.

A better control over dimensionality and size of perovskite NCs during synthesis is one of the critical issues for understanding structure–property relationships as well as advancing their applications. The formation of perovskites occurs extremely rapidly, making a study of the intermediate particles difficult, which could aid in understanding the growth mechanisms. While there have been some studies showing an ability to control the shape of the perovskite particles to some degree, a full understanding of their nucleation and growth mechanisms is still lacking.

**Acknowledgement** Part of the contents was adapted and modified from "Colloidal lead halide perovskite nanocrystals: synthesis, optical properties and applications" by He Huang et al. NPG Asia Mater 8, e328 (2016). https://doi.org/10.1038/am.2016.167 with CC BY 4.0.

## References


1. J. Burschka, N. Pellet, S.J. Moon, R. Humphry-Baker, P. Gao, M.K. Nazeeruddin, M. Gratzel, Sequential deposition as a route to high-performance perovskite-sensitized solar cells. Nature **499**(7458), 316–319 (2013). https://doi.org/10.1038/nature12340
2. Y. Li, X. Zhang, H. Huang, S.V. Kershaw, A.L. Rogach, Advances in metal halide perovskite nanocrystals: Synthetic strategies, growth mechanisms, and optoelectronic applications. Mater. Today **32**, 204–221 (2020). https://doi.org/10.1016/j.mattod.2019.06.007
3. M.V. Kovalenko, L. Protesescu, M.I. Bodnarchuk, Properties and potential optoelectronic applications of lead halide perovskite nanocrystals. Science **358**(6364), 745–750 (2017). https://doi.org/10.1126/science.aam7093
4. H. Huang, L. Polavarapu, J.A. Sichert, A.S. Susha, A.S. Urban, A.L. Rogach, Colloidal lead halide perovskite nanocrystals: synthesis, optical properties and applications. NPG. Asia. Mater **8**(11), e328 (2016). https://doi.org/10.1038/am.2016.167
5. A. Kojima, K. Teshima, Y. Shirai, T. Miyasaka, Organometal halide perovskites as visible-light sensitizers for photovoltaic cells. J. Am. Chem. Soc **131**(17), 6050–6051 (2009). https://doi.org/10.1021/ja809598r
6. M.M. Lee, J. Teuscher, T. Miyasaka, T.N. Murakami, H.J. Snaith, Efficient hybrid solar cells based on meso-superstructured organometal halide perovskites. Science **338**(6107), 643–647 (2012). https://doi.org/10.1126/science.1228604
7. M. Liu, M.B. Johnston, H.J. Snaith, Efficient planar heterojunction perovskite solar cells by vapour deposition. Nature **501**(7467), 395–398 (2013). https://doi.org/10.1038/nature12509
8. Z.K. Tan, R.S. Moghaddam, M.L. Lai, P. Docampo, R. Higler, F. Deschler, M. Price, A. Sadhanala, L.M. Pazos, D. Credgington, F. Hanusch, T. Bein, H.J. Snaith, R.H. Friend, Bright light-emitting diodes based on organometal halide perovskite. Nat. Nanotechnol. **9**(9), 687–692 (2014). https://doi.org/10.1038/nnano.2014.149
9. G. Xing, N. Mathews, S.S. Lim, N. Yantara, X. Liu, D. Sabba, M. Gratzel, S. Mhaisalkar, T.C. Sum, Low-temperature solution-processed wavelength-tunable perovskites for lasing. Nat. Mater. **13**(5), 476–480 (2014). https://doi.org/10.1038/nmat3911
10. H. Zhou, Q. Chen, G. Li, S. Luo, T.B. Song, H.S. Duan, Z. Hong, J. You, Y. Liu, Y. Yang, Interface engineering of highly efficient perovskite solar cells. Science **345**(6196), 542–546 (2014). https://doi.org/10.1126/science.1254050
11. National Renewable Energy Laboratory (2020) Best Research-Cell Efficiencies. https://www.nrel.gov/ncpv/images/efficiency_chart.jpg. Accessed 11 Mar 2020
12. D. Weber, The perovskite system $CH_3NH_3[Pb_n, Sn_{1-n}X3]$ (X=Cl, Br, I). Z. Naturforsch. B. **34**(7), 939–941 (1979)
13. D. Weber, $CH_3NH_3SnBr_xI_{3-x}$ (X = 0–3)—Sn(II)-System with cubic perovskite structure. Z. Naturforsch. B. **33**(8), 862–865 (1978)
14. D. Weber, $CH_3NH_3PbX_3$, a Pb(II)-System with cubic perovskite structure. Z. Naturforsch. B. **33**(12), 1443–1445 (1978)
15. D.B. Mitzi, C.A. Feild, W.T.A. Harrison, A.M. Guloy, Conducting tin halides with a layered organic-based perovskite structure. Nature **369**(6480), 467–469 (1994). https://doi.org/10.1038/369467a0
16. D.B. Mitzi, K. Chondroudis, C.R. Kagan, Organic-inorganic electronics. IBM. J. Res. Dev. **45**(1), 29–45 (2001)
17. D.B. Mitzi, C.D. Dimitrakopoulos, L.L. Kosbar, Structurally tailored organic-inorganic perovskites: optical properties and solution-processed channel materials for thin-film transistors. Chem. Mater. **13**(10), 3728–3740 (2001)
18. A. Kojima, M. Ikegami, K. Teshima, T. Miyasaka, Highly luminescent lead bromide perovskite nanoparticles synthesized with porous alumina media. Chem. Lett **41**(4), 397–399 (2012). https://doi.org/10.1246/cl.2012.397
19. J.A. Christians, P.A. Miranda Herrera, P.V. Kamat, Transformation of the excited state and photovoltaic efficiency of $CH_3NH_3PbI_3$ perovskite upon controlled exposure to humidified air. J. Am. Chem. Soc. **137**(4), 1530–1538 (2015). https://doi.org/10.1021/ja511132a

20. G.E. Eperon, S.D. Stranks, C. Menelaou, M.B. Johnston, L.M. Herz, H.J. Snaith, Formamidinium lead trihalide: a broadly tunable perovskite for efficient planar heterojunction solar cells. Energ. Environ. Sci. **7**(3), 982 (2014). https://doi.org/10.1039/c3ee43822h
21. F. Zhang, H. Zhong, C. Chen, X.G. Wu, X. Hu, H. Huang, J. Han, B. Zou, Y. Dong, Brightly luminescent and color-tunable colloidal $CH_3NH_3PbX_3$ (X = Br, I, Cl) quantum dots: potential alternatives for display technology. ACS Nano **9**(4), 4533–4542 (2015). https://doi.org/10.1021/acsnano.5b01154
22. L. Protesescu, S. Yakunin, M.I. Bodnarchuk, F. Krieg, R. Caputo, C.H. Hendon, R.X. Yang, A. Walsh, M.V. Kovalenko, Nanocrystals of cesium lead halide perovskites ($CsPbX_3$, X = Cl, Br, and I): novel optoelectronic materials showing bright emission with wide color gamut. Nano. Lett. **15**(6), 3692–3696 (2015). https://doi.org/10.1021/nl5048779
23. J.A. Sichert, Y. Tong, N. Mutz, M. Vollmer, S. Fischer, K.Z. Milowska, R. Garcia Cortadella, B. Nickel, C. Cardenas-Daw, J.K. Stolarczyk, A.S. Urban, J. Feldmann, Quantum size effect in organometal halide perovskite nanoplatelets. Nano. Lett. **15**(10), 6521–6527 (2015). https://doi.org/10.1021/acs.nanolett.5b02985
24. S. Gonzalez-Carrero, R.E. Galian, J. Pérez-Prieto, Maximizing the emissive properties of $CH_3NH_3PbBr$ 3 perovskite nanoparticles. J. Mater. Chem. A. **3**(17), 9187–9193 (2015). https://doi.org/10.1039/c4ta05878j
25. H. Huang, A.S. Susha, S.V. Kershaw, T.F. Hung, A.L. Rogach, Control of Emission Color of High Quantum Yield CH3NH3PbBr 3 Perovskite Quantum Dots by Precipitation Temperature. Adv Sci **2**(9), 1500194 (2015). https://doi.org/10.1002/advs.201500194
26. L.C. Schmidt, A. Pertegas, S. Gonzalez-Carrero, O. Malinkiewicz, S. Agouram, G. Minguez Espallargas, H.J. Bolink, R.E. Galian, J. Perez-Prieto, Nontemplate synthesis of $CH_3NH_3PbBr$ 3 perovskite nanoparticles. J. Am. Chem. Soc. **136**(3), 850–853 (2014). https://doi.org/10.1021/ja4109209
27. H. Huang, Y. Li, Y. Tong, E.P. Yao, M.W. Feil, A.F. Richter, M. Doblinger, A.L. Rogach, J. Feldmann, L. Polavarapu, Spontaneous crystallization of perovskite nanocrystals in nonpolar organic solvents: a versatile approach for their shape-controlled synthesis. Angew. Chem. Int. Ed. Engl. **58**(46), 16558–16562 (2019). https://doi.org/10.1002/anie.201906862
28. S. Sun, D. Yuan, Y. Xu, A. Wang, Z. Deng, Ligand-mediated synthesis of shape-controlled cesium lead halide perovskite nanocrystals via reprecipitation process at room temperature. ACS Nano **10**(3), 3648–3657 (2016). https://doi.org/10.1021/acsnano.5b08193
29. C. Muthu, S.R. Nagamma, V.C. Nair, Luminescent hybrid perovskite nanoparticles as a new platform for selective detection of 2,4,6-trinitrophenol. RSC. Adv. **4**(99), 55908–55911 (2014). https://doi.org/10.1039/c4ra07884e
30. B. Luo, Y.-C. Pu, Y. Yang, S.A. Lindley, G. Abdelmageed, H. Ashry, Y. Li, X. Li, J.Z. Zhang, Synthesis, optical properties, and exciton dynamics of organolead bromide perovskite nanocrystals. J. Phys. Chem. C. **119**(47), 26672–26682 (2015). https://doi.org/10.1021/acs.jpcc.5b08537
31. P. Tyagi, S.M. Arveson, W.A. Tisdale, Colloidal organohalide perovskite nanoplatelets exhibiting quantum confinement. J. Phys. Chem. Lett. **6**(10), 1911–1916 (2015). https://doi.org/10.1021/acs.jpclett.5b00664
32. K. Tanaka, T. Takahashi, T. Ban, T. Kondo, K. Uchida, N. Miura, Comparative study on the excitons in lead-halide-based perovskite-type crystals $CH_3NH_3PbBr$ 3 $C_H3N_H3Pb_I3$. Solid. State. Commun. **127**(9–10), 619–623 (2003). https://doi.org/10.1016/s0038-1098(03)00566-0
33. A. Buin, R. Comin, A.H. Ip, E.H. Sargent, Perovskite quantum dots modeled using ab initio and replica exchange molecular dynamics. J. Phys. Chem. C. **119**(24), 13965–13971 (2015). https://doi.org/10.1021/acs.jpcc.5b03613
34. X. Li, Y. Wu, S. Zhang, B. Cai, Y. Gu, J. Song, H. Zeng, $CsPbX_3$ quantum dots for lighting and displays: room-temperature synthesis, photoluminescence superiorities, underlying origins and white light-emitting diodes. Adv. Funct. Mater. **26**(15), 2435–2445 (2016). https://doi.org/10.1002/adfm.201600109
35. Y. Ling, Z. Yuan, Y. Tian, X. Wang, J.C. Wang, Y. Xin, K. Hanson, B. Ma, H. Gao, Bright light-emitting diodes based on organometal halide perovskite nanoplatelets. Adv. Mater. **28**(2), 305–311 (2016). https://doi.org/10.1002/adma.201503954

36. O. Vybornyi, S. Yakunin, M.V. Kovalenko, Polar-solvent-free colloidal synthesis of highly luminescent alkylammonium lead halide perovskite nanocrystals. Nanoscale **8**(12), 6278–6283 (2016). https://doi.org/10.1039/c5nr06890h
37. Z. Yuan, Y. Shu, Y. Tian, Y. Xin, B. Ma, A facile one-pot synthesis of deep blue luminescent lead bromide perovskite microdisks. Chem. Commun. **51**(91), 16385–16388 (2015). https://doi.org/10.1039/c5cc06750b
38. S. Pathak, N. Sakai, F.W.R. Rivarola, S.D. Stranks, J.W. Liu, G.E. Eperon, C. Ducati, K. Wojciechowski, J.T. Griffiths, A.A. Haghighirad, A. Pellaroque, R.H. Friend, H.J. Snaith, Perovskite crystals for tunable white light emission. Chem. Mater. **27**(23), 8066–8075 (2015). https://doi.org/10.1021/acs.chemmater.5b03769
39. Q.A. Akkerman, S.G. Motti, A.R. Srimath Kandada, E. Mosconi, V. D'Innocenzo, G. Bertoni, S. Marras, B.A. Kamino, L. Miranda, F. De Angelis, A. Petrozza, M. Prato, L. Manna, Solution synthesis approach to colloidal cesium lead halide perovskite nanoplatelets with monolayer-level thickness control. J. Am. Chem. Soc. **138**(3), 1010–1016 (2016). https://doi.org/10.1021/jacs.5b12124
40. K.H. Wang, L. Wu, L. Li, H.B. Yao, H.S. Qian, S.H. Yu, Large-scale synthesis of highly luminescent perovskite-related $cspb_2br$ 5 nanoplatelets and their fast anion exchange. Angew. Chem. Int. Ed. **55**(29), 8328–8332 (2016). https://doi.org/10.1002/anie.201602787
41. T.C. Jellicoe, J.M. Richter, H.F. Glass, M. Tabachnyk, R. Brady, S.E. Dutton, A. Rao, R.H. Friend, D. Credgington, N.C. Greenham, M.L. Bohm, Synthesis and optical properties of lead-free cesium tin halide perovskite nanocrystals. J. Am. Chem. Soc. **138**(9), 2941–2944 (2016). https://doi.org/10.1021/jacs.5b13470
42. Q. Zhong, M. Cao, Y. Xu, P. Li, Y. Zhang, H. Hu, D. Yang, Y. Xu, L. Wang, Y. Li, X. Zhang, Q. Zhang, L-type ligand-assisted acid-free synthesis of cspbbr 3 nanocrystals with near-unity photoluminescence quantum yield and high stability. Nano. Lett. **19**(6), 4151–4157 (2019). https://doi.org/10.1021/acs.nanolett.9b01666
43. H. Huang, F. Zhao, L. Liu, F. Zhang, X.G. Wu, L. Shi, B. Zou, Q. Pei, H. Zhong, Emulsion synthesis of size-tunable $CH_3NH_3PbBr$ 3 quantum dots: an alternative route toward efficient light-emitting diodes. ACS. Appl. Mater. Interfaces. **7**(51), 28128–28133 (2015). https://doi.org/10.1021/acsami.5b10373
44. R. Naphade, S. Nagane, G.S. Shanker, R. Fernandes, D. Kothari, Y. Zhou, N.P. Padture, S. Ogale, Hybrid perovskite quantum nanostructures synthesized by electrospray antisolvent-solvent extraction and intercalation. ACS. Appl. Mater. Interfaces. **8**(1), 854–861 (2016). https://doi.org/10.1021/acsami.5b10208
45. P. Audebert, G. Clavier, A.-R. Vr, E. Deleporte, S. Zhang, J.S. Lauret, G. Lanty, C. Boissière, Synthesis of new perovskite luminescent nanoparticles in the visible range. Chem Mater **21**(2), 210–214 (2009). https://doi.org/10.1021/cm8020462
46. G.C. Papavassiliou, G. Pagona, N. Karousis, G.A. Mousdis, I. Koutselas, A. Vassilakopoulou, Nanocrystalline/microcrystalline materials based on lead-halide units. J. Mater. Chem. **22**(17), 8271 (2012). https://doi.org/10.1039/c2jm15783g
47. I. Saikumar, S. Ahmad, J.J. Baumberg, G. Vijaya Prakash, Fabrication of excitonic luminescent inorganic-organic hybrid nano- and microcrystals. Scripta. Mater. **67**(10), 834–837 (2012). https://doi.org/10.1016/j.scriptamat.2012.07.048
48. F. Zhu, L. Men, Y. Guo, Q. Zhu, U. Bhattacharjee, P.M. Goodwin, J.W. Petrich, E.A. Smith, J. Vela, Shape evolution and single particle luminescence of organometal halide perovskite nanocrystals. ACS Nano **9**(3), 2948–2959 (2015). https://doi.org/10.1021/nn507020s
49. L. Protesescu, S. Yakunin, S. Kumar, J. Bar, F. Bertolotti, N. Masciocchi, A. Guagliardi, M. Grotevent, I. Shorubalko, M.I. Bodnarchuk, C.J. Shih, M.V. Kovalenko, Dismantling the "Red Wall" of colloidal perovskites: highly luminescent formamidinium and formamidinium-cesium lead iodide nanocrystals. ACS Nano **11**(3), 3119–3134 (2017). https://doi.org/10.1021/acsnano.7b00116
50. I. Lignos, S. Stavrakis, G. Nedelcu, L. Protesescu, A.J. deMello, M.V. Kovalenko, Synthesis of cesium lead halide perovskite nanocrystals in a droplet-based microfluidic platform: fast parametric space mapping. Nano. Lett. **16**(3), 1869–1877 (2016). https://doi.org/10.1021/acs.nanolett.5b04981

51. M. Koolyk, D. Amgar, S. Aharon, L. Etgar, Kinetics of cesium lead halide perovskite nanoparticle growth; focusing and de-focusing of size distribution. Nanoscale **8**(12), 6403–6409 (2016). https://doi.org/10.1039/c5nr09127f
52. E. Yassitepe, Z. Yang, O. Voznyy, Y. Kim, G. Walters, J.A. Castañeda, P. Kanjanaboos, M. Yuan, X. Gong, F. Fan, J. Pan, S. Hoogland, R. Comin, O.M. Bakr, L.A. Padilha, A.F. Nogueira, E.H. Sargent, Amine-free synthesis of cesium lead halide perovskite quantum dots for efficient light-emitting diodes. Adv. Func. Mater. **26**(47), 8757–8763 (2016). https://doi.org/10.1002/adfm.201604580
53. M. Imran, V. Caligiuri, M. Wang, L. Goldoni, M. Prato, R. Krahne, L. De Trizio, L. Manna, Benzoyl halides as alternative precursors for the colloidal synthesis of lead-based halide perovskite nanocrystals. J. Am. Chem. Soc. **140**(7), 2656–2664 (2018). https://doi.org/10.1021/jacs.7b13477
54. D. Zhang, S.W. Eaton, Y. Yu, L. Dou, P. Yang, Solution-phase synthesis of cesium lead halide perovskite nanowires. J. Am. Chem. Soc **137**(29), 9230–9233 (2015). https://doi.org/10.1021/jacs.5b05404
55. D. Zhang, Y. Yu, Y. Bekenstein, A.B. Wong, A.P. Alivisatos, P. Yang, Ultrathin colloidal cesium lead halide perovskite nanowires. J. Am. Chem. Soc. **138**(40), 13155–13158 (2016). https://doi.org/10.1021/jacs.6b08373
56. M. Imran, F. Di Stasio, Z. Dang, C. Canale, A.H. Khan, J. Shamsi, R. Brescia, M. Prato, L. Manna, Colloidal synthesis of strongly fluorescent cspbbr 3 nanowires with width tunable down to the quantum confinement regime. Chem. Mater. **28**(18), 6450–6454 (2016). https://doi.org/10.1021/acs.chemmater.6b03081
57. J. Song, L. Xu, J. Li, J. Xue, Y. Dong, X. Li, H. Zeng, Monolayer and few-layer all-inorganic perovskites as a new family of two-dimensional semiconductors for printable optoelectronic devices. Adv. Mater. **28**(24), 4861–4869 (2016). https://doi.org/10.1002/adma.201600225
58. Y. Bekenstein, B.A. Koscher, S.W. Eaton, P. Yang, A.P. Alivisatos, Highly luminescent colloidal nanoplates of perovskite cesium lead halide and their oriented assemblies. J. Am. Chem. Soc. **137**(51), 16008–16011 (2015). https://doi.org/10.1021/jacs.5b11199
59. W. Liu, Q. Lin, H. Li, K. Wu, I. Robel, J.M. Pietryga, V.I. Klimov, Mn2+-doped lead halide perovskite nanocrystals with dual-color emission controlled by halide content. J. Am. Chem. Soc. **138**(45), 14954–14961 (2016). https://doi.org/10.1021/jacs.6b08085
60. D. Parobek, B.J. Roman, Y. Dong, H. Jin, E. Lee, M. Sheldon, D.H. Son, Exciton-to-dopant energy transfer in mn-doped cesium lead halide perovskite nanocrystals. Nano. Lett. **16**(12), 7376–7380 (2016). https://doi.org/10.1021/acs.nanolett.6b02772
61. G. Pan, X. Bai, D. Yang, X. Chen, P. Jing, S. Qu, L. Zhang, D. Zhou, J. Zhu, W. Xu, B. Dong, H. Song, Doping lanthanide into perovskite nanocrystals: highly improved and expanded optical properties. Nano. Lett. **17**(12), 8005–8011 (2017). https://doi.org/10.1021/acs.nanolett.7b04575
62. A. Wang, Y. Guo, F. Muhammad, Z. Deng, Controlled synthesis of lead-free cesium tin halide perovskite cubic nanocages with high stability. Chem. Mater. **29**(15), 6493–6501 (2017). https://doi.org/10.1021/acs.chemmater.7b02089
63. J. Pal, S. Manna, A. Mondal, S. Das, K.V. Adarsh, A. Nag, Colloidal synthesis and photophysics of M3 Sb2 I9 (M=Cs and Rb) nanocrystals: lead-free perovskites. Angew. Chem. Int. Ed. Engl. **56**(45), 14187–14191 (2017). https://doi.org/10.1002/anie.201709040
64. B. Pradhan, G.S. Kumar, S. Sain, A. Dalui, U.K. Ghorai, S.K. Pradhan, S. Acharya, Size tunable cesium antimony chloride perovskite nanowires and nanorods. Chem. Mater. **30**(6), 2135–2142 (2018). https://doi.org/10.1021/acs.chemmater.8b00427
65. L. Zhou, Y.F. Xu, B.X. Chen, D.B. Kuang, C.Y. Su, Synthesis and photocatalytic application of stable lead-free cs2 agbibr 6 perovskite nanocrystals. Small **14**(11), e1703762 (2018). https://doi.org/10.1002/smll.201703762
66. S.E. Creutz, E.N. Crites, M.C. De Siena, D.R. Gamelin, Colloidal nanocrystals of lead-free double-perovskite (elpasolite) semiconductors: synthesis and anion exchange to access new materials. Nano. Lett. **18**(2), 1118–1123 (2018). https://doi.org/10.1021/acs.nanolett.7b04659

67. M. Imran, P. Ijaz, L. Goldoni, D. Maggioni, U. Petralanda, M. Prato, G. Almeida, I. Infante, L. Manna, Simultaneous cationic and anionic ligand exchange for colloidally stable CsPbBr 3 nanocrystals. ACS. Energy. Lett. **4**(4), 819–824 (2019). https://doi.org/10.1021/acsenergylett.9b00140
68. H. Huang, Y. Li, Y. Tong, E.-P. Yao, M.W. Feil, A.F. Richter, M. Döblinger, A.L. Rogach, J. Feldmann, L. Polavarapu, Spontaneous crystallization of perovskite nanocrystals in nonpolar organic solvents: a versatile approach for their shape-controlled synthesis. Angew. Chem. Int. Ed. (2019). https://doi.org/10.1002/anie.201906862
69. H. Huang, L. Wu, Y. Wan, A.F. Richter, M. Doblinger, J. Feldmann, Facile synthesis of FAPbI3 nanorods. Nanomaterials **10**(1) (2019). doi:10.3390/nano10010072
70. Y. Tong, E. Bladt, M.F. Ayguler, A. Manzi, K.Z. Milowska, V.A. Hintermayr, P. Docampo, S. Bals, A.S. Urban, L. Polavarapu, J. Feldmann, Highly luminescent cesium lead halide perovskite nanocrystals with tunable composition and thickness by ultrasonication. Angew. Chem. Int. Ed. **55**(44), 13887–13892 (2016). https://doi.org/10.1002/anie.201605909
71. Y. Tong, B.J. Bohn, E. Bladt, K. Wang, P. Muller-Buschbaum, S. Bals, A.S. Urban, L. Polavarapu, J. Feldmann, From precursor powders to CsPbX3 perovskite nanowires: one-pot synthesis, growth mechanism, and oriented self-assembly. Angew. Chem. Int. Ed. **56**(44), 13887–13892 (2017). https://doi.org/10.1002/anie.201707224
72. Y. Tong, M. Fu, E. Bladt, H. Huang, A.F. Richter, K. Wang, P. Muller-Buschbaum, S. Bals, P. Tamarat, B. Lounis, J. Feldmann, L. Polavarapu, Chemical cutting of perovskite nanowires into single-photon emissive low-aspect-ratio CsPbX3 (X=Cl, Br, I) nanorods. Angew. Chem. Int. Ed. **57**(49), 16094–16098 (2018). https://doi.org/10.1002/anie.201810110
73. H. Huang, Q. Xue, B. Chen, Y. Xiong, J. Schneider, C. Zhi, H. Zhong, A.L. Rogach, Top-down fabrication of stable methylammonium lead halide perovskite nanocrystals by employing a mixture of ligands as coordinating solvents. Angew. Chem. Int. Ed. **56**(32), 9571–9576 (2017). https://doi.org/10.1002/anie.201705595
74. M. Chen, Y. Zou, L. Wu, Q. Pa, D. Yang, H. Hu, Y. Ta, Q. Zhon, Y. Xu, H. Liu, B. Sun, Q. Zhang (2017) Solvothermal synthesis of high-quality all-inorganic cesium lead halide perovskite nanocrystals: from nanocube to ultrathin nanowire. Adv. Funct. Mater. 1701121. doi:10.1002/adfm.201701121
75. D. Chen, G. Fang, X. Chen, L. Lei, J. Zhong, Q. Mao, S. Zhou, J. Li (2018) Mn-doped CsPbCl3 perovskite nanocrystals: solvothermal synthesis, dual-color luminescence and improved stability. doi:10.1039/C8TC03139H
76. Q. Pan, H. Hu, Y. Zou, M. Chen, L. Wu, D. Yang, X. Yuan, J. Fan, B. Sun, Q. Zhang, Microwave-assisted synthesis of high-quality "all-inorganic" CsPbX3 (X = Cl, Br, I) perovskite nanocrystals and their application in light emitting diodes. J. Mater. Chem. C. **5**(42), 10947–10954 (2017). https://doi.org/10.1039/c7tc03774k
77. Y. Li, H. Huang, Y. Xiong, S.V. Kershaw, A.L. Rogach, Revealing the formation mechanism of cspbbr 3 perovskite nanocrystals produced via a slowed-down microwave-assisted synthesis. Angew. Chem. Int. Ed. **57**(20), 5833–5837 (2018). https://doi.org/10.1002/anie.201713332
78. L. Protesescu, S. Yakunin, O. Nazarenko, D.N. Dirin, M.V. Kovalenko, Low-cost synthesis of highly luminescent colloidal lead halide perovskite nanocrystals by wet ball milling. ACS. Appl. Nano. Mater. **1**(3), 1300–1308 (2018). https://doi.org/10.1021/acsanm.8b00038
79. F. Palazon, Y. El Ajjouri, P. Sebastia-Luna, S. Lauciello, L. Manna, H.J. Bolink, Mechanochemical synthesis of inorganic halide perovskites: evolution of phase-purity, morphology, and photoluminescence. J. Mater. Chem. C. **7**(37), 11406–11410 (2019). https://doi.org/10.1039/c9tc03778k

# Chapter 2
# Strongly Quantum Confined Metal Halide Perovskite Nanocrystals

**Daniel Rossi, David Parobek, and Dong Hee Son**

**Abstract** Metal halide perovskite nanocrystals with the chemically tunable bandgap and superb optical properties are promising candidates for a number of high performance optoelectronic and photonic applications. The majority of studies on perovskites have focused on the large perovskite nanoparticles in the weak confinement regime, however recent synthetic advances have allowed for the preparation of high quality 0D, 1D, and 2D confined perovskites, both hybrid organic-inorganic and all-inorganic. These new synthetic methods have enabled the study of confinement effects on various photophysical properties of perovskites nanocrystals previously not possible. In the first section of this chapter, we discuss the synthetic methods for the preparation of both hybrid and all-inorganic perovskite nanocrystals in the strong confinement regime, including the synthetic protocols as well as the mechanistic aspects of the nanocrystal growth. The second section discusses the confinement effects on various static and dynamic photophysical properties of exciton experiencing varying degree and dimensionality of quantum confinement.

## 2.1 Introduction

Colloidal semiconductor nanocrystals (NCs) were studied in recent decades as the sources of photons and charge carriers for photovoltaics, optoelectronics [1–3], and photonics, as well as biological applications [4–6]. Perhaps one of the most widely utilized and interesting function of the colloidal semiconductor NCs is their size-tunable exciton absorption and emission due to quantum confinement [7]. For NCs with size smaller than the twice of exciton Bohr radius, the exciton experiences significant spatial confinement of the wavefunction, and various properties of

D. Rossi
Center for Nanomedicine, Institute for Basic Science (IBS), and Graduate Program of Nano Biomedical Engineering, Advanced Science Institute, Yonsei University, Seoul 03722, Republic of Korea

D. Parobek · D. H. Son (✉)
Department of Chemistry, Texas A&M University, College Station, Texas 777843, USA
e-mail: dhson@tamu.edu; dhson@chem.tamu.edu

 
Y. Zhou and Y. Wang (eds.), *Perovskite Quantum Dots*, Springer Series in Materials Science 303, https://doi.org/10.1007/978-981-15-6637-0_2

NCs derived from exciton can be tailored by varying the NC size and morphology [8–10]. In addition, it is well-known that adding impurities into the semiconductor NCs could alter the optical properties, often introducing new emission pathways, emitting photons at different wavelength from the exciton of the host NCs [11, 12]. For example, quantum dots doped with paramagnetic transition metal ions such as $Mn^{2+}$ could exhibit bright phosphorescence resulting from the sensitization of the dopant transition via exciton of the host quantum dots [13–15].

In recent years, metal halide perovskite (MHP) NCs with $ABX_3$ composition (A: organic or metal monovalent cation, B: divalent metal cation, X: halide) were introduced into the family of colloidal semiconductor NCs [16, 17]. Fig. 2.1 illustrates the structure of MHP NCs that exhibit near-cubic lattice structure, however the NCs often exhibit small distortion to adopt orthorhombic structure [18, 19]. Although the research of bulk MHP has the longer history, the colloidal nanocrystals synthesis was realized only in recent years. MHP NCs with organic cation at A cation site (organic-inorganic hybrid perovskite NCs) were first synthesized in 2014 and the application in light emitting diodes (LED) was demonstrated. [16] However, the relatively poor

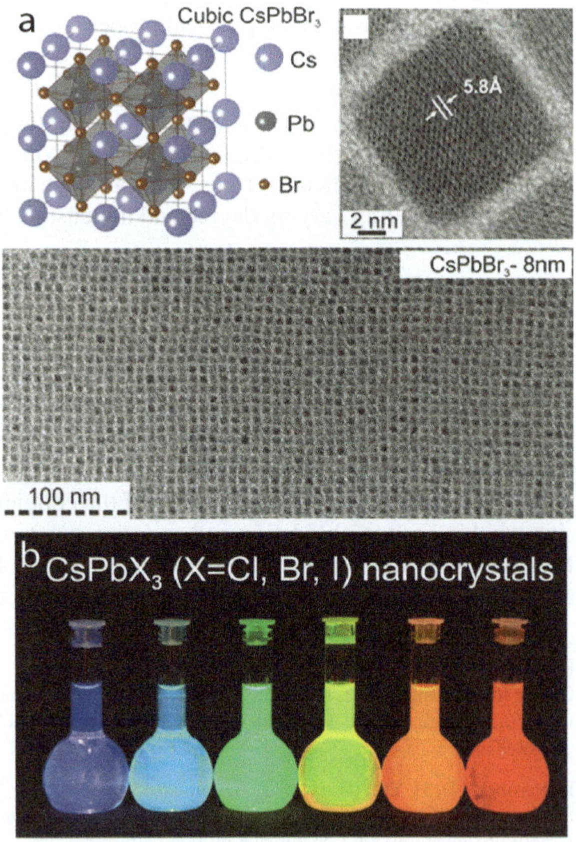


**Fig. 2.1** **a** Low and high resolution TEM and a schematic of the $CsPbBr_3$ crystal structure. Photograph of $CsPbX_3$ and mixed halide NC solutions under UV irrigation showing PL over the entire visible spectrum. Reprinted (adapted) with permission from Ref. [17] Copyright (2015) American Chemical Society

stability of organic-inorganic hybrid perovskite NCs limited their wider utilization in the applications. In 2015, fully inorganic MHP NCs, where $Cs^+$ and $Pb^{2+}$ ions occupy A and B sites respectively, exhibiting significantly higher stability were synthesized through a hot injection method. [17] Following this work, studies on $CsPbX_3$ NCs exploring the utility of these new semiconductor NCs in the applications, such as solar cell, LED and lasing, where more traditional II–VI and IV–VI semiconductor NCs have been extensively utilized [20–22].

For both hybrid and fully inorganic MHP NCs, the superb tolerance to charge carrier-trapping defects provides an important advantages compared to other semiconductor NCs used as the source of photons and charges. For this reason, MHP NCs typically showed high luminescence quantum yields (QY) of up to 90% without surface treatment and QY approaching unity with appropriate surface ligand or (and) with doping. [23–28] Furthermore, bandgap tuning of MHP NCs was readily achieved through facile halide exchange or control of the composition of mixed halide without resorting to quantum confinement, providing color tunable emission throughout the entire visible spectrum. [29, 30] Therefore, the majority of research on MHP NCs and their application as a source of color-tunable light has focused on weakly or non-confined NCs with a chemically tunable bulk-like bandgap, rather than size tunable bandgap in quantum confinement regime. Part of the reason for the domination of the weakly and non-confined MHP NCs in the earlier research and the application of MHP NCs is the slower development of reliable methods to control the size of NCs in the quantum confinement regime with sufficiently high uniformity.

More recently, synthetic methods for producing highly uniform MHP nanocrystals with strong quantum confinement and varying dimensionality of confinement were developed, enabling the exploration and utilization of properties of MHP NCs tuned by the quantum confinement. [31–39] Quantum confinement not only alters the energetics and dynamics of exciton, but also enhances the coupling of exciton with other degrees of freedom including the lattice phonon, electronic and vibrational degrees of freedom of the ligands, and spin-active impurities doped within the lattice. Interparticle electronic coupling can also be enhanced in strongly confined NCs by facilitating the spatial overlap of the exciton wavefunctions between coupled NCs. [40] The enhanced coupling of exciton with other degrees of freedom in strongly confined semiconductor NCs is known to introduce new optical, electronic and optomagnetic properties. [41, 42] Similar benefits of quantum confinement are also anticipated in MHP NCs, which will expand the range of controllable material properties beyond what can be obtained from non-confined systems. In this chapter, we discuss recent progresses made in the synthesis of MHP NCs with tunable quantum confinement and morphology, and the exploration of their photophysical properties reflecting the effect of quantum confinement.

## 2.2 Synthesis of Quantum-Confined Perovskite Nanocrystals

Since the first synthesis of colloidal all-inorganic perovskite ($CsPbX_3$) NCs, significant advancements have been made in the synthetic methodologies and in the understanding of the formation mechanism of the NCs. Now there exists a library of methods for the preparation of MHP NCs with varying A and B site cations, halide composition, size, phase, and morphology [30, 31, 37, 43–46]. The currently available synthesis methods for MHP NCs can be largely grouped into two categories: high-temperature hot injection synthesis and room-temperature reprecipitation.

Hot injection synthesis of colloidal NCs was first developed for the preparation of CdSe quantum dots, and has since become a valuable tool for NC synthesis in many semiconductor material systems. Hot injection methods provide a good compromise between simplicity and sample quality; typically requiring only a Schlenk line, and yielding samples with narrow size distributions and high PL quantum yields [47–49]. Furthermore, the use of surface coordinating ligands in the synthesis allows NCs to be dispersed in both organic and inorganic solvents through the proper choice of ligand molecules and enables facile post-synthesis chemical functionalization of the NCs. The large and easily tunable parameter space (reaction times and temperature, precursor composition and concentration, ligand choice etc.) also makes hot injection synthesis a valuable tool for understanding NC growth mechanisms. A number of studies have identified the fast growth rate of MHP NCs under hot injection condition as a barrier to the control of the NC size, especially in the smaller size regime imposing strong quantum confinement [31, 32, 45]. Nevertheless, the insight obtained from the study of kinetic and mechanistic aspects of hot injection MHP NC synthesis has been instrumental in the development of hot injection methods for the synthesis of quantum confined MHP NCs with varying morphology and dimensionality of confinement, including quantum dots (QDs), nanowires (NWs) and nanoplatelets (NPLs).

While the room temperature reprecipitation methods have not received as much attention as hot injection methods, they have proven remarkably useful for the preparation of 1D and 2D perovskites e.g. NWs and NPLs [43, 50]. Reprecipitation typically involves the addition antisolvent to a concentrated salt precursor solutions, which induces NC nucleation and growth from the reduced ion solubility. One significant advantage of the reprecipitation method over hot injection method is the relative simplicity, typically requiring only simple air free conditions. As such the reprecipitation method is more easily scalable and implementable for the industrial scale preparation of perovskites. On the other hand, sample quality and the size control are inferior to the hot injection methods. In this section, the development and implementation of the hot injection and reprecipitation methods for the synthesis of quantum confined QDs, NWs and NPLs of all-inorganic and organic-inorganic hybrid MHP NCs are described.

### 2.2.1 All-Inorganic Metal Halide Perovskites Quantum Dots

The first preparation of all-inorganic MHP NCs was demonstrated by Protesescu et al. in $CsPbX_3$ system employing the hot injection method [17]. In a typical hot injection synthesis condition of $CsPbX_3$ NCs, Pb salt is dissolved in organic solvent at high temperature (100–200 °C) in the presence of organic ligands, and Cs oleate is injected subsequently to initiate the reaction. After the injection, NCs grow rapidly within a few to ten seconds forming generally cube-shaped NCs. Detailed examination of the growth kinetics was performed in a study employing a microfluidic reactor, which showed that the growth of NCs was completed in less than 5 s within the range of reaction temperature examined (130–220 °C). [51] Compared to II–VI or III–V QDs, the nucleation and growth kinetics are much faster, posing a significant challenge in controlling the size of the NCs via timed quenching of the reaction achieved by rapid cooling of the reactor. Because of the fast reaction kinetics forming $CsPbX_3$ NCs under typical hot injection synthesis condition, the resulting NCs had relatively large size (>10 nm) in weakly confined regime and a broad size distribution. For this reason, earlier efforts to explore the size-dependent optical properties of MHP NCs in the quantum confined regime resorted to post-synthesis size selection of the NC samples to improve the size uniformity of the QDs or rely on single-particle spectroscopy [17].

Recently significant advancements have been made to extend the synthesis method of $CsPbX_3$ NCs into the quantum confined regime. One strategy used to synthesize $CsPbX_3$ NCs in the quantum confined regime is using the lower reaction temperatures to slow down the reaction kinetics, which would make size control via control of the growth kinetics more feasible. Generally, the size of the NCs decreased with decreasing reaction temperature giving an access to the QDs in strongly confined regime [43, 46, 52, 53]. Figure 2.2a shows the absorption and photoluminescence spectra of $CsPbBr_3$ NCs of varying sizes obtained by using different reaction temperature, which exhibit blueshift of excitonic absorption and emission peak from quantum confinement below 9 nm. However, the size distribution of the obtained NCs determined from the TEM images was significantly larger compared to what

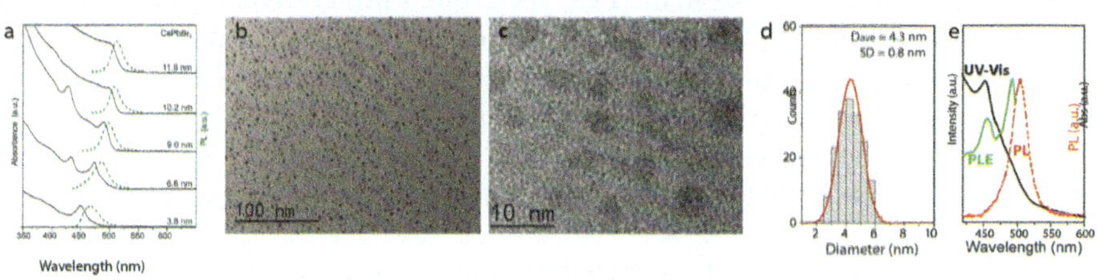


**Fig. 2.2** **a** Absorption and PL spectra of $CsPbBr_3$ QDs made by decreasing the reaction temperature. **b** and **c** TEM images of QDs made by the room temperature reprecipitation methods. **d** Size distribution of QDs showed in (**b**) and (**c**). **e** Absorption, PL and PLE spectra of QDs made with ligand-mediated method showing multiple features resulted from ensemble inhomogeneity. Reprinted (adapted) with permission from Ref. [52]. Copyright (2015) American Chemical Society

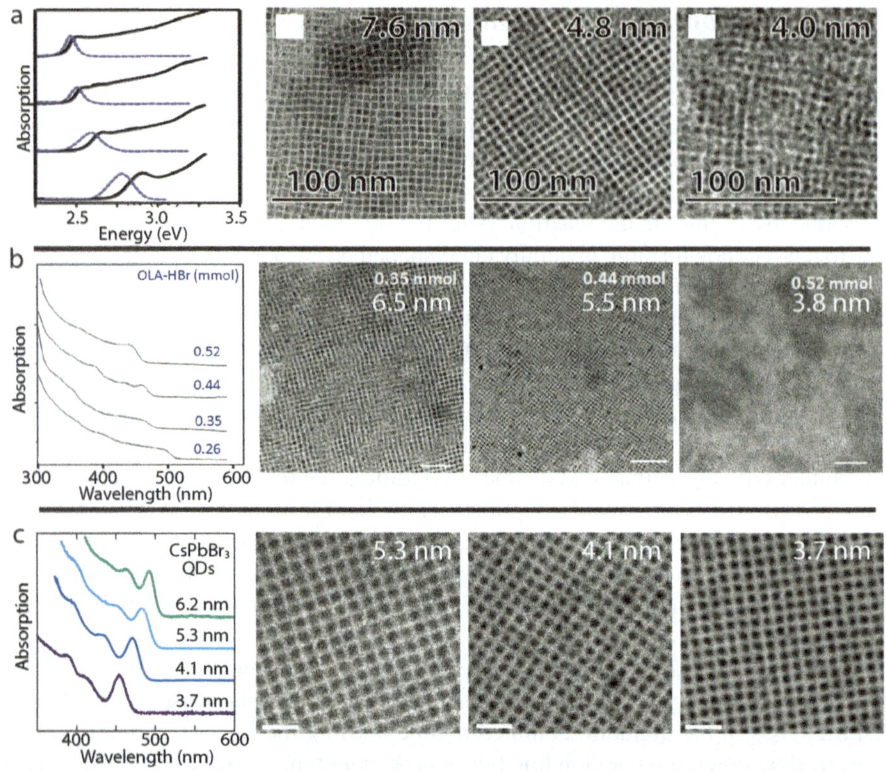


**Fig. 2.3** **a**, **b** and **c** Absorption spectra and corresponding TEM images for QDs made in Ref [31, 37, 44], respectively. Reprinted (adapted) with permission from Refs. [31, 37, 44]. Copyright (2015) American Chemical Society

has been achieved in other semiconductor QDs such as II–VI metal chalcogenide QDs. (Figure 2.2b–d) Often, the heterogeneity of size could be revealed more clearly in the photoluminescence excitation (PLE) spectrum as shown in Fig. 2.2e, which shows the presence two sizes in the PLE spectrum. Furthermore, contamination by NCs of other morphologies was another shortcoming of this approach. For instance, exciton absorption peak from the nanoplatelets was seen prominently (e.g., peak near 430 nm in Fig. 2.2a), superimposed on the absorption from the nanocubes, while the nanoplatelets leave less distinct spectroscopic signature in the luminescence spectrum due to their relatively low photoluminescence quantum yield compared to highly bright nanocubes.

Another approach to synthesize $CsPbX_3$ QDs of varying sizes in the strongly confined regime is varying the amount or type of organic ligands. Manna and coworkers reported that the size of $CsPbBr_3$ NCs can be reduced to ca. 4 nm by varying the composition or the amount of the ligands they used, specifically the ratio of oleic acid to oleylamine [44]. Fig. 2.3a shows the absorption and emission spectra, and TEM images of $CsPbBr_3$ NCs of different sizes obtained by this approach. The reduction of the size was achieved by suppressing the rapid ripening rate of the perovskite nanocrystals during the growth phase by increasing the acid concentration or reducing the temperature, which the authors considered responsible for ripening of the nanocrystals. Because of the difference in the rate of ripening for the $CsPbX_3$ NCs with different halide, the size control via modification of the ligand could not be extended to $CsPbI_3$ NCs. Pradhan and coworkers reported that the size of $CsPbBr_3$ NCs can be reduced by adding oleylamine-HBr mixture to the reactant. The size of $CsPbBr_3$ QDs could be more continuously tuned by varying the amount of oleylamine-HBr mixture [37]. Fig. 2.3b shows the absorption spectra and TEM images of $CsPbBr_3$ QDs of varying sizes produced in this way. They proposed that the mechanism of size control is the competition between $Cs^+$ and protonated ligands during the growth of the NCs. All these methods relied on the extra means to control the growth kinetics, therefore still had an intrinsic size heterogeneity that smears out the size-dependent exciton absorption and emission spectra. Furthermore, absorption spectra in Fig. 2.3a and b indicate that the contamination from other morphology is diminished compared to those shown in Fig. 2.2, a significant improvement in the purity of the sample.

More recently, Son and coworkers developed a new synthesis approach that can produce strongly quantum confined $CsPbX_3$ QDs with precise control of the size enabling the synthesis of QD samples with very high ensemble uniformity [31]. In contrast to the majority of other synthesis methods that attempt to control the reaction kinetics, their approach used the control of the thermodynamic equilibrium of halide ions between the QD lattice and solution phase during the synthesis as a means to control the size as explained in detail below. Figure 2.3c shows the absorption spectra and TEM images of $CsPbBr_3$ QDs of varying sizes produced employing their methods. Compared to Fig. 2.3a, b, the exciton absorption feature in the optical spectra in Fig. 2.3c is much better defined, similarly to those observed in highly monodisperse II–VI QDs. [54] Absorption peaks from the higher-energy exciton transitions, which are not clearly visible in Fig. 2.3a, b, are also seen more clearly. They showed that the same approach based on equilibrium of halide is generally applicable to the synthesis of $CsPbX_3$ QDs of all three halides (Cl, Br and I). Figure 2.4 shows the optical spectra and TEM images of the QDs of $CsPbCl_3$ and $CsPbI_3$ prepared employing the same halide equilibrium-based size control.

In general, thermodynamic equilibrium is not considered as a viable mechanism to control the size of nanocrystalline solids for the majority of materials. However, lead halide perovskite is a unique case, where the equilibrium approach can be used. One crucial property of $CsPbX_3$ NCs and other related lead halide perovskite materials that allow the use of thermodynamic equilibrium to control the size is the highly labile halide ions in the crystals. It is well known that the lead halide perovskite NCs can undergo facile anion exchange even at room temperature, due to their low kinetic barrier for the diffusion in the lattice. [29, 30, 55] The low kinetic barrier, especially at the elevated temperatures of the hot injection condition (typically 100–200 °C), makes the thermodynamic equilibrium of halide a feasible mechanism to control the size. The other crucial property of $CsPbX_3$ NCs that enabled the thermodynamic equilibrium-based size control is the size-dependent composition of halide, resulting

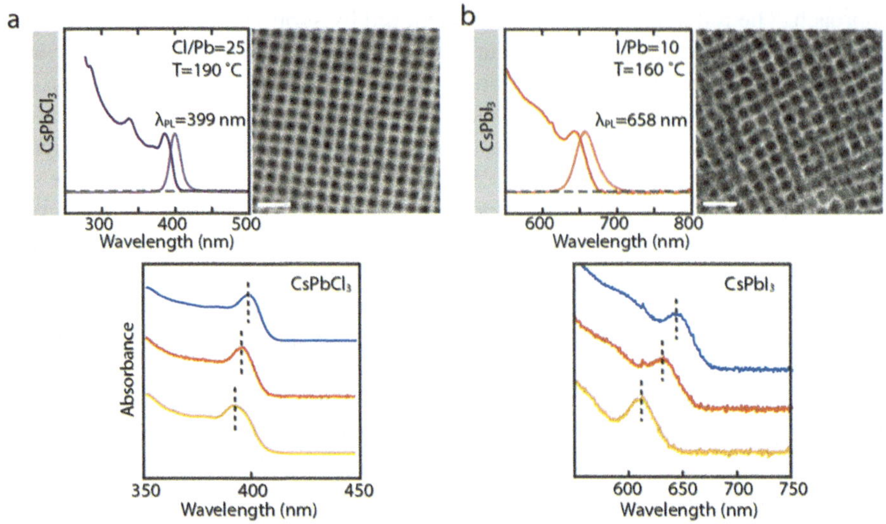


**Fig. 2.4** **a** Absorption and PL spectra (left) and TEM image (right) of $CsPbCl_3$ QDs prepared via halide equilibrium control method. Bottom is the absorptions spectra of the of $CsPbCl_3$ QDs of different sizes. **b** Absorption and PL spectra (left) and TEM image (right) of $CsPbI_3$ QDs prepared via halide equilibrium control method. Bottom is the absorptions spectra of the of $CsPbI_3$ QDs of different sizes. Reprinted (adapted) with permission from Refs. [31]. Copyright (2015) American Chemical Society

from halide-rich surface of the NCs formed under the employed reaction condition. The relative composition of halide is higher for the QDs of the smaller size, which allows the equilibrium of halide ions at the interface of NCs and solvent medium, a critical factor that can determine the size of the NCs during the growth.

The size control was achieved either by varying the amount of halide in the reactant mixture for a given amount of Cs and Pb at a fixed reaction temperature or by varying the temperature for a fixed amount of the precursors of all three elements constituting the $CsPbX_3$ QDs. Figure 2.5a shows the dependence of the size of $CsPbBr_3$ QDs with the variation of temperature and molar ratio of Br/Pb in the reactant mixture. At a fixed reaction temperature, the smaller QDs were obtained when the concentration of $Br^-$ increased (i.e., higher Br/Pb ratio) through the law of mass action as will be explained in detail below. For the fixed concentration of $Br^-$ (i.e., fixed Br/Pb ratio), lowering the temperature resulted in the smaller QDs. Using this approach, $CsPbBr_3$ QDs with very high ensemble uniformity were produced in the strongly confined regime (3.7–7 nm)

The mechanism of size control through halide equilibrium between the QD lattice and solvent medium is illustrated by the simple model shown in Fig. 2.5b. In this model, the control of the QD size ($l$) with $Br^-$ concentration ($[Br^-]$) can be viewed as satisfying the following equilibrium condition.

$$\mu_{Br^-,sol}\left(\left[\mathrm{Br}^-\right], \mathrm{T}\right) = \mu_{Br^-,QD}(l, \mathrm{T}) \tag{2.1}$$

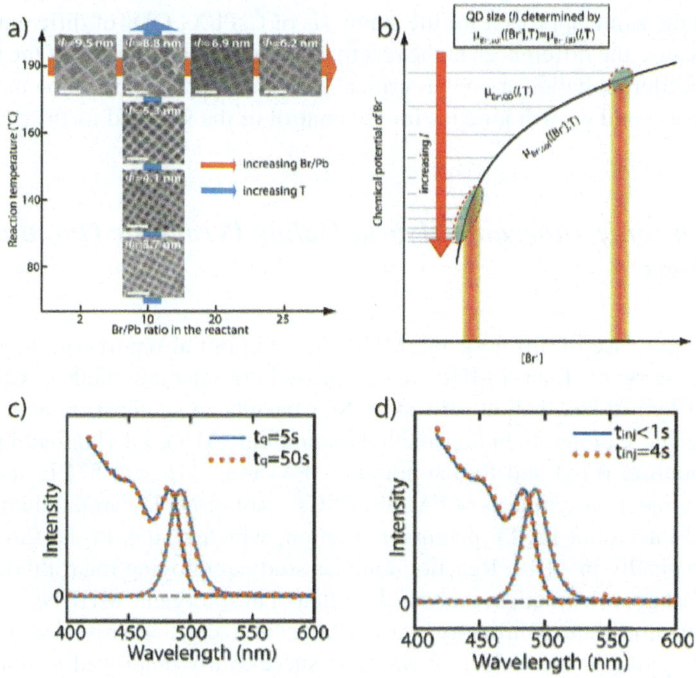


**Fig. 2.5** **a** reaction conditions that could produce QDs with various sizes. **b** Proposed mechanism of halide equilibrium control of quantum confinement in Cs based MHP QD synthesis. **c** Absorption and PL spectra of $CsPbBr_3$ NCs quenched at 5 s (blue line) and 50 s (orange dashed line), **d** Absorption and PL spectra of $CsPbBr_3$ NCs with cesium oleate injected swiftly (blue line) and over 4 s dropwise (orange dashed line) Images were modified with permission from Ref [31]. Reprinted (adapted) with permission from Ref. [31]. Copyright (2015) American Chemical Society

$\mu_{Br^-,sol}([Br^-], T)$ is the chemical potential of $Br^-$ at a given $[Br^-]$ and temperature T in the solution medium. $\mu_{Br^-,QD}(l, T)$ is the chemical potential of $Br^-$ in the lattice of QD with size $l$ at temperature T, which depends on the relative stoichiometric composition of $Br^-$. $\mu_{Br^-,sol}([Br^-], T)$ should increase approximately logarithmically with $[Br^-]$ following $\mu_{Br^-,sol}([Br^-], T) = \mu^{\circ}_{Br^-,sol} + RTln([Br^-])$, where $\mu^{\circ}_{Br^-,sol}$, $R$, and $\gamma$ are the chemical potential at standard state, gas constant and activity coefficient respectively. The larger slope of $\mu_{Br^-,sol}([Br^-], T)$ at the lower ($[Br^-]$) and decreasing size dependence of $\mu_{Br^-,QD}(l, T)$ for the larger QDs explains the poorer size uniformity of the larger NCs synthesized under relatively low $[Br^-]$, i.e., Br/Pb ratio of 2. Since the size control is achieved via thermodynamic equilibrium, the size of the QDs is independent of nucleation and growth kinetics, which is one of the most distinct and advantages of the equilibrium-based synthesis of size-controlled QDs. Figure 2.5c–d show the identical absorption and emission spectra of $CsPbBr_3$ QDs synthesized with varying reaction quenching time and precursor injection rate that proves that the size control is via thermodynamic equilibrium not the kinetic control. Such independence of the QD size on the reaction kinetics made it easier to expand the same approach for the synthesis of $CsPbX_3$ QDs of different halides. This is because the differences in the reactive chemistry and kinetics of the reactions involving different halide are not as critical as in the approach based on modulating the nucleation and growth kinetics for the control of the size and its distribution.

### 2.2.2 *Organic-Inorganic Hybrid Halide Perovskite Quantum Dots*

Similar to the case of all-inorganic MHP NCs, the initial reports on the synthesis of organic-inorganic hybrid MHP NCs employed hot injection method. Currently a number of hybrid lead halide perovskite NCs have been synthesized with different A-site organic cations including methylammonium (MA), ethylammonium (EA), octylammonium (OA) and formamidinium (FA) ions. [16, 56, 57] In the case of typical hot injection synthesis of $FAPbBr_3$ NCs, a solution oleylammonium bromide is injected into a hot FA-Pb precursor solution, which results in the formation of the NCs typically in ~10 s. Reaction kinetics study employing microfluidic reactor revealed fast growth kinetics comparable to that of all-inorganic MHP NCs, therefore sharing the same issue of difficulty in controlling the size with narrow size distribution [58]. Post-synthesis size selection was also successfully employed to improve the size distribution [59]. Fig. 2.6a–c show the TEM mages, size distribution and optical spectra of $MAPbBr_3$ QDs of different sizes obtained from this approach.

Compared to all-inorganic MHP NCs, the ability to tune the size of the NCs with ligand choice, reactant concentration, and temperature is more limited for a number of reasons. Notably, the usable temperature range is smaller and the morphology of the NCs formed is more sensitive to the choice of surfactant and cation concentration.

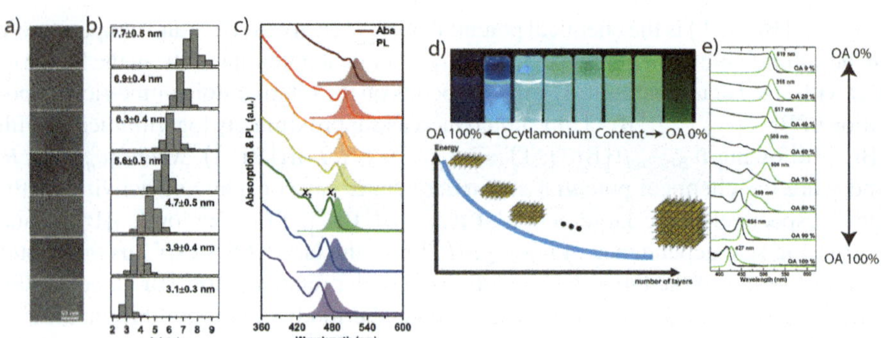


**Fig. 2.6** **a–c** TEM, absorption and PL spectra of $FAPbBr_3$ quantum dots synthesized using hot injection. The size was controlled via size selective precipitation. **d–e** Structural evolution of $OAPbBr_3$ as a function of ligand concentration, smaller QDs are made with increasing ligand concentration eventually leading to the formation of monolayer NPLs in the limit of 100% OA. Reprinted (adapted) with permission from Refs. [39, 59]. Copyright (2015) American Chemical Society

For the synthesis of hybrid MHP NCs, both the surface ligand and organic cation require protonation for the growth of NCs, and the NC size and shape morphology are sensitive to changes in the concentration of either species. Similar to all-inorganic NCs, the increase of protonated species reduces the NC size and in high concentrations induces anisotropic particle growth, resulting in the formation of surfactant rich $L_2(FAPbI_3)_{n-1}PbX_4$ layered structure. [38] This narrows the useable parameter space for using kinetics to tune the size of hybrid MHP QDs, however provides a route to prepare of layered hybrid MHP NCs by varying the oleic acid concentration as shown in Fig. 2.6d–e for $OAPbBr_3$ NPLs and QDs [39].

Utilizing the hot injection methods adapted from the synthesis of $CsPbX_3$ QDs, the similar attempts to control the NC size via temperature-dependent reaction kinetic were made. However, the resulting QDs were often contaminated by NCs with different morphologies, such as nanoplatelets for the reasons discussed above. Within the temperature range where anisotropic structure formation was minimized, the size distribution of the QDs was still relatively poor. However, post-synthesis size selection could be successfully applied to improve the uniformity of the QD samples sufficiently to reveal size-dependent and strongly confined excitonic feature in the absorption spectra as shown in Fig. 2.6c. While the size control based on the equilibrium of halide should in principle be applicable to the hybrid MHP NCs, so far successful synthesis of size controlled organic-inorganic hybrid perovskite QDs in strongly quantum confined regime has not been reported.

### 2.2.3 Synthesis of Quantum Confined Nanowires and Nanoplatelets of Perovskites

Although the cube-shaped lead halide perovskite NCs are commonly formed under typical hot injection synthesis condition due to the near-cubic lattice symmetry, synthesis methods of MHP NCs with anisotropic morphology exhibiting quantum confinement in 1D or 2D have also been developed. Currently, a significant library of both hot injection and reprecipitation methods for the preparation of nanowires (NWs) with 1D confinement and nanoplatelets (NPLs) with 2D confinement are available for both hybrid and all-inorganic MHP, with the ability to tune the thickness at unit cell precision and lateral dimension/length of NPLs and NWs.

There are several different methods that have been developed for the synthesis of perovskite monolayer and multilayer NPLs and NWs in colloidal solution form and in thin films. These include solid-state crystallization, chemical exfoliation, non-solvent crystallization, and hot injection [34]. These methods have been used to synthesize both NPLs and NWs of the organic-inorganic hybrid and all-inorganic based perovskites. The solid-state crystallization method was first developed in the 90's by creating layered perovskite thin films through evaporating a polar solvent (i.e. DMF) resulting in the crystallization of the metal salts forming monolayer and multilayered perovskite NPLs [60, 61]. As the solvent evaporates, a 2D sheet of

$PbBr_4^{2-}$ forms and charge compensation is accomplished by the protonated amine ligand in solution. The number of layers is controlled through the concentration of the metal salts and the evaporation rate of the solvent. The chemical exfoliation of the bulk perovskite in a nonpolar solvent will also result in NPLs in a variety of thicknesses that can be isolated by centrifugation [62].

Non-solvent crystallization methods have been heavily used to create nanoplatelets down to monolayer thicknesses for both MHP and organic-inorganic perovskites. This was first used to synthesize $CH_3NH_3PbBr_3$ nanocrystals by Schmidt et al. and was adapted for use in NPL synthesis shortly after [16, 39]. The metal salts and ligands are dissolved into a polar solvent (DMF) and then added dropwise to a nonpolar solvent (toluene) which results in the crystallization of the NCs. Feldmann and coworkers found that increasing the concentration of the amine ligand resulted in its increased binding to the A site of the perovskite structure which caused thinner nanoplatelets from the arrested growth in the thickness direction. Weidman et al. used this method to synthesize monolayer NPLs with the A site cation being replaced with the ligand, using the notation $L_2PbX_4$ (L = ligand). Without the presence of the A site cation they received a monodisperse sample since no other thicknesses could form with only the ligands present in solution [63]. There are also several examples of adaptations of the non-solvent crystallization being used for the all-inorganic MHP to form NPLs and NWs [43, 46, 64, 65].

Figure 2.7a, b show two samples of room temperature reprecipitation methods, similar to non-solvent crystallization, for the growth of anisotropic perovskites. Notably, Stratakis and coworkers prepared $CsPbBr_3$ NWs via supersaturated precipitation [64]. Likewise, Deng and coworkers prepared $CsPbBr_3$ NPLs with variable thickness, tunable via ligand choice, via supersaturated precipitation [46]. While there is less mechanistic insight into the NC growth mechanism for the NPLs, they are superior to the hot injection synthesis as the low temperatures promote anisotropic particle growth and they provide samples with superior size and shape distributions compared to NPLs or NWs prepared with hot injection methods. Regardless, there are well defined recipes for both thickness controlled NPLs as well as NWs as shown in Fig. 2.7a, b.

Hot injection methods to synthesize NPLs and NWs have primarily centered around all-inorganic MHP with several mechanistic studies to determine the effects of temperature and ligand concentration on the NC anisotropy. The competition between the reactivity of surfactant and cesium was proven critical to promote anisotropic growth in perovskites. In this case, the NPLs are capped with protonated amine (R-AMH) where R is an organic ligand (oleylamine, octylamine, dodecylamie and ect.), which becomes protonated during the synthesis, typically from oleic acid. The concentration and reactivity of protonated R-AM, which can be controlled via acid concentration or temperature, becomes the critical parameter to promote anisotropic particle growth. Almeida et al. demonstrated this, preparing $CsPbBr_3$ NPLs down to a single unit cell thickness, by increasing the R-AMH concentration with excess Oleic acid. [44] It was further noted the importance of the temperature, where both the R-AMH concentration as well as the binding affinity to the NC surface are increased with decreasing temperatures. Figure 2.7a, b shows the evolution of NP size and

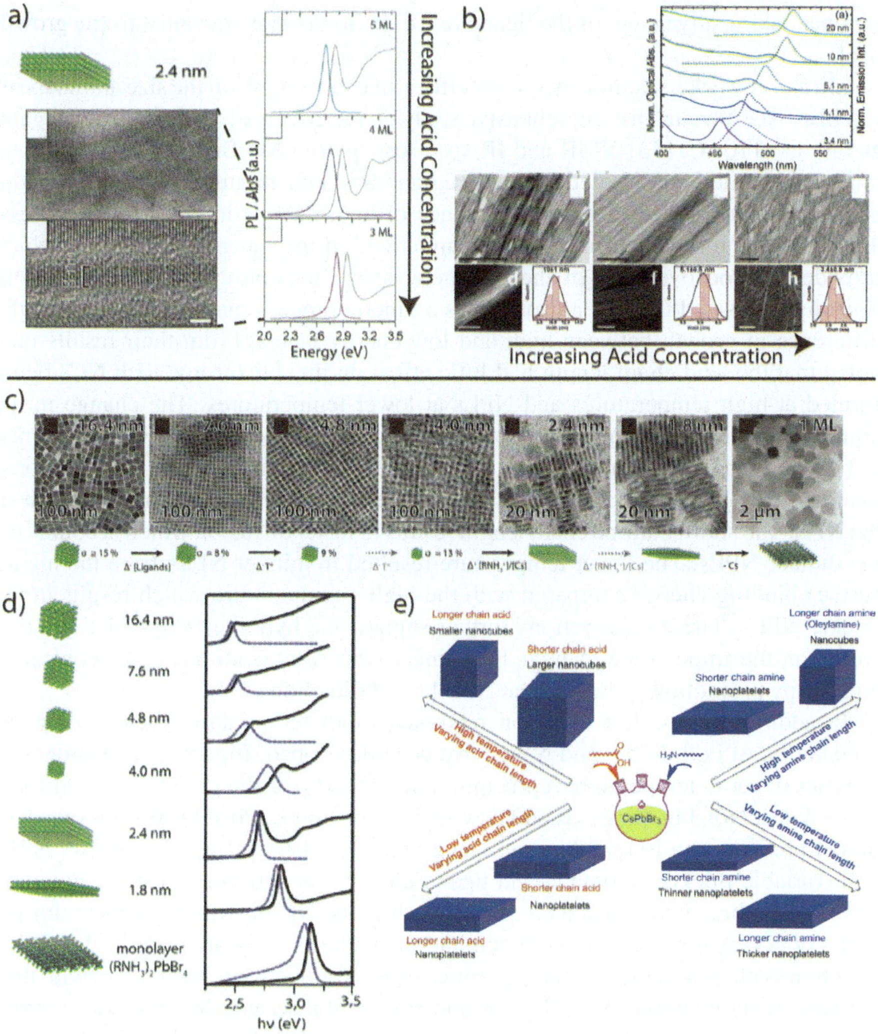

**Fig. 2.7** **a–b** TEM, absorption, and PL from $CsPbBr_3$ platelets and NWs nanowires. **c–d** TEM, absorption and PL from $CsPbBr_3$ QDs and NPLs synthesized under varying reaction conditions. Reduction of the temperature and increase of acid concentration results in QD and NPL formation. **e** Schematic depicting the effects of reaction conditions on the growth of $CsPbBr_3$ NPLs. Reprinted (adapted) with permission from Refs. [44, 46, 53, 65]. Copyright (2015) American Chemical Society

shape morphology for $CsPbBr_3$ NPs explored by Manna by varying the size and temperature. By decreasing the temperature and increasing the R-AMH concentration, smaller QDs were produced. Pushing to lower temperatures and higher R-AMH concentrations, NPLs were formed down to single unit cell thickness. Similar hot injection methods for the preparation of NWs have also been developed with similar rational, taking advantage of the ligand reactivity to promote the anisotropic growth [36].

Liu and coworkers demonstrated the effects of chain length on the size and shape of $CsPbBr_3$ NCs through a comprehensive study on the effects of acid and amine ligand at 140 and 170 °C. [53] NMR and IR spectroscopy of $CsPbBr_3$ NPLs prepared via hot injection method revealed the surface ligands to be a mixture of both carboxylic acid and amine, suggesting the importance of both ligands in the growth process. The results of these experiments are summarized in the schematic Fig. 2.7c which provides a good overview of the parameter space for confined $CsPbBr_3$ growth. They hypothesized that the anisotropy is a kinetic process that is controlled by the difference in growth between high and low energy facets. From their results they noted that the acid chain length had little effect on the anisotropy, with NCs being formed at high temperatures and NPLs at lower temperatures. The change in the amine chain length had a greater impact on the morphology with NPLs being formed at both low and high temperatures. They believed the amine had a greater affect, competitively binding to the A site location, resulting in the anisotropic growth of the NPL. The shorter amine chain length results in faster diffusion which accounts for the thinner NPLs. The lower temperature resulted in thinner NPLs from the higher surface binding energy compared with the higher temperature which results in the thicker NPLs. These observations further support the hypothesis provided in other works on the importance of ligand binding to the A site cation for the control of anisotropy promoting growth in the NPL lateral directions.

In addition to the hot injection methods, room temperature methods for the preparation of both NPLs and NWs have been developed. Figure 2.7d–e show two samples of room temperature reprecipitation methods for the growth of anisotropic perovskites. Notably, Stratakis and coworkers prepared $CsPbBr_3$ NWs via supersaturated precipitation [64]. Likewise, Deng and coworkers prepared $CsPbBr_3$ NPLs with variable thickness, tunable via ligand choice, via supersaturated precipitation [46]. While there is less mechanistic insight into the NC growth mechanism for the NPLs or NWs grown with the reprecipitation synthesis, they are superior to the hot injection synthesis as the low temperatures promote anisotropic particle growth, they are significantly simpler than the hot injection methods, and they provide samples with superior size and shape distributions compared to NPLs or NWs prepared with hot injection methods. Regardless, there are well defined recipes for both thickens controlled NPLs as well as NWs as shown in Fig. 2.7d–e.

Some of the synthetic methods for strongly confined 1D and 2D MHP NCs are hindered by poor size distributions due to the fast growth kinetics, requiring size selective precipitation to obtain NWs and NPLs with high size and shape uniformity. Recently, Son and coworkers developed a new synthesis method that can produce NWs and NPLs with highly uniform thickness by combining the thermodynamic equilibrium of halide discussed in the previous section for the synthesis of $CsPbX_3$ QDs and the kinetic anisotropy that becomes more pronounced at lower reaction temperatures. This approach utilizes thermodynamics to control the NC thickness, and the kinetics to control the lateral dimension and length [32]. As shown in Fig. 2.8a, b, this allows for the isolation of colloidal NPLs and NWs with precise control over the

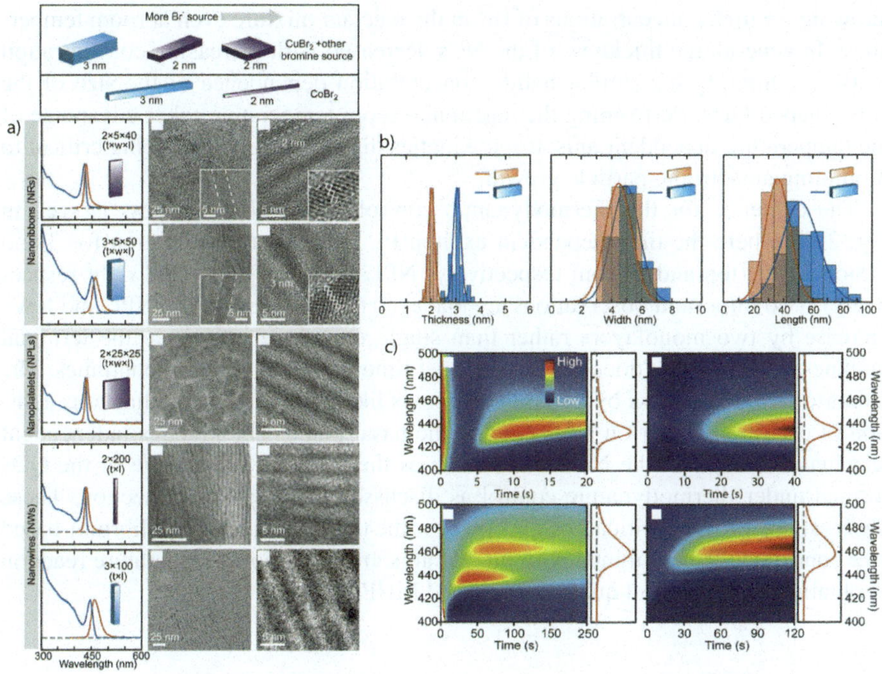


**Fig. 2.8** **a** Absorption (blue) and photoluminescence (PL) spectra (red) of NRs, NPLs, and NWs. The dimension of each nanocrystal is indicated in each panel. Wide-view transmission electron microscopy (TEM) images of the nanocrystal corresponding to the optical spectra. The insets show the NRs lying flat on the substrate, revealing the width different from the thickness. TEM images at the higher magnification showing the thickness of each nanocrystal sample. The insets of clearly show the number of $PbBr_6^{4-}$ octahedral units n = 3 and 5, respectively. **b** Histograms showing the distribution of thickness (i), width (ii), and length (iii) of NRs with t = 2 nm (red) and t = 3 nm (blue). **c** In situ measurement of photoluminescence (PL) during the synthesis of NRs and NWs. NRs with t = 2 and 3 nm, top left and right respectively. NWs with t = 2 and 3 nm, bottom left and right respectively. The PL spectra of the final product are shown on the right side of the contour plots. Reprinted (adapted) with permission from Refs. [32]. Copyright (2015) American Chemical Society

thickness down to three unit cells. The reactions are performed at room temperature where the non-reactive precursors are combined, and the reaction is initiated via a sharp change in solvent polarity. Since it is not possible to control the kinetics via temperature, the kinetics were controlled via halide precursor ($CuBr_2$, $NiBr_2$, $CoBr_2$, etc.) choice which modulates the halide release rate.

In this reaction scheme, the sharp increase in solvent polarity activates the metal ions, which immediately initiates nucleation and growth of the NCs. As previously discussed for the control of the QD size, the concentration of halide plays the critical role in determining the size of the NCs in the confined dimension, i.e., thickness of NWs and NPLs. Additional metal bromide salts, such as $CoBr_2$ and $CuBr_2$, were used as the source of halide for their relatively high solubility in non-polar solvent, allowing for high concentrations of $Br^-$ in the reactant mixture even at room temperature. In general, the thickness of the NCs decreased with increasing concentration of $Br^-$, exhibiting the similar halide concentration dependence of the size of the cube-shaped QDs. Performing the reaction at room temperature takes advantage of the temperature dependent anisotropic kinetics discussed before as being critical to promoting anisotropic particle growth.

The evidence for the thermodynamic control of the NC thickness is seen in Fig. 2.8c, where the time-dependent exciton PL after reaction initiation for 3 and 5 monolayer (top and bottom respectively) NPLs and NWs (left and right respectively). Two important observations are made; (i) the thickness of the NPL and NWs increase by two monolayers rather than single monolayer. (ii) once the terminal thickness is reached, it does not change over time while the reaction continues. The increase of the thickness by two monolayers is likely due to the simultaneous addition of each monolayer on both sides of the basal plane. The kinetics independent terminal thickness of the NCs is the same as the behavior of the size of the QDs formed under thermodynamic control as discussed in the previous section. These observations provide critical insights on how the thermodynamic equilibrium-based size control can be combined with kinetic anisotropy at room-temperature reaction to obtain anisotropic and quantum confined MHP NCs.

## 2.3 Photophysical Properties of Quantum-Confined MHP NCs

Quantum confined MHP NCs exhibit significantly different static and dynamic photophysical properties from those of weakly-confined NCs and bulk due to the stronger interaction of electron and hole within the confined space altering not only the energetics of exciton but also its dynamic relaxation pathways. To date, the characterization and understanding of the properties of quantum-confined MHP NCs, directly influenced by the size or morphology-dependent quantum confinement, has been lagging behind compared to the development of the synthetic methodologies of various MHP NCs with controllable quantum confinement and dimensionality. One of the limiting factors in particular with optical characterization of the size-dependent properties of the quantum confined MHP NCs has been the relatively lower stability of the NCs, especially for the smaller NCs in the strongly confined regime compared to other semiconductor NCs such as II–VI QDs that have been extensively studied as the archetypal QD system. For instance, photoluminescence quantum yield of quantum confined MHP NCs including QDs, NWs and NPLs are more prone to degradation when exposed to the ambient environment and under continued photoexcitation condition even when compared to the large MHP NCs in weakly confined regime [55]. Photoreaction between the MHP NCs and halogenated solvent commonly used to disperse the colloidal semiconductor QDs, such as dichloromethane and chloroform, was also reported. The highly labile halide anion in MHP coupled with interfacial electron transfer-induced dissociation of halogenated solvent resulted in halide exchange upon above-gap photoexcitation altering the halide composition of the MHP NCs when the solvent and NCs have different halogen elements [55]. These illustrate the additional challenges in accurate and reliable characterization of the MHP NCs compared to other semiconductor NCs. Nevertheless, significant progress has been made in understanding the photophysical properties of MHP NCs that reflect the influence of the size and morphology-dependent quantum confinement in recent years. In this section, several static and dynamic photophysical properties of quantum confined MHP NCs with focus on QDs will be discussed.

### 2.3.1 *Size-Dependent Exciton Level Structure and Absorption Cross Section of MHP QDs*

The size-dependent bandgap that dictates the exciton absorption and emission energy is the most widely utilized property of semiconductor QDs in various applications [9, 66]. However, in the early studies of MHP NCs, the tunability of exciton absorption and emission was obtained mostly by chemical modification of the bandgap, taking advantage of facile exchange of halides (Cl, Br, and I), enabling continuous tuning of the bandgap across the entire visible spectrum. With the availability of synthetic and post-synthetic methods that can produce MHP QDs with controlled size in the quantum confinement regime, more reliable characterization of the size-dependent exciton transition energy, Stokes shift, and spectral linewidth became possible. While both chemical tuning and size tuning of the bandgap can vary the color of the luminescence from MHP NCs for the applications requiring color tunable light source, the two approached do not have the same effect on all the characteristics of exciton luminescence. For instance, linewidth of the exciton luminescence is sensitive to the size of the QDs, while not very sensitive to the change of halide composition.

Figure 2.9a shows the size-dependent absorption and emission spectra of $CsPbBr_3$ QDs and Stokes shift. The size-dependent exciton transition energy studied by several groups were close to each other and calculated exciton absorption peak based on effective mass approximation was in reasonable agreement with the experiment [17, 67]. Kuno et al. studied the size-dependent Stokes shift in $CsPbBr_3$ QDs, and explained the observed increase in stokes shift with decreasing particle size in terms of the size-dependent energy gap between the valence band edge level and delocalized hole level. The observed stokes shift, extracted from the fluorescence line narrowing experiments, ranged from 20–120 meV for NCs with edge lengths from 4–12 nm. From the electronic structure calculation, they suggested that the Stokes shift results from the existence of a confined hole state above the valence band edge [67]. They argued that emission in $CsPbBr_3$ occurs from a bulk-like electron state and delocalized hole state which is dark in the absorption. The resulting stokes shift was interpreted as the energy difference between the bulk-like bandedge hole state and delocalized hole states. The different orbital character of these states results

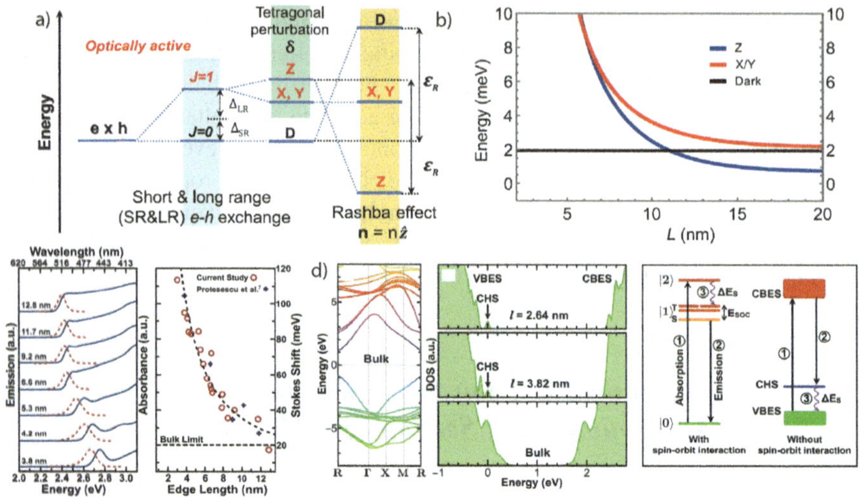


**Fig. 2.9** **a–b** Schematic of the band edge exciton fine structure in weakly confined $CsPbX_3$ NCs. Calculated size dependence of band edge exciton fine structure from the effective mass approximation for $CsPbBr_3$ QDs. **c** Size-dependent absorption, PL, and Stokes shift from $CsPbBr_3$ QDs, **d** Electronic structure calculations predicting the existence of a confined hole state above the valence band edge used to explain the size-dependent Stokes shift. Reprinted (adapted) with permission from Refs. [67, 72]. Copyright (2015) American Chemical Society

in different confinement effects for the delocalized hole state and band edge hole state, resulting in the size dependent stokes shift. There are no other current models or experiments which attempt to understand the size dependent stokes shift in the $CsPbBr_3$ however given the relative importance it will likely be revisited in more detail.

Another characteristic of exciton in $CsPbBr_3$ QDs that may exhibit size dependence is the exciton fine structure, particularly on the ordering of dipole-allowed bright exciton and dipole-forbidden dark exciton. A recent study in $CsPbBr_3$ NCs in weakly confined system indicated that the lowest-energy exciton state is bright exciton, which is different from the majority of other semiconductor NCs that have dark exciton as the lowest-energy exciton state. In typical semiconductor NCs, the lowest-energy dark exciton is located a few meV below the bright exciton level, therefore the luminescence property of the NCs change drastically at low temperatures where the thermal energy becomes smaller than the bright-dark exciton splitting. At sufficiently low temperature, where thermal excitation from the lowest to the upper exciton level is suppressed, long-lived dark exciton dominates the luminescence. However, this was not the case in $CsPbBr_3$ NCs, where only short-lived bright exciton emission (~500 ps) was observed in the 4 K PL, suggesting its unique exciton fine structure. The authors explained this with the Rashba effect; perturbations to the electronic structure resulting from the asymmetric lattice distortion which inverts the bright and dark level ordering as shown in Fig. 2.10a [68]. On the other hand, another study in $MAPbBr_3$ NCs showed an evidence indicating that the lowest-energy exciton

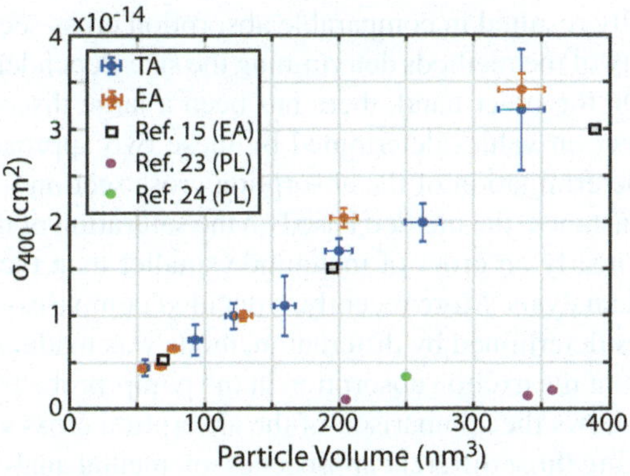


**Fig. 2.10** Comparison of the size-dependent absorption cross section values determined by different methods adapted from the data in Ref [62]. Results from four different studies employing three different methods are compared. EA, PL, TA each indicate the method based on elemental analysis, transient PL and transient absorption respectively. Reprinted from Ref. [62], with permission from AIP Publishing

state is dark state ~2.5 meV below the bright state, indicating potential complexity of the exciton fine structure in MHP NCs [69]. Recently, it was predicted that the ordering of bright and dark exciton, which are influenced by both Rashba effect and size-dependent electron-hole exchange energy will result in switching of the bright and dark exciton level order as a function of NC size, (Fig. 2.9d) which predicts the dark lowest-energy exciton state for strongly quantum confined $CsPbBr_3$ QDs, opposite to what has been reported for very weakly confined $CsPbBr_3$ NCs. While further research is necessary, experimental evidence supporting the existence of the low energy dark exciton in strongly confined $CsPbBr_3$ is emerging [70–72].

Another fundamental size-dependent property important for the characterization of various photophysical behavior of MHP QDs is the absorption cross section, which determines the density of exciton or charge carriers created at a given photoexcitation condition. Because various nonlinear optical properties of the photoexcited MHP QDs are dependent on the density of photoexcited exciton, accurate determination of the size-dependent absorption cross section is important.

Determination of the absorption cross section is commonly achieved by employing a few different approaches. On is through careful correlated analysis of the absorption intensities of the colloidal solution of the NCs to the concentrations of the NCs of known size using Beer's Law. Often, the concentration of colloidal NCs is determined from elemental analysis in conjunction with the information on the size (volume) of NCs, obtained from the electron microscopy. Another approach that has been frequently employed is examining the saturation behavior of the transient absorption or photoluminescence intensity as a function of photoexcitation fluence under the condition the Poisson distribution is applicable for the number of photons absorbed in each QDs. For relatively small NCs experiencing quantum confinement, such assumption should be valid. In the case of II–VI and III–V QDs, these two approaches generally resulted in comparable absorption cross section values, establishing the reliability of the methods determining the size-dependent absorption cross section [73, 74]. On the other hand, there has been a large discrepancy among the absorption cross section values determined by these two approaches in the earlier literatures on the determination of the absorption cross section in the case of MHP NCs [75–77]. For instance, the method based on the saturation of the transient photoluminescence was nearly an order of magnitude smaller than the value determined from the elemental analysis. More recently, a detailed comparison of the absorption cross section values determined by different methods was made, including the saturation of the bleach of the exciton absorption in the pump-probe transient absorption [78]. Figure. 2.10 shows the comparison of the absorption cross section of $CsPbBr_3$ NCs determined using three different approaches: elemental analysis (EA), transient photoluminescence (PL), transient absorption (TA). This comparison concluded that the values determined by elemental analysis, which are in good agreement with those determined by transient absorption, are more reliable than those determined by analyzing the saturation of the transient photoluminescence intensity. Compared to other colloidal binary semiconductor QDs, MHP QDs present additional challenges in determining the absorption cross section accurately. For instance, the significant departure of the Cs:Pb:X stoichiometric ratio in $CsPbX_3$ QDs from the bulk value of 1:1:3 that depends on the size of the QD and on the reaction condition can complicate the determination of the absorption cross section via elemental analysis. Because of the labile anion and the sensitivity of the crystal phase to the surrounding environment, MHP QDs are more susceptible to change their stoichiometric composition or structure in response to changes of ligand and solvent environment [79]. The Cs/Pb ratio in the range of 0.62–1.5 that varies depending on the QD size, synthesis method, and aging of the sample was reported [31, 77, 79–81]. A large deviation of the Cs/Pb ratio from 1 that varies with the thickness was also observed in 1D and 2D structures [32]. The Br/Pb ratio is also shown to be size dependent, exhibiting increasing Br/Pb ratio with decreasing QD size in the Br-terminated QDs [31]. In the case of $CsPbCl_3$ QDs, post-synthesis self-anion exchange resulted in a significant increase of the absorption intensity with a concomitant increase of PL quantum yield without a noticeable change in the particle size, presumably by removing the existing Cl vacancies in the QDs [55]. Therefore, careful enumeration of the stoichiometric ratio of the constituting elements in the QDs is important for accurate determination of the absorption cross section.

With the establishment of reliable method to determine the absorption cross section of colloidal MHP NCs, the size-dependent absorption cross section of $CsPbBr_3$ QDs in strongly quantum confined regime has been also determined. At the wavelengths significantly above the bandgap where the density of states in the conduction and valence bands are less discrete and more bulk-like, the absorption cross section values are linear to the volume of the NCs as expected. Interestingly, $CsPbBr_3$ QDs exhibit very similar absorption cross section values to those of CdSe QDs of the comparable size at 400 nm, contrary to earlier perception that MHP NCs possess comparatively large absorption cross section and functions as the more efficient light absorber.

### 2.3.2 *Size-Dependent Exciton Dynamics in MHP QDs*

Upon photoexcitation of QDs above the bandgap, the electron and hole in conduction and valance bands will cool down to the bandedge and undergo recombination. In QDs experiencing quantum confinement, the rates of cooling of hot charge carriers and relaxation of bandedge excitons are affected by the size-dependent energy level structure and electron-hole interaction. Furthermore, when multiple excitons are prepared simultaneously, the relaxation of exciton modified by Auger-type multi-exciton interaction is sensitive to the spatial confinement, exhibiting size-dependent relaxation dynamics. The stronger interaction among the charge carriers can also lead to the enhancement of the spectroscopic features influenced by the stronger quantum confinement. This section discusses recent progress made in understanding the size-dependent dynamics of hot carrier cooling, Auger relaxation of excitons, and pump-induce nonlinear absorption in the photoexcited MHP QDs in quantum confined regime.

Taking advantage of the recently developed synthetic techniques, several studies have explored the effect of varying the QD size on the hot carrier cooling and Auger recombination rates in $CsPbBr_3$ and $FAPbBr_3$ perovskite QDs. Auger recombination wherein the energy of an exciton or excited charge carrier is transferred to another exciton or charge carrier is a critical dynamic process for a number of applications. Understanding Auger recombination mechanisms and rates is important for processes such as multi exciton generation (MEG) and lasing where the preparation or relaxation of multiexcitons dictates applications feasibility. For example, the utilization of multi-excitons for MEG enhanced solar cells or lasing applications depends strongly on the ability to sustain multiexciton states. The size-dependent Auger recombination of biexcitons was recently studied in $CsPbBr_3$ QDs, the lifetime scaled linearly with QD volume which is the typical behavior observed in other QD systems [82]. Furthermore, Auger recombination rates were about 10 times faster than in CdSe and PbS suggesting the utility of $CsPbB_3$ for applications where biexcitons are detrimental like single photon source. It is interesting to compare the Auger recombination rates of $CsPbBr_3$ NC with different composition or shape morphology. Compared with $CsPbBr_3$ QDs exhibiting the Auger recombination rate scaling linearly with volume, the Auger recombination rate in $CsPbBr_3$ NWs and NPLs scaled linearly with the lateral area or length. This departure from the expected linear volume scaling behavior is understood in terms of the exciton-exciton collisional frequency where. Unlike in QDs, where the multiple excitons are forced to occupy the same volume, excitons in NPLs and NWs, with one or two non-confined dimensions, are not forced into the same volume, and must undergo a collision to experience Auger recombination [33].

Hot carrier extraction from semiconductor NCs is a rapidly growing field, and information of hot carrier lifetimes is not only of fundamental interest but also critical for the design of more efficient hot carrier harvesting devices. The hot carrier cooling mechanism in QDs such as CdSe and PbSe have been studied for over a decade, nevertheless, new cooling pathways are still being studied. The hot carrier relaxation of both hybrid and all-inorganic MHP QDs has been studied mostly from transient absorption measurements, and a number of cooling mechanisms have been proposed including exciton-phonon energy transfer, Auger mediated cooling, and direct energy transfer from exciton to ligand. While hot carrier cooling in bulk perovskites is affected by the polaron formation, there is no evidence to support polaron formation affecting the cooling rates in perovskite NCs. The most current model suggests cooling mechanisms similar to CdSe QDs, where the hot charge carriers cool via coupling with the vibrational modes of the surface ligands. This mechanism proposes that hot electrons and holes relax via two-step process involving the Auger mediated relaxation of hot electrons, followed by the non-adiabatic ligand mediated relaxation of hot holes. Ultrafast transient absorption experiments in $CsPbBr_3$ and $FAPbBr_3$ QDs revealed the carrier cooling times are relatively unaffected by the particle size (300–400 fs for $CsPbBr_3$ and 120–150 fs $FAPbBr_3$) for QDs sized 2.5–10 nm [59, 82]. In comparison, the energy loss rates dE/dt showed a strong dependence on the QD size, which is consistent with non-adiabatic ligand mediated carrier relaxation involving the vibrational states of the surface ligands. The size dependence of dE/dt comes from the differing wavefunction overlap between the exciton and the ligand, increasing with decreasing particle size as shown in Fig. 2.11b. Furthermore, the cooling rates showed relatively no influence from ligand choice or at temperatures down to 80 K, suggesting the absence of phonon emission or ligand assisted cooling via carrier-ligand Auger recombination. This model may also explain the enhanced hot carrier cooling in the $FAPbBr_3$, where the vibrational modes of the organic cations may be used to assist in the hot carrier cooling.

The effects of lattice distortion following photoexcitation has been of high interest in the perovskite community given the important role of polaron in determining various optical electronic and transport properties MHP materials [83, 84]. Recent studies in single crystal $MAPbBr_3$ and $CsPbBr_3$ discussed the effect of polaron formation by pulsed excitation light on the dynamic optical properties that reflect the dynamics of polaron formation. For instance, in a time-resolved optical Kerr effect study by Zhu and coworkers, the formation of a large polaron was observed in 300 and 700 fs for $MAPbBr_3$ and $CsPbBr_3$ respectively, involving exclusively the deformation of the $PbBr_3^-$ octahedra. More recently, transient absorption study by Son and coworkers in strongly confined $CsPbBr_3$ QDs reported the effect of polaron on activating the parity forbidden transition, which is prominently observed only in strongly confined QDs in contrast to the larger NCs with weak quantum confinement [85]. The formation of polaron in the excited state is expected to provide a large perturbation to the lattice, providing a significant distortion to the electronic structure and providing oscillator strength to the formally forbidden transition, Fig. 2.12b. In general, the TA spectra under single exciton condition consists of a bleach signal of the first exciton transition (B1), and two induced absorption signals (IA1, IA2) blueshifted and redshifted from B1. The similar behavior is observed across a range of sizes; Fig. 2.12c shows the evolution of the TA spectra from $CsPbBr_3$ QDs (4–10 nm). The exciton bleach feature results from the state filling of the band edge transition, while the identity of IA1 and IA2 are more complicated. Initially IA2 on the red side of B1 was assigned to the excitation of biexcitons by the probe pulse, however a number or recent papers suggest it results from polaronic effects

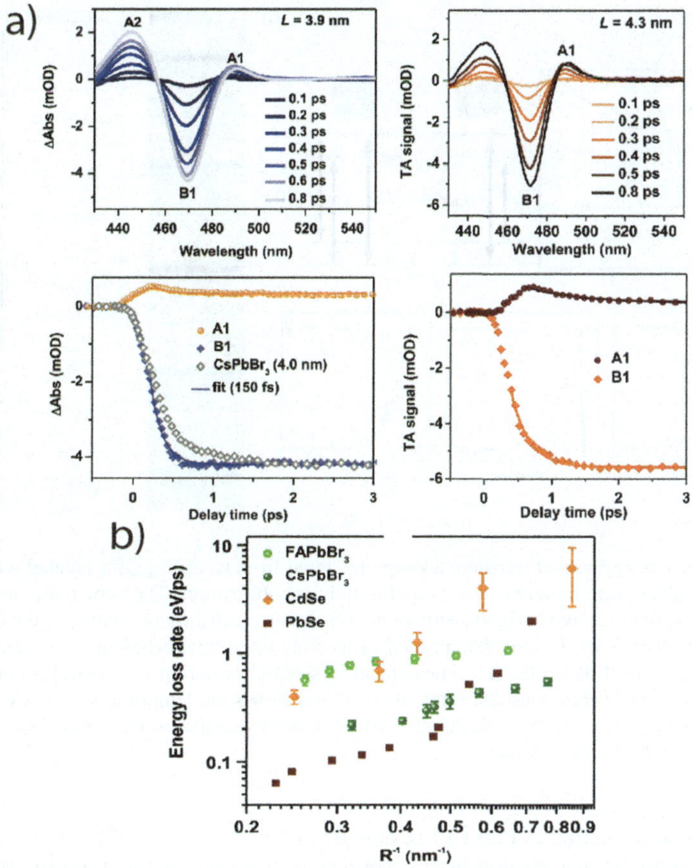


**Fig. 2.11** Size-dependent hot carrier cooling and Auger recombination rates in $CsPbBr_3$ and $FAPbBr_3$ QDs. **a** Early-time dynamics in strongly quantum confined $CsPbBr_3$ (left) and $FAPbBr_3$ (right) QDs with the corresponding bleach and induced absorption kinetics plotted below. **b** Comparison of the size dependent energy loss rate in $FAPbBr_3$, $CsPbBr_3$, CdSe, and PbSe. Ref [82]—Published by The Royal Society of Chemistry. Reprinted (adapted) with permission from Ref. [59]. Copyright (2015) American Chemical Society

as explained below. IA1 on the blue side of B1 is observed at an energy between the first and second exciton transition and results from the activation of parity forbidden transitions. The bleach recovery dynamics from these transitions show that IA2 is a short-lived state, decay time of roughly 600 fs, showing little size dependence. B1 and IA1 on the other hand are long lived with ~ 5 ns lifetime, similar to the PL decay, reflecting the exciton population, Fig. 2.12c.

IA features are common in the excited state absorption spectra for QDs, the most common of which is the biexcitons absorption which results in a derivative shaped feature from the energy shift between the exciton and biexciton absorption. These features are absent in the transient absorption data for $CsPbBr_3$ QDs, Fig. 2.12c. A

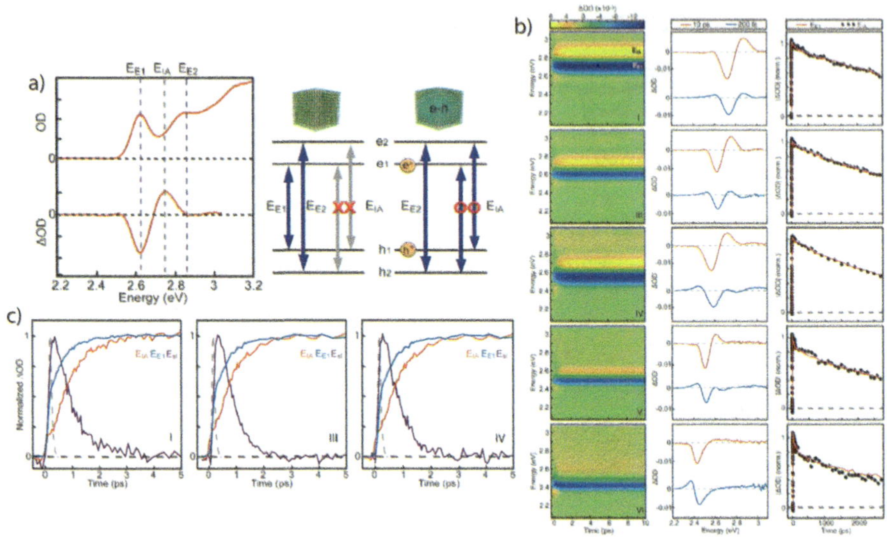


**Fig. 2.12** **a** Absorption and transient absorption signal from $CsPbBr_3$ QDs labeled with the first and second absorption features as well as the induced absorption. The schematic describes the origin of these features including the activation of the induced absorption feature in the QD excited state. **b** Size dependent transient absorption data showing the similar behavior across the entire size range, the right panel shows the full time dependent spectral evolution, the central panel shows the early and late time spectra, and the right hand column shows the long time kinetics. **c** Early time dynamics of $E_{E1}$, $E_{E2}$, and $E_{IA}$. Reprinted (adapted) with permission from Ref. [85]. Copyright (2015) American Chemical Society

more consistent picture is that IA1 results from the absorption of formally forbidden transitions which are activated in the excited state outlined in the diagram, Fig. 2.12b. This is supported by considering the short time dynamics, where the rise of B1, IA1 and IA2 have different dynamics. IA2 has an IRF limited rise time (~ 75 fs), while the rise time of IA1 is much longer is identical to the decay of IA2 (~ 600 fs), reflecting the polaron formation time which is similar to that observed by Zhu et all. The relative intensity of IA1 scales with the particle size, where the symmetry breaking perturbation appears more intense for the smaller particles as the particle size becomes comparable to the polaron size. In the nonconfined NPs, IA1 is low intensity and difficult resolve, likely resulting from the smaller perturbation induced by the polaron formation, nevertheless the rise and decay times of IA1 and IA2 present a similar picture. Recently, the similar spectral signature of polaron activated transitions were observed in $FAPbBr_3$ QDs [59].

## 2.4 Applications and Future Outlook

Considering the relatively short time strongly confined MHP NCs have been synthetically available, they have become well integrated into the current applications-based researches, notably in LEDs. Compared with bulk MHP materials, the NCs have several advantages. Bulk MHP materials are known to have relatively small exciton binding energy (2–6 meV), long lifetimes, and long diffusion lengths, all of which limit their ability to achieve high external quantum efficiency (EQE) of light emission in LED platform. However, quantum confined NC structures, especially 1D and 2D confined structure, can have exciton binding energies up to 250 meV, and the shorter radiative lifetimes relative to the bulk, making them better candidates for the source of photons in LED applications. $CsPbBr_3$ NCs were integrated into high quantum efficiency green LED (up to 21.3% EQE), however the low QY of chloride and mixed halide NCs made it difficult to prepared efficient blue emitting LED. As such, the confined $CsPbBr_3$ takes advantage of the defect tolerant nature and high QY of $CsPbBr_3$, potentially extending it to high efficiency blue emitting LED taking advantage of the size-tunable bandgap.

Quantum confined $CsPbBr_3$ NCs were demonstrated as efficient blue light emitters, with emission wavelength down to 460 nm for the smallest reported QDs and ~ 420 nm for monolayer NPLs or NWs [31, 43]. While the QDs can retain their high QY (up to 80%) the strongly confined platelets and wires have markedly smaller QY, however post synthesis modifications suggest it can be repaired. Shih et all recently prepared LED based on blue emitting $MAPbBr_3$ NPLs with variable thickens from 1–7 unit cells [86]. Electroluminescence was observed at room temperature from 440–510 nm with efficiencies ranging from 0.1–1.2%. Stranks and coworkers demonstrated 464 nm emitting LED with 0.3% EQE and more importantly identified the issues which need to be addressed to enhance the perovskite NPL LED [87]. They suggest that the most significant issue is the hole injection, which they suggest can be tuned via proper ligand and matrix choice. Nevertheless, before the efficient LED can be made, to take full advantage of the perovskite NPLs, the major issues that needs to be addressed is the intrinsically low PLQY.

The defect tolerant nature and facile control over the bandgap, coupled with confinement enhanced properties, such as exciton binding energies and carrier cooling dynamics, makes confined perovskites NCs desirable for a number of applications such as single photon source and high-efficiency LED. Towards this end, the synthesis of both hybrid and all-inorganic QDs, NPLs, and NWs with highly uniform size distributions has been well developed. In addition, further improvement of PL QY has been achieved via various post synthetic surface treatments or doping. While relatively lower stability of the MHP NCs still remains as a major problem to be resolved, this new family of quantum confined semiconductor NCs is a promising superior alterative to the existing semiconductor NCs as the source of photons and charges in various applications.

## References


1. J.-H. Choi, H. Wang, S.J. Oh, T. Paik, P. Sung, J. Sung, X. Ye, T. Zhao, B.T. Diroll, C.B. Murray, C.R. Kagan, Exploiting the colloidal nanocrystal library to construct electronic devices. Science **352**(6282), 205–208 (2016). https://doi.org/10.1126/science.aad0371
2. C.R. Kagan, E. Lifshitz, E.H. Sargent, D.V. Talapin, Building devices from colloidal quantum dots. Science **353** (6302) (2016). https://doi.org/10.1126/science.aac5523
3. S.A. McDonald, G. Konstantatos, S. Zhang, P.W. Cyr, E.J.D. Klem, L. Levina, E.H. Sargent, Solution-processed PbS quantum dot infrared photodetectors and photovoltaics. Nat. Mater. **4**, 138 (2005). https://doi.org/10.1038/nmat1299
4. X. Michalet, F.F. Pinaud, L.A. Bentolila, J.M. Tsay, S. Doose, J.J. Li, G. Sundaresan, A.M. Wu, S.S. Gambhir, S. Weiss, Quantum dots for live cells, in vivo imaging, and diagnostics. Science **307**(5709), 538–544 (2005). https://doi.org/10.1126/science.1104274
5. I.L. Medintz, H.T. Uyeda, E.R. Goldman, H. Mattoussi, Quantum dot bioconjugates for imaging, labelling and sensing. Nat. Mater. **4**(6), 435–446 (2005). https://doi.org/10.1038/nmat1390
6. X.H. Gao, Y.Y. Cui, R.M. Levenson, L.W.K. Chung, S.M. Nie, In vivo cancer targeting and imaging with semiconductor quantum dots. Nat. Biotechnol. **22**(8), 969–976 (2004). https://doi.org/10.1038/nbt994
7. A.P. Alivisatos, Semiconductor clusters, nanocrystals, and quantum dots. Science **271**(5251), 933–937 (1996). https://doi.org/10.1126/science.271.5251.933
8. T. Vossmeyer, L. Katsikas, M. Giersig, I.G. Popovic, K. Diesner, A. Chemseddine, A. Eychmueller, H. Weller, CdS nanoclusters: synthesis, characterization, size dependent oscillator strength, temperature shift of the excitonic transition energy, and reversible absorbance shift. J. Phys. Chem. **98**(31), 7665–7673 (1994). https://doi.org/10.1021/j100082a044
9. D.J. Norris, M.G. Bawendi, Measurement and assignment of the size-dependent optical spectrum in CdSe quantum dots. Phys. Rev. B **53**(24), 16338–16346 (1996). https://doi.org/10.1103/PhysRevB.53.16338
10. H.J. Yun, T. Paik, B. Diroll, M.E. Edley, J.B. Baxter, C.B. Murray, Nanocrystal size-dependent efficiency of quantum dot sensitized solar cells in the strongly coupled CdSe nanocrystals/$TiO_2$ system. ACS Appl. Mater. Interfaces **8**(23), 14692–14700 (2016). https://doi.org/10.1021/acsami.6b05552
11. D.J. Norris, A.L. Efros, S.C. Erwin, Doped nanocrystals. Science **319**(5871), 1776–1779 (2008). https://doi.org/10.1126/science.1143802
12. D.J. Norris, N. Yao, F.T. Charnock, T.A. Kennedy, High-quality manganese-doped ZnSe nanocrystals. Nano Lett. **1**(1), 3–7 (2001). https://doi.org/10.1021/nl005503h
13. N. Pradhan, X.G. Peng, Efficient and color-tunable Mn-doped ZnSe nanocrystal emitters: control of optical performance via greener synthetic chemistry. J. Am. Chem. Soc. **129**(11), 3339–3347 (2007). https://doi.org/10.1021/ja068360v
14. B.B. Srivastava, S. Jana, N.S. Karan, S. Paria, N.R. Jana, D.D. Sarma, N. Pradhan, Highly luminescent Mn-doped ZnS nanocrystals: gram-scale synthesis. J. Phys. Chem. Lett. **1**(9), 1454–1458 (2010). https://doi.org/10.1021/jz100378w
15. Y.A. Yang, O. Chen, A. Angerhofer, Y.C. Cao, On doping CdS/ZnS core/shell nanocrystals with Mn. J. Am. Chem. Soc. **130**(46), 15649–15661 (2008). https://doi.org/10.1021/ja805736k
16. L.C. Schmidt, A. Pertegás, S. González-Carrero, O. Malinkiewicz, S. Agouram, G. Mínguez Espallargas, H.J. Bolink, R.E. Galian, J. Pérez-Prieto, Nontemplate synthesis of $CH_3NH_3PbBr_3$ perovskite nanoparticles. J. Am. Chem. Soc. **136**(3), 850–853 (2014). https://doi.org/10.1021/ja4109209
17. L. Protesescu, S. Yakunin, M.I. Bodnarchuk, F. Krieg, R. Caputo, C.H. Hendon, R.X. Yang, A. Walsh, M.V. Kovalenko, Nanocrystals of cesium lead halide perovskites ($CsPbX_3$, X = Cl, Br, and I): Novel optoelectronic materials showing bright emission with wide color gamut. Nano Lett. **15**(6), 3692–3696 (2015). https://doi.org/10.1021/nl5048779

18. M.C. Brennan, M. Kuno, S. Rouvimov, Crystal structure of individual $CsPbBr_3$ perovskite nanocubes. Inorg. Chem. **58**(2), 1555–1560 (2019). https://doi.org/10.1021/acs.inorgchem.8b03078
19. M.S. Kirschner, B.T. Diroll, P. Guo, S.M. Harvey, W. Helweh, N.C. Flanders, A. Brumberg, N.E. Watkins, A.A. Leonard, A.M. Evans, M.R. Wasielewski, W.R. Dichtel, X. Zhang, L.X. Chen, R.D. Schaller, Photoinduced, reversible phase transitions in all-inorganic perovskite nanocrystals. Nat Commun. **10**(1), 504 (2019). https://doi.org/10.1038/s41467-019-08362-3
20. A. Swarnkar, A.R. Marshall, E.M. Sanehira, B.D. Chernomordik, D.T. Moore, J.A. Christians, T. Chakrabarti, J.M. Luther, Quantum dot–induced phase stabilization of $\alpha$-$CsPbI_3$ perovskite for high-efficiency photovoltaics. Science **354**(6308), 92–95 (2016). https://doi.org/10.1126/science.aag2700
21. X. Zhang, H. Lin, H. Huang, C. Reckmeier, Y. Zhang, W.C.H. Choy, A.L. Rogach, Enhancing the brightness of cesium lead halide perovskite nanocrystal based green light-emitting devices through the interface engineering with perfluorinated ionomer. Nano Lett. **16**(2), 1415–1420 (2016). https://doi.org/10.1021/acs.nanolett.5b04959
22. S. Yakunin, L. Protesescu, F. Krieg, M.I. Bodnarchuk, G. Nedelcu, M. Humer, G. De Luca, M. Fiebig, W. Heiss, M.V. Kovalenko, Low-threshold amplified spontaneous emission and lasing from colloidal nanocrystals of caesium lead halide perovskites. Nat. Commun. **6**, 8056 (2015). https://doi.org/10.1038/ncomms9056
23. M.V. Kovalenko, L. Protesescu, M.I. Bodnarchuk, Properties and potential optoelectronic applications of lead halide perovskite nanocrystals. Science **358**(6364), 745–750 (2017). https://doi.org/10.1126/science.aam7093
24. Q.A. Akkerman, G. Rainò, M.V. Kovalenko, L. Manna, Genesis, challenges and opportunities for colloidal lead halide perovskite nanocrystals. Nat. Mater. **17**(5), 394–405 (2018). https://doi.org/10.1038/s41563-018-0018-4
25. B.A. Koscher, J.K. Swabeck, N.D. Bronstein, A.P. Alivisatos, Essentially trap-free $CsPbBr_3$ colloidal nanocrystals by postsynthetic thiocyanate surface treatment. J. Am. Chem. Soc. **139**(19), 6566–6569 (2017). https://doi.org/10.1021/jacs.7b02817
26. J.-S. Yao, J. Ge, B.-N. Han, K.-H. Wang, H.-B. Yao, H.-L. Yu, J.-H. Li, B.-S. Zhu, J.-Z. Song, C. Chen, Q. Zhang, H.-B. Zeng, Y. Luo, S.-H. Yu, Ce3 + -doping to modulate photoluminescence kinetics for efficient $CsPbBr_3$ nanocrystals based light-emitting diodes. J. Am. Chem. Soc. **140**(10), 3626–3634 (2018). https://doi.org/10.1021/jacs.7b11955
27. D. Parobek, B.J. Roman, Y. Dong, H. Jin, E. Lee, M. Sheldon, D.H. Son, Exciton-to-dopant energy transfer in Mn-doped cesium lead halide perovskite nanocrystals. Nano Lett. **16**(12), 7376–7380 (2016). https://doi.org/10.1021/acs.nanolett.6b02772
28. Y. Cheng, C. Shen, L. Shen, W. Xiang, X. Liang, $Tb^{3+}$, $Eu^{3+}$ Co-doped $CsPbBr_3$ QDs glass with highly stable and luminous adjustable for white LEDs. ACS Appl. Mater. Interfaces. **10**(25), 21434–21444 (2018). https://doi.org/10.1021/acsami.8b05003
29. G. Nedelcu, L. Protesescu, S. Yakunin, M.I. Bodnarchuk, M.J. Grotevent, M.V. Kovalenko, Fast anion-exchange in highly luminescent nanocrystals of cesium lead halide perovskites ($CsPbX_3$, X = Cl, Br, I). Nano Lett. **15**(8), 5635–5640 (2015). https://doi.org/10.1021/acs.nanolett.5b02404
30. Q.A. Akkerman, V. D'Innocenzo, S. Accornero, A. Scarpellini, A. Petrozza, M. Prato, L. Manna, Tuning the optical properties of cesium lead halide perovskite nanocrystals by anion exchange reactions. J. Am. Chem. Soc. **137**(32), 10276–10281 (2015). https://doi.org/10.1021/jacs.5b05602
31. Y. Dong, T. Qiao, D. Kim, D. Parobek, D. Rossi, D.H. Son, Precise control of quantum confinement in cesium lead halide perovskite quantum dots via thermodynamic equilibrium. Nano Lett. **18**(6), 3716–3722 (2018). https://doi.org/10.1021/acs.nanolett.8b00861
32. Y. Dong, T. Qiao, D. Kim, D. Rossi, S.J. Ahn, D.H. Son, Controlling anisotropy of quantum-confined $CsPbBr_3$ nanocrystals by combined use of equilibrium and kinetic anisotropy. Chem. Mater. **31**(15), 5655–5662 (2019). https://doi.org/10.1021/acs.chemmater.9b01515
33. Q. Li, Y. Yang, W. Que, T. Lian, Size- and morphology-dependent auger recombination in $CsPbBr_3$ perovskite two-dimensional nanoplatelets and one-dimensional nanorods. Nano Lett. **19**(8), 5620–5627 (2019). https://doi.org/10.1021/acs.nanolett.9b02145

34. M.C. Weidman, A.J. Goodman, W.A. Tisdale, Colloidal halide perovskite nanoplatelets: an exciting new class of semiconductor nanomaterials. Chem. Mater. **29**(12), 5019–5030 (2017). https://doi.org/10.1021/acs.chemmater.7b01384
35. Z. Liang, S. Zhao, Z. Xu, B. Qiao, P. Song, D. Gao, X. Xu, Shape-controlled synthesis of all-inorganic $CsPbBr_3$ perovskite nanocrystals with bright blue emission. ACS Appl. Mater. Interfaces. **8**(42), 28824–28830 (2016). https://doi.org/10.1021/acsami.6b08528
36. D. Zhang, Y. Yu, Y. Bekenstein, A.B. Wong, A.P. Alivisatos, P. Yang, Ultrathin colloidal cesium lead halide perovskite nanowires. J. Am. Chem. Soc. **138**(40), 13155–13158 (2016). https://doi.org/10.1021/jacs.6b08373
37. A. Dutta, S.K. Dutta, S. Das Adhikari, N. Pradhan, Tuning the size of $CsPbBr_3$ nanocrystals: All at one constant temperature. ACS Energy Lett. **3**(2), 329–334 (2018). https://doi.org/10.1021/acsenergylett.7b01226
38. S.K. Ha, C.M. Mauck, W.A. Tisdale, Toward stable deep-blue luminescent colloidal llead halide perovskite nanoplatelets: systematic photostability investigation. Chem. Mater. **31**(7), 2486–2496 (2019). https://doi.org/10.1021/acs.chemmater.8b05310
39. J.A. Sichert, Y. Tong, N. Mutz, M. Vollmer, S. Fischer, K.Z. Milowska, R. García Cortadella, B. Nickel, C. Cardenas-Daw, J.K. Stolarczyk, A.S. Urban, J. Feldmann, Quantum size effect in organometal halide perovskite nanoplatelets. Nano Lett. **15**(10), 6521–6527 (2015). https://doi.org/10.1021/acs.nanolett.5b02985
40. C.B. Murray, C.R. Kagan, M.G. Bawendi, Self-organization of CdSe nanocrystallites into three-dimensional quantum dot superlattices. Science **270**(5240), 1335–1338 (1995). https://doi.org/10.1126/science.270.5240.1335%JScience
41. M. Nirmal, D.J. Norris, M. Kuno, M.G. Bawendi, A.L. Efros, M. Rosen, Observation of the "dark exciton" in CdSe quantum dots. Phys. Rev. Lett. **75**(20), 3728–3731 (1995). https://doi.org/10.1103/PhysRevLett.75.3728
42. J. Lim, Y.-S. Park, V.I. Klimov, Optical gain in colloidal quantum dots achieved with direct-current electrical pumping. Nat. Mater. **17**(1), 42–49 (2018). https://doi.org/10.1038/nmat5011
43. Q.A. Akkerman, S.G. Motti, A.R. Srimath Kandada, E. Mosconi, V. D'Innocenzo, G. Bertoni, S. Marras, B.A. Kamino, L. Miranda, F. De Angelis, A. Petrozza, M. Prato, L. Manna, Solution synthesis approach to colloidal cesium lead halide perovskite nanoplatelets with monolayer-level thickness control. J. Am. Chem. Soc. **138**(3), 1010–1016 (2016). https://doi.org/10.1021/jacs.5b12124
44. G. Almeida, L. Goldoni, Q. Akkerman, Z. Dang, A.H. Khan, S. Marras, I. Moreels, L. Manna, Role of acid–base equilibria in the size, shape, and phase control of cesium lead bromide nanocrystals. ACS Nano **12**(2), 1704–1711 (2018). https://doi.org/10.1021/acsnano.7b08357
45. M. Koolyk, D. Amgar, S. Aharon, L. Etgar, Kinetics of cesium lead halide perovskite nanoparticle growth; focusing and de-focusing of size distribution. Nanoscale **8**(12), 6403–6409 (2016). https://doi.org/10.1039/C5NR09127F
46. S. Sun, D. Yuan, Y. Xu, A. Wang, Z. Deng, Ligand-mediated synthesis of shape-controlled cesium lead halide perovskite nanocrystals via reprecipitation process at room temperature. ACS Nano **10**(3), 3648–3657 (2016). https://doi.org/10.1021/acsnano.5b08193
47. C.B. Murray, D.J. Norris, M.G. Bawendi, Synthesis and characterization of nearly monodisperse CdE (E = sulfur, selenium, tellurium) semiconductor nanocrystallites. J. Am. Chem. Soc. **115**(19), 8706–8715 (1993). https://doi.org/10.1021/ja00072a025
48. A.B. Greytak, P.M. Allen, W. Liu, J. Zhao, E.R. Young, Z. Popović, B.J. Walker, D.G. Nocera, M.G. Bawendi, Alternating layer addition approach to CdSe/CdS core/shell quantum dots with near-unity quantum yield and high on-time fractions. Chem. Sci. **3**(6), 2028–2034 (2012). https://doi.org/10.1039/C2SC00561A
49. R.C. Page, D. Espinobarro-Velazquez, M.A. Leontiadou, C. Smith, E.A. Lewis, S.J. Haigh, C. Li, H. Radtke, A. Pengpad, F. Bondino, E. Magnano, I. Pis, W.R. Flavell, P. O'Brien, D.J. Binks, Near-unity quantum yields from chloride treated CdTe colloidal quantum dots. Small **11**(13), 1548–1554 (2015). https://doi.org/10.1002/smll.201402264
50. Y. Liu, M. Guo, S. Dong, X. Jiao, T. Wang, D. Chen, Room temperature colloidal synthesis of $CsPbBr_3$ nanowires with tunable length, width and composition. J. Mater. Chem. **6**(29), 7797–7802 (2018). https://doi.org/10.1039/C8TC02636J

51. I. Lignos, S. Stavrakis, G. Nedelcu, L. Protesescu, A.J. deMello, M.V. Kovalenko, Synthesis of cesium lead halide perovskite nanocrystals in a droplet-based microfluidic platform: Fast parametric space mapping. Nano Lett. **16**(3), 1869–1877 (2016). https://doi.org/10.1021/acs.nanolett.5b04981
52. Y. Bekenstein, B.A. Koscher, S.W. Eaton, P. Yang, A.P. Alivisatos, Highly luminescent colloidal nanoplates of perovskite cesium lead halide and their oriented assemblies. J. Am. Chem. Soc. **137**(51), 16008–16011 (2015). https://doi.org/10.1021/jacs.5b11199
53. A. Pan, B. He, X. Fan, Z. Liu, J.J. Urban, A.P. Alivisatos, L. He, Y. Liu, Insight into the ligand-mediated synthesis of colloidal $CsPbBr_3$ perovskite nanocrystals: the role of organic acid, base, and cesium precursors. ACS Nano **10**(8), 7943–7954 (2016). https://doi.org/10.1021/acsnano.6b03863
54. D.V. Talapin, A.L. Rogach, A. Kornowski, M. Haase, H. Weller, Highly luminescent monodisperse CdSe and CdSe/ZnS nanocrystals synthesized in a Hexadecylamine – Trioctylphosphine Oxide – Trioctylphospine mixture. Nano Lett. **1**(4), 207–211 (2001). https://doi.org/10.1021/nl0155126
55. D. Parobek, Y. Dong, T. Qiao, D. Rossi, D.H. Son, Photoinduced anion exchange in cesium lead halide perovskite nanocrystals. J. Am. Chem. Soc. **139**(12), 4358–4361 (2017). https://doi.org/10.1021/jacs.7b01480
56. F. Zhang, H. Zhong, C. Chen, X-g Wu, X. Hu, H. Huang, J. Han, B. Zou, Y. Dong, Brightly luminescent and color-tunable colloidal $CH_3NH_3PbX_3$ ($X = Br, I, Cl$) quantum dots: potential alternatives for display technology. ACS Nano **9**(4), 4533–4542 (2015). https://doi.org/10.1021/acsnano.5b01154
57. L. Protesescu, S. Yakunin, M.I. Bodnarchuk, F. Bertolotti, N. Masciocchi, A. Guagliardi, M.V. Kovalenko, Monodisperse formamidinium lead bromide nanocrystals with bright and stable green photoluminescence. J. Am. Chem. Soc. **138**(43), 14202–14205 (2016). https://doi.org/10.1021/jacs.6b08900
58. R.M. Maceiczyk, K. Dümbgen, I. Lignos, L. Protesescu, M.V. Kovalenko, A.J. deMello, Microfluidic reactors provide preparative and mechanistic insights into the synthesis of formamidinium lead halide perovskite nanocrystals. Chem. Mater. **29**(19), 8433–8439 (2017). https://doi.org/10.1021/acs.chemmater.7b02998
59. Y. Li, T. Ding, X. Luo, Y. Tian, X. Lu, K. Wu, Synthesis and spectroscopy of monodispersed, quantum-confined $FAPbBr_3$ perovskite nanocrystals. Chem. Mater. **32**(1), 549–556 (2020). https://doi.org/10.1021/acs.chemmater.9b04297
60. L. Dou, A.B. Wong, Y. Yu, M. Lai, N. Kornienko, S.W. Eaton, A. Fu, C.G. Bischak, J. Ma, T. Ding, Atomically thin two-dimensional organic-inorganic hybrid perovskites. Science **349**(6255), 1518–1521 (2015)
61. T. Ishihara, J. Takahashi, T. Goto, Optical properties due to electronic transitions in two-dimensional semiconductors $(CnH_{2n+1}NH_3)_2PbI_4$. Phys. Rev. B **42**(17), 11099 (1990)
62. V.A. Hintermayr, A.F. Richter, F. Ehrat, M. Döblinger, W. Vanderlinden, J.A. Sichert, Y. Tong, L. Polavarapu, J. Feldmann, A.S. Urban, Tuning the optical properties of perovskite nanoplatelets through composition and thickness by ligand-assisted exfoliation. Adv. Mater. **28**(43), 9478–9485 (2016)
63. M.C. Weidman, M. Seitz, S.D. Stranks, W.A. Tisdale, Highly tunable colloidal perovskite nanoplatelets through variable cation, metal, and halide composition. ACS Nano **10**(8), 7830–7839 (2016). https://doi.org/10.1021/acsnano.6b03496
64. A. Kostopoulou, M. Sygletou, K. Brintakis, A. Lappas, E. Stratakis, Low-temperature benchtop-synthesis of all-inorganic perovskite nanowires. Nanoscale **9**(46), 18202–18207 (2017)
65. M. Imran, F. Di Stasio, Z. Dang, C. Canale, A.H. Khan, J. Shamsi, R. Brescia, M. Prato, L. Manna, Colloidal synthesis of strongly fluorescent $CsPbBr_3$ nanowires with width tunable down to the quantum confinement regime. Chem. Mater. **28**(18), 6450–6454 (2016). https://doi.org/10.1021/acs.chemmater.6b03081
66. I. Moreels, K. Lambert, D. Smeets, D. De Muynck, T. Nollet, J.C. Martins, F. Vanhaecke, A. Vantomme, C. Delerue, G. Allan, Z. Hens, Size-dependent optical properties of colloidal PbS quantum dots. ACS Nano **3**(10), 3023–3030 (2009). https://doi.org/10.1021/nn900863a

67. M.C. Brennan, J.E. Herr, T.S. Nguyen-Beck, J. Zinna, S. Draguta, S. Rouvimov, J. Parkhill, M. Kuno, Origin of the size-dependent stokes shift in $CsPbBr_3$ perovskite nanocrystals. J. Am. Chem. Soc. **139**(35), 12201–12208 (2017). https://doi.org/10.1021/jacs.7b05683
68. M.A. Becker, R. Vaxenburg, G. Nedelcu, P.C. Sercel, A. Shabaev, M.J. Mehl, J.G. Michopoulos, S.G. Lambrakos, N. Bernstein, J.L. Lyons, T. Stöferle, R.F. Mahrt, M.V. Kovalenko, D.J. Norris, G. Rainò, A.L. Efros, Bright triplet excitons in caesium lead halide perovskites. Nature **553**, 189 (2018). https://doi.org/10.1038/nature25147
69. P. Tamarat, M.I. Bodnarchuk, J.-B. Trebbia, R. Erni, M.V. Kovalenko, J. Even, B. Lounis, The ground exciton state of formamidinium lead bromide perovskite nanocrystals is a singlet dark state. Nat. Mater. **18**(7), 717–724 (2019). https://doi.org/10.1038/s41563-019-0364-x
70. L. Chen, B. Li, C. Zhang, X. Huang, X. Wang, M. Xiao, Composition-dependent energy splitting between bright and dark excitons in lead halide perovskite nanocrystals. Nano Lett. **18**(3), 2074–2080 (2018). https://doi.org/10.1021/acs.nanolett.8b00184
71. X.L. Daniel Rossi, Y. Lee, M. Khurana, J. Puthenpurayil, K. Kim, A. Akimov, J. Cheon, D.H. Son (2020) Intense dark exciton emission from strongly quantum confined $CsPbBr_3$ nanocrystals. arXiv:200208589
72. P.C. Sercel, J.L. Lyons, D. Wickramaratne, R. Vaxenburg, N. Bernstein, A.L. Efros, Exciton fine structure in perovskite nanocrystals. Nano Lett. **19**(6), 4068–4077 (2019). https://doi.org/10.1021/acs.nanolett.9b01467
73. M. Ji, S. Park, S.T. Connor, T. Mokari, Y. Cui, K.J. Gaffney, Efficient multiple exciton generation observed in colloidal PbSe quantum dots with temporally and spectrally resolved intraband excitation. Nano Lett. **9**(3), 1217–1222 (2009). https://doi.org/10.1021/nl900103f
74. I. Moreels, K. Lambert, D. De Muynck, F. Vanhaecke, D. Poelman, J.C. Martins, G. Allan, Z. Hens, Composition and size-dependent extinction coefficient of colloidal PbSe quantum dots. Chem. Mater. **19**(25), 6101–6106 (2007). https://doi.org/10.1021/cm071410q
75. J.A. Castañeda, G. Nagamine, E. Yassitepe, L.G. Bonato, O. Voznyy, S. Hoogland, A.F. Nogueira, E.H. Sargent, C.H.B. Cruz, L.A. Padilha, Efficient biexciton interaction in perovskite quantum dots under weak and strong confinement. ACS Nano **10**(9), 8603–8609 (2016). https://doi.org/10.1021/acsnano.6b03908
76. N.S. Makarov, S. Guo, O. Isaienko, W. Liu, I. Robel, V.I. Klimov, Spectral and dynamical properties of single excitons, biexcitons, and trions in cesium–lead-halide perovskite quantum dots. Nano Lett. **16**(4), 2349–2362 (2016). https://doi.org/10.1021/acs.nanolett.5b05077
77. J. Maes, L. Balcaen, E. Drijvers, Q. Zhao, J. De Roo, A. Vantomme, F. Vanhaecke, P. Geiregat, Z. Hens, Light absorption coefficient of $CsPbBr_3$ perovskite nanocrystals. J. Phys. Chem. Lett. **9**(11), 3093–3097 (2018). https://doi.org/10.1021/acs.jpclett.8b01065
78. J. Puthenpurayil, O.H.-C. Cheng, T. Qiao, D. Rossi, D.H. Son, On the determination of absorption cross section of colloidal lead halide perovskite quantum dots. J. Chem. Phys. **151**(15), 154706 (2019). https://doi.org/10.1063/1.5126039
79. M.I. Bodnarchuk, S.C. Boehme, S. ten Brinck, C. Bernasconi, Y. Shynkarenko, F. Krieg, R. Widmer, B. Aeschlimann, D. Günther, M.V. Kovalenko, I. Infante, Rationalizing and controlling the surface structure and electronic passivation of cesium lead halide nanocrystals. ACS Energy Lett **4**(1), 63–74 (2019). https://doi.org/10.1021/acsenergylett.8b01669
80. X. Du, G. Wu, J. Cheng, H. Dang, K. Ma, Y.-W. Zhang, P.-F. Tan, S. Chen, High-quality $CsPbBr_3$ perovskite nanocrystals for quantum dot light-emitting diodes. RSC Adv. **7**(17), 10391–10396 (2017). https://doi.org/10.1039/C6RA27665B
81. J. De Roo, M. Ibáñez, P. Geiregat, G. Nedelcu, W. Walravens, J. Maes, J.C. Martins, I. Van Driessche, M.V. Kovalenko, Z. Hens, Highly dynamic ligand binding and light absorption coefficient of cesium lead bromide perovskite nanocrystals. ACS Nano **10**(2), 2071–2081 (2016). https://doi.org/10.1021/acsnano.5b06295
82. Y. Li, R. Lai, X. Luo, X. Liu, T. Ding, X. Lu, K. Wu, On the absence of a phonon bottleneck in strongly confined $CsPbBr_3$ perovskite nanocrystals. Chem. Sci. **10**(23), 5983–5989 (2019). https://doi.org/10.1039/C9SC01339C
83. T.J.S. Evans, K. Miyata, P.P. Joshi, S. Maehrlein, F. Liu, X.Y. Zhu, Competition between hot-electron cooling and large polaron screening in $CsPbBr_3$ perovskite single crystals. J. Phys. Chem. C **122**(25), 13724–13730 (2018). https://doi.org/10.1021/acs.jpcc.8b00476

84. K. Miyata, T.L. Atallah, X. –Y. Zhu (2017) Lead halide perovskites: crystal-liquid duality, phonon glass electron crystals, and large polaron formation. Science **3**(10):e1701469. https://doi.org/10.1126/sciadv.1701469%JScienceAdvances
85. D. Rossi, H. Wang, Y. Dong, T. Qiao, X. Qian, D.H. Son, Light-induced activation of forbidden exciton transition in strongly confined perovskite quantum dots. ACS Nano **12**(12), 12436–12443 (2018). https://doi.org/10.1021/acsnano.8b06649
86. S. Kumar, J. Jagielski, S. Yakunin, P. Rice, Y.-C. Chiu, M. Wang, G. Nedelcu, Y. Kim, S. Lin, E.J.G. Santos, M.V. Kovalenko, C.-J. Shih, Efficient blue electroluminescence using quantum-confined two-dimensional perovskites. ACS Nano **10**(10), 9720–9729 (2016). https://doi.org/10.1021/acsnano.6b05775
87. R.L.Z. Hoye, M.-L. Lai, M. Anaya, Y. Tong, K. Gałkowski, T. Doherty, W. Li, T.N. Huq, S. Mackowski, L. Polavarapu, J. Feldmann, J.L. MacManus-Driscoll, R.H. Friend, A.S. Urban, S.D. Stranks, Identifying and reducing interfacial losses to enhance color-pure electroluminescence in blue-emitting perovskite nanoplatelet light-emitting diodes. ACS Energy Lett. **4**(5), 1181–1188 (2019). https://doi.org/10.1021/acsenergylett.9b00571

# Chapter 3
# All-Inorganic Perovskite Quantum Dots: Ligand Modification, Surface Treatment and Other Strategies for Enhanced Stability and Durability

**Zhigang Zang and Dongdong Yan**

**Abstract** All-inorganic perovskite $CsPbX_3$ (X = Cl, Br, and I) QDs exhibit outstanding optical performance, including high photoluminescent quantum yield (PLQY), high stability, low Auger recombination loss and large exciton binding energy. However, they suffer from problems related poor humidity and poor thermal stability, which restrict their practical application. In this Chapter, we will focus on the recent advances on the stability of inorganic $CsPbX_3$ QDs. Our review on understanding of the origins of instability and the intrinsic factors affecting their stability are summarized. The various strategies for enhancing the stability of $CsPbBr_3$ perovskite QDs are introduced. Finall, we propose our thoughts on the future development of this field.

## 3.1 Introduction

Perovskite is an unremarkable calcium titanium oxide or calcium titanate mineral, which has a chemical formula of $CaTiO_3$. This crystallographic material was firstly discovered in the Ural Mountains by Gustav Rose in 1839 and named after Russian mineralogist Count Lev Alekseevich Perovski [1]. The perovskite materials have the formula of $ABX_3$, where A represents usually a monovalent organic or metal cation (such as $Cs^+$, $CH_3NH_3^+$ (MA in short)), the B represents a divalent metal cation (typically $Pb^{2+}$ or $Sn^{2+}$), and the X is a halide anion (typically $Cl^-$, $Br^-$, and $I^-$) [2]. Additionally, the perovskite materials can be divided into two general families: organic-inorganic hybrid perovskites (A site is an organic cation, e.g., MA) and all-inorganic perovskite (A site is an inorganic cation, e.g., $Cs^+$, $Rb^+$). Significant attention and extremely rapid progress in perovskite materials have been aroused and realized due to their excellent optoelectronic properties, such as low exciton binding energy, [3, 4] high light absorption coefficient [5–7] and long charge carrier diffusion lengths, [8–10]. In

Z. Zang (✉) · D. Yan
Key Laboratory of Optoelectronic Technology & Systems (Ministry of Education), Chongqing University, Chongqing 400044, China
e-mail: zangzg@cqu.edu.cn



Y. Zhou and Y. Wang (eds.), *Perovskite Quantum Dots*, Springer Series in Materials Science 303, https://doi.org/10.1007/978-981-15-6637-0_3

2009, the organic-inorganic hybrid perovskites (OIHP) structural materials were first utilized into solar cells [11]. After that, the perovskite materials started to rapidly emerge as a vigorous researching field of optoelectronics. Besides the solar cells, the OIHP materials have many other optical applications, such as light-emitting diodes (LEDs), [12, 13] photodetectors [14, 15] and microlasers [16, 17]. Despite exhibiting great progress and exciting potentials, the OIHP materials face serious challenges in the commercial application, which is due to the instability of organic groups against water, heat, oxygen and light irradiation in the ambient environment [18, 19]. Although numerous efforts have been devoted to solving the issues, the stability of OIHP materials has not been improved significantly. In contrast, all-inorganic ($CsPbX_3$, X = Cl, Br, I) perovskite exhibited enhanced stability in the ambient environment, facilitating to be the more suitable candidates in the field of optical applications. All-inorganic perovskite was first synthesized though the hot-injection method with wide color gamut and 50–85% photoluminescence quantum yields (PLQY) [20]. These $CsPbX_3$ perovskites were commonly prepared from solutions of metal salts (e.g., $PbX_2$ and CsX). The crystal structures of $CsPbX_3$ include tetragonal/orthorhombic and cubic symmetry structure, which have been validated by X-ray crystallography [21, 22]. In the characteristic structure, a $Pb^{2+}$ occupies in the middle of the octahedrally coordinated structure ($[PbX_6]^{4-}$ octahedra), in which six halide anions lie on the corners. As a stable "A" atom, a $Cs^+$ cation resides in the center of the cube crystal. For a typical three-dimensional (3D) halide perovskite, the $CsPbX_3$ perovskites can exhibit different crystalline structures due to the high flexibility of the $PbX_6$ octahedral unit with different synthetic temperature [23]. In comparison with the OIHP materials, the $CsPbX_3$ materials exhibit superior environmental stability, which makes them more suitable candidates in the field of optical applications [24, 25]. The basic properties of $CsPbX_3$ perovskite materials are also excellent, such as high photoluminescence quantum yields (PLQYs, up to 90%), wide color gamut (up to 140%), high optical gain and narrow photoluminescence emission line-widths. Since the first reported on colloidal $CsPbX_3$ quantum dots (QDs), sustainability efforts have been focused on the preparation and applications of $CsPbX_3$ QDs with various sizes, shapes and compositions. For the traditional II–VI and III–V QDs, e.g. CdSe, CdS, and InP, their various interesting optical performance has been reported, which was due to the change of QDs sizes in different synthetic temperature. In fact, the optical properties of $CsPbX_3$ QDs depend on not only the halide composition but the size. With these properties and advantages mentioned above, $CsPbX_3$ QDs have been acknowledged as promising materials in the optical field, resulting in great researching achievement in the past several years [26]. However, compared to the classical II–VI and III–V QDs, [27] $CsPbX_3$ QDs exhibit poor stability during the purification process and readily precipitated from the crude solution [28]. Up to now, previous chemical reactions and surface chemistry researches on $CsPbX_3$ QDs have suggested that the facile proton exchange between the common oleate and amine surfactant ligands may be one of the dominant causes for the instability [29, 30]. This phenomenon can lead to a dynamic equilibrium of bound (unbound) ligands on the surface, which enables a facile ligand loss, leading to the instability of $CsPbX_3$ QDs.

In this outlook, we focus on the strategies of the enhanced stability and durability about inorganic $CsPbX_3$ QDs and summarize the recent advances. We address the various strategies of surface chemistry for $CsPbBr_3$ perovskite QDs, including post-synthetic ligand treatment, coating and compositional engineering, followed by our thoughts on the future directions of the field. We seek to connect the surface treatments and the distinctive optical properties of the $CsPbX_3$ perovskites. Finally, we demonstrate the application of inorganic perovskite quantum dots with enhanced stability and provide a forward look on the approaches to furtherly improve the stability of the inorganic perovskite quantum dots.

## 3.2 Structure, Synthesis and Morphological Control of $CsPbX_3$ QDs

### *3.2.1 Crystal Structure*

The high-temperature phase of $CsPbX_3$ QDs, cubic crystallinity, is the usual product in the hot-inject method at 140–200 °C [20]. However, the lower symmetry phases (including orthorhombic phase, tetragonal phase, monoclinic phase, and rhombohedral phase) are found to exist at lower temperatures. The typical symmetries of the above phases can be observed in the $CsPbX_3$ QDs [31]. Notably, the concept of Goldschmidt tolerance factor (t) [32] has been widely used to predict the stability of perovskite structures based only on the chemical formula $ABX_3$, in which, the phase diversity of perovskites can be qualitatively explained.

$$t = \frac{r_A + r_X}{\sqrt{2}(r_B + r_X)} \tag{3.1}$$

where $r_A$, $r_B$, and $r_X$ are the ionic radii for the components A, B, and X atoms, respectively. Empirically, most of the 3D perovskite structures are favored for the t value in the range of 0.8–1 [33], that is, the stable 3D structures are formed when t is within the range. Beyond this range, other perovskite-related structures are usually unstable (see Fig. 3.1a, b and c).

Some perovskites at the boundary of the tolerance factor range, such as $CsPbI_3$ (t ~ 0.8) and $FAPbI_3$ (t ~ 1), readily undergo a structural phase transition to become more stable hexagonal/orthorhombic phases at room temperature [34]. A second constraint for perovskite structure is known as the octahedral factor (μ), which is defined as $\mu = r_B/r_X$ a typically lying in the range of $0.442 \leq \mu \leq 0.895$ [35]. This $\mu$ is a measure of octahedral stability and it depends on the radii of both the B and X ions. The octahedral and the tolerance factor are widely used to predict the perovskite formability [36–38]. As shown in Fig. 3.1d–f, the perovskites are classified as 3D, 2D, 1D and 0D structure, in which the dimension of perovskite depends on the $BX_6$ octahedra corner-shared.

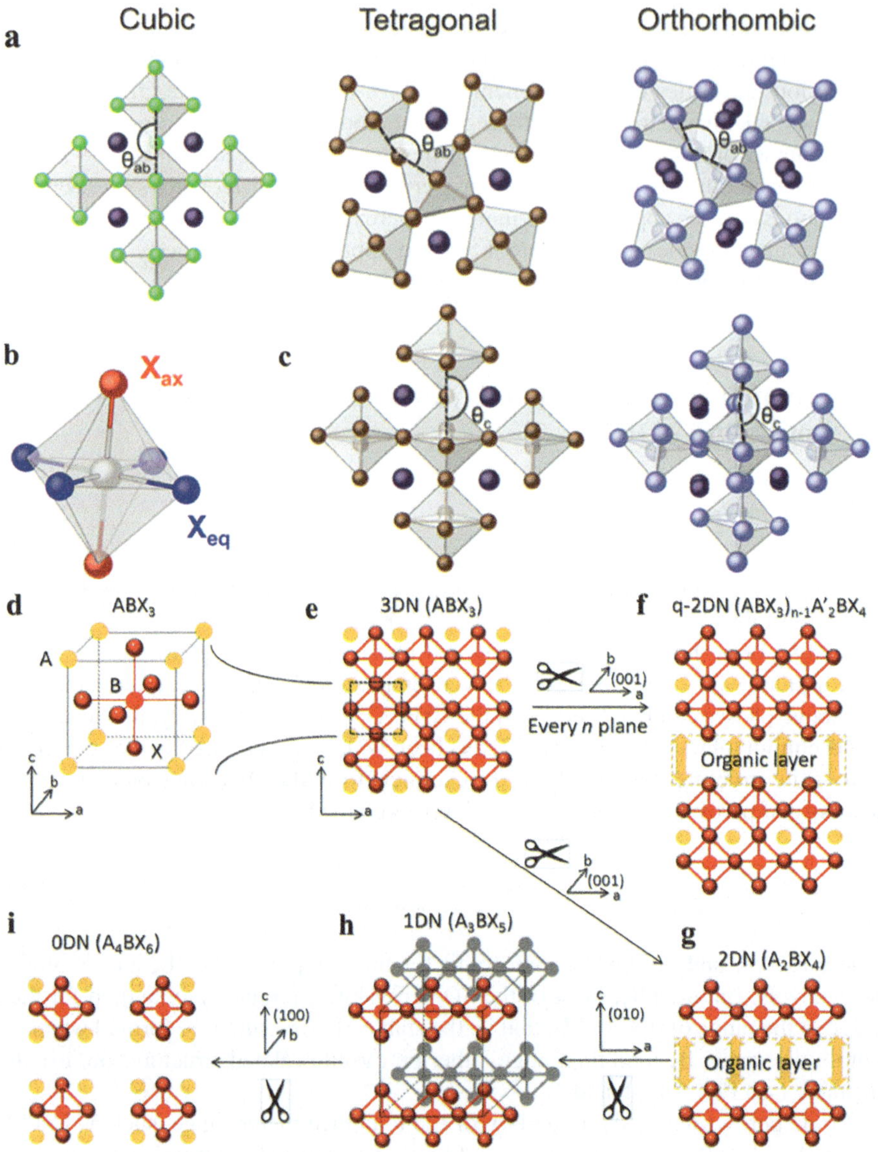


**Fig. 3.1** **a–c** Schematic diagram of $PbX_6$ octahedral 3D arrangement in $CsPbX_3$ perovskites unit cell. Reproduced from Ref. [23], Copyright 2017 American Chemical Society. **d–i** Schematic diagram of $BX_6$ octahedra (in the view of the crystallographic planes) [38]

### 3.2.2 Synthetic Strategies

In 2015, the Kovalenko group [20] reported the first successful preparation of high-quality $CsPbX_3$ perovskite QDs using a hot-injection method, which announces the launch of a new field into optical nanomaterials.

Figure 3.2a shows the typical schematic diagram of the hot-injection method for colloidal $CsPbX_3$ QDs, [39] which are obtained by reacting the Cs-oleate with predissolved $PbX_2$ in high boiling ODE solvent at 140–200 °C. Several seconds later, the reaction system was rapidly cooled down to room temperature using an ice-bath. In addition, the size of the $CsPbX_3$ QDs is tuned by the reaction temperature rather than the growth time. The $CsPbX_3$ QDs show excellent optical properties, such as high PLQY (~90%), narrow emission line-widths, widely tunable emission wavelength in the visible spectrum and wide color gamuts (up to 140%). In addition, this hot-injection method has been modified in different ways to synthesize $CsPbX_3$ QDs with various shapes, compositions, and sizes [44, 45]. However, the hot-injection method usually requires high reaction temperature and inert atmosphere, which may severely limit the practical application. To overcome these disadvantages, in 2016, Zeng's group firstly report a ligand assisted reprecipitation method at room temperature and made it possible to obtain high-quality $CsPbBr_3$ QDs (Fig. 3.2b) [40]. In the room temperature reprecipitation method, CsX and $PbX_2$ (X = Cl, Br, I) were firstly pre-dissolved in a polar solvent (dimethyl formamide (DMF) or dimethyl sulfoxide (DMSO)) together with oleic acid (OA) and oleylamine (OAm) as precursors, followed by a quick injection into toluene (a nonpolar solvent), as shown in

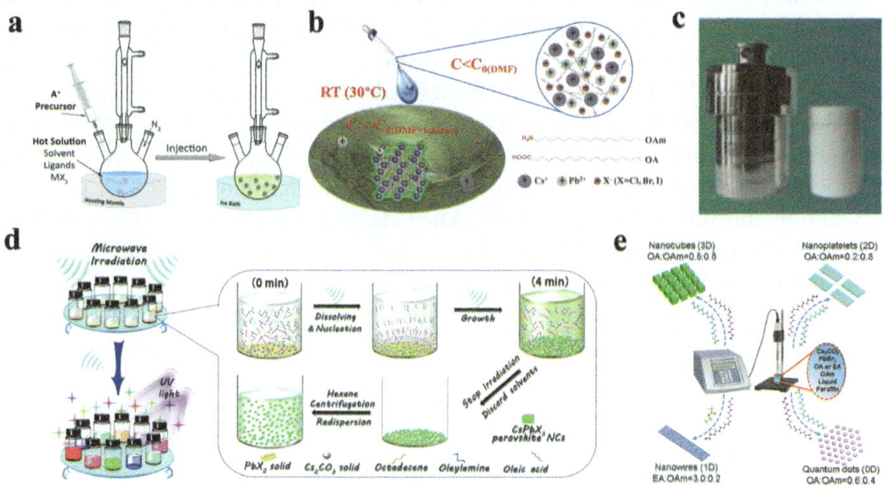


**Fig. 3.2** Schematic illustration of the methods for the synthesis of $CsPbX_3$ perovskite nanocrystals: **a** hot-injection method, **b** room temperature reprecipitation method, (**c**) solvothermal method, **d** microwave-assisted method and **e** ultrasonication-assisted method. The images were modified with permission from refs (**a**) [39], (**b**) [40], (**c**) [41], (**d**) [42], and (**e**) [43]

Fig. 3.2b. This research has been followed immediately by many other reports, which aimed to develop more facile and effective synthetic strategies to obtain high-quality $CsPbX_3$ NCs. In addition, solvothermal [41] (Fig. 3.2c), microwave-assisted [42] (Fig. 3.2d) and ultrasonication-assisted method [43] (Fig. 3.2e) have been proposed to synthesize $CsPbX_3$ QDs efficiently in air. The methods are still under development to synthesize stable, effective and high-quality $CsPbX_3$ QDs (Table 3.1).

### *3.2.3 Phase Transformation*

Recently, other types of the cesium lead halide perovskite, such as $CsPb_2X_5$ and $Cs_4PbX_6$, have been synthesized by researchers [57–62]. Although their component elements are the same as the classic $CsPbX_3$ perovskite, they are found to possess entirely different structures. Compared with the $CsPbX_3$, the $Cs_4PbX_6$ exhibit completely decoupled and $Cs^+$ cations fill the holes formed by four adjacent $[PbX_6]^{4-}$ octahedra. For the $CsPb_2X_5$ structure, the $PbBr_6$ octahedrons to $PbBr_8$ capped-triangular prisms to form metastable tetragonal structures can exhibit a sandwich structure consisting of $Pb_2X_5$-layers and intercalated $Cs^+$ ions [63]. In terms of the dimension of materials, $CsPbX_3$, $CsPb_2X_5$ and $Cs_4PbX_6$ can be found as 3D, 2D and 0D structures, respectively. In addition, it has been found the structures can transform among the cesium lead halide perovskite family arbitrarily. Manna et al. regarded the 0D $Cs_4PbX_6$ NCs, which were synthesized by treating $Cs_4PbX_6$ with excessive $PbX_2$, as a less-lead perovskite material (see Fig. 3.3d) [58]. Alivisatos et al. later studied the role of ligands in the synthesis process and proposed a single-step method to transform from $CsPbBr_3$ to $Cs_4PbX_6$ QDs after being treated with an excessive amine treatment, as shown in Scheme 3.1 and Fig. 3.3a [59]. Meanwhile, Yin's group [57] showed that $Cs_4PbX_6$ QDs could be defined as CsX-rich structural material (see Fig. 3.3b). It has been demonstrated that the nonluminescent $Cs_4PbX_6$ QDs can be converted into highly luminescent $CsPbX_3$ QDs when $Cs_4PbX_6$ QDs were treated with water via a CsX-stripping process owing to the high solubility of CsX in water. In addition, the 3D $CsPbBr_3$ NCs perovskite structure could be reported to transform from the 3D structure into 2D $CsPb_2Br_5$ nanosheets with dodecyl dimethylammonium bromide (DDAB) ligand treatment (see Fig. 3.3c) [61]. Banerjee's group [64] has demonstrated the Cs-Pb-Br system with respect to structural and morphological dimensionality as dictated by concentration, denticity and steric bulk of added ligands (see Fig. 3.3e).

### *3.2.4 Degradation Mechanism of $CsPbX_3$ QDs*

Although the $CsPbX_3$ QDs have exhibited excellent optical performances, the degradation of $CsPbX_3$ QDs is still a big hurdle for their further applications [65–68]. Various synethesis strategies and passivation mechanisms have been proposed to

**Table 3.1** Summaries of PL properties and stability of all-inorganic $CsPbX_3$ QDs

| QDs | Synthetic method | Emission peak (nm) | FWHM (nm) | PL QYs (%) | Stability | References. |
|---|---|---|---|---|---|---|
| $CsPbBr_3$ | LARP | 513 | 20 | 95 | 90% (30 d, air) | [40] |
| $CsPbCl_3$ | LARP | 405 | 12 | 10 | | [40] |
| $CsPbBr_{1.5}I_{1.5}$ | LARP | 600 | 38 | 72 | | [40] |
| $CsPbBr_{1.5}Cl_{1.5}$ | LARP | 455 | 16 | 37 | | [40] |
| $CsPbCl_3$ | Ultrasonication | | | 10 | 8% (4 months, air) | [46] |
| $CsPbBr_3$ | Ultrasonication | | | 92 | 89% (4 months, air) | [46] |
| $CsPbI_3$ | Ultrasonication | | | 90 | 0% (2 months, air) | [46] |
| $CsPbCl_3$ | Microwave irradiation | 410 | 14 | 7 | | [47] |
| $CsPbBr_3$ | Microwave irradiation | 517 | 17 | 90 | | [47] |
| $CsPbI_3$ | Microwave irradiation | 691 | 35 | 70 | | [47] |
| $CsPbCl_3$: Ni | Hot-injection (210 °C) | 407 | | 96.5 | >90% (6 d, air) | [48] |
| $CsPb(Br/I)_3$: Al | Hot-injection (150 °C) | 536 | | 40 | | [49] |
| $CsPbBr_3$: Al | Hot-injection (150 °C) | 456 | 16 | 42 | | [49] |
| $CsPbBr_3$ | Hot-injection (160 1C) | 509 | 16 | 92 ± 2 | | [50] |
| $CsPb_{1-x}Sn_xBr_3$ | Post-synthesis | 479–512 | | 62 | | [51] |
| $CsPbBr_3$: Sn(IV) | Hot-injection (180 1C) | 517 | 20 | 83 | | [52] |
| $CsPbBr_3$: Ce | Hot-injection (185 1C) | 510 | | 89 | 60% (30 d, air) | [53] |
| $CsPbBr_3$: Mn | Hot-injection (150 1C) | 514–517 | 20 | 90 | 60% (120 d, air) | [54] |
| $CsPbCl_3$: Mn | Hot-injection (150 °C) | 589 | | 27 | | [55] |
| $CsPbI_3$: Mn | Hot-injection (150 1C) | 680 | 40 | 82 ± 9 | Stable over a month | [44] |
| $CsPb_{0.73}Mn_{0.27}Cl_3$ | Hot-injection (170 1C) | 580 | | 54 | 40% (60 min, UV) | [56] |

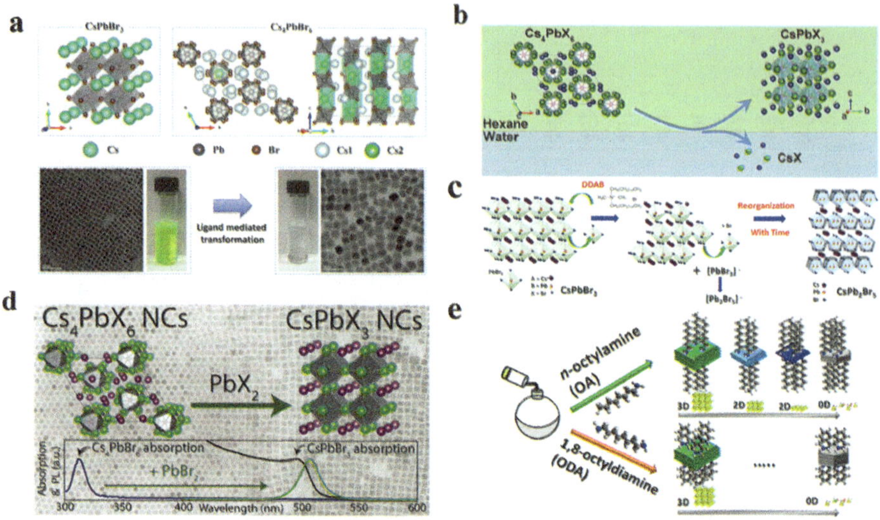


**Fig. 3.3** **a** The schematic crystal structure of perovskite ($CsPbBr_3$, $Cs_4PbX_6$) and their corresponding TEM images, respectively [59]. **b** The transformation process of $Cs_4PbX_6$ QDs to $CsPbBr_3$ QDs [57]. **c** The schematic of cubic $CsPbBr_3$ perovskite nanocrystals to tetragonal $CsPb_2Br_5$ nanosheets [61]. **d** The reaction of $Cs_4PbX_6$ with $PbBr_2$ forming $CsPbBr_3$ [58]. **e** Schematic of phase transformation with increasing the ligand concentration [64]

$$4CsPbBr_3 \xrightarrow{\text{OLA+Thiol}} Cs_4PbBr_6 + 3PbBr_2$$

**Scheme 3.1** The transformation process from $CsPbBr_3$ to $Cs_4PbBr_6$ [59]

solve this problem. As we all known, the degradation of $CsPbX_3$ QDs can be attributed to environmental factors, including temperature, UV light, moisture, and oxygen, etc., as well as intrinsic factors of ion migration. Recently, Ju and coworker [69] summarized the degradation mechanisms for all kinds of halide perovskites. It has been found that hydration, decomposition, oxidation and polymorphic transition, which are induced by moisture, light, heat and oxygen, may degrade the $CsPbX_3$ critically (Fig. 3.4). Therefore, the researcher may facus on the factor above to suppress the degradation of perovskite.

Kovalenko and coworker [26] have summarized the long-term stabilities of lead halide perovskite (LHP) QDs. They addressed the four significant characteristics of instability for LHP QDs. Firstly, all LHPs are partially or highly soluble in polar solvents. Secondly, the internal and QD-ligand binding in LHP QDs is highly ionic in solutions [30, 70]. Thirdly, the long-term stability of $CsPbX_3$ may still be limited in the presence of ambient conditions (such as one or a combination of moisture, oxygen, and light) [71]. Lastly, the low melting points of LHPs render densely packed arrays of LHP QDs in high temperatures, which is undesired for the devices operating at elevated temperatures.

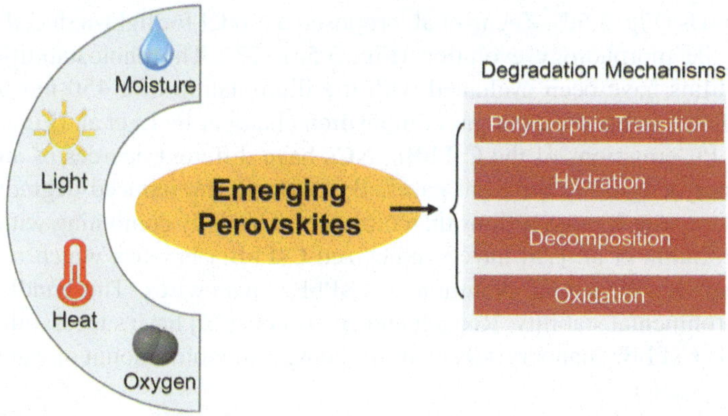


**Fig. 3.4** Illustration of degradation mechanisms of halide perovskites under vital environmental factors (moisture, light, heat, and oxygen). Adapted from Ref. [69] with permission requested from Elsevier

Sargent's group [28] investigated the effects of the polar solvents on the as-synthesized $CsPbBr_3$ NCs (Fig. 3.5a–c). The results demonstrated that the $CsPbBr_3$ QDs were strongly affected by the solvents of dimethyl sulfoxide, dimethylformamide, methanol, and ethanol, leading to the photoluminescent decrease and quenching of can make the $CsPbBr_3$ NCs QDs. Schaller et al. [72] systematically investigated the dependence of the synthesis temperature and the fluorescence of

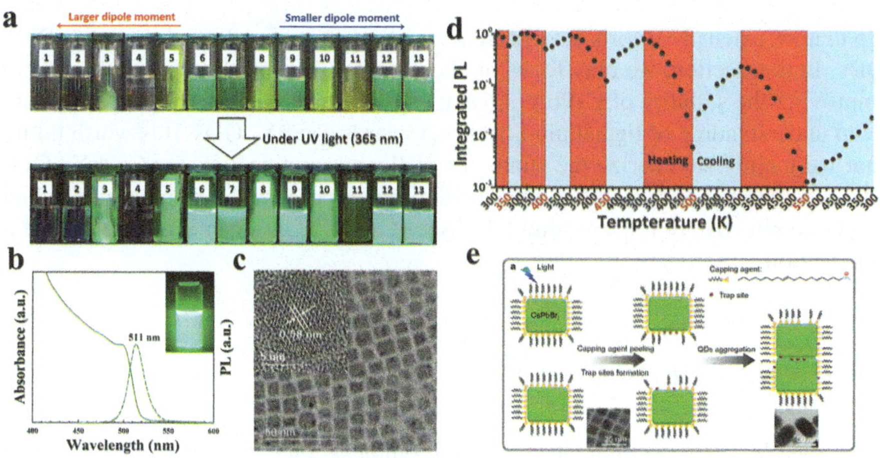


**Fig. 3.5** **a**. The as-synthesized $CsPbBr_3$ NCs were subjected to different polar solvents: (1) dimethyl sulfoxide, (2) dimethylformamide, (3) acetonitrile, (4) methanol, (5) acetone, (5) ethyl acetate, (6) tertbutanol, (7) 1-butanol, (8) tetrahydrofuran, (9) isopropyl alcohol, (10) ethanol, (11) chloroform, and (12) dichloromethane, respectively [28]. **b**. UV—is absorbance and PL spectra of the once-purified $CsPbBr_3$ NCs [28]. **c**. HR-TEM image [28]. **d**. The performance of $CsPbBr_3$ PL under thermal cycling measurements [72]. **e**. Schematic process of the photo-degradation [73]

$CsPbBr_3$ QDs (Fig. 3.5d). Zheng et al. proposed a model for light-induced regrowth based on the morphology evolution (Fig. 3.5e) [73]. The photostabilities of the $CsPbBr_3$ films have been evaluated with the illumination of a 450 nm blue LED (175 mW/cm$^2$) in the different gas composition chamber by Li et al. (Figure 3.6a–i) [74]. The PL emissions of the $CsPbBr_3$ NCs have different degrees of attenuation and enhancement in the different gases. By contrast, the induced degradation and PL loss of oxygen is more serious than nitrogen, especially combining with the light and water. Sahin et al. [75] have synthesized $CsPbBr_3$ bundles which result from the water-driven structural evolution of $CsPbBr_3$ nanowires. The bundles exhibit good environmental stability. Rogach and coworkers [76] have successfully synthesized stable $CsPbBr_3$ nanocrystals by introducing a suitable amount of water into the reaction precursor.

As mentioned above, the instabilities of the $CsPbX_3$ QDs are reportedly accelerated by light, oxygen, moisture, heat, and their synergistic effects. Moreover, the intrinsic chemical instability of $CsPbX_3$ QDs is much greater than what we could imagine, $CsPbI_3$ NCs even decompose at room temperature. The crystals of $CsPbX_3$ perovskite can readily grow into large crystals under moisture and light conditions, and the tendency can be accelerated in the presence of oxygen. Hence, to obtain stable $CsPbX_3$ QDs, the degradation of perovskite should be eliminated by focusing on the relevant mechanism.

## 3.3 Surface Ligand Modification

Up to now, extensive efforts have been devoted to improving the stability of $CsPbX_3$ QDs. In this section, we may focus on the progress of recently reported studies on improving the stability of $CsPbX_3$ QDs using ligand modification approaches, and deep understanding of ligand modification towards $CsPbX_3$ QDs. It is worth noting that we mainly summarize the strategies of ligand modification for $CsPbX_3$ QDs. The other strategies to stabilize perovskite films or perovskite crystal grains, which might be effective to improve the stability of $CsPbX_3$ QDs, can also be included in this section.

### *3.3.1 Surface Engineering*

Kovalenko's group accomplished the first synthesis of $CsPbX_3$ in 2015, and they adopted a pair of oleic acid (OA) and oleylamine (OAm) capping ligands to these ligands' protection, morphology control and stabilization for the NCs [20, 77, 78]. These ligands can affect the thermodynamic process and modify the surface energy of perovskite shapes during the chemical reaction [79]. Following the pioneering report of Kovalenko's group, OA and OAm ligands have been widely used for the synthesis of $CsPbX_3$ QDs. However, the proton exchange between coordinated OA and OAm may easily facilitate them to desorb from the as-prepared $CsPbX_3$ QDs surface,

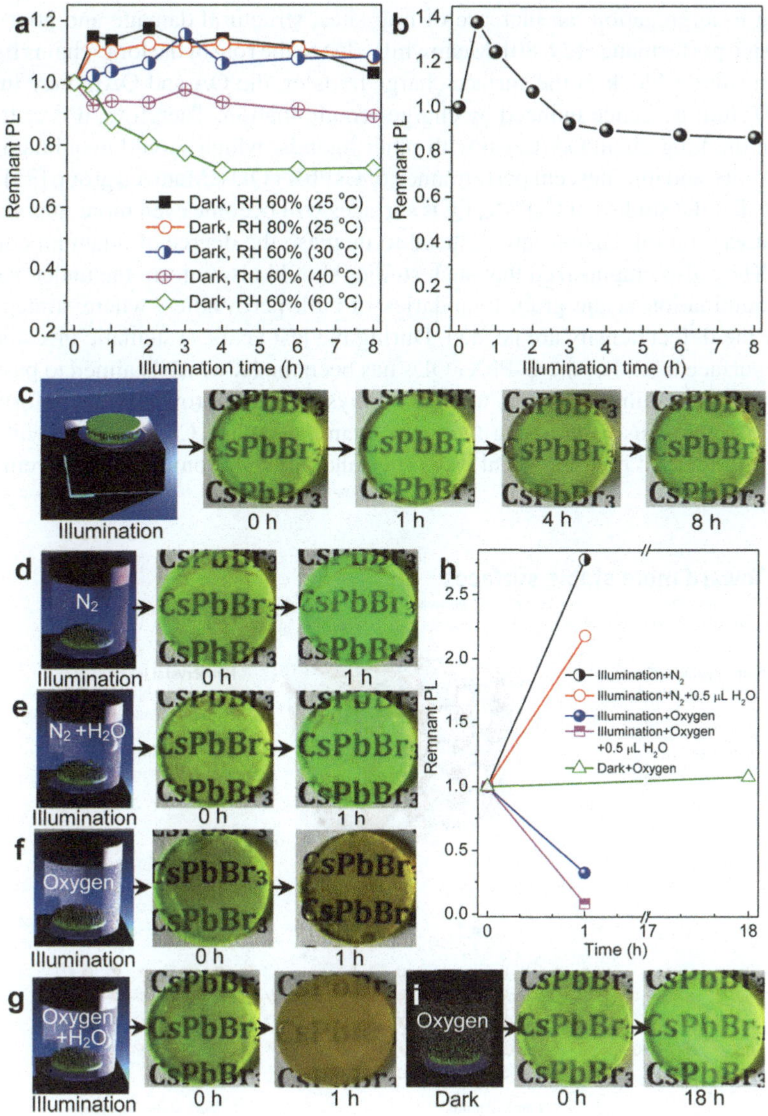


**Fig. 3.6** Factors affecting the photostabilities of the $CsPbBr_3$ films. **a** The conditions of different RH levels and temperatures. **b** The performance of remnant PL emissions. **c** Optical images of the CPB film sandwiched between two quartz coverslips as a function of the illumination time (175 mW/cm$^2$, RH 60%). **d–i** Optical images of the CPB films exposed to various conditions (inset the figure) [74]

leading to aggregation, an increase of trap sites, structural damage and poor optoelectronic performance [29, 80]. Meanwhile, due to the role of the long-chains ligands on an insulator block of the surface charge transfer, the OA and OAm may impede the QDs luminescence induced by charge-recombination. Therefore, it is critical to replace the long-chain OA/OAm with other ligands, which should avoid damaging the stability and luminescent performance of $CsPbX_3$ QDs. Manna's group [81] highlighted that the studies of $CsPbX_3$ QDs surfaces will become even more important as the nanocrystal dimensions are reduced to increase the degree of quantum confinement. They also emphasized that such studies should also inform the understanding of recombination at the grain boundaries in bulk perovskites, where strategies to reduce the defect density are needed. During the last years, systematic investigation on the surface ligand of the $CsPbX_3$ QDs has been made, which is aimed to passivate surface traps, minimize defects, tune photophysics and improve device performance (Fig. 3.7). Obviously, choosing the right ligands to make $CsPbX_3$ NCs with high stability, excellent photophysical properties and optoelectronic efficiency remains a critical challenge.

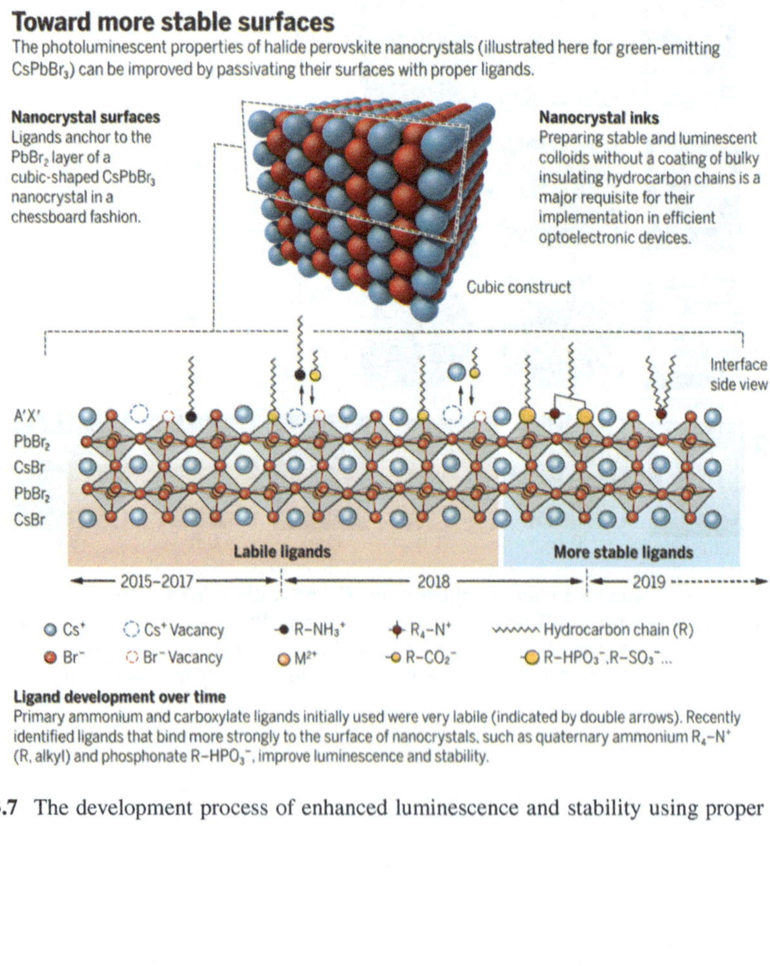


**Fig. 3.7** The development process of enhanced luminescence and stability using proper ligands [81]

### 3.3.2 Categories of the Common Ligands

Recently, researches have proven that the covalent bond classification (CBC) is an established concept in organometallic chemistry to classify molecules by identifying the number and types of bonds ligands. The CBC provided a general, unambiguous, and consistent concept and was therefore adopted by the nanocrystal material field to classify the interactions between QD surface and ligands. The types of the surface ligands can be conveniently described by the CBC, as developed by Green [82]. According to the Jonathan De Roo's [30] report, ligands were labeled as L-, X-, or Z-type, depending on the number of electrons (2, 1 or 0, respectively) when the neutral ligand contributes to the bond between the $CsPbX_3$ NCs surface and the ligand. L-type ligands are regarded as Lewis bases which are typically coordination to the surface metal atoms, while Z-type ligands are Lewis acids which usually coordinate to the surface nonmetal atoms, respectively (see Fig. 3.8). In addition, the X-type ligands can interact with either nonmetal or metal atoms, depending on the affinity. Therefore, the perovskite-type NCs could be constructed different binding motifs due to the combinations of QD stoichiometries and ligand types.

Talapin's group [77] reported that ligand chemistries of conventional QDs relied on long-alkyl-chain ligands bound to the QD surface via a single anchoring group, or on carboxylate or phosphonate (X-type) ligands attached at surface metal atoms. Similarly, the results can be applied to $CsPbX_3$ QDs. Kovalenko and his coworker [83] pointed out that each of oleylammonium bromide and/or oleylammonium carboxylate on the surface may act as an ionic pair, i.e. ammonium groups displace a surface

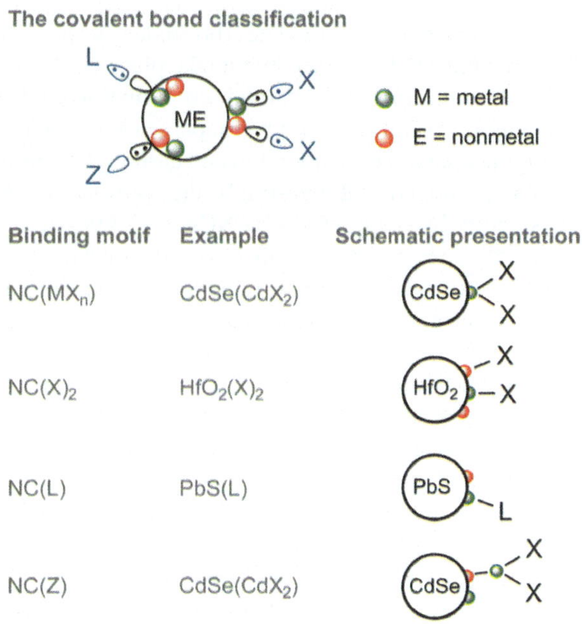


**Fig. 3.8** Schematic diagram of the CBC. The tye ligands in the picture: X-type, L-type and Z-type is one-electron donors, Lewis bases and Lewis acids, respectively [30]

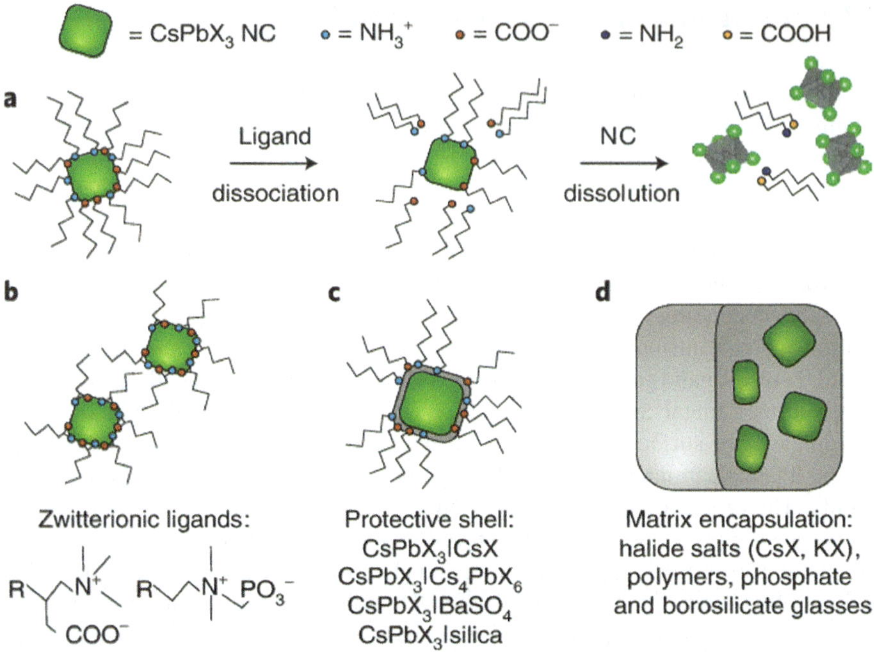


**Fig. 3.9** **a** The process of the $CsPbX_3$ NCs lost their stability or structural integrity. **b** using zwitterionic long-chain molecules; **c** overcoating strategy using shell; and **d** matrix encapsulation into stable materials (including polymers, amorphous inorganic matrixes) [83]

A-site cation, while carboxylates (bromides) act as surface anions. The desorptions of these ligands on the surface might take place easily, as shown in Fig. 3.9a. They also pointed out that more static surface coordination might be confirmed from the ligands containing several anchoring groups, such as zwitterionic molecules (Fig. 3.9b), including quaternary ammonium and carboxylate groups (or sulfonate, phosphonate or other groups) [83]. Interestingly, they presented a classical approach to improving the stability by coating $CsPbX_3$ NCs (Fig. 3.9c, d).

In order to better understand the role of amino and carboxylic groups on the surface of perovskite QDs, Pan et al. studied the impact of chain lengths of acid and amine on the surface of IHP QDs systematically [84]. Fig. 3.10a shows clearly the relation between the morphologies of as-prepared QDs and the chain length of the ligands in the synthesis reaction. It was found that the carboxylates presumably chelate with lead atoms on the QDs surface, while alkylammonium interacts with the bromine atoms on the surface via hydrogen-bonding interactions. It proved that, compared with the ammonium, the carboxylate ligands are more strongly attached to the surface than the ammonium ions do. Deng's group reported a facile ligand-assisted, room-temperature reprecipitation strategy to synthesize $CsPbX_3$ QDs with well-defined morphology, including 0D spherical quantum dots, nanocubes, 1D nanorods, and 2D few-unit-cell thick nanoplatelets, as shown in Fig. 3.10b [62]. It was found that

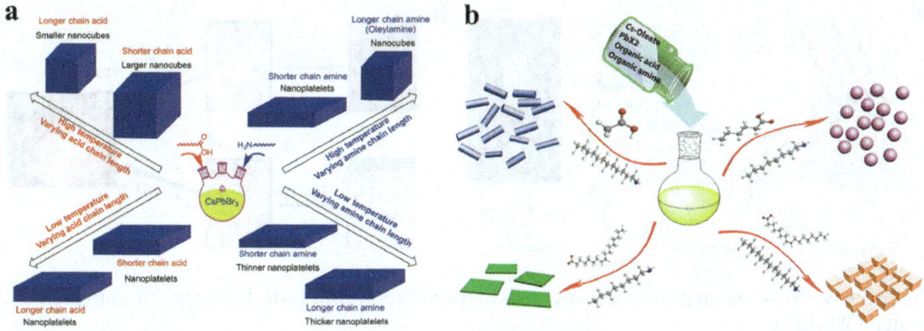


**Fig. 3.10** **a** Presentation of the structures dependence on the ligand length. Reproduced with permission [84]. Copyright 2016, American Chemical Society. **b** Schematic diagram of the formation process for different $CsPbX_3$ NCs [62]

the use of crosslinking surface ligands could strengthen the interactions of perovskite QDs and minimize ligand loss. Tan and coworkers [85] explored a trimethylaluminum (TMA) as a crosslinking agent and achieve the formation of $CsPbX_3$ QDs film. In another study, Manna et al. [86] deposited OA and OLA capped $CsPbX_3$ PQDs on a silicon substrate. The formation of intermolecular $C = C$ bonds, which linked the adjacent $CsPbX_3$ QDs, resulted from X-ray irradiation. Overall, based on the various crosslinking protocols, the obtained $CsPbX_3$ QDs could exhibit high PLQYs for a long period of time under ambient conditions due to the reduction of the ligand loss.

### *3.3.3 Ligand Modification*

For the synthesis of $CsPbX_3$ QDs, OA and OAm are the most popular and successful ligands. However, the QDs aggregation, degradation, and structural destruction of $CsPbX_3$ QDs, which resulting from the high dynamic in surface ligand banding to $CsPbX_3$ surface, [30] could limit their further application. Ligand modification strategy is one of the effective approaches to enhance the stability of $CsPbX_3$ QDs. The $CsPbX_3$ QDs could be effectively passivated and protected by Zwitterionic ligands that possess several anchoring groups at the same end of the molecule (Fig. 3.11a) [80]. Compared with classic OA and OAm, the zwitterionic ligands are capable of coordinating simultaneously to the surface cations and anions in $CsPbX_3$ QDs via special chelate effect, resulting in stronger adhesion on the surface of $CsPbX_3$ QDs followed by improved stability and durability of $CsPbX_3$ QDs. Ligand exchange on presynthesized perovskite QDs is another effective way to passivate the surface of QDs. Bakr et al. reported that both the PLQY and stability of $CsPbX_3$ QDs could be dramatically improved after ligand exchange with didodecyl dimethylammonium bromide and bidentate 2,2′-iminodibenzoic acid (Fig. 3.11b) [87].

In addition, replacing the long-chain, and insulating organic ligands with conductive inorganic ligands has been proposed to increase charge carrier mobility, which

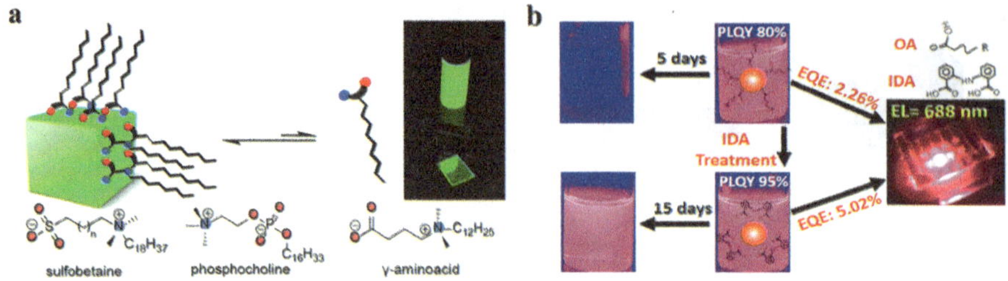


**Fig. 3.11** **a** Novel strategy usinga single zwitterionic molecule [80]. **b** Images of untreated and treated QDs [87]

has worked well in QD-based photovoltaics, LEDs, and field-effect transistor devices [88, 89]. Zang's group [90] have proposed a simple and effective ligand-modification strategy by introducing 2-hexyldecanoic acid (DA) with two short branched chains to replace long-chains OA using the modified hot-injection method (see Fig. 3.12a, b). These two short branched chains not only maintained their colloidal stability, but also contributed to efficient radiative recombination. Due to the strong binding energy between DA ligands and QDs, DA ligand modified $CsPbBr_3$ QDs exhibited no aggregation phenomenon even after stored in air for more than 70 days, and the corresponding films could maintain 94.3% of initial PL intensity after stored in air for 28 days, while $CsPbBr_3$ QDs films based on OA ligand occured a rapid degradation of PL intensity.

Using a similar approach, Zeng's group [91] reported a method to obtain efficient $CsPbBr_3$ QDs. They proposed a concept of equivalent ligand, which the strong ionic sulfonate heads of benzenesulfonic acid could firmly bind to the exposed Pb ions, resulting in forming a steady binding state. This equivalent ligand could effectively eliminate the exciton trapping probability and provide surface defect passivation. The high PLQY (above 90%) was facilely obtained without any amine-related ligands during the synthesis process. In addition, the PLQYs are well maintained even after eight purification cycles, storage over five months and light irradiation.

Recently, Sun et al. replaced the OA/OAm ligands with octylphosphonic acid (OPA) for synthesis of $CsPbX_3$ QDs [92]. They found that the OPA ligand enabled to modify the QDs surface and improve the stability of $CsPbX_3$ QDs. Because of the strong interaction between $Pb^{2+}$ ions and OPA ligands, the OPA ligands modified $CsPbX_3$ (OPA-$CsPbBr_3$) QDs exhibit a highly PLQY (above 90%). As a result, the green LED device based on OPA-$CsPbBr_3$ NCs shows an external quantum efficiency of 6.5% (Fig. 3.12e). Manna and co-workers demonstrated that the addition of trioctylphosphine oxide (TOPO) into the lead halide precursor could obtain monodisperse $CsPbBr_3$ QDs with robust colloidal stability [93].

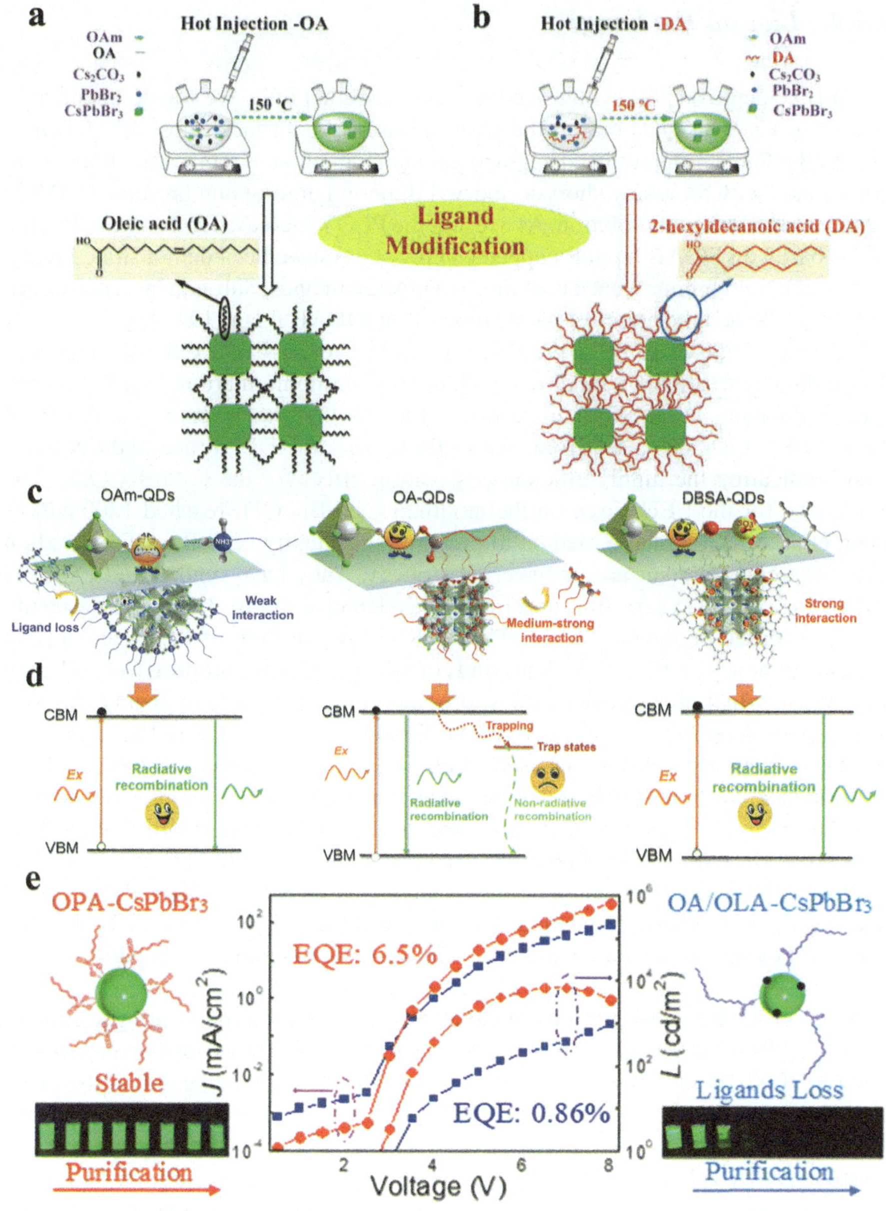


**Fig. 3.12** **a**, **b** Schematic presentation of OA/DA ligand modification [90]. **c**, **d** Comparison of different ligand strategies [91]. **e** Schematic structure of OPA-capped $CsPbBr_3$ QDs [92]

### 3.3.4 Ligand Exchange

Ligand exchange on post synthesized process is another effective way to obtain high quality of $CsPbX_3$ QDs due to the passivation of the surface traps. As shown in Fig. 3.13a, Pan's group utilized a ligand-exchange method to replace the long-chain OA ligands with relatively short di-dodecyl dimethyl ammonium bromide (DDAB) ligand into $CsPbX_3$ QD solution. As a result, the PLQY increased from 49 to 71%, the LEDs based on DDAB ligands capped $CsPbX_3$ QDs have been demonstrated much higher EQE and luminescence than those of OA/OAm ones. Subsequent reports have further confirmed the effective passivation of the DDAB ligand [94, 95].

Recently, Tian et al. [97] proposed a method to passivate $CsPbBr_3$ quantum dots (QDs) by treatment with di-dodecyldimethylammonium bromide (DDAB) and sodium thiocyanate (NaSCN), as shown in Fig. 3.13b. In their report, the PLQY of the $CsPbBr_3$ QDs increased from 73 to ~100% after the DDAB and NaSCN treatment, indicating the highly efficient passivation effect for the $CsPbBr_3$ QDs. The luminance for the LED-based on the modified $CsPbBr_3$ QDs reached 1200 cd/m$^2$. Zeng et al. [98] reported a method to prepare high-quality $CsPbBr_3$ QDs based on a room-temperature surface engineering strategy. They have synthesized $CsPbBr_3$ QDs using several synergistic short ligands, as depicted in Fig. 3.13c–g. As a result, the LEDs based on the modified $CsPbBr_3$ NCs show an EQE of 11.6% and a peak current efficiency of 45.4 cd/A. Wang and coworkers [99] have proposed an exchange strategy, in which they replaced the short-chain ligands with long-chain OA/OAm ligand pairs (see Fig. 3.14a–c). With the ligand exchange, it was found that the surface defects of $CsPbBr_3$ QDs have been passivated and generated the best-LED performance. Many researchers demonstrated that replacing the common ligand (OA or OAm) by other ligand is an effective method to enhance the stability of $CsPbX_3$ QDs [100–102]. Manna's group [103] used propionic acid (PrAc) to synthesize the $CsPbBr_3$ QDs under room temperature, as shown in Fig. 3.14d–i. They also point out that they could synthesize high optical qualities $CsPbBr_3$ close to those QDs obtained from the hot-injection method and the easily scale up to gram-scale production (see Fig. 3.14i).

Sun's group has chosen thionyl halides ($SOCl_2$, $SOBr_2$) as passivating agents for $CsPbCl_3$. [104] They achieved high-performance blue LED based on the treated QDs. Song et al. [105] investigated the effect of the chain length of ligand on the properties and stability of $CsPbBr_3$ QDs by replacing OAm with other suitable ligands (see Fig. 3.15b). In addition, based on the exchange of didecyldimethyl ammonium bromide (DDAB) ligands, they achieved a highly efficient green emissive LED with a current efficiency of 31.7 cd/A and EQE of 9.7%, which were enhanced 16-fold compared to those of untreated $CsPbBr_3$ LEDs. Li's group [106] used a short conjugation molecular ligand phenethylamine (PEA) as ligands to synthesize $CsPbX_3$ QDs and then treated the $CsPbX_3$ QD films with phenethylammonium bromide (PEABr) or phenethylammonium iodide (PEAI) (see Fig. 3.15c). They pointed out that the

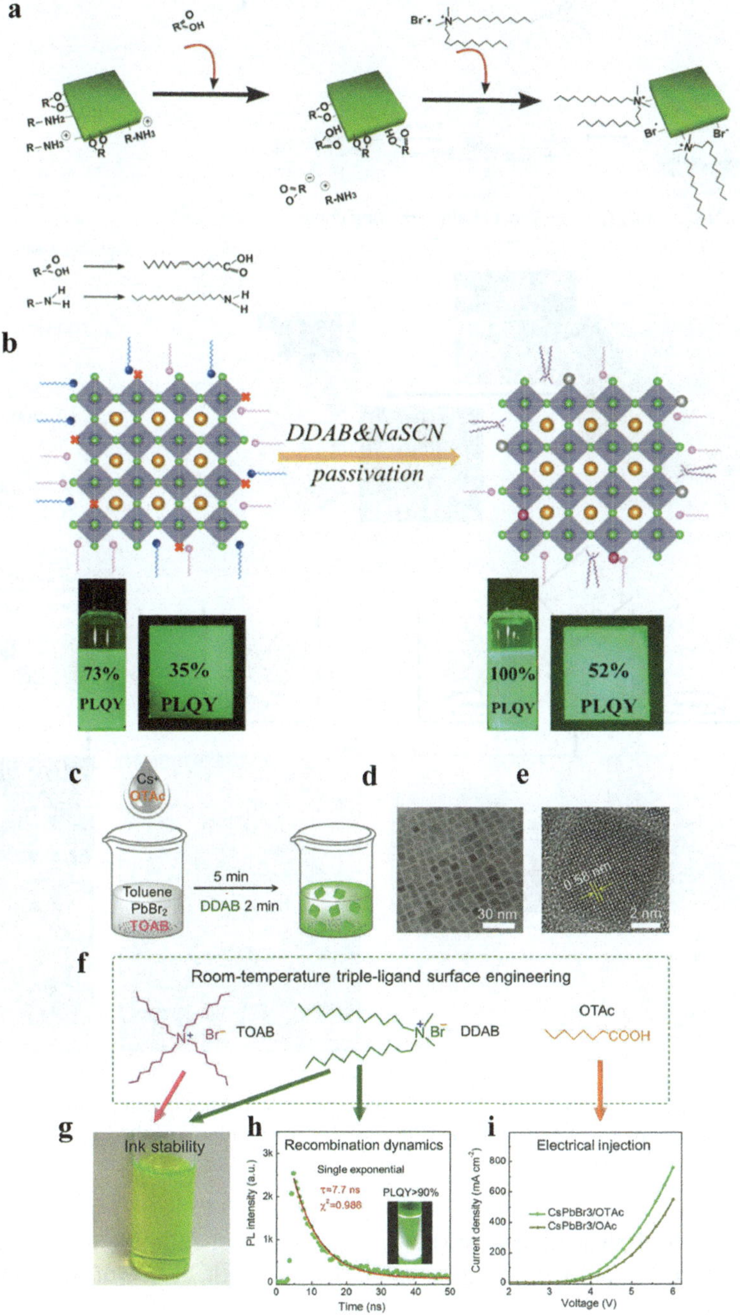


**Fig. 3.13** **a** The diagram of the ligand-exchange process [96]. **b** Surface passivations [97]. **c**–**g** Triple-ligand surface engineering under room-temperature [98]

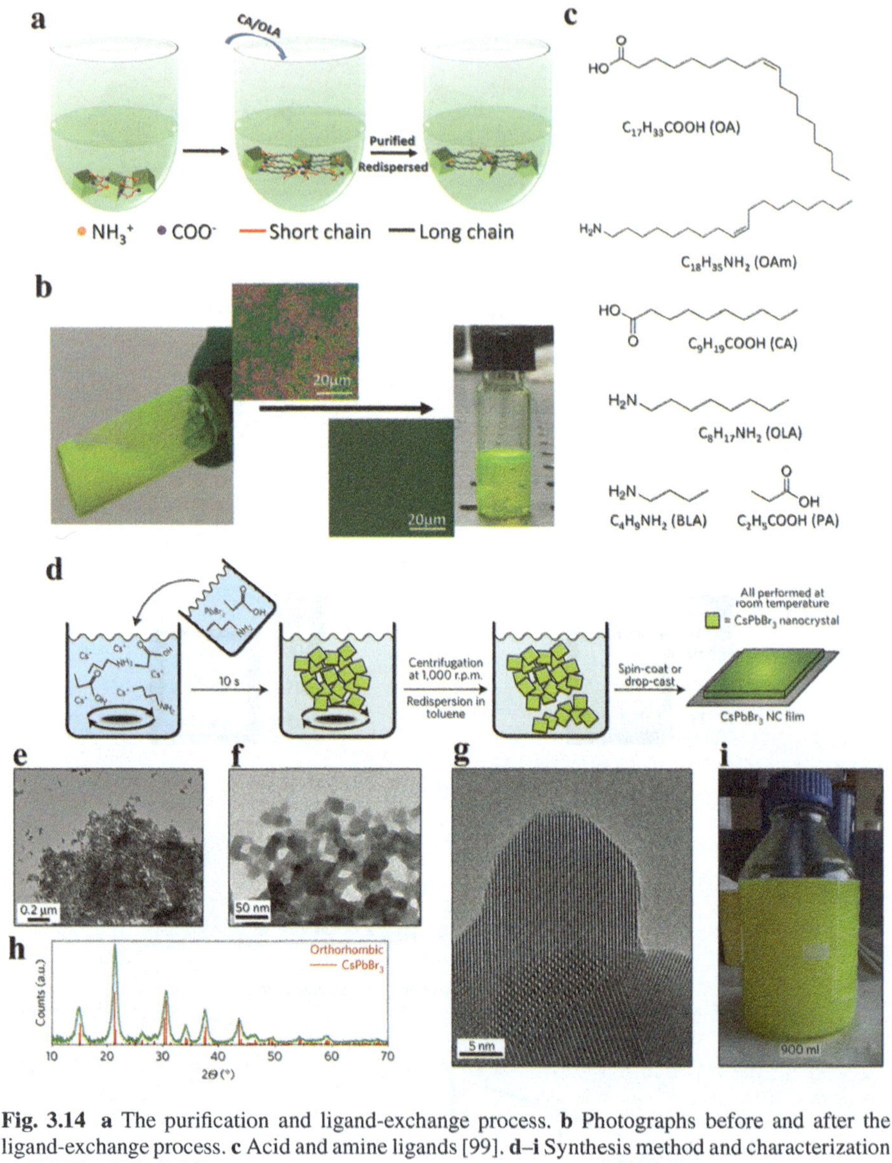


**Fig. 3.14** **a** The purification and ligand-exchange process. **b** Photographs before and after the ligand-exchange process. **c** Acid and amine ligands [99]. **d–i** Synthesis method and characterization of $CsPbBr_3$ QDs [103]

incorporation of phenethylamine (PEA) as ligand resulted in the fewer trap states in both $CsPbBr_3$ and CsPbI3 QDs, exhibiting the high PLQY of 93% and 95%, respectively.

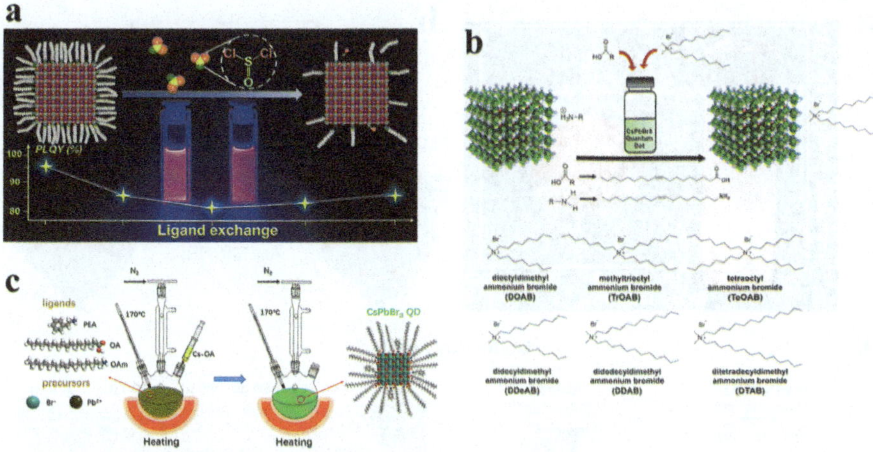


**Fig. 3.15** Schematic of **a** the mild reaction and ligand-exchange process and **b** ligand exchange process [105] **c** synthesis process of PEA ligand modified $CsPbBr_3$ NCs [106]

## 3.4 Post-Synthetic Ligand Treatments

Various post-synthetic ligand treatments have also been developed to produce high-quality $CsPbX_3$ QDs, such as with thiocyanates, $PbBr_2$ or trioctylphosphine $PbI_2$, in which is the PLQY of QDs is close to 100% [50, 107]. As shown in Figs. 3.16a, b, one of the key factors should be due to the improved passivation of the defects resulting from under-coordinated Pb ions on the $CsPbX_3$ surface.

Colvin's group [108] have demonstrated that trioctylphosphine (TOP) can effectively recover the luminescent emission of aged red perovskite QDs to their initial value and improve the emission intensity of the newly synthesized $CsPbX_3$ QDs, as shown in Fig. 3.16c–e. In addition, the TOP can enhance the stability of perovskite QDs against various ambient conditions, such as temperature, UV irradiation, and polar solvents. This may be attributed to the fact that surface defects are passivated in the presence of TOP. Wang et al. have developed a strategy to synthesize stable alkyl phosphate (TDPA) coated $CsPbBr_3$ QDs ($CsPbBr_3$/TDPA QDs) under ambient conditions [109]. In their report, the 1-Tetradecylphosphonic acid has been chosen both as the precursors in the formation of alkyl phosphate and the necessary ligands for the synthesis of $CsPbBr_3$ QDs. Based on the as-prepared $CsPbBr_3$/TDPA QDs with excellent stability, the WLEDs exhibit a wide color gamut (122% of NTSC) and high luminous efficiency of 63 lm/W. Li et al. [110] reported a highly luminescent $CsPbBr_3$@$NH_4Br$ core-shell composites using an ion exchange of CsPbI3 PQDs strategy. They proposed that the rich-bromide environments can sufficiently passivate the $CsPbBr_3$ PQDs surface. In comparison with the regular $CsPbBr_3$ QDs, the obtained $CsPbBr_3$ QDs with a core-shell structure can avoid PL quenching, enhance stability and maintain the excellent optical. When water contact with the $CsPbBr_3$@$NH_4Br$ composites, the $NH_4Br$ framework firstly dissolved in water and

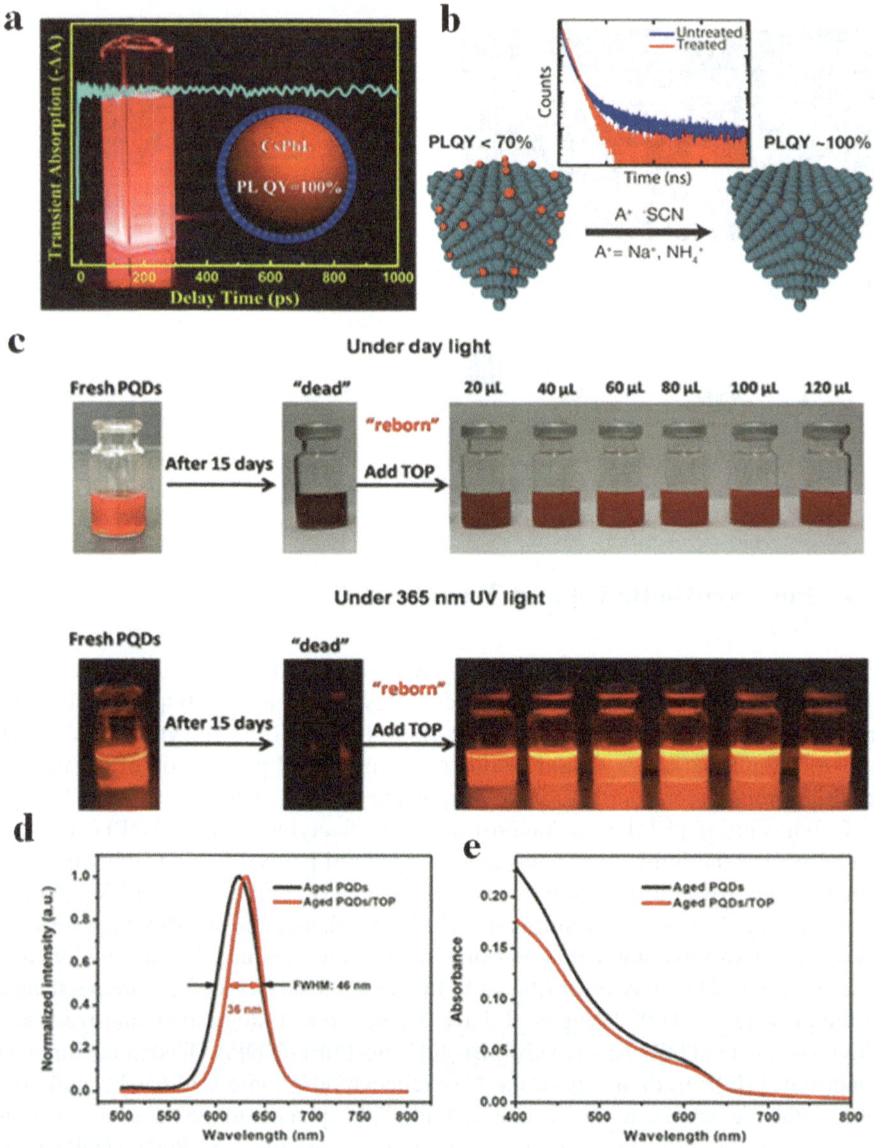


**Fig. 3.16** **a** Highly luminescent phase-stable $CsPbI_3$ NCs [107]. **b** Postsynthetic thiocyanate surface treatment [50]. **c–e** Pictures of various PQDs (fresh/aged PQDs) with different amounts of TOP. Reprinted with permission from Ref. [108]

then released numerous ammonium and Br ions, which protected the $CsPbBr_3$ QDs isolated from water. Jasieniak et al. [111] have proposed that the structural distortion of $CsPbI_3$ QDs arising from the ligand–surface interactions (see Fig. 3.17). They have obtained stable $CsPbI_3$ QDs using a ligand modification method by replacing regular OA ligand with bis-(2,2,4-trimethylpentyl) phosphinic acid (TMPPA) in the synthesis process. The PL performance of TMPPA modified $CsPbI_3$ PQDs ($CsPbI_3$–TMPPA) was similar to that of common $CsPbI_3$ PQDs ($CsPbI_3$–OA). Compared with $CsPbI_3$–OA decomposing within 3 days, the PL intensity of $CsPbI_3$–TMPPA almost retained the initial value during storage over 20 days. Therefore, replacing OA with TMPPA could effectively enhance the phase stability of $CsPbI_3$ QDs.

Chan and coworkers [112] have demonstrated that the performance of $CsPbBr_3$ nanoparticles using surface treatment of with a mixture of oleic acid, oleylamine and $PbBr_2$ not only improved their PLQY (up to unity) and prolonged their stability from days to months in the air, but it also observably increased the photoluminescence lifetime from ~0.9 to ~1.6 ns, as shown in Fig. 3.18. They emphasized that the intensity of the PL intensity was drastically improved by the $PbBr_2$/OA/OLAM treatment and the emission peak and FWHM remained almost unchanged (Fig. 3.18a and the inset). These results indicated that surface treatment not affected the intrinsic optical

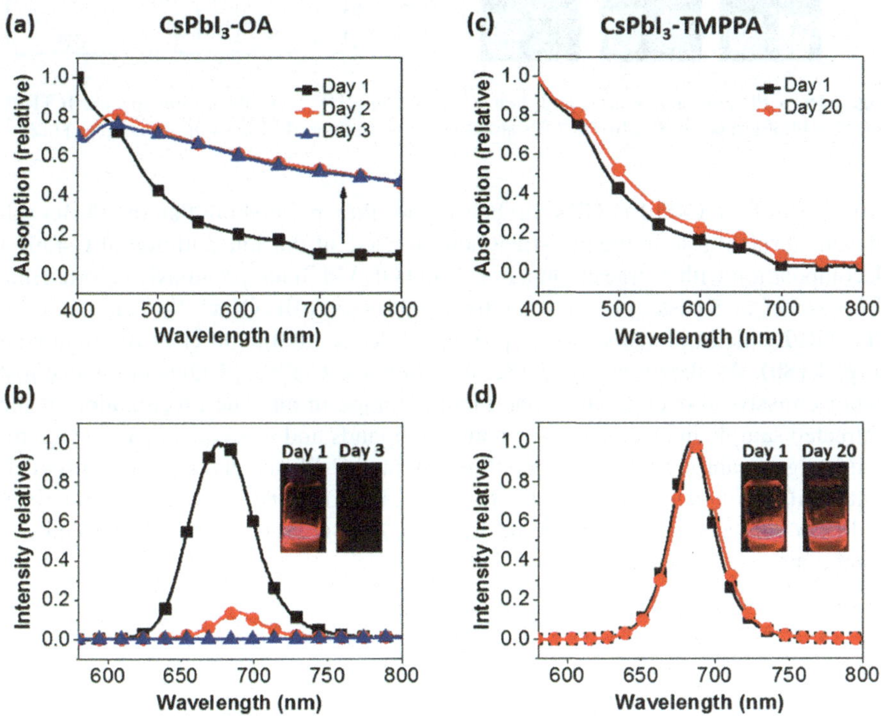


**Fig. 3.17** UV-Vis absorption and PL spectra for $CsPbI_3$–OA (**a** and **b**) and $CsPbI_3$–TMPPA (**c** and **d**). Insets of (**b** and **d**): solutions of the respective washed QDs under UV light [111]

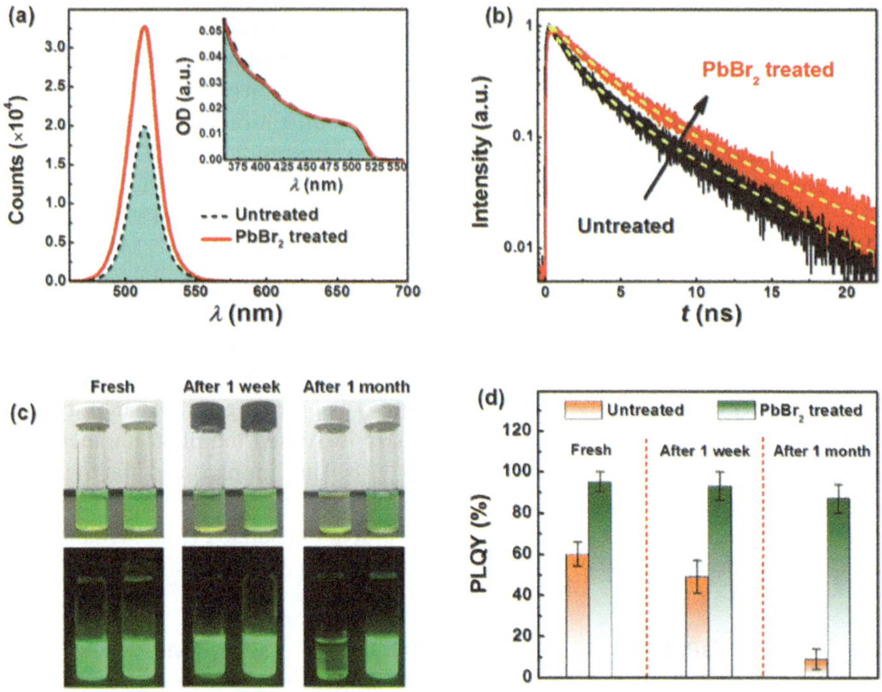


**Fig. 3.18** **a** PL emission spectra and their corresponding UV—vis absorption spectra. **b** TRPL traces. **c** Photographs of $CsPbBr_3$ CQDs stored along with time. **d** PLQYs of the samples [112]

properties of the $CsPbBr_3$ QDs. In their report, they pointed out that the $PbBr_2$ salt played a crucial role in the post-treatment because of the sole addition of $CsPbBr_3$ decomposition with a large amount of OA and OLAM. It has previously reported that the presence of excess amine structurally degrades $CsPbBr_3$ QDs [30]. Time-resolved PL (TRPL) measurements show a prolonged decay lifetime after $PbBr_2$ treatment (Fig. 3.18b). As shown in Fig. 3.18c, the untreated $CsPbBr_3$ CQDs in toluene lost major emissive properties after one-month storage in air. The precipitation of the untreated sample induced the loss of surface ligands and subsequent particle aggregation significantly. By contrast, the $PbBr_2$-treated $CsPbBr_3$ QDs exhibited an excellent clarification in the solvent with brightness. Figure 3.18d showed the comparison of PLQY between the untreated and treated $CsPbBr_3$ QDs samples under ambient conditions.

## 3.5 Coating Strategies

### 3.5.1 *Silica Coating for Stable $CsPbX_3$ QDs*

The dynamic nature and low crystal lattice energy of perovskite QDs lead to the soluble in nearly all polar solvents and subsequently result in the disintegration under such solvents conditions. This problem remained difficult to mitigate until a strategy for the full encapsulation of perovskite QDs with an inert shell material. Up to now, the $SiO_2$ coating method has been widely used in the traditional QDs, [113] such as lanthanide-doped QDs [114] and magnetic nanocomposites [115]. Remarkably, $SiO_2$ is an inorganic oxide with chemical stability and optical transparency in the whole visible spectrum. The coating perovskite QDs with $SiO_2$ has been found to not change the optical properties of luminescent materials and can protect materials against polar solvents dissolution. The silica coating for $CsPbX_3$ QDs is a typical process using Stöberor [116] or reverse microemulsion methods [117]. Liu et al. [118] reported a method that they obtained $CsPbBr_3$ QDs with silica spheres with the majority in the 12–15 nm range, as shown in Fig. 3.19a. They have demonstrated that the mesoporous $CsPbX_3$ QDs could prevent the ion-exchange effect and show enhanced thermal stability and photostability. Based on the mixture of green MP-$CsPbBr_3$ QDs and red perovskite QDs in silicon, they fabricated an on-chip LED device which exhibited an NTSC value of 113%.

Kovalenko's group and Yamauchi's group also have adopted the method of embedding the LHP QDs inside the mesoporous silica [119, 121]. They further demonstrated a common method of the growth of perovskite nanocrystals inside mesoporous silica templates, as shown in Fig. 3.19b. The dimensions of the perovskites can be well-controlled by altering the channels of mesoporous silica templates. Zeng et al. [120] successfully grew $CsPbBr_3$ QDs on the silica spheres based on a presynthesized aminated silica sphere as the substrate. The isolation of the QDs on the surface of the silica spheres prevents the QDs from contacting each other, which limits the possibility of photoinduced regrowth and deterioration of QDs (Fig. 3.19c). The obtained $CsPbBr_3$–$SiO_2$ composites show excellent stability against light irradiation, moisture, and air [96]. Additionally, Rogach et al. and Liu et al. presented a method to prepare stable solid-state perovskite by introducing polyhedral oligomeric silsesquioxane (POSS) to coat the LHPs NCs [122, 123]. The POSS-$CsPbBr_3$ nanocomposites showed high stability, including the resistance to water and UV light. Meanwhile, the POSS coating layers could efficiently prevent the anion exchange after mixing two perovskite QDs with different halide compositions.

In recent years, Zhang's group. [124, 125] have prepared monodisperse $CsPbX_3/SiO_2$ and $CsPbX_3/Ta_2O_5$ Janus nanoparticles based on the water-triggered transformation process and a sol-gel protocol. Taking $CsPbX_3/SiO_2$ as an example (see Fig. 3.20), in their report, the $CsPbBr_3/SiO_2$ composite showed robust stability in the different surroundings, such as water, high humid air (40 °C and humidity of 75%) and at 375 nm UV light. Leng et al. [126] presented a facile method for the in situ growth of silica shells onto $CsPbX_3$ QDs without using any water. The amorphous

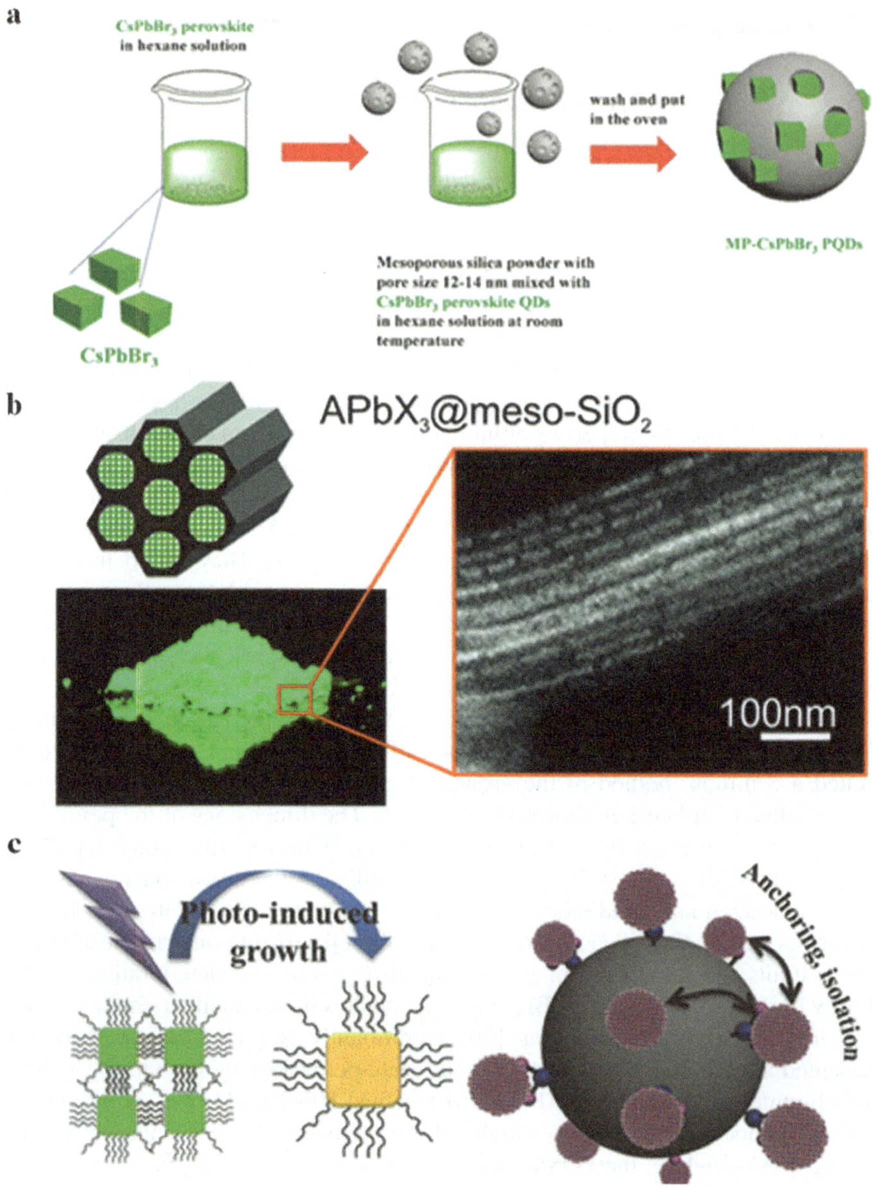


**Fig. 3.19** **a** The synthesis process [118]. **b**. Schematic, photographs, and TEM. Adapted from Ref. [119]. **c** Schematic diagram of photoinduced regrowth [120]

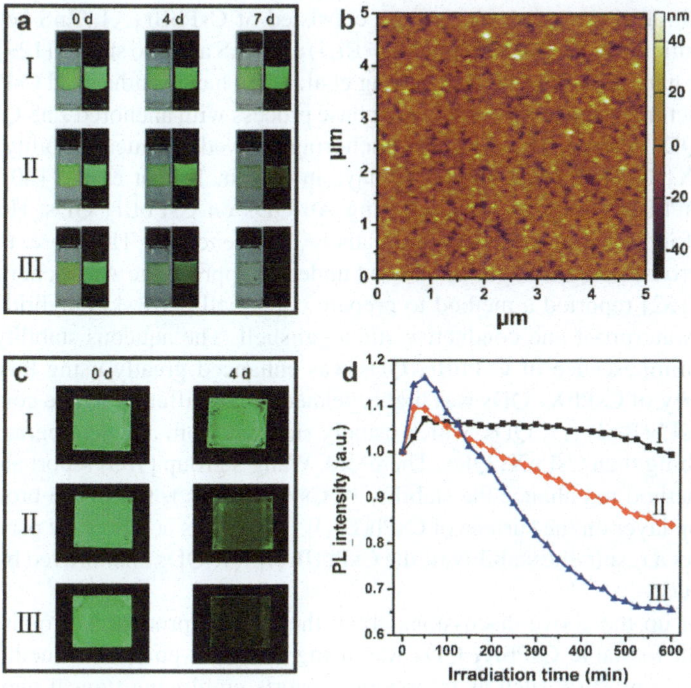


**Fig. 3.20** **a** The stability against water of (I) $CsPbBr_3/SiO_2$ QDs, (II) WT-$CsPbBr_3$ QDs and (III) HI-$CsPbBr_3$ QDs. **b** AFM image for $CsPbBr_3/SiO_2$ thin film. **c** The evolution of the thin films in air. **d** The photostabilities for different thin films [124]

$SiO_2$ layers were formed on the PQDs rapidly when the silica precursor tetraethoxysilane (TEOS) was injected during the synthesis process under high temperature. In contrast with the common $CsPbBr_3$ QDs, the obtained $CsPbBr_3/SiO_2$ composites presented a dramatically enhanced stability, such as improved PL intensity, nonblinking optical properties, and photostability. Jang et al. [127] have employed the perhydropolysilazane (PHPS) as a silica precursor during the synthesis process. The silica layer formed on the $CsPbBr_3$ PQDs under the moisture condition. The silica-coated $CsPbX_3$ QDs showed good thermal stability and chemical stability compared to the untreated $CsPbX_3$ QDs. In recent years, monodisperse $CsPbBr_3@SiO_2$ core-shell QDs have been synthesized by varying synthetic conditions, such as temperature, pH, and precursors [125, 128]. In comparison with the universal phenomenon of aggregates, the stable $CsPbBr_3@SiO_2$ QDs exhibited excellently dispersed in the solvent, leading to a potential in optoelectronic applications, such as high-quality film, cell imaging, LED, etc.

In addition to the silica coating strategy, recently, some reports commented that anchoring $CsPbBr_3$ QDs onto a substrate can effectively improve their stability. Importantly, the PL intensity has a little decrease even after 40 days of storage in ambient air. The proposed architecture consisted of $CsPbBr_{3-x}I_x$/ZnS heterostructures. Atomic structures of $CsPbX_3$ (X = Br, I) and ZnS are also shown [129]. A novel anchoring nanocrystal architecture, Tang et al. [129] have synthesized $CsPbX_3$/ZnS heterostructures using a facile solution-phase process with anchored ZnS QDs on the $CsPbX_3$ QDs surface. The ZnS QDs anchoring showed chemical stability in which the $CsPbX_3$ QDs remained up to 12 days in the air. Kamat et al. [130] reported an Au–$CsPbBr_3$ composite by anchoring Au QDs on $CsPbBr_3$ QDs. However, it has been found that an excess of Au leads to PL quenching. Therefore, the brightness of perovskite QDs might be showed under an appropriate Au-anchoring [131]. Li et al. [132] reported a method to prepare core-shell perovskite hybrid by introducing a waterproof and conductive silica gel shell. The aqueous stability of electrochemiluminescence of $CsPbBr_3$ QDs was enhanced greatly using this method. The stability of $CsPbX_3$ QDs was highly related to the different halide components, such as $CsPb(Br_xI_{1-x})_3$ QDs, which usually suffered from a faster degradation and PL quenching than $CsPbX_3$ QDs. Therefore, Wang's group [133] report an acetone-etching method to enhance the stability of $CsPb(Br_xI_{1-x})_3$ QDs. The bromine-rich passivation layer on the surface of $CsPb(Br_xI_{1-x})_3$ QDs is achieved by partial iodine etching. As a result, the stability of the $CsPb(Br_xI_{1-x})_3$ QDs is improved by 3 orders of magnitude.

To sum up the above discussions about the coating process, a promising alternative path to stable $CsPbX_3$ QDs was using the glass-embedded method during the synthesis process, such as in previous reports employing typical phosphate or borosilicate glasses [134–137]. The size of the obtained $CsPbX_3$ QDs was tunable by melt quenching and by adjusting the subsequent thermal conditions, eventually, they were surrounded by robust glass matrix. Such stable $CsPbX_3$ NCs could be easily integrated into the application of backlighting in displays or as an effective material in lighting.

### *3.5.2 Other Materials Coating*

As can be seen from the discussion above, many efforts have been devoted to silica coating CsPbX3 NCs. In addition, introducing $CsPbX_3$ QDs into presynthesized porous materials was a feasible strategy for improving stability. In the previous report, it has been demonstrated that embedding perovskites into mesoporous materials is an effective approach to improving their stability [138, 139]. Ye et al. [140] presented a facile two-step synthesis process for luminescent and color-tunable $CsPbX_3$-zeolite-Y composite, as shown in Fig. 3.21a. First, they infused $Cs^+$ into zeolites by the ion-exchange reaction. Then, $Cs^+$ cations diffused and reacted with the $PbBr_2$ solution, and finally formed $CsPbX_3$ QDs inside the zeolite pores. Cheng et al. [141] integrated $CsPbX_3$ QDs into $CaF_2$ hierarchical nanospheres and then obtained stable $CsPbX_3$ QDs. In comparison with the regular $CsPbX_3$ perovskite materials, the treatment of $CsPbX_3$ QDs has improved the stability and remained the excellent optical

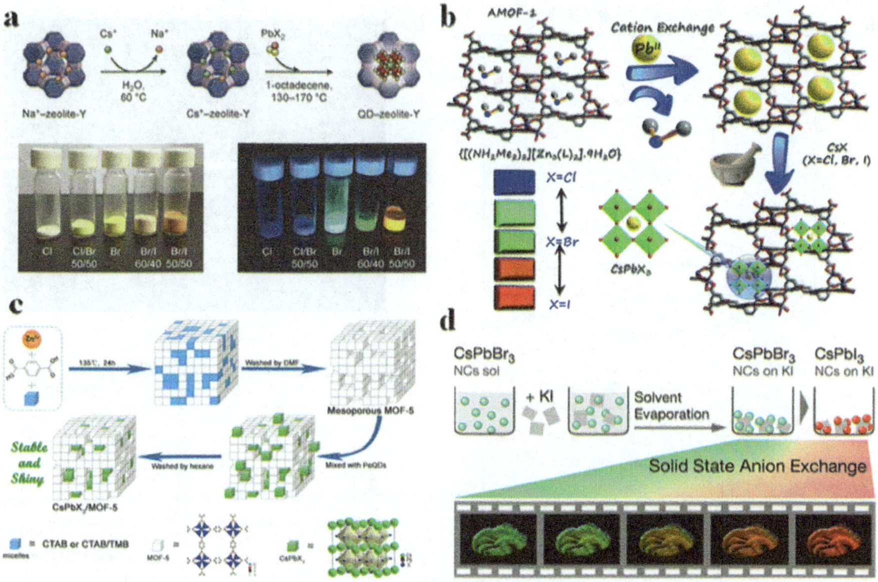


**Fig. 3.21** **a** Schematic illustration of the two-step synthesis process and photograph under visible illumination or 365 nm [140]. **b** The synthesis of PQD@MOF composites [142]. **c** Synthesis strategy of the composites [143]. **d** Scheme of the embedding $CsPbBr_3$ NCs into different KX (X = Cl, Br, or I) salts [144]

properties, such as high color purity, high brightness and so on. In addition, by co-doping of $Ce^{3+}$/$Tb^{3+}$ into $CaF_2$ matrices, the obtained samples exhibited strong green brightness and reached a high PLQY of 77%. They fabricated white LEDs with a power efficiency of 62.7 lm/W based on the green $CaF_2$-$CsPbBr_3$ composites and red commercial phosphors.

Among the porous materials, metal-organic frameworks (MOFs) with tunable pore size have been regarded as a great potential for encapsulation of perovskites. Bhattacharyya and coworkers [142] synthesized a $CsPbX_3$@MOF composite using a mechanochemical method, as shown in Fig. 3.21b. The obtained $CsPbX_3$ @MOF exhibited narrow and intense emission bands. Based on a similar method, Wang et al. [143] prepared a stable $CsPbX_3$ NCs by embedding $CsPbX_3$ PeQDs into the mesostructure zinc-based MOF (MOF-5) crystals, as shown in Fig. 3.21c. In their report, the mesoporous structured MOF-5 crystals protected the $CsPbX_3$ QDs against the various ambient conditions. Based on the $CsPbX_3$/MOF-5 composites, they fabricated a warm white LED that exhibited a color temperature of 3607 K and a wide color gamut. Gaponik et al. [144] have incorporated $CsPbBr_3$ QDs into different KX (X = Cl, Br, I) salts under high pressure of 2.2 GPa and showed that the $CsPbX_3$/KX pellets with bright fluorescence could be tuned in the entire visible light region (see Fig. 3.21d). The obtained $CsPbX_3$/KX QDs showed excellent robustness and stability under ambient conditions. Most inorganic oxides are a class of highly robust that has

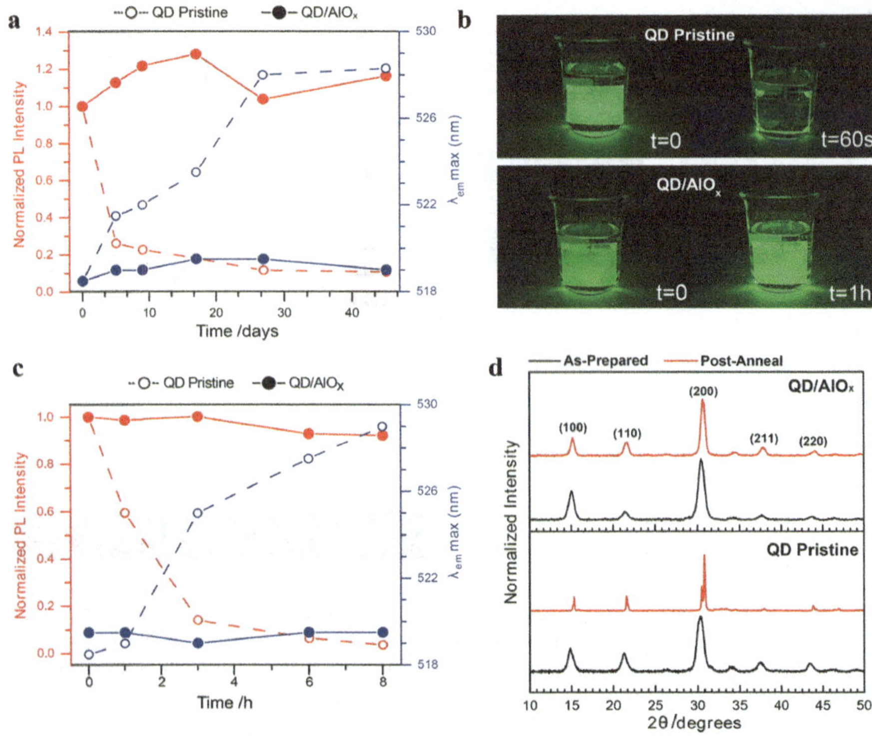


**Fig. 3.22** Stability studies with $CsPbBr_3$ QD/AlOx nanocomposites. **a** PL properties. **b** Photographs under UV illumination. **c** PL properties after 8 h photosoaking under solar spectrum irradiation. **d** XRD spectra [145]

application as a coating material. Oxides used to protect $CsPbX_3$ NCs are a reasonable design strategy. Buonsanti et al. [145] prepared $CsPbBr_3$/AlOx composites via an atomic layer deposition (ALD) technique, as shown in Fig. 3.22. The growth of the amorphous alumina (AlOx) matrix was using the virtue of the low-temperature and fast ALD process. The obtained $CsPbBr_3$/AlOx QDs have been tested under various conditions, including air (Fig. 3.22a), water (Fig. 3.22b), light (Fig. 3.22c), and temperature (Fig. 3.22d). The results of the stabilities have been demonstrated that the AlOx layer leads to protection for $CsPbBr_3$ PQDs, resulting in the preserved the optoelectronic properties as well as enhanced stability towards the air, high-energy light irradiation, and thermal and water.

Zheng et al. [146] proposed a method for obtaining monodispersed $CsPbBr_3/TiO_2$ core/shell NCs. The schematic illustration of the fabrication process is shown in Fig. 3.23a. The $CsPbBr_3/TiO_2$ core/shell QDs were synthesized using the encapsulation of colloidal $CsPbBr_3$ QDs with titanium precursor, and then calcined samples at 300 °C for 5 h. The pristine $CsPbBr_3$ NCs have diameters of 12 $\pm$ 2 nm (see Fig. 3.23b) and the obtained $CsPbBr_3/TiO_2$ core/shell NCs present a $TiO_2$ shell thickness of 5 $\pm$ 3 nm (see Fig. 3.23c). The resulting monodispersed $CsPbBr_3/TiO_2$

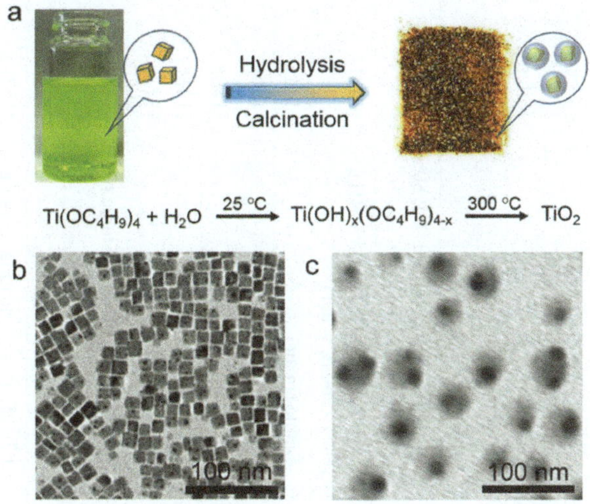


**Fig. 3.23** **a** Schematic illustration of the fabrication process. TEM images of **b** $CsPbBr_3$ QDs and **c** $CsPbBr_3/TiO_2$ core/shell QDs [146]

nanocomposites showed excellent water stability and improved PL intensity. In addition, the $TiO_2$ shell could improve the chemical stability and photostability of $CsPbBr_3$ QDs (Table 3.2).

## 3.6 Compositional Engineering

Zeng et al. [161] summarised the doping perovskite $ABX_3$ with targeted ions. It has been demonstrated that the doping cations of A-site were always monovalent metal cations or organic cations; B-site cations or doped cations covered various bivalent or heterovalent metal cations, whereas X-site anions were mainly single or mixed halides as well as anion group SCN. They summarized the applied ions as shown in Fig. 3.24.

Among the family of $CsPbX_3$ QDs, $CsPbCl_3$ QDs has the most appropriate band gap for the $M^{2+}$ doping [162]. Dong and Klimov groups demonstrated that $Mn^{2+}$ doping can be accomplished using the hot-injection method, as shown in Fig. 3.25a [55, 163]. In their reports, the precursors ($Mn^{2+}$ and $Pb^{2+}$) were completely dissolved under high temperatures, and then a measured amount of cesium oleate was swiftly injected into the mixture precursors for obtaining Mn-doped QDs. They found that the Mn-doped $CsPbCl_3$ QDs exhibited a bright yellow emission under the UV light, which results from the Mn d-d transition. Bekr et al. [164] adopted $Bi^{3+}$ ions into $CsPbBr_3$ QDs via a hot-injection method, as shown in Fig. 3.25b. Figure 3.25c illustrates the fluorescent tunability of $CsPb_xMn_{1-x}Cl_3$ QDs by alerting the doping condition [56]. Figure 3.25d shows the $CsPbX_3$ crystal after doping the $Mn^{2+}$ ions [54]. Donega and coworker [51] reported a method that adopted partial cation exchange in $CsPbBr_3$ QDs resulting in favourable preservation of the original structure (see

**Table 3.2** Summary of PL properties and stability of all inorganic perovskites NCs@glass

| Glass composition | All-inorganic perovskite | Emission (nm) | PLQY (%) | Stability | References |
|---|---|---|---|---|---|
| $B_2O_3$–$SiO_2$–ZnO–$SrCO_3$ | $CsPbI_3$ | 630–698 | 4.2 | 91.1% (200 h, 25 °C, 85% RH) 30% | [135] |
| $2O_3$–$SiO_2$–$Na_2O$–CaO | $CsPbBr_3$ | 460–620 | 28 | ($CsPb(Cl_{0.5}Br_{0.5})_3$, 190 °C); 80% ($CsPbBr_3$, 12 h, UV) | [147] |
| $2O_5$–$SiO_2$–ZnO | $CsPb(Br/I)_3$ | 523–693 | 40.5 | 95.9% (500 h, 25 °C, 85% RH); 90% (960 h, water) | [148] |
| $P_2O_5$–$SiO_2$–ZnO | $CsPbBr_3$ | 497–523 | 42 | Stable (in air) | [134] |
| $B_2O_3$–$SiO_2$–ZnO | $CsPb(BrI)_3$ | 520–705 | 42.5 | | [149] |
| $B_2O_3$–$SiO_2$–ZnO–$Al_2O_3$ | $CsPb_2Br_5$ | 500–519 | 43.8 | | [150] |
| $P_2O_5$–$SiO_2$–$SrCO_3$–$Al_2O_3$ | $CsPbBr_3$ | 432–521 | 51.5 | | [136] |
| $B_2O_3$–$SiO_2$–ZnO | $CsPbBr_{1.5}I_{1.5}$: Eu | 501–672 | 64.7 | | [151] |
| $TeO_2$–$H_3BO_3$–$Al_2O_3$–ZnO–$Na_2CO_3$ | $CsPbBr_3$ | 522 | ~70 | 60% (45 days, water); Stable (70 h, UV); | [152] |
| $GeO_2$–$B_2O_3$–ZnO–CaO–$K_2O$ | $CsPbX_3$ | 478–698 | 80 | 97% (7 days, water); | [144] |
| $P_2O_5$–$SiO_2$–SrO–$Al_2O_3$ | $CsPbBr_3$ | 516 | | 97.12% (for 10 heating–cooling cycles at 483 K) | [153] |
| $B_2O_3$–$SiO_2$–ZnO–$Al_2O_3$ | $CsPb_2Br_5$ | 499–506 | | 30.6% (433 K) | [154] |
| $B_2O_3$–$SiO_2$–ZnO–$SrCO_3$ | $CsPbBr_xI_{3-x}$ | 520–698 | | 95.8% (500 h, 85 °C, 85% RH) | [155] |
| $B_2O_3$–$SiO_2$–ZnO | $CsPbBr_3$: $Sn^{2+}$ | 506–518 | | Stable (100 days, 85% RH) | [156] |
| $B_2O_3$–$SiO_2$–ZnO | $CsPbBr_3$: $Eu^{3+}$ | 480–550, 580–700 | | | [157] |
| $B_2O_3$–$SiO_2$–ZnO | $CsPb(Cl/Br)_3$: $Mn^{2+}$ | 505–750 | | 90.0% (80 h, 25 °C, 85% RH) | [158] |
| $B_2O_3$–$SiO_2$–ZnO | $CsPbBr_3$: $Tb^{3+}$,$Eu^{3+}$ | 400–580, 600–750 | | 67% (245 °C) | [159] |
| $B_2O_3$–$SiO_2$–ZnO | $CsPb(Br_xI_{1-x})_3$: $Mn^{2+}$ | 572–708 | | over 45% (358 K) | [160] |

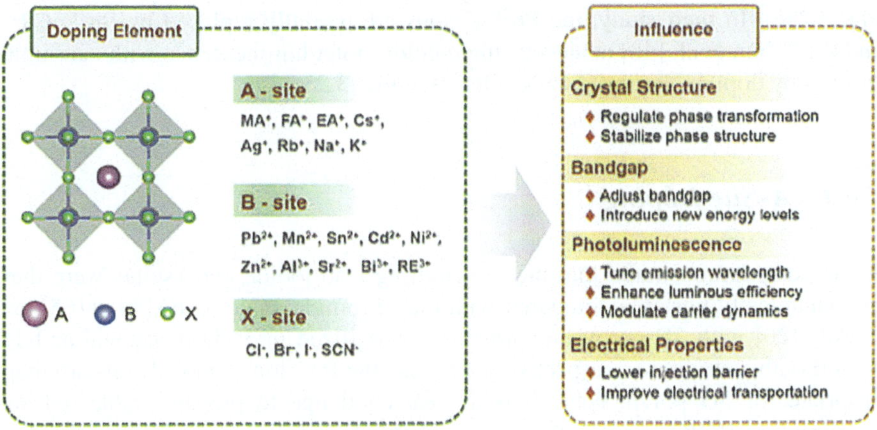


**Fig. 3.24** Schematic diagram of the doped ions and influence in $ABX_3$ perovskite QDs [161]

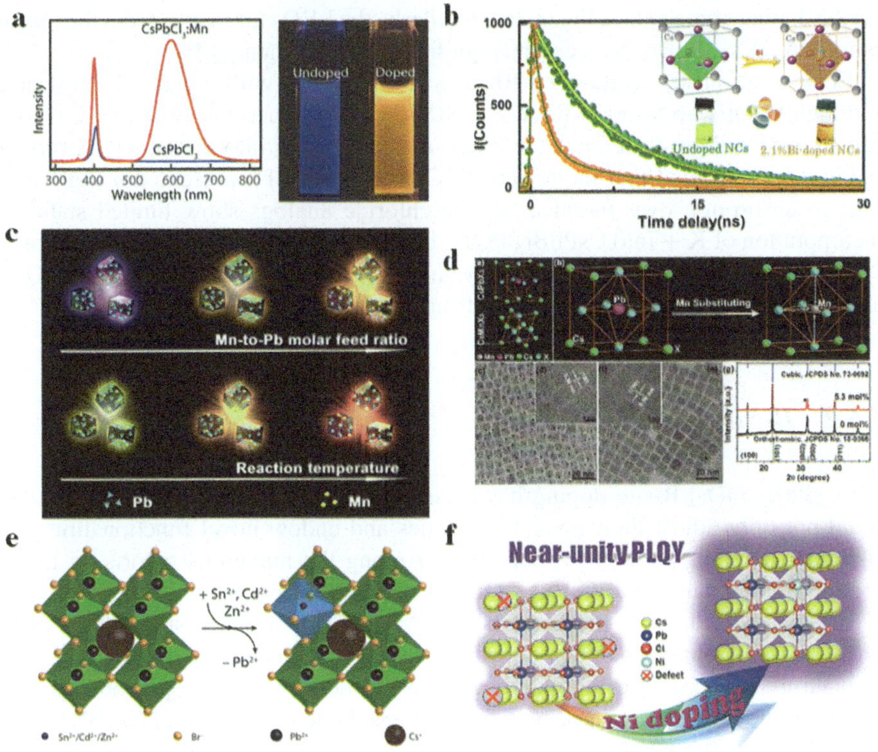


**Fig. 3.25** **a** Mn-doped $CsPbCl_3$ [163]. **b** $Bi^{3+}$−doped $CsPbBr_3$[164]. **c** Mn doped [56]. **d** Schematic illustration showing the lattice contraction [54]. **e** Schematic overview of partial cation exchange in $CsPbBr_3$ nanocrystals [51]. **f** $Ni^{3+}$ doped $CsPbCl_3$ NCs [48]

Fig. 3.25e). In their study, the $Pb^{2+}$ cations are partially replaced by $Sn^{2+}$, $Cd^{2+}$, and $Zn^{2+}$. Sun et al. [48] achieved single-color violet luminescence with near-unity PLQYs by doping Ni ions $CsPbX_3$ QDs (see Fig. 3.25f).

### 3.6.1 A-Site Doping

It has been demonstrated that the red-emitting lead iodide perovskites were thermodynamically unstable compared with their bromide analogs. $FAPbI_3$ [165] and $CsPbI_3$ [20] with 3D perovskite structure crystallized into 1D hexagonal and 1D orthorhombic structures, respectively, because the $FA^+$ ions and $Cs^+$ ions are inappropriate for 3D polymorphs. It is a great challenge to prepare stable red and near-infrared (NIR) nano emitters, termed the "perovskite red wall". It could be addressed by rational doping of $FA^+$ into the $CsPbI_3$ lattice [37]. The as-designed $FA_{0.1}Cs_{0.9}PbI_3$ PQDs showed stable and bright red-emission with PLQYs over 70% via the doping strategy (see Fig. 3.26). Similarly, the LEDs fabricated from the mixed cation $FA_{0.8}Cs_{0.2}PbBr_3$ NCs exhibit significant enhancement [13].

Etgar et al. [166] synthesized $Rb_xCs_{1-x}PbX_3$ PQDs with tunable fluorescence and high PLQEs up to 60%. $Rb_xCs_{1-x}PbX_3$ PQDs retained the structural integrity and showed a slight difference upon $Rb^+$ cations doping into the lattice. A further stability test revealed that the mixed Rb/Cs bromide PQDs presented a slight redshift in absorption over months, but the chloride analogs show limited stability. Incorporation of K + into $CsPbBr_3$ PQDs by the postmodification method improves PL QYs from 65 to 83% [167]. Similarly, the incorporation of $K^+$ into $CsPbCl_3$ PQDs brings in the improvement of PL QYs from 3.2 to 10.3% [168].

### 3.6.2 B-Site Doping

For $CsPbX_3$ QDs, B-site doping has been regarded as an effective technique that could not only adjust their optical properties and endow novel functionalities but also improve their stability [49, 66, 169]. Among the numerous reports on B-site doping, doping $Mn^{2+}$ into $CsPbX_3$ QDs has received great attention [55, 56, 170–175]. Shedon et al. [163] prepared the $Mn^{2+}$ into $CsPbCl_3$ and $CsPb(Cl/Br)_3$ QDs. The successful synthesis of Mn-doped perovskite nanocrystals and the observation of intense sensitized Mn luminescence showed the highly homogeneous spectral characteristics of Mn luminescence from an ensemble of Mn-doped $CsPbX_3$ QDs. Chen and coworkers [54] reported a $Mn^{2+}$-doped strategy to improve the environmental stability and PL properties of QDs. $CsPbBr_3$: Mn QDs preserved about 60% of the initial PL brightness upon exposure to ambient air for 120 days (see Fig. 3.27). Similarly, Meggiolar et al. [44] doped the $Mn^{2+}$ in the lattice of $\alpha$–$CsPbI_3$ QDs, the obtained $CsPb_xMn_{1-x}I_3$ QDs could be stable in films and in solution during storage over 1 month.

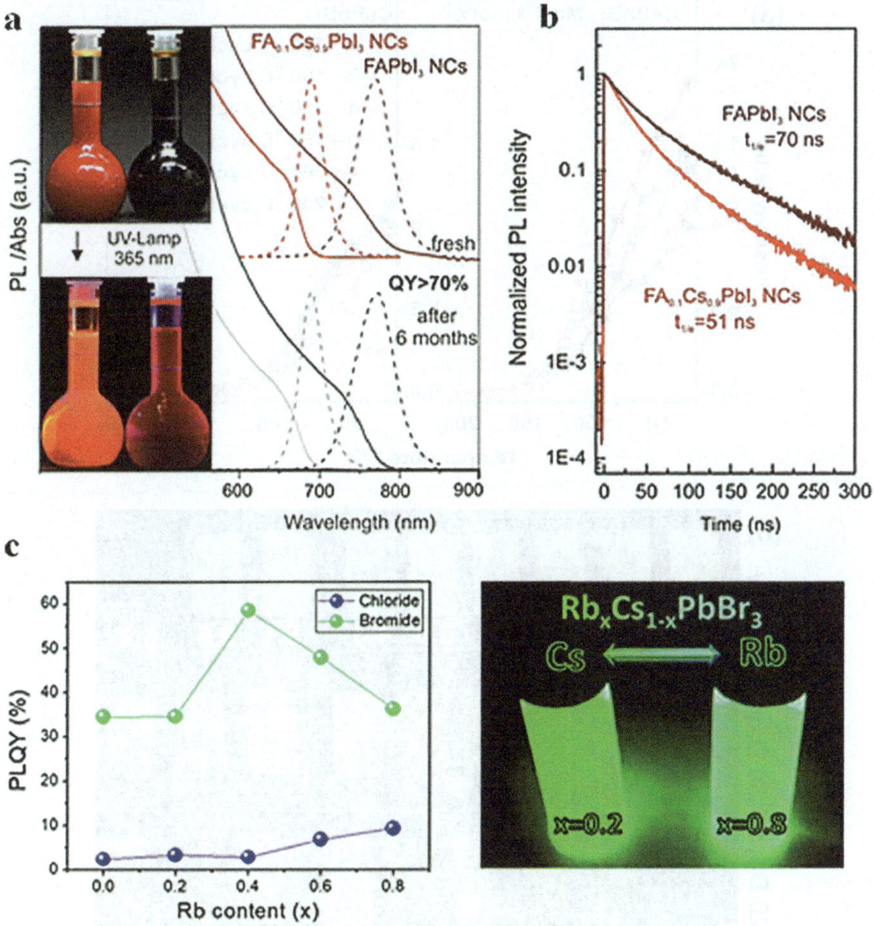


**Fig. 3.26** **a** The photographs of the $FAPbI_3$ NCs and $FA_{0.1}Cs_{0.9}PbI_3$ NCs and **b** PL decay traces for $FAPbI_3$ and $FA_{0.1}Cs_{0.9}PbI_3$ NCs. [37]. **c** PLQYs [166]

Until now, there are many metal ions have been incorporated into $CsPbX_3$ QDs. Most investigations are focused on the effect of phase structure or optical properties based on the doping strategies. The optical properties of B-site doping with various metal ions in $CsPbX_3$ QDs are summarized in Table 3.3.

## 3.7 Polymer Encapsulation

Tüysüz et al. [184] proposed that the $CsPbI_3$ perovskite QDs can be regarded as a photocatalyst under light irradiation and polymerize monomers. In their report,

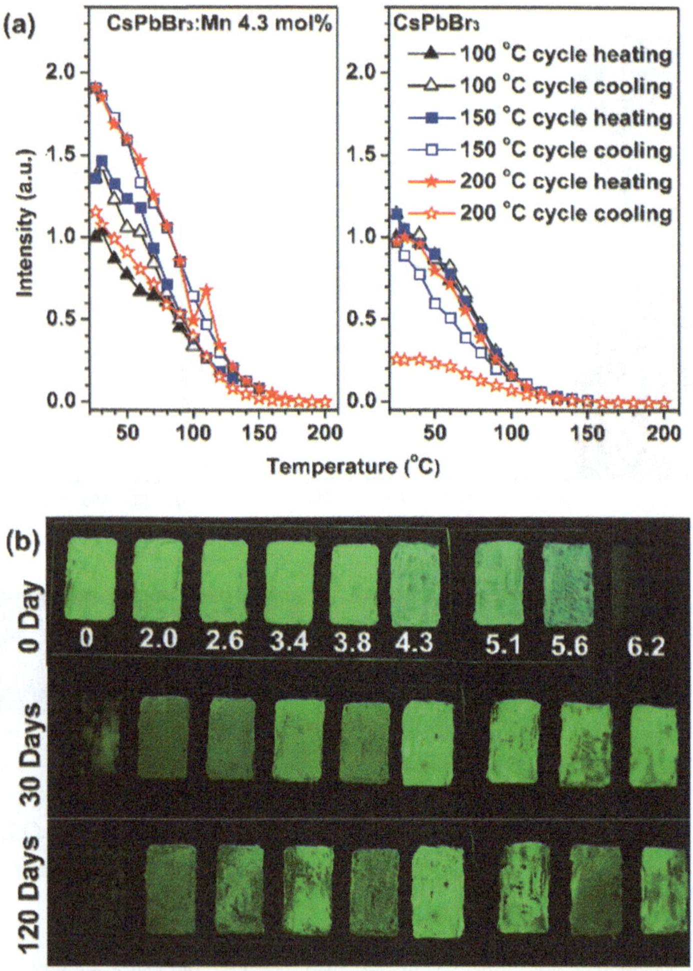


**Fig. 3.27** **a** PL intensities. **b** PL emission photographs for $CsPbBr_3$:Mn QDs [54]

they used 2, 2, 5′,2″-ter-3, 4-Ethylenedioxythiophene (TerEDOT) polymerized over $CsPbI_3$ PQDs with the help of 1,4-benzoquinone. Finally, $CsPbI_3$ QDs have been encapsulated in the poly(3,4-ethylenedioxythiophene) (PEDOT) networks and the initial cubic structure is retained. Tan et al. [185] reported a method to synthesize the luminescent perovskite-polymer nanocomposites, in which they adopted $CsPbBr_3$ PQDs as the photoinitiator in the polymerization of vinyl monomers (see Fig. 3.28a). Upon white light illumination, the in situ growth of polymer chains grew on the surface of $CsPbX_3$ QDs, resulting in the individually encapsulated perovskite-polymer nanoparticles with monodisperse core-shell structure. Thus, the perovskite–polymer nanocomposites showed a significant enhancement in PLQY and excellent stability against moisture. Moreover, the obtained nanocomposites provided the

**Table 3.3** Summaries of B-site doping with various metal ions

| B-site doped | PL (nm) | PLQY (%) | References |
|---|---|---|---|
| $CsPbCl_3$: Mn | 420 | 67 | [176] |
| $CsPbCl_3$: Mn | 405/600 | 39 | [177] |
| $CsPbX_3$: Mn | 430/525 | 56 | [178] |
| $CsPb_{0.75}Mn_{0.25}Cl_3$ | 392/607 | 59.3 | [52] |
| $CsPbBr_3$: Mn | 520 | 90 | [54] |
| $CsPb_xMn_{1-x}Cl_3$ | 580 | 54 | [56] |
| $CsPbI_3$: Bi | | | [179] |
| $CsPbBr_3$: Bi | 517 | 55 | [164] |
| $CsPb_{0.93}Cu_{0.07}X_3$ | 453 | 80 | [180] |
| $CsPbBr_3$: $Sn^{2+}$, $Zn^{2+}$, $Cd^{2+}$ | 462–506 | >50 | [51] |
| $CsPbBr_3$: Al | 456 | 42 | [49] |
| $CsPbBr_3$: $Ce^{3+}$ | 515 | 89 | [53] |
| $CsPbI_3$: Ag | | 80 | [181] |
| $CsPbCl_3$: $Ni^{2+}$ | 405 | 96.5 | [48] |
| $CsPbX_3$: Fe | 585/530 | – | [182] |
| CsPbCl3: $Yb^{3+}$ | 986 | 127.80 | [183] |

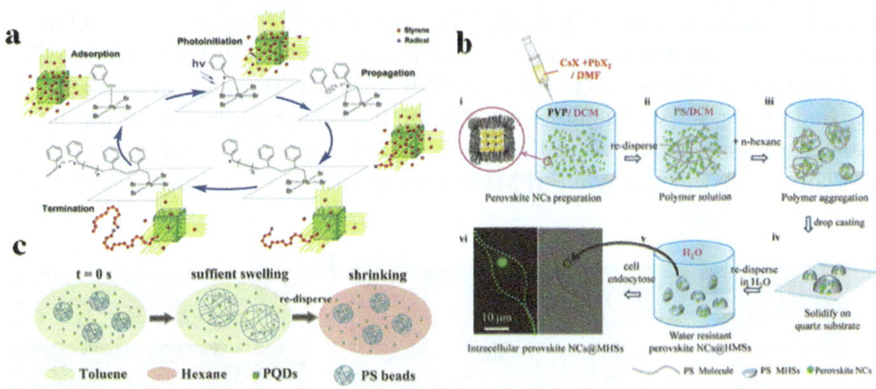


**Fig. 3.28** **a** Reaction mechanism. Reprinted with permission from Ref. [185]. **b** Schematic diagram of $CsPbX_3$ NCs@MHSs fabrication process. Reprinted with permission from Ref. [186]. **c** PQDs@PS composites [187]

possible application for functional downconversion films, displays and lighting. Fu and co-workers [186] developed an approach to synthesize polyvinylpyrrolidone (PVP)-capped colloidal $CsPbX_3$ QDs (see Fig. 3.28b). Under the protection of the PVP, the obtained composite-tunable $CsPbX_3$ QDs exhibited a high QY and narrow emission line widths of 12–34 nm. In addition, they embedded the QDs in polystyrene microhemispheres (MHSs). As a result, the obtained luminescent NCs@MHSs could be employed as luminescent probes for cell imaging due to the high water stability and nontoxicity in live cells.

Lin et al. [187] proposed a strategy to synthesize the uniform morphology of PQDs@PS composite beads with high luminescence (see Fig. 3.28c). In their report, the obtained composite exhibited the high luminescence superior stability against the water. The composite still had a strong luminescence over nine months period. Li and coworkers [188] fabricated monodisperse $CsPbBr_3$@PS microspheres with uniform morphology. PS layer provided protection against the anion-exchange and water. Meanwhile, the as-prepared $CsPbBr_3$@PS exhibited excellent luminescence and superior stability. Zheng et al. [189] proposed a general strategy to obtain a $CsPbX_3$@polymer with highly uniform size and spatial distribution via a situ growth of $CsPbX_3$ QDs in polymer fibers (see Fig. 3.29a). They pointed out that $CsPbX_3$ QDs can be uniformly encapsulated within the polymer fibers by rationally adjusting the ratios of $PbX_2$ and CsX during the synthesized process.

Zhang et al. [191] integrated the formation of $CsPbX_3$ QDs and the polymer matrix in a one-pot reaction to obtain the composites. This technique avoided the tedious separation and preparation of perovskites. In their report, they have demonstrated that it was a universal method for most polymers, such as poly(methyl methacrylate), poly(butyl methacrylate), and polystyrene. The resulted perovskite–polymer composites could maintain their quantum yield for more than 1 month under the ambient conditions. Chen and coworker [190] synthesized the $CsPbBr_3$ perovskite quantum dots (PQDs)/ethylene vinyl acetate (EVA) composite films via a one-step method. Figure 3.29b exhibited the photographs of the $CsPbBr_3$ PQD–EVA solution in toluene and $CsPbBr_3$ PQDs/EVA films. They demonstrated that the prepared films possessed long-term stable luminescent properties under the air and water conditions. Chen et al. [192] proposed a facile approach to fabricate $CsPbBr_3$ NCs/poly(methyl)methacrylate (PMMA) composites with excellent PL stability.

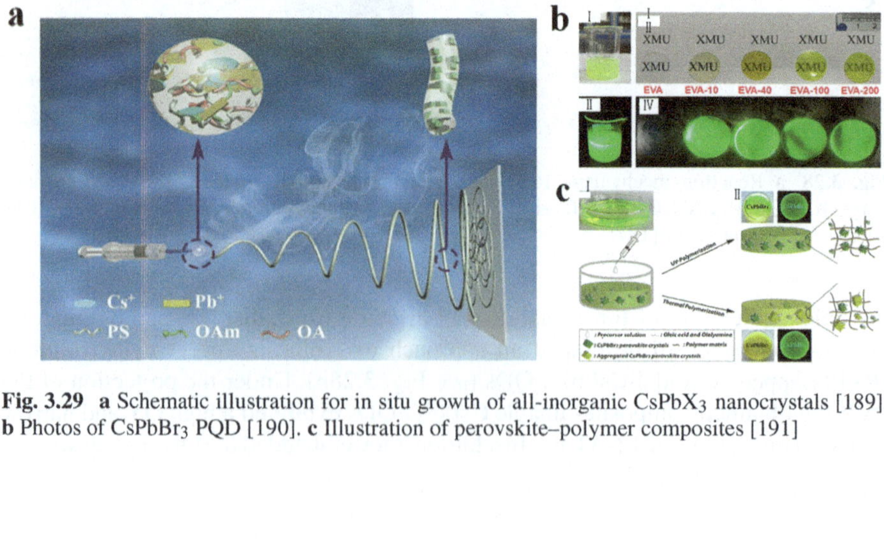


**Fig. 3.29** **a** Schematic illustration for in situ growth of all-inorganic $CsPbX_3$ nanocrystals [189]. **b** Photos of $CsPbBr_3$ PQD [190]. **c** Illustration of perovskite–polymer composites [191]

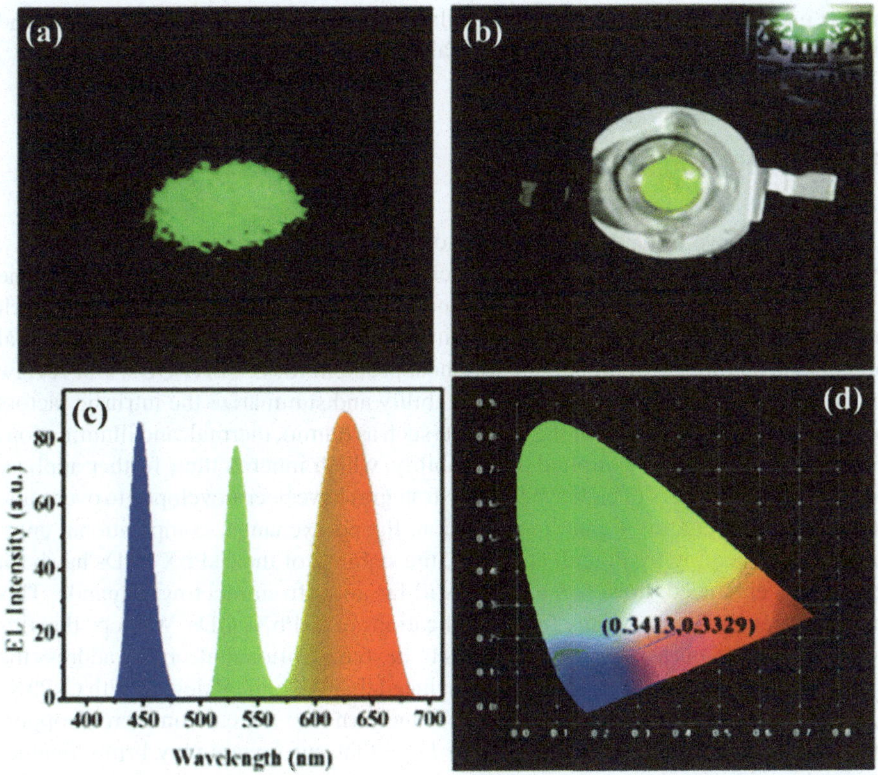


**Fig. 3.30** **a** Picture of $CsPbBr_3$/PMMA powders. **b**. A picture of the WLED device prepared by mixing $CsPbBr_3$/PMMA powders with CdSe QDs and a photograph of the as-prepared WLED device under the operation of 5 mA (inset). **c** EL spectrum. **d** Chromatic diagram [192]

Based on the resulted $CsPbBr_3$/PMMA, they prepared the WLED with a rendering index up to 89.2 (Fig. 3.30).

## 3.8 Application in Pc-LEDs

It has been found that LEDs based on the perovskite were promising solid-state light sources and widely used in the fields of photoelectricity because of their high electro-optical conversion efficiency, low cost, and environmentally friendly properties [193, 194]. Overall, $CsPbX_3$ QDs can be employed as promising optical materials in the LEDs. But the poor stability hindered their further application. Particularly, with the development of the stable $CsPbX_3$ QDs, there are many useful strategies for the $CsPbX_3$ NCs based LED, such as compositional engineering, surface engineering, matrix encapsulation, and device encapsulation. The reported improved stability and device performance are summarized in Table 3.4.

## 3.9 Conclusion and Outlook

In conclusion, we have introduced all-inorganic metal halide perovskite ($CsPbX_3$, X = Cl, Br, and I) nanocrystals as a special class of semiconductor materials. The $CsPbX_3$ QDs achieved superior performances for practical optical applications, such as photodetectors, solar cells, light-emitting diodes, and lasers. However, structural and optical stability will remain the foremost problem for $CsPbX_3$ QDs. Our review on an understanding of the origins of instability and summarize the intrinsic factors affecting their stability. Under the ambient (such as humid, thermal, and illumination) condition the LHP has exhibited an instability, which inherits their further application. After recent years of endeavor, many strategies have been developed to overcome this problem, including ligand modification, ligand exchange, compositional engineering, and coating treatment, etc. Thus, the stability of the $CsPbX_3$ QDs has been effectively enhanced, however, they are still far away from meeting demands. This poses a great challenge for any future applications of $CsPbX_3$ QDs. We hope that this review could encourage more researchers to devote additional efforts to address the stability of $CsPbX_3$ QDs. And we are convinced that the optical device with $CsPbX_3$ QDs will gain increasing scientific interest and turn into a great commercial opportunity and change in the near future. We hope that our review may bring together researchers from various areas to explore broad attention and experimental for this new promising material.

**Table 3.4** Summary of the PL properties and stability of perovskite composites, performance in pc-WLEDs

| Composites | PL QYs | Stability | Device structure | Luminous efficacy and device lifetime | References |
|---|---|---|---|---|---|
| $CsPbX_3$@SBS | 11% (blue) 23% (green) 12.2% (red) | 100% (10 min, water) | Blue chip (450 nm)/$CsPb(Br/I)_3$@SBS/$CsPbBr_3$@SBS | 9 lm/W | [195] |
| $CsPbBr_3$/EC | 37.20% | 95% (150 h, air) | Blue chip/$CsPbBr_3$/EC, $Sr_2Si_5N_8$: $Eu^{2+}$ (silicone) | 65.78 lm/W (20 mA) | [196] |
| $CsPbX_3$–PLA | 33–90% | 80% (50 h, working on LED) | Blue chip/$CsPbBr_3$–PLA (PMMA)/(Ba, Ca, $Sr)_3SiO_5$: Eu | 62.93 lm/W | [197] |
| $CsPbBr_3$/EVA | 40.50% | 100% (192 h, air) ~ 100% (720 h, water) | Blue chip (460 nm)/$CsPbBr_3$/EVA/(Sr, Ca) $AlSiN_3$: $Eu^{2+}$ | 37.7 lm/W (20 mA) | [190] |
| $CsPb(Br/I)_3$@anthracene | 41.90% | 50% (100 1C) | UV chip (385 nm)/$CsPb(Br_{0.6}I_{0.4})_3$@anthracene (PMMA) | | [198] |
| $CsPbX_3$–PS | 44% (green) | 82% (24 h, water) | Blue chip/$CsPbBr_{1.2}I_{1.8}$–PS/$CsPbBr_3$–PS | | [185] |
| $CsPbX_3$/PMMA | 45% | 75% (3 d, 30 1C air with 70% humidity) | Blue chip (460 nm)/$CsPbX_3$/PMMA, CdSe QDs (silicone resin) | | [192] |
| $CsPbX_3$/PS | 48% (green) | 70% (192 h, water) 50% (2 h, 80 1C) | Blue chip (410 nm)/CdSe QDs (ethoxyline resin)/$CsPbBr_3$/PS fiber | | [189] |
| $CsPbX_3$@POSS | 62% (green) 45% (red) | Stable in water over 10 weeks | Blue chip (455 nm)/$CsPbBr_3$@POSS, $CsPbBr_{1.2}I_{1.8}$ (silicone resin) | 14.1 lm/W | [123] |

(continued)

**Table 3.4** (continued)

| Composites | PL QYs | Stability | Device structure | Luminous efficacy and device lifetime | References |
|---|---|---|---|---|---|
| $CsPbBr_3$–phospho–silicate glass | 63% | 85% (5 d, UV) | Blue chip (460 nm)/$CsPbBr_3$ glass/$CaAlSiN_3$: $Eu^{2+}$ (epoxy resin) | 50.5 (20 mA) | [134] |
| $CsPbBr_3$–SDDA/$SiO_2$/PMMA | 63% | ~100% (cooling from 100 °C to RT) | Blue chip (458 nm)/$CsPbBr_3$–SDDA/$SiO_2$/PMMA, KSF (silicone resin) | | [199] |
| $CsPbBr_3$@$NH_4Br$ | 64.21% | 40% (3.5 h, water) 96% (cooling from 100 1C to RT) | Blue chip/$CsPbBr_3$@$NH_4Br$, KSF (silicone resin) | | [110] |
| $CsPbBrI_2$ | 65% | | Blue chip (460 nm)/YAG: Ce PiG/$CsPbBrI_2$ (PMMA) (remote-type) | 58 (20 mA) | [200] |
| $CsPbX_3$@PS | 68% (green) | 60% (16 d, UV) | Blue chip (450 nm)/$CsPbBr_3$@$SiO_2$, $CsPb(Br_{0.4}I_{0.6})_3$ | | [187] |
| $CsPbBr_3$–TDPA | 68% | 80% (300 min, water) | Blue chip (453 nm)/$CsPbBr_3$–TDPA, KSF (silicone resin) | 63 lm/W (20 mA) 57 lm/W (Working for 15 h) | [109] |
| $CsPbBr_3$–$SiO_2$ | 71.80% | 63.5% (80 1C) | Blue chip (450 nm)/$CsPbBr_3$–$SiO_2$, $CsPbBr_{1.2}I_{1.8}$ (silicone resin | 35.4 lm/W (20 mA) | [127] |
| $CsPbBr_3$/$SiO_2$ | 80% | 98% (10 h, UV) | Blue chip/CdSe NCs (PMMA)/$CsPbBr_3$/$SiO_2$ (PMMA) | 56 lm/W (5 mA) Unchanged after working for 1 h (1 mA) | [124] |
| $CsPbX_3$/$SiO_2$ | 11.2% (blue) 82% (green) | 93% (30 d, air) | Blue chip (441 nm)/$CsPbBr_3$/$SiO_2$, $CsPb(Br_{0.3}/I_{0.7})_3$/$SiO_2$ (PMMA | 35.32 lm/W Little variation after working for 40 h | [128] |

(continued)

**Table 3.4** (continued)

| Composites | PL QYs | Stability | Device structure | Luminous efficacy and device lifetime | References |
|---|---|---|---|---|---|
| $CsPbX_3$–$CaF_2$ | 50% (blue) 82% (green) | 56% (80 h, UV) 60% (2 d, air with 100% humidity) | Blue chip (460 nm)/KSF (silicone resin)/$CsPbBr_3$– $CaF_2$ (PMMA) | 62.7 lm/W (20 mA) | [141] |
| $CsPbX_3/SiO_2$ | 85% (green) 88% (red) | 95% (90 d, air) Stable (5 d, UV) | Blue chip (458 nm)/$CsPbBr_3/SiO_2$, red QDs (PMMA) (remote-type) | 61.2 lm/W (20 mA) Little variation after working for 10 h | [201] |
| $CsPbBr_3$/PSZ | 81.70% | >60% (70 h, UV) ~ 100% (60 d, air) | Blue chip (455 nm)/$CsPbBr_3$/PSZ, KSF (PMMA) | 138.6 (60 mA) | [202] |
| $CsPbX_3$–PMPOPNC | 67% (blue) 85% (green) 69% (red) | 95% (60 d, water) >80% (120 1C) 81% (156 h, UV) | Blue chip (460 nm)/red YAG (silicone resin)/$CsPbX_3$–PMPOPNC (silicone resin) | 26.3 (20 mA) | [122] |
| $CsPbBr_3/SiO_2$–Al2$O_3$ | 90% | 90% (34 h, air) 90% (300 h, UV) 70% (1 h, 100 1C) | Blue chip (455 nm)/$CsPbBr_3$, $CsPbBr_3/SiO_2$–$Al_2O_3$ (PDMS) | 80.77 | [203] |
| $CsPbBr_3/Cs_4PbX_6$ | 95% | 93% (80 min, blue light irradiation) 80% (14 d, air) | $CsPbBr_3/Cs_4PbX_6$, KSF | 73.8 lm/W) (5 mA) | [204] |
| $CsPbBr_3/Cs_4PbX_6$ | 97% | | Blue chip/KSF (silicone resin)/$CsPbBr_3/Cs_4PbX_6$ | 151 lm/W (20 mA) | [205] |

(continued)

**Table 3.4** (continued)

| Composites | PL QYs | Stability | Device structure | Luminous efficacy and device lifetime | References |
|---|---|---|---|---|---|
| $CsPbCl_3$: Ce, Eu | | 40% (35 h, UV) | UV chip (365 nm)/$CsPbCl_3$: Ce, Eu (PMMA) | 24 lm/W Unchanged (working over 288 h) | [206] |
| $CsPbX_3$–CB | | 80% (30 min, UV) | Blue chip/$CsPbBr_3$–CB, $CsPb(Br_{0.4}I_{0.6})_3$–CB | | [207] |
| $CsPbBr_3$–$SiO_2$ | | B100% (cooling from 100 °C to RT) 80% (96 h, UV) | Blue chip (450 nm)/$CsPbBr_3$–$SiO_2$, $CsPb(Br_{0.4}I_{0.6})_3$ (silicone resin) | 30 lm/W | [118] |
| $CsPbX_3$–zeolites | | 50% (80 °C) | Blue chip (460 nm)/$CsPb(Br_{0.4}I_{0.6})_3$, $CsPbBr_3$–zeolites (silicone resin) | 2 lm/W (20 mA) | [140] |

## References


1. G. Rose, De Novis Quibusdam Fossilibus Quae In Montibus Uraliis Inveniuntur (1839)
2. M. Abdi-Jalebi, M. Pazoki, B. Philippe et al., Dedoping of lead halide perovskites incorporating monovalent cations. ACS Nano **12**, 7301–7311 (2018). https://doi.org/10.1021/acsnano.8b03586
3. S. Sun, T. Salim, N. Mathews et al., The origin of high efficiency in low-temperature solution-processable bilayer organometal halide hybrid solar cells. Energy Environ. Sci. **7**, 399–407 (2014). https://doi.org/10.1039/C3EE43161D
4. V. D'Innocenzo, G. Grancini, M.J.P. Alcocer et al., Excitons versus free charges in organo-lead tri-halide perovskites. Nat. Commun. **5**, 3586 (2014). https://doi.org/10.1038/ncomms4586. https://www.nature.com/articles/ncomms4586#supplementary-information
5. G. Xing, N. Mathews, S. Sun et al., Long-range balanced electron- and hole-transport lengths in organic-inorganic $CH_3NH_3PbI_3$. Science **342**, 344–347 (2013). https://doi.org/10.1126/science.1243167
6. Z. Chen, Q. Dong, Y. Liu et al., Thin single crystal perovskite solar cells to harvest below-bandgap light absorption. Nat. Commun. **8**, 1890 (2017). https://doi.org/10.1038/s41467-017-02039-5
7. C. Motta, F. El-Mellouhi, S. Kais et al., Revealing the role of organic cations in hybrid halide perovskite $CH_3NH_3PbI_3$. Nat. Commun. **6**, 7026 (2015). https://doi.org/10.1038/ncomms8026, https://www.nature.com/articles/ncomms8026#supplementary-information
8. N.J. Jeon, J.H. Noh, Y.C. Kim et al., Solvent engineering for high-performance inorganic–organic hybrid perovskite solar cells. Nat. Mater. **13**, 897 (2014). https://doi.org/10.1038/nmat4014, https://www.nature.com/articles/nmat4014#supplementary-information
9. Q. Jiang, L. Zhang, H. Wang et al., Enhanced electron extraction using $SnO_2$ for high-efficiency planar-structure $HC(NH_2)_2PbI_3$-based perovskite solar cells. Nat. Energy **2**, 16177 (2016). https://doi.org/10.1038/nenergy.2016.177, https://www.nature.com/articles/nenergy2016177#supplementary-information
10. C.S. Ponseca, T.J. Savenije, M. Abdellah et al., Organometal halide perovskite solar cell materials rationalized: ultrafast charge generation, high and microsecond-long balanced mobilities, and slow recombination. J. Am. Chem. Soc. **136**, 5189–5192 (2014). https://doi.org/10.1021/ja412583t
11. A. Kojima, K. Teshima, Y. Shirai et al., Organometal halide perovskites as visible-light sensitizers for photovoltaic cells. J. Am. Chem. Soc. **131**, 6050–6051 (2009). https://doi.org/10.1021/ja809598r
12. Y.-H. Kim, H. Cho, J.H. Heo et al., Multicolored organic/inorganic hybrid perovskite light-emitting diodes. Adv. Mater. **27**, 1248–1254 (2015). https://doi.org/10.1002/adma.201403751
13. X. Zhang, H. Liu, W. Wang et al., Hybrid perovskite light-emitting diodes based on perovskite nanocrystals with organic-inorganic mixed cations. Adv. Mater. **29**, 1606405 (2017). https://doi.org/10.1002/adma.201606405
14. J. Zhou, J. Huang, Photodetectors based on organic-inorganic hybrid lead halide perovskites. Adv. Sci. **5**, 1700256 (2018). https://doi.org/10.1002/advs.201700256
15. L. Dou, Y. Yang, J. You et al., Solution-processed hybrid perovskite photodetectors with high detectivity. Nat. Commun. **5**, 5404 (2014). https://doi.org/10.1038/ncomms6404, https://www.nature.com/articles/ncomms6404#supplementary-information
16. P. Liu, X. He, J. Ren et al., Organic-inorganic hybrid perovskite nanowire laser arrays. ACS Nano **11**, 5766–5773 (2017). https://doi.org/10.1021/acsnano.7b01351
17. H. Yu, K. Ren, Q. Wu et al., Organic–inorganic perovskite plasmonic nanowire lasers with a low threshold and a good thermal stability. Nanoscale **8**, 19536–19540 (2016). https://doi.org/10.1039/C6NR06891J
18. A. Amat, E. Mosconi, E. Ronca et al., Cation-induced band-gap tuning in organohalide perovskites: interplay of spin-orbit coupling and octahedra tilting. Nano Lett. **14**, 3608–3616 (2014). https://doi.org/10.1021/nl5012992

19. G.Y. Kim, A. Senocrate, T.-Y. Yang et al., Large tunable photoeffect on ion conduction in halide perovskites and implications for photodecomposition. Nat. Mater. **17**, 445–449 (2018). https://doi.org/10.1038/s41563-018-0038-0
20. L. Protesescu, S. Yakunin, M.I. Bodnarchuk et al., Nanocrystals of cesium lead halide perovskites ($CsPbX_3$, X = Cl, Br, and I): novel optoelectronic materials showing bright emission with wide color Gamut. Nano Lett. **15**, 3692–3696 (2015). https://doi.org/10.1021/nl5048779
21. C.K. Møller, Crystal structure and photoconductivity of Cæsium plumbohalides. Nature **182**, 1436 (1958)
22. C.K. Møller, A phase transition in Cæsium plumbochloride. Nature **180**, 981 (1957)
23. F. Bertolotti, L. Protesescu, M.V. Kovalenko et al., Coherent nanotwins and dynamic disorder in cesium lead halide perovskite nanocrystals. ACS Nano **11**, 3819–3831 (2017). https://doi.org/10.1021/acsnano.7b00017
24. Q. Sun, W.-J. Yin, thermodynamic stability trend of cubic perovskites. J. Am. Chem. Soc. **139**, 14905–14908 (2017). https://doi.org/10.1021/jacs.7b09379
25. G. Divitini, S. Cacovich, F. Matteocci et al., In situ observation of heat-induced degradation of perovskite solar cells. Nat. Energy **1**, 15012 (2016). https://doi.org/10.1038/nenergy.2015.12
26. M.V. Kovalenko, L. Protesescu, M.I. Bodnarchuk, Properties and potential optoelectronic applications of lead halide perovskite nanocrystals. Science **358**, 745–750 (2017). https://doi.org/10.1126/science.aam7093
27. X. Bai, G. Caputo, Z. Hao et al., Efficient and tuneable photoluminescent boehmite hybrid nanoplates lacking metal activator centres for single-phase white LEDs. Nat. Commun. **5**, 5702 (2014). https://doi.org/10.1038/ncomms6702
28. Y. Kim, E. Yassitepe, O. Voznyy et al., Efficient luminescence from perovskite quantum dot solids. ACS Appl. Mater. Inter. **7**, 25007–25013 (2015). https://doi.org/10.1021/acsami.5b09084
29. E. Yassitepe, Z. Yang, O. Voznyy et al., Amine-free synthesis of cesium lead halide perovskite quantum dots for efficient light-emitting diodes. Adv. Func. Mater. **26**, 8757–8763 (2016). https://doi.org/10.1002/adfm.201604580
30. J. De Roo, M. Ibáñez, P. Geiregat et al., Highly dynamic ligand binding and light absorption coefficient of cesium lead bromide perovskite nanocrystals. ACS Nano **10**, 2071–2081 (2016). https://doi.org/10.1021/acsnano.5b06295
31. G. Almeida, L. Goldoni, Q. Akkerman et al., Role of acid-base equilibria in the size, shape, and phase control of cesium lead bromide nanocrystals. ACS Nano **12**, 1704–1711 (2018). https://doi.org/10.1021/acsnano.7b08357
32. V.M. Goldschmidt, Die Gesetze der Krystallochemie. Naturwissenschaften **14**, 477–485 (1926). https://doi.org/10.1007/bf01507527
33. B. Saparov, D.B. Mitzi, Organic-inorganic perovskites: structural versatility for functional materials design. Chem. Rev. **116**, 4558–4596 (2016). https://doi.org/10.1021/acs.chemrev.5b00715
34. J.-P. Correa-Baena, M. Saliba, T. Buonassisi et al., Promises and challenges of perovskite solar cells. Science **358**, 739–744 (2017). https://doi.org/10.1126/science.aam6323
35. W. Travis, E.N.K. Glover, H. Bronstein et al., On the application of the tolerance factor to inorganic and hybrid halide perovskites: a revised system. Chem. Sci. **7**, 4548–4556 (2016). https://doi.org/10.1039/C5SC04845A
36. C. Li, X. Lu, W. Ding et al., Formability of ABX3 (X = F, Cl, Br, I) halide perovskites. Acta Crystallogr. B **64**, 702–707 (2008). https://doi.org/10.1107/s0108768108032734
37. L. Protesescu, S. Yakunin, S. Kumar et al., Dismantling the "Red Wall" of Colloidal Perovskites: Highly Luminescent Formamidinium and Formamidinium-Cesium Lead Iodide Nanocrystals. ACS Nano **11**, 3119–3134 (2017). https://doi.org/10.1021/acsnano.7b00116
38. M.I. Saidaminov, O.F. Mohammed, O.M. Bakr, Low-dimensional-networked metal halide perovskites: the next big thing. ACS Energy Lett. **2**, 889–896 (2017). https://doi.org/10.1021/acsenergylett.6b00705

39. J. Shamsi, A.S. Urban, M. Imran et al., Metal halide perovskite nanocrystals: synthesis, post-synthesis modifications, and their optical properties. Chem. Rev. **119**, 3296–3348 (2019). https://doi.org/10.1021/acs.chemrev.8b00644
40. X. Li, Y. Wu, S. Zhang et al., $CsPbX_3$ quantum dots for lighting and displays: room-temperature synthesis, photoluminescence superiorities, underlying origins and white light-emitting diodes. Adv. Func. Mater. **26**, 2435–2445 (2016). https://doi.org/10.1002/adfm.201600109
41. M. Chen, Y. Zou, L. Wu et al., Solvothermal synthesis of high-quality all-inorganic cesium lead halide perovskite nanocrystals: from nanocube to ultrathin nanowire. Adv. Func. Mater. **27**, 1701121 (2017). https://doi.org/10.1002/adfm.201701121
42. Z. Long, H. Ren, J. Sun et al., High-throughput and tunable synthesis of colloidal $CsPbX_3$ perovskite nanocrystals in a heterogeneous system by microwave irradiation. Chem. Commun. **53**, 9914–9917 (2017). https://doi.org/10.1039/C7CC04862A
43. L. Rao, Y. Tang, C. Song et al., Polar-solvent-free synthesis of highly photoluminescent and stable CsPbBr 3 nanocrystals with controlled shape and size by ultrasonication. Chem. Mater. **31**, 365–375 (2019). https://doi.org/10.1021/acs.chemmater.8b03298
44. Q.A. Akkerman, D. Meggiolaro, Z. Dang et al., Fluorescent alloy CsPbxMn1−xI3 perovskite nanocrystals with high structural and optical stability. ACS Energy Lett. **2**, 2183–2186 (2017). https://doi.org/10.1021/acsenergylett.7b00707
45. L. Peng, A. Dutta, R. Xie et al., Dot–Wire–Platelet–Cube: step growth and structural transformations in CsPbBr 3 perovskite nanocrystals. ACS Energy Lett. **3**, 2014–2020 (2018). https://doi.org/10.1021/acsenergylett.8b01037
46. Y. Tong, B.J. Bohn, E. Bladt et al., From precursor powders to $CsPbX_3$ perovskite nanowires: one-pot synthesis, growth mechanism, and oriented self-assembly. Angew. Chem. Int. Ed. **56**, 13887–13892 (2017). https://doi.org/10.1002/anie.201707224
47. H. Liu, Z. Wu, H. Gao et al., One-Step preparation of cesium lead halide $CsPbX_3$ ($X =$ Cl, Br, and I) perovskite nanocrystals by microwave irradiation. ACS Appl. Mater. Inter. **9**, 42919–42927 (2017). https://doi.org/10.1021/acsami.7b14677
48. Z.-J. Yong, S.-Q. Guo, J.-P. Ma et al., Doping-enhanced short-range order of perovskite nanocrystals for near-unity violet luminescence quantum yield. J. Am. Chem. Soc. **140**, 9942–9951 (2018). https://doi.org/10.1021/jacs.8b04763
49. M. Liu, G. Zhong, Y. Yin et al., Aluminum-doped cesium lead bromide perovskite nanocrystals with stable blue photoluminescence used for display backlight. Adv. Sci. **4**, 1700335 (2017). https://doi.org/10.1002/advs.201700335
50. B.A. Koscher, J.K. Swabeck, N.D. Bronstein et al., Essentially trap-free CsPbBr 3 colloidal nanocrystals by postsynthetic thiocyanate surface treatment. J. Am. Chem. Soc. **139**, 6566–6569 (2017). https://doi.org/10.1021/jacs.7b02817
51. W. van der Stam, J.J. Geuchies, T. Altantzis et al., Highly emissive divalent-ion-doped colloidal CsPb1−xMxBr 3 perovskite nanocrystals through cation exchange. J. Am. Chem. Soc. **139**, 4087–4097 (2017). https://doi.org/10.1021/jacs.6b13079
52. H.-C. Wang, W. Wang, A.-C. Tang et al., High-performance CsPb1 − xSnxBr 3 perovskite quantum dots for light-emitting diodes. Angew. Chem. Int. Ed. **56**, 13650–13654 (2017). https://doi.org/10.1002/anie.201706860
53. J.-S. Yao, J. Ge, B.-N. Han et al., Ce3 + -Doping to modulate photoluminescence kinetics for efficient CsPbBr 3 nanocrystals based light-emitting diodes. J. Am. Chem. Soc. **140**, 3626–3634 (2018). https://doi.org/10.1021/jacs.7b11955
54. S. Zou, Y. Liu, J. Li et al., Stabilizing cesium lead halide perovskite lattice through Mn(II) substitution for air-stable light-emitting diodes. J. Am. Chem. Soc. **139**, 11443–11450 (2017). https://doi.org/10.1021/jacs.7b04000
55. W. Liu, Q. Lin, H. Li et al., $Mn^{2+}$ -Doped lead halide perovskite nanocrystals with dual-color emission controlled by halide content. J. Am. Chem. Soc. **138**, 14954–14961 (2016). https://doi.org/10.1021/jacs.6b08085
56. H. Liu, Z. Wu, J. Shao et al., CsPbxMn1−xCl3 perovskite quantum dots with high Mn substitution ratio. ACS Nano **11**, 2239–2247 (2017). https://doi.org/10.1021/acsnano.6b08747

57. L. Wu, H. Hu, Y. Xu et al., From Nonluminescent Cs4PbX6 (X = Cl, Br, I) nanocrystals to highly luminescent $CsPbX_3$ nanocrystals: water-triggered transformation through a CsX-Stripping mechanism. Nano Lett. **17**, 5799–5804 (2017). https://doi.org/10.1021/acs.nanolett.7b02896
58. Q.A. Akkerman, S. Park, E. Radicchi et al., Nearly monodisperse insulator Cs4PbX6 (X = Cl, Br, I) nanocrystals, their mixed halide compositions, and their transformation into $CsPbX_3$ nanocrystals. Nano Lett. **17**, 1924–1930 (2017). https://doi.org/10.1021/acs.nanolett.6b05262
59. Z. Liu, Y. Bekenstein, X. Ye et al., Ligand mediated transformation of cesium lead bromide perovskite nanocrystals to lead depleted Cs4PbBr 6 nanocrystals. J. Am. Chem. Soc. **139**, 5309–5312 (2017). https://doi.org/10.1021/jacs.7b01409
60. G. Li, H. Wang, Z. Zhu et al., Shape and phase evolution from CsPbBr 3 perovskite nanocubes to tetragonal CsPb2Br5 nanosheets with an indirect bandgap. Chem. Commun. **52**, 11296–11299 (2016). https://doi.org/10.1039/C6CC05877A
61. S.K. Balakrishnan, P.V. Kamat, Ligand assisted transformation of cubic CsPbBr 3 nanocrystals into two-dimensional CsPb2Br5 nanosheets. Chem. Mater. **30**, 74–78 (2018). https://doi.org/10.1021/acs.chemmater.7b04142
62. S. Sun, D. Yuan, Y. Xu et al., Ligand-mediated synthesis of shape-controlled cesium lead halide perovskite nanocrystals via reprecipitation process at room temperature. ACS Nano **10**, 3648–3657 (2016). https://doi.org/10.1021/acsnano.5b08193
63. K.-H. Wang, L. Wu, L. Li et al., Large-scale synthesis of highly luminescent perovskite-related CsPb2Br 5 nanoplatelets and their fast anion exchange. Angew. Chem. Int. Ed. **55**, 8328–8332 (2016). https://doi.org/10.1002/anie.201602787
64. J. Cho, S. Banerjee, Ligand-directed stabilization of ternary phases: synthetic control of structural dimensionality in solution-grown cesium lead bromide nanocrystals. Chem. Mater. **30**, 6144–6155 (2018). https://doi.org/10.1021/acs.chemmater.8b02730
65. R.E. Beal, D.J. Slotcavage, T. Leijtens et al., Cesium lead halide perovskites with improved stability for tandem solar cells. J. Phys. Chem. Lett. **7**, 746–751 (2016). https://doi.org/10.1021/acs.jpclett.6b00002
66. A. Swarnkar, W.J. Mir, A. Nag, Can B-site doping or alloying improve thermal- and phase-stability of all-inorganic $CsPbX_3$ (X = Cl, Br, I) Perovskites? ACS Energy Lett. **3**, 286–289 (2018). https://doi.org/10.1021/acsenergylett.7b01197
67. X. Yuan, X. Hou, J. Li et al., Thermal degradation of luminescence in inorganic perovskite CsPbBr 3 nanocrystals. Phys. Chem.Chem. Phys. **19**, 8934–8940 (2017). https://doi.org/10.1039/C6CP08824D
68. X. Yuan, S. Ji, M.C. De Siena et al., Photoluminescence temperature dependence, dynamics, and quantum efficiencies in $Mn^{2+}$ -Doped CsPbCl3 perovskite nanocrystals with varied dopant concentration. Chem. Mater. **29**, 8003–8011 (2017). https://doi.org/10.1021/acs.chemmater.7b03311
69. M.-G. Ju, M. Chen, Y. Zhou et al., Toward Eco-friendly and stable perovskite materials for photovoltaics. Joule **2**, 1231–1241 (2018). https://doi.org/10.1016/j.joule.2018.04.026
70. J. Owen, The coordination chemistry of nanocrystal surfaces. Science **347**, 615–616 (2015). https://doi.org/10.1126/science.1259924
71. N. Aristidou, C. Eames, I. Sanchez-Molina et al., Fast oxygen diffusion and iodide defects mediate oxygen-induced degradation of perovskite solar cells. Nat. Commun. **8**, 15218 (2017). https://doi.org/10.1038/ncomms15218
72. B.T. Diroll, G. Nedelcu, M.V. Kovalenko et al., High-temperature photoluminescence of $CsPbX_3$ (X = Cl, Br, I) nanocrystals. Adv. Func. Mater. **27**, 1606750 (2017). https://doi.org/10.1002/adfm.201606750
73. J. Chen, D. Liu, M.J. Al-Marri et al., Photo-stability of CsPbBr 3 perovskite quantum dots for optoelectronic application. Sci. China Mater. **59**, 719–727 (2016). https://doi.org/10.1007/s40843-016-5123-1
74. S. Huang, Z. Li, B. Wang et al., Morphology evolution and degradation of CsPbBr 3 nanocrystals under blue light-emitting diode illumination. ACS Appl. Mater. Inter. **9**, 7249–7258 (2017). https://doi.org/10.1021/acsami.6b14423

75. B. Akbali, G. Topcu, T. Guner et al., CsPbBr 3 perovskites: Theoretical and experimental investigation on water-assisted transition from nanowire formation to degradation. Phys. Rev. Mater. **2**, 034601 (2018). https://doi.org/10.1103/PhysRevMaterials.2.034601
76. X. Zhang, X. Bai, H. Wu et al., Water-assisted size and shape control of CsPbBr 3 perovskite nanocrystals. Angew. Chem. Int. Ed. **57**, 3337–3342 (2018). https://doi.org/10.1002/anie.201710869
77. M.A. Boles, D. Ling, T. Hyeon et al., The surface science of nanocrystals. Nat. Mater. **15**, 141–153 (2016). https://doi.org/10.1038/nmat4526
78. M. Imran, P. Ijaz, D. Baranov et al., Shape-pure, nearly monodispersed CsPbBr 3 nanocubes prepared using secondary aliphatic amines. Nano Lett. **18**, 7822–7831 (2018). https://doi.org/10.1021/acs.nanolett.8b03598
79. X. Yin, J. Wu, P. Li et al., Self-heating approach to the fast production of uniform metal nanostructures. Chem. Nano Mater. **2**, 37–41 (2016). https://doi.org/10.1002/cnma.201500123
80. F. Krieg, S.T. Ochsenbein, S. Yakunin et al., Colloidal $CsPbX_3$ (X = Cl, Br, I) nanocrystals 2.0: zwitterionic capping ligands for improved durability and stability. ACS Energy Let. **3**, 641–646 (2018). https://doi.org/10.1021/acsenergylett.8b00035
81. G. Almeida, I. Infante, L. Manna, Resurfacing halide perovskite nanocrystals. Science **364**, 833–834 (2019). https://doi.org/10.1126/science.aax5825
82. M.L.H. Green, G. Parkin, Application of the covalent bond classification method for the teaching of inorganic chemistry. J. Chem. Educ. **91**, 807–816 (2014). https://doi.org/10.1021/ed400504f
83. Q.A. Akkerman, G. Rainò, M.V. Kovalenko et al., Genesis, challenges and opportunities for colloidal lead halide perovskite nanocrystals. Nat. Mater. **17**, 394–405 (2018). https://doi.org/10.1038/s41563-018-0018-4
84. A. Pan, B. He, X. Fan et al., Insight into the ligand-mediated synthesis of colloidal CsPbBr 3 perovskite nanocrystals: the role of organic acid, base, and cesium precursors. ACS Nano **10**, 7943–7954 (2016). https://doi.org/10.1021/acsnano.6b03863
85. G. Li, F.W.R. Rivarola, N.J.L.K. Davis et al., highly efficient perovskite nanocrystal light-emitting diodes enabled by a universal crosslinking method. Adv. Mater. **28**, 3528–3534 (2016). https://doi.org/10.1002/adma.201600064
86. F. Palazon, Q.A. Akkerman, M. Prato et al., X-ray lithography on perovskite nanocrystals films: from patterning with anion-exchange reactions to enhanced stability in air and water. ACS Nano **10**, 1224–1230 (2016). https://doi.org/10.1021/acsnano.5b06536
87. J. Pan, Y. Shang, J. Yin et al., Bidentate ligand-passivated $CsPbI_3$ perovskite nanocrystals for stable near-unity photoluminescence quantum yield and efficient red light-emitting diodes. J. Am. Chem. Soc. **140**, 562–565 (2018). https://doi.org/10.1021/jacs.7b10647
88. H. Zhang, J. Jang, W. Liu et al., Colloidal nanocrystals with inorganic halide, pseudohalide, and halometallate ligands. ACS Nano **8**, 7359–7369 (2014). https://doi.org/10.1021/nn502470v
89. X. Li, Y.-B. Zhao, F. Fan et al., Bright colloidal quantum dot light-emitting diodes enabled by efficient chlorination. Nat. Photonics **12**, 159–164 (2018). https://doi.org/10.1038/s41566-018-0105-8
90. D. Yan, T. Shi, Z. Zang et al., Ultrastable CsPbBr 3 perovskite quantum dot and their enhanced amplified spontaneous emission by surface ligand modification. Small **15**, 1901173 (2019). https://doi.org/10.1002/smll.201901173
91. Yang D, Li X, Zhou Wet al., $CsPbBr_3$ quantum dots 2.0: benzenesulfonic acid equivalent ligand awakens complete purification. Adv. Mater. **31**, 1900767 (2019). https://doi.org/10.1002/adma.201900767
92. Y. Tan, Y. Zou, L. Wu et al., Highly luminescent and stable perovskite nanocrystals with octylphosphonic acid as a ligand for efficient light-emitting diodes. ACS Appl. Mater. Inter. **10**, 3784–3792 (2018). https://doi.org/10.1021/acsami.7b17166
93. G. Almeida, O.J. Ashton, L. Goldoni et al., The phosphine oxide route toward lead halide perovskite nanocrystals. J. Am. Chem. Soc. **140**, 14878–14886 (2018). https://doi.org/10.1021/jacs.8b08978

94. H. Wu, Y. Zhang, M. Lu et al., Surface ligand modification of cesium lead bromide nanocrystals for improved light-emitting performance. Nanoscale **10**, 4173–4178 (2018). https://doi.org/10.1039/C7NR09126E
95. W. Chen, X. Tang, P. Wangyang et al., Surface-passivated cesium lead halide perovskite quantum dots: toward efficient light-emitting diodes with an inverted sandwich structure. Adv. Opt. Mater. **6**, 1800007 (2018). https://doi.org/10.1002/adom.201800007
96. J. Pan, L.N. Quan, Y. Zhao et al., Highly efficient perovskite-quantum-dot light-emitting diodes by surface engineering. Adv. Mater. **28**, 8718–8725 (2016). https://doi.org/10.1002/adma.201600784
97. C. Zheng, C. Bi, F. Huang et al., Stable and strong emission CsPbBr 3 quantum dots by surface engineering for high-performance optoelectronic films. ACS Appl. Mater. Inter. **11**, 25410–25416 (2019). https://doi.org/10.1021/acsami.9b07818
98. J. Song, J. Li, L. Xu et al., Room-temperature triple-ligand surface engineering synergistically boosts ink stability, recombination dynamics, and charge injection toward EQE-11.6% perovskite QLEDs. Adv. Mater. **30**, 1800764 (2018). https://doi.org/10.1002/adma.201800764
99. F. Ye, H. Zhang, W. Li et al., Ligand-exchange of low-temperature synthesized CsPbBr 3 perovskite toward high-efficiency light-emitting diodes. Small Methods **3**, 1800489 (2019). https://doi.org/10.1002/smtd.201800489
100. Y.S. Shin, Y.J. Yoon, K.T. Lee et al., Vivid and fully saturated blue light-emitting diodes based on ligand-modified halide perovskite nanocrystals. ACS Appl. Mater. Inter. **11**, 23401–23409 (2019). https://doi.org/10.1021/acsami.9b04329
101. H. Wang, X. Zhang, Q. Wu et al., Trifluoroacetate induced small-grained CsPbBr 3 perovskite films result in efficient and stable light-emitting devices. Nat. Commun. **10**, 665 (2019). https://doi.org/10.1038/s41467-019-08425-5
102. Y. Lou, Y. Niu, D. Yang et al., Rod-shaped thiocyanate-induced abnormal band gap broadening in SCN − doped CsPbBr 3 perovskite nanocrystals. Nano Res. **11**, 2715–2723 (2018). https://doi.org/10.1007/s12274-017-1901-z
103. Q.A. Akkerman, M. Gandini, F. Di Stasio et al., Strongly emissive perovskite nanocrystal inks for high-voltage solar cells. Nat. Energy **2**, 16194 (2016). https://doi.org/10.1038/nenergy.2016.194
104. B.-B. Zhang, S. Yuan, J.-P. Ma et al., General mild reaction creates highly luminescent organic-ligand-lacking halide perovskite nanocrystals for efficient light-emitting diodes. J. Am. Chem. Soc. **141**, 15423–15432 (2019). https://doi.org/10.1021/jacs.9b08140
105. J.H. Park, A.-Y. Lee, J.C. Yu et al., Surface Ligand engineering for efficient perovskite nanocrystal-based light-emitting diodes. ACS Appl. Mater. Inter. **11**, 8428–8435 (2019). https://doi.org/10.1021/acsami.8b20808
106. G. Li, J. Huang, H. Zhu et al., Surface ligand engineering for near-unity quantum yield inorganic halide perovskite QDs and high-performance QLEDs. Chem. Mater. **30**, 6099–6107 (2018). https://doi.org/10.1021/acs.chemmater.8b02544
107. F. Liu, Y. Zhang, C. Ding et al., Highly luminescent phase-stable $CsPbI_3$ perovskite quantum dots achieving near 100% absolute photoluminescence quantum yield. ACS Nano **11**, 10373–10383 (2017). https://doi.org/10.1021/acsnano.7b05442
108. H. Wang, N. Sui, X. Bai et al., Emission recovery and stability enhancement of inorganic perovskite quantum dots. J. Phys. Chem. Lett. **9**, 4166–4173 (2018). https://doi.org/10.1021/acs.jpclett.8b01752
109. T. Xuan, X. Yang, S. Lou et al., Highly stable CsPbBr 3 quantum dots coated with alkyl phosphate for white light-emitting diodes. Nanoscale **9**, 15286–15290 (2017). https://doi.org/10.1039/C7NR04179A
110. S. Lou, T. Xuan, C. Yu et al., Nanocomposites of CsPbBr 3 perovskite nanocrystals in an ammonium bromide framework with enhanced stability. J. Mater. Chem. C **5**, 7431–7435 (2017). https://doi.org/10.1039/C7TC01174A
111. C. Wang, A.S.R. Chesman, J.J. Jasieniak, Stabilizing the cubic perovskite phase of $CsPbI_3$ nanocrystals by using an alkyl phosphinic acid. Chem. Commun. **53**, 232–235 (2017). https://doi.org/10.1039/C6CC08282C

112. Y. Wang, M. Zhi, Y.-Q. Chang et al., Stable, ultralow threshold amplified spontaneous emission from CsPbBr 3 nanoparticles exhibiting trion gain. Nano Lett. **18**, 4976–4984 (2018). https://doi.org/10.1021/acs.nanolett.8b01817
113. D. Gerion, F. Pinaud, S.C. Williams et al., Synthesis and properties of biocompatible water-soluble silica-coated CdSe/ZnS semiconductor quantum dots. J. Phys. Chem. B **105**, 8861–8871 (2001). https://doi.org/10.1021/jp0105488
114. J.-N. Liu, W.-B. Bu, J.-L. Shi, Silica coated upconversion nanoparticles: a versatile platform for the development of efficient theranostics. Acc. Chem. Res. **48**, 1797–1805 (2015). https://doi.org/10.1021/acs.accounts.5b00078
115. D.K. Yi, S.T. Selvan, S.S. Lee et al., Silica-coated nanocomposites of magnetic nanoparticles and quantum dots. J. Am. Chem. Soc. **127**, 4990–4991 (2005). https://doi.org/10.1021/ja0428863
116. W. Stöber, A. Fink, E.Bohn, Controlled growth of monodisperse silica spheres in the micron size range. J. Colloid Interface Sci. **26**, 62–69 (1968). https://doi.org/10.1016/0021-9797(68)90272-5
117. J. Ziegler, S. Xu, E. Kucur et al., Silica-coated InP/ZnS nanocrystals as converter material in white LEDs. Adv. Mater. **20**, 4068–4073 (2008). https://doi.org/10.1002/adma.200800724
118. H.-C. Wang, S.-Y. Lin, A.-C. Tang et al., Mesoporous silica particles integrated with all-inorganic CsPbBr 3 perovskite quantum-dot nanocomposites (MP-PQDs) with high stability and wide color gamut used for backlight display. Angew. Chem. Int. Ed. **55**, 7924–7929 (2016). https://doi.org/10.1002/anie.201603698
119. D.N. Dirin, L. Protesescu, D. Trummer et al., Harnessing defect-tolerance at the nanoscale: highly luminescent lead halide perovskite nanocrystals in mesoporous silica matrixes. Nano Lett. **16**, 5866–5874 (2016). https://doi.org/10.1021/acs.nanolett.6b02688
120. X. Li, Y. Wang, H. Sun et al., Amino-mediated anchoring perovskite quantum dots for stable and low-threshold random lasing. Adv. Mater. **29**, 1701185 (2017). https://doi.org/10.1002/adma.201701185
121. V. Malgras, S. Tominaka, J.W. Ryan et al., Observation of quantum confinement in monodisperse methylammonium lead halide perovskite nanocrystals embedded in mesoporous silica. J. Am. Chem. Soc. **138**, 13874–13881 (2016). https://doi.org/10.1021/jacs.6b05608
122. A. Pan, J. Wang, M.J. Jurow et al., General strategy for the preparation of stable luminous nanocomposite inks using chemically addressable $CsPbX_3$ peroskite nanocrystals. Chem. Mater. **30**, 2771–2780 (2018). https://doi.org/10.1021/acs.chemmater.8b00587
123. H. Huang, B. Chen, Z. Wang et al., Water resistant $CsPbX_3$ nanocrystals coated with polyhedral oligomeric silsesquioxane and their use as solid state luminophores in all-perovskite white light-emitting devices. Chem. Sci. **7**, 5699–5703 (2016). https://doi.org/10.1039/C6SC01758D
124. H. Hu, L. Wu, Y. Tan et al., Interfacial synthesis of highly stable $CsPbX_3$/Oxide janus nanoparticles. J. Am. Chem. Soc. **140**, 406–412 (2018). https://doi.org/10.1021/jacs.7b11003
125. Q. Zhong, M. Cao, H. Hu et al., One-pot synthesis of highly stable CsPbBr 3@$SiO_2$ Core-shell nanoparticles. ACS Nano **12**, 8579–8587 (2018). https://doi.org/10.1021/acsnano.8b04209
126. Z. Hu, Z. Liu, Y. Bian et al., Enhanced two-photon-pumped emission from in situ synthesized nonblinking CsPbBr 3/$SiO_2$ nanocrystals with excellent stability. Adv. Optical Mater. **6**, 1700997 (2018). https://doi.org/10.1002/adom.201700997
127. D.H. Park, J.S. Han, W. Kim et al., Facile synthesis of thermally stable $CsPbBr_3$ perovskite quantum dot-inorganic $SiO_2$ composites and their application to white light-emitting diodes with wide color gamut. Dyes Pigm **149**, 246–252 (2018). https://doi.org/10.1016/j.dyepig.2017.10.003
128. N. Ding, D. Zhou, X. Sun et al., Highly stable and water-soluble monodisperse $CsPbX_3/SiO_2$ nanocomposites for white-LED and cells imaging. Nanotechnology **29**, 345703 (2018). https://doi.org/10.1088/1361-6528/aac84d
129. W. Chen, J. Hao, W. Hu et al., Enhanced stability and tunable photoluminescence in perovskite $CsPbX_3$/ZnS quantum dot heterostructure. Small **13**, 1604085 (2017). https://doi.org/10.1002/smll.201604085

130. S.K. Balakrishnan, P.V. Kamat, Au–CsPbBr 3 hybrid architecture: anchoring gold nanoparticles on cubic perovskite nanocrystals. ACS Energy Lett. **2**, 88–93 (2017). https://doi.org/10.1021/acsenergylett.6b00592
131. B.J. Roman, J. Otto, C. Galik et al., au exchange or au deposition: dual reaction pathways in Au–CsPbBr 3 Heterostructure nanoparticles. Nano Lett. **17**, 5561–5566 (2017). https://doi.org/10.1021/acs.nanolett.7b02355
132. Z. Li, Q. Kang, L. Chen et al., Enhancing aqueous stability and radiative-charge-transfer efficiency of $CsPbBr_3$ perovskite nanocrystals via conductive silica gel coating. Electrochim. Acta **330**, 135332 (2020). https://doi.org/10.1016/j.electacta.2019.135332
133. Q. Jing, M. Zhang, X. Huang et al., Surface passivation of mixed-halide perovskite CsPb(BrxI1 − x)3 nanocrystals by selective etching for improved stability. Nanoscale **9**, 7391–7396 (2017). https://doi.org/10.1039/C7NR01287J
134. X. Di, Z. Hu, J. Jiang et al., Use of long-term stable CsPbBr 3 perovskite quantum dots in phospho-silicate glass for highly efficient white LEDs. Chem. Commun. **53**, 11068–11071 (2017). https://doi.org/10.1039/C7CC06486A
135. S. Liu, Y. Luo, M. He et al. Novel $CsPbI_3$ QDs glass with chemical stability and optical properties. J. Eur Ceram. Soc. **38**, 1998–2004 (2018). https://doi.org/10.1016/j.jeurceramsoc.2017.10.012
136. B. Ai, C. Liu, J. Wang et al., Precipitation and optical properties of CsPbBr 3 quantum dots in phosphate glasses. J. Am. Ceram. Soc. **99**, 2875–2877 (2016). https://doi.org/10.1111/jace.14400
137. B. Ai, C. Liu, Z. Deng et al., Low temperature photoluminescence properties of CsPbBr 3 quantum dots embedded in glasses. Phys. Chem. Chem. Phys. **19**, 17349–17355 (2017). https://doi.org/10.1039/C7CP02482G
138. S. Guarnera, A. Abate, W. Zhang et al., Improving the long-term stability of perovskite solar cells with a porous $Al_2O_3$ buffer layer. J. Phys. Chem. Lett. **6**, 432–437 (2015). https://doi.org/10.1021/jz502703p
139. T. Leijtens, B. Lauber, G.E. Eperon et al., The importance of perovskite pore filling in organometal mixed halide sensitized $TiO_2$-based solar cells. J. Phys. Chem. Lett. **5**, 1096–1102 (2014). https://doi.org/10.1021/jz500209g
140. J.-Y. Sun, F.T. Rabouw, X.-F. Yang et al., Facile two-step synthesis of all-inorganic perovskite $CsPbX_3$ (X = Cl, Br, and I) zeolite-Y composite phosphors for potential backlight display application. Adv. Func. Mater. **27**, 1704371 (2017). https://doi.org/10.1002/adfm.201704371
141. Y. Wei, H. Xiao, Z. Xie et al., Highly luminescent lead halide perovskite quantum dots in hierarchical $CaF_2$ matrices with enhanced stability as phosphors for white light-emitting diodes. Adv. Opt. Mater. **6**, 1701343 (2018). https://doi.org/10.1002/adom.201701343
142. S. Bhattacharyya, D. Rambabu, T.K. Maji, Mechanochemical synthesis of a processable halide perovskite quantum dot–MOF composite by post-synthetic metalation. J. Mater. Chem. A **7**, 21106–21111 (2019). https://doi.org/10.1039/C9TA05977F
143. J. Ren, T. Li, X. Zhou et al., Encapsulating all-inorganic perovskite quantum dots into mesoporous metal organic frameworks with significantly enhanced stability for optoelectronic applications. Chem. Eng. J. **358**, 30–39 (2019). https://doi.org/10.1016/j.cej.2018.09.149
144. C. Guhrenz, A. Benad, C. Ziegler et al., Solid-state anion exchange reactions for color tuning of $CsPbX_3$ perovskite nanocrystals. Chem. Mater. **28**, 9033–9040 (2016). https://doi.org/10.1021/acs.chemmater.6b03980
145. A. Loiudice, S. Saris, E. Oveisi et al., CsPbBr 3 QD/AlOx inorganic nanocomposites with exceptional stability in water, light, and heat. Angew. Chem. Int. Ed. **56**, 10696–10701 (2017). https://doi.org/10.1002/anie.201703703
146. Z.-J. Li, E. Hofman, J. Li et al., Photoelectrochemically active and environmentally stable CsPbBr 3/TiO2 Core/shell nanocrystals. Adv. Func. Mater. **28**, 1704288 (2018). https://doi.org/10.1002/adfm.201704288
147. Y. Han, J. Sun, S. Ye et al., A stimuli responsive material of perovskite quantum dots composited nano-porous glass. J. Mater. Chem. C **6**, 11184–11192 (2018). https://doi.org/10.1039/C8TC04383C

148. J. Jiang, G. Shao, Z. Zhang et al., Ultrastability and color-tunability of CsPb(Br/I)3 nanocrystals in P-Si–Zn glass for white LEDs. Chem. Commun. **54**, 12302–12305 (2018). https://doi.org/10.1039/C8CC06442C
149. R. Yuan, L. Ding, G. Shao et al., Suitable medium for $CsPbBr_3$ quantum dots toward light-emitting-diodes fabrication. Mater. Lett. **234**, 275–278 (2019). https://doi.org/10.1016/j.matlet.2018.09.065
150. P. Li, C. Hu, L. Zhou et al., Novel synthesis and optical characterization of CsPb2Br5 quantum dots in borosilicate glasses. Mater. Lett. **209**, 483–485 (2017). https://doi.org/10.1016/j.matlet.2017.08.079
151. R. Yuan, J. Liu, H. Zhang et al., $Eu^{3+}$-doped CsPbBr1.5I1.5 quantum dots glasses: a strong competitor among red fluorescence solid materials. J. Am. Ceram. Soc. **101,** 4927–4932 (2018). https://doi.org/10.1111/jace.15933
152. S. Yuan, D. Chen, X. Li et al., In situ crystallization synthesis of CsPbBr 3 perovskite quantum dot-embedded glasses with improved stability for solid-state lighting and random upconverted lasing. ACS Appl. Mater. Inter. **10**, 18918–18926 (2018). https://doi.org/10.1021/acsami.8b05155
153. C. Wang, H. Lin, X. Xiang et al., CsPbBr 3/EuPO4 dual-phase devitrified glass for highly sensitive self-calibrating optical thermometry. J. Mater. Chem. C **6**, 9964–9971 (2018). https://doi.org/10.1039/C8TC03457E
154. P. Li, Y. Cheng, L. Zhou et al., Photoluminescence properties and device application of $CsPb_2Br_5$ quantum dots in glasses. Mater. Res. Bull. **105**, 63–67 (2018). https://doi.org/10.1016/j.materresbull.2018.04.028
155. S. Liu, M. He, X. Di et al., Precipitation and tunable emission of cesium lead halide perovskites ($CsPbX_3$, X = Br, I) QDs in borosilicate glass. Ceram. Int. **44**, 4496–4499 (2018). https://doi.org/10.1016/j.ceramint.2017.12.012
156. S. Liu, G. Shao, L. Ding et al., Sn-doped $CsPbBr_3$ QDs glasses with excellent stability and optical properties for WLED. Chem. Eng. J. **361,** 937–944 (2019). https://doi.org/10.1016/j.cej.2018.12.147
157. R. Yuan, L. Shen, C. Shen et al., CsPbBr 3:xEu3 + perovskite QD borosilicate glass: a new member of the luminescent material family. Chem. Commun. **54**, 3395–3398 (2018). https://doi.org/10.1039/C8CC00243F
158. M. He, Y. Cheng, L. Shen et al., Doping manganese into $CsPb(Cl/Br)_3$ quantum dots glasses: dual-color emission and super thermal stability. J. Am. Ceram. Soc. **102**, 1090–1100 (2019). https://doi.org/10.1111/jace.15945
159. Y. Cheng, C. Shen, L. Shen et al., $Tb^{3+}$, $Eu^{3+}$ Co-doped CsPbBr 3 QDs glass with highly stable and luminous adjustable for white LEDs. ACS Appl. Mater. Inter. **10**, 21434–21444 (2018). https://doi.org/10.1021/acsami.8b05003
160. He M, Ding L, Liu S, et al. Superior fluorescence and high stability of B-Si-Zn glasses based on Mn-doped CsPbBrxI3-x nanocrystals. J. Alloys Compound. **780**, 318–325 (2019). https://doi.org/10.1016/j.jallcom.2018.11.382
161. L. Xu, S. Yuan, H. Zeng, et al., A comprehensive review of doping in perovskite nanocrystals/quantum dots: evolution of structure, electronics, optics, and light-emitting diodes. Mater. Today Nano **6**, 100036 (2019). https://doi.org/10.1016/j.mtnano.2019.100036
162. Y. Zhou, J. Chen, O.M. Bakr et al., Metal-doped lead halide perovskites: synthesis, properties, and optoelectronic applications. Chem. Mater. **30**, 6589–6613 (2018). https://doi.org/10.1021/acs.chemmater.8b02989
163. D. Parobek, B.J. Roman, Y. Dong et al., Exciton-to-dopant energy transfer in mn-doped cesium lead halide perovskite nanocrystals. Nano Lett. **16**, 7376–7380 (2016). https://doi.org/10.1021/acs.nanolett.6b02772
164. R. Begum, M.R. Parida, A.L. Abdelhady et al., Engineering interfacial charge transfer in CsPbBr 3 perovskite nanocrystals by heterovalent doping. J. Am. Chem. Soc. **139**, 731–737 (2017). https://doi.org/10.1021/jacs.6b09575
165. N.J. Jeon, J.H. Noh, W.S. Yang et al., Compositional engineering of perovskite materials for high-performance solar cells. Nature **517**, 476–480 (2015). https://doi.org/10.1038/nature14133

166. D. Amgar, T. Binyamin, V. Uvarov et al., Near ultra-violet to mid-visible band gap tuning of mixed cation RbxCs1 − xPbX3 (X = Cl or Br) perovskite nanoparticles. Nanoscale **10**, 6060–6068 (2018). https://doi.org/10.1039/C7NR09607K
167. S. Huang, B. Wang, Q. Zhang et al., Postsynthesis potassium-modification method to improve stability of CsPbBr 3 perovskite nanocrystals. Adv. Opt. Mater. **6**, 1701106 (2018). https://doi.org/10.1002/adom.201701106
168. Y. Liu, G. Pan, R. Wang et al., Considerably enhanced exciton emission of CsPbCl3 perovskite quantum dots by the introduction of potassium and lanthanide ions. Nanoscale **10**, 14067–14072 (2018). https://doi.org/10.1039/C8NR03581D
169. F. Yang, D. Hirotani, G. Kapil et al., All-Inorganic CsPb1 − xGexI2Br perovskite with enhanced phase stability and photovoltaic performance. Angew. Chem. Int. Ed. **57**, 12745–12749 (2018). https://doi.org/10.1002/anie.201807270
170. G. Fang, D. Chen, S. Zhou et al., Reverse synthesis of CsPbxMn1 − x(Cl/Br)3 perovskite quantum dots from CsMnCl3 precursors through cation exchange. J. Mater. Chem. C **6**, 5908–5915 (2018). https://doi.org/10.1039/C8TC01426D
171. P.K. Santra, P.V. Kamat, Mn-doped quantum dot sensitized solar cells: a strategy to boost efficiency over 5%. J. Am. Chem. Soc. **134**, 2508–2511 (2012). https://doi.org/10.1021/ja211224s
172. A.K. Guria, S.K. Dutta, S.D. Adhikari et al., Doping Mn2 + in lead halide perovskite nanocrystals: successes and challenges. ACS Energy Lett. **2**, 1014–1021 (2017). https://doi.org/10.1021/acsenergylett.7b00177
173. G. Huang, C. Wang, S. Xu et al., Postsynthetic doping of MnCl2 molecules into preformed CsPbBr 3 perovskite nanocrystals via a halide exchange-driven cation exchange. Adv. Mater. **29**, 1700095 (2017). https://doi.org/10.1002/adma.201700095
174. S. Das Adhikari, S.K. Dutta, A. Dutta et al., Chemically tailoring the dopant emission in manganese-doped CsPbCl3 perovskite nanocrystals. Angew. Chem. Int. Ed. **56**, 8746–8750 (2017). https://doi.org/10.1002/anie.201703863
175. W.J. Mir, M. Jagadeeswararao, S. Das et al., Colloidal Mn-doped cesium lead halide perovskite nanoplatelets. ACS Energy Lett. **2**, 537–543 (2017). https://doi.org/10.1021/acsenergylett.6b00741
176. W. Xu, F. Li, F. Lin et al., Synthesis of CsPbCl3-Mn nanocrystals via cation exchange. Adv. Opt. Mater. **5**, 1700520 (2017). https://doi.org/10.1002/adom.201700520
177. K. Xu, C.C. Lin, X. Xie et al., Efficient and stable luminescence from Mn2 + in core and cCore-Isocrystalline shell CsPbCl3 perovskite nanocrystals. Chem. Mater. **29**, 4265–4272 (2017). https://doi.org/10.1021/acs.chemmater.7b00345
178. W.J. Mir, Y. Mahor, A. Lohar et al., Postsynthesis doping of Mn and Yb into $CsPbX_3$ (X = Cl, Br, or I) perovskite nanocrystals for downconversion emission. Chem. Mater. **30**, 8170–8178 (2018). https://doi.org/10.1021/acs.chemmater.8b03066
179. Y. Hu, F. Bai, X. Liu et al., Bismuth incorporation stabilized $\alpha$-$CsPbI_3$ for fully inorganic perovskite solar cells. ACS Energy Lett. **2**, 2219–2227 (2017). https://doi.org/10.1021/acsenergylett.7b00508
180. C. Bi, S. Wang, Q. Li et al., Thermally stable copper(II)-doped cesium lead halide perovskite quantum dots with strong blue emission. J. Phys. Chem. Lett. **10**, 943–952 (2019). https://doi.org/10.1021/acs.jpclett.9b00290
181. M. Lu, X. Zhang, X. Bai et al., Spontaneous silver doping and surface passivation of $CsPbI_3$ perovskite active layer enable light-emitting devices with an external quantum efficiency of 11.2%. ACS Energy Lett. **3**, 1571–1577 (2018). https://doi.org/10.1021/acsenergylett.8b00835
182. S. Zou, G. Yang, T. Yang et al., Template-free synthesis of high-yield fe-doped cesium lead halide perovskite ultralong microwires with enhanced two-photon absorption. J. Phys. Chem. Lett. **9**, 4878–4885 (2018). https://doi.org/10.1021/acs.jpclett.8b02127
183. X. Zhang, Y. Zhang, X. Zhang et al., $Yb^{3+}$ and $Yb^{3+}/Er^{3+}$ doping for near-infrared emission and improved stability of $CsPbCl_3$ nanocrystals. J. Mater. Chem. C **6**, 10101–10105 (2018). https://doi.org/10.1039/C8TC03957G

184. K. Chen, X. Deng, G. Dodekatos et al., Photocatalytic polymerization of 3,4-Ethylenedioxythiophene over cesium lead iodide perovskite quantum dots. J. Am. Chem. Soc. **139**, 12267–12273 (2017). https://doi.org/10.1021/jacs.7b06413
185. Y.-C. Wong, Ng J. De Andrew, Z.-K. Tan, Perovskite-initiated photopolymerization for singly dispersed luminescent nanocomposites. Adv. Mater. **30**, 1800774 (2018). https://doi.org/10.1002/adma.201800774
186. H. Zhang, X. Wang, Q. Liao et al., Embedding perovskite nanocrystals into a polymer matrix for tunable luminescence probes in cell imaging. Adv. Func. Mater. **27**, 1604382 (2017). https://doi.org/10.1002/adfm.201604382
187. Y. Wei, X. Deng, Z. Xie et al., Enhancing the stability of perovskite quantum dots by encapsulation in crosslinked polystyrene beads via a swelling-shrinking strategy toward superior water resistance. Adv. Func. Mater. **27**, 1703535 (2017). https://doi.org/10.1002/adfm.201703535
188. X. Yang, T. Xu, Y. Zhu et al., Preparation of CsPbBr 3@PS composite microspheres with high stability by electrospraying. J. Mater. Chem. C **6**, 7971–7975 (2018). https://doi.org/10.1039/C8TC01408F
189. H. Liao, S. Guo, S. Cao et al., A general strategy for in situ growth of all-inorganic $CsPbX_3$ (X = Br, I, and Cl) perovskite nanocrystals in polymer fibers toward significantly enhanced water/thermal stabilities. Adv. Opt. Mater. **6**, 1800346 (2018). https://doi.org/10.1002/adom.201800346
190. Y. Li, Y. Lv, Z. Guo et al., One-step preparation of long-term stable and flexible CsPbBr 3 perovskite quantum dots/ethylene vinyl acetate copolymer composite films for white light-emitting diodes. ACS Appl Mater Inter **10**, 15888–15894 (2018). https://doi.org/10.1021/acsami.8b02857
191. Y. Xin, H. Zhao, J. Zhang, Highly stable and luminescent perovskite-polymer composites from a convenient and universal strategy. ACS Appl. Mater. Inter. **10**, 4971–4980 (2018). https://doi.org/10.1021/acsami.7b16442
192. K. Ma, X.-Y. Du, Y.-W. Zhang et al., In situ fabrication of halide perovskite nanocrystals embedded in polymer composites via microfluidic spinning microreactors. J. Mater. Chem. C **5**, 9398–9404 (2017). https://doi.org/10.1039/C7TC02847D
193. Z. Xia, A. Meijerink, $Ce^{3+}$-Doped garnet phosphors: composition modification, luminescence properties and applications. Chem. Soc. Rev. **46**, 275–299 (2017). https://doi.org/10.1039/C6CS00551A
194. G. Li, Y. Tian, Y. Zhao et al., Recent progress in luminescence tuning of $Ce^{3+}$ and Eu2 + -activated phosphors for pc-WLEDs. Chem. Soc. Rev. **44**, 8688–8713 (2015). https://doi.org/10.1039/C4CS00446A
195. C.C. Lin, D.-H. Jiang, C.-C. Kuo et al., Water-resistant efficient stretchable perovskite-embedded fiber membranes for light-emitting diodes. ACS Appl. Mater. Inter. **10**, 2210–2215 (2018). https://doi.org/10.1021/acsami.7b15989
196. Y.H. Song, J.S. Yoo, B.K. Kang et al., Long-term stable stacked CsPbBr 3 quantum dot films for highly efficient white light generation in LEDs. Nanoscale **8**, 19523–19526 (2016). https://doi.org/10.1039/C6NR07410C
197. L. Rao, Y. Tang, C. Yan et al., Tuning the emission spectrum of highly stable cesium lead halide perovskite nanocrystals through poly(lactic acid)-assisted anion-exchange reactions. J. Mat. Chem. C **6**, 5375–5383 (2018). https://doi.org/10.1039/C8TC00582F
198. X. Shen, C. Sun, X. Bai et al., Efficient and Stable $CsPb(Br/I)_3$@Anthracene composites for white light-emitting devices. ACS Appl. Mater. Inter. **10**, 16768–16775 (2018). https://doi.org/10.1021/acsami.8b03158
199. X. Zhang, H.-C. Wang, A.-C. Tang et al., Robust and stable narrow-band green emitter: an option for advanced wide-color-gamut backlight display. Chem. Mater. **28**, 8493–8497 (2016). https://doi.org/10.1021/acs.chemmater.6b04107
200. J. Zhou, F. Huang, H. Lin et al., Inorganic halide perovskite quantum dot modified YAG-based white LEDs with superior performance. J. Mater. Chem. C **4**, 7601–7606 (2016). https://doi.org/10.1039/C6TC02405J

201. C. Sun, Y. Zhang, C. Ruan et al., Efficient and stable white LEDs with silica-coated inorganic perovskite quantum dots. Adv. Mater. **28**, 10088–10094 (2016). https://doi.org/10.1002/adma.201603081
202. H.C. Yoon, S. Lee, J.K. Song et al., Efficient and stable CsPbBr 3 quantum-dot powders passivated and encapsulated with a mixed silicon nitride and silicon oxide inorganic polymer matrix. ACS Appl. Mater. Inter. **10**, 11756–11767 (2018). https://doi.org/10.1021/acsami.8b01014
203. Z. Li, L. Kong, S. Huang et al., Highly luminescent and ultrastable CsPbBr 3 perovskite quantum dots incorporated into a silica/alumina monolith. Angew. Chem. Int. Ed. **56**, 8134–8138 (2017). https://doi.org/10.1002/anie.201703264
204. Y.-M. Chen, Y. Zhou, Q. Zhao et al., Cs4PbBr 6/CsPbBr3 perovskite composites with near-unity luminescence quantum yield: large-scale synthesis, luminescence and formation mechanism, and white light-emitting diode application. ACS Appl. Mater. Inter. **10**, 15905–15912 (2018). https://doi.org/10.1021/acsami.8b04556
205. X. Chen, F. Zhang, Y. Ge et al., Centimeter-sized Cs4PbBr 6 crystals with embedded CsPbBr3 nanocrystals showing superior photoluminescence: nonstoichiometry induced transformation and light-emitting applications. Adv. Func. Mater. **28**, 1706567 (2018). https://doi.org/10.1002/adfm.201706567
206. G. Pan, X. Bai, D. Yang et al., Doping lanthanide into perovskite nanocrystals: highly improved and expanded optical properties. Nano Lett. **17**, 8005–8011 (2017). https://doi.org/10.1021/acs.nanolett.7b04575
207. W. Xu, Z. Cai, F. Li et al., Embedding lead halide perovskite quantum dots in carboxybenzene microcrystals improves stability. Nano Res. **10**, 2692–2698 (2017). https://doi.org/10.1007/s12274-017-1471-0

# Chapter 4
# Perovskite Quantum Dots Based Light-Emitting Diodes

**Yun-Fei Li, Jing Feng, and Hong-Bo Sun**

**Abstract** Beyond the extraordinary success fulfilled in solid-state solar cells, perovskite materials have demonstrated significant potentials in other photonic and optoelectronic devices since 2014. In addition to the perovskite thin films, nanometer-scale perovskite quantum dots (PQDs) with unique optical properties, such as PLQYs approaching unity, high color purity, tunable wavelength, narrow emission bandwidth, and high defect tolerance, have been hotly pursued as promising alternative phosphors in white-light down-converted light-emitting diodes (DC-LEDs) and emitting materials in AM-QLEDs for next-generation lighting sources and displays. In this chapter, the advances in perovskite QDs-based LEDs are reviewed with strategies of material composition design, surface ligand engineering, and device optimization. Besides, the major hurdles for perovskite QLEDs of short lifetime and toxicity (lead-based) are investigated and highlighted; while the solutions undertaken related to surface coating and treatment, device encapsulation, and free-lead perovskite QDs are identified. Moreover, perspectives for the perovskite QDs applied in flexible/stretchable devices and their further development for commercial requirements are also demonstrated.

Y.-F. Li (✉)
Tianjin Key Laboratory of Electronic Materials and Devices, School of Electronics and Information Engineering, Hebei University of Technology, 5340 Xiping Road, Beichen District, Tianjin 300401, People's Republic of China
e-mail: yfli@hebut.edu.cn

J. Feng · H.-B. Sun
State Key Laboratory of Integrated Optoelectronics, College of Electronic Science and Engineering, Jilin University, 2699 Qianjin Street, Changchun 130012, China

H.-B. Sun
State Key Lab of Precision Measurement Technology and Instruments, Department of Precision Instrument, Tsinghua University, Haidian, Beijing 100084, China



Y. Zhou and Y. Wang (eds.), *Perovskite Quantum Dots*, Springer Series in Materials Science 303, https://doi.org/10.1007/978-981-15-6637-0_4

## 4.1 Introduction

Stimulated by the high photoluminescence quantum yields (PLQYs), narrow emission wavelength, high color-rendering index, wide color tenability, and compatibility with flexible devices of colloidal quantum dots (QDs), the quantum dots light-emitting diodes (QLEDs) have aroused a worldwide concern for decades [1–7]. The emission wavelength of QDs can be precisely controlled in relation to content [8] and size [9, 10] with quantum confinement effect [11–14]. Due to these advantages, QDs are potential candidates for white-light LEDs as light converters [15, 16] and active-mode QLEDs (QLEDs) as emitters, [17] which hold great promise in next-generation commercial lighting sources and displays. So far, the external quantum efficiency (EQE) of inorganic Cd-based QLEDs has surpassed 20%, [18] which is comparable to that of the state-of-the-art organic light-emitting diodes (OLEDs).

The booming development of QLEDs requires the exploration of more efficient emitters with low cost [19]. Perovskite QDs show high radiative recombination and favorable optical properties arising from their defect-tolerate nature, where intrinsic defects do not function as trap states [20–25]. Promoted by the extraordinary opto-electronic characteristics of perovskite QDs, numerous studies have devoted into perovskite QLEDs to enhance the performance through composition design, surface engineering, and device structure modification [26–29]. To date, rapid advances of perovskite QLEDs have been achieved within the past decade. It is noteworthy that the first perovskite nanocrystals-based television demon has been demonstrated at CES 2018 exhibition.

Here, we address the device performance evolution of down-converted perovskite QLEDs and perovskite AM-QLEDs and investigate their applications in high-quality lighting sources and vivid color displays. Besides, we demonstrate the major hurdles for perovskite QLEDs of instability and toxicity (lead-based) and summary the effective strategies to circumvent these issues. Moreover, perspectives for the perovskite QDs applied in flexible/stretchable electronics and their further development for commercial requirements are also presented.

## 4.2 Down-Converted Perovskite QLEDs

At present, down-converted perovskite QLEDs are still the mainstream for applications in lighting sources and displays, due to their superior luminance efficiency, high electro-optical conversion efficiency, and relatively long operational lifetime. In the following section, we will discuss the device evolution of white-light down-converted perovskite QLEDs and the strategies towards enhancing the device stability in detail.

### 4.2.1 Solid-State Lighting Sources

Solid-state lighting sources have drawn a significant concern in the general lighting market because of their extended lifetime, energy saving, and eco-friendly properties. In recent decades, down-converted QLEDs function as attractive solid-state lighting sources and are widely employed as alternative to conventional indoor incandescent light, outdoor stage/architectural lighting, due to their high luminescence, simple production process, cost effective, and long lifetime [30]. These white-light QLEDs are consisted of LED chips exe cutting electro-optical conversion, and phosphor down-converting the chip light [31–34]. InGaN chip has been regarded as an optimal light excitation source due to its cost-effective and highly reliable characteristics. Therefore, incorporation of a suitable phosphor to down-convert the chip light can significantly determine the performance of devices.

In the case of indoor and outdoor lighting applications, two major performance metrics, circadian and color performance of the down-converted LEDs (DC-LEDs) should be addressed. Specifically, luminous efficiency (LE) has been regarded as the primary parameter for DC-LEDs. High LE is pursued since it reflects the electro-optical conversion efficiency and reduces the energy consumption. Color rendering index (CRI) is coordinated to the spectrum of the lighting source, requiring the value exceeding 80 [35–37]. However, trade-off always happens between LE and CRI in devices, which needs a good balance in practical applications [38]. Color temperature (CT) describes the color appearance of white LEDs, namely, the absolute temperature of an ideal black-body that radiates light of a color comparable to that of the lighting source. Although CT can effectively evaluate Planckian lighting sources, such as sunlight and tungsten lamp, it is not suitable for not Planckian lighting sources, including DC-LEDs. Correlated color temperature (CCT) is brought up to characterize these kinds of lighting sources, distributing in the range of 2500–6500 K. Specifically, CCT around 2700 K is defined as warm white light, around 4000 K is called neutral white light, while 5000 K or more moves to cool white light. Generally, warm white light facilitates relaxation, benefiting for lighting sources. Besides, lighting sources also demand Commission Internationale de l'Eclairage (CIE) referring to the black body radiator. Optimization of the four major performance metrics, namely LE, CRI, CCT and CIE, plays a critical role in boosting the applications of DC-LEDs in lighting and displays; while integrating an appropriate phosphor is significantly beneficial to improve the device performance.

Compared to the traditional lighting systems, white-light DC-LEDs exhibit enhanced device efficiency and tunable emission characteristics. Effective white-light DC-LEDs can be fulfilled by associating chips with one or more phosphors based on the additive color mixing rules. Basically, the renovation of phosphor materials constructs the foundation of light generation in white-light DC-LEDs. Figure 4.1 shows the evolution for five generations of general white-light DC-LEDs, specifically, the choice of phosphor materials and corresponding color-mixing tactics [39]. In the first generation, InGaN chip with blue emission and $Ce^{3+}$:YAG phosphors with broadband yellow emission were commonly chosen to form the white-light

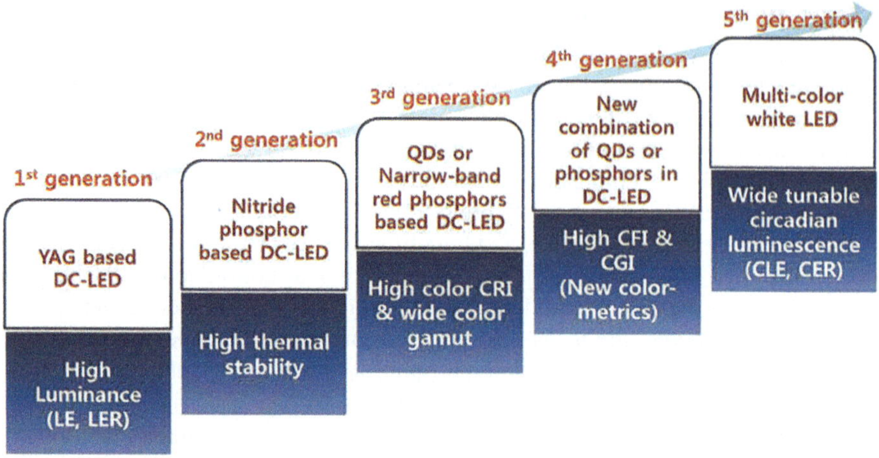


**Fig. 4.1** Schematic of the development strategies of the first through fifth generations for white-light LEDs. Reproduced with permission from Ref. [39]. Copyright 2017, Nature Publishing Group

systems, which had been commercialized. Although the first-generation YAG-based DC-LEDs is easy fabrication and cost-effective, combined with high luminance and high LE characteristics, they can only emit cool white light with CCT above 5000 K, along with CRI values ranging within 70–80 due to the innate shortage of red components [40, 41]. Besides, the first-generation YAG-based DC-LEDs produce light with unstable color, which leads to the halo phenomenon under various current. The other one is the detrimentally poor thermal stability of the organic binders, causing spectral drifting and luminous decay of the phosphor materials under long-term radiation. To settle the above limitations and drawbacks, second-generation nitride phosphors with red emission based LEDs were investigated. $Eu^{2+}$ or $Ce^{3+}$ doped nitrides were commonly used red phosphors in this period [42, 43]. These devices exhibited high thermal stability. However, the LEs of these LEDs were a bit sacrificed owing to the broader FWHM and thick phosphor layer. Specifically, thick phosphor layer was employed to decrease blue-emission to fulfil a low CCT of the DC-LED with blue chip. For the sake of the prerequisites for high-quality backlight display, third-generation narrow-band phosphor ($K_2SiF_6$) combined with red-emission QDs was explored. The LEDs based on red QDs and YAG phosphor exhibit wider color gamut and higher color CRI compared to the previous generations of DC-LEDs. For example, Liu and Chen demonstrated the highly efficient non-rare-earth phosphors with red emission, namely, $K_2TiF_6{:}Mn^{4+}$, with a photoluminescence quantum yield (PLQY) up to 98% [44]. The warm white DC-LED realized superior performance with luminous efficiency of 116 lm/W, high CRI of 81, and low CCT of 3556 K. High luminous efficiency and high color quality (CRI) serve as the critical performance metrics in the evolution from the first to third generation of white-light DC-LEDs. Subsequently, the Illuminating Engineering Society of North America came up with a new color metric, TM-30-2015, with a color gamut index (CGI) and color fidelity index (CF) for the accurate assessment of fourth-generation DC-LEDs. New combination of QDs or phosphors in the fourth-generation DC-LEDs influences the color metrics of trichromatic warm white-light DC-LEDs. However, it is substantiated that three-primary-color warm white-light DC-LEDs with a CRI value over 90 are difficult to achieve a preferable CF score exceeding 90. Actually, DC-LEDs should emit high-quality white colors spanning warm to cool white-light. In the fifth-generation lighting system, healthy and color-reproducible white-light DC-LEDs were the major pursuits. These smart DC-LEDs demonstrate wide tunable circadian luminescence and daily discoloration in a natural circadian rhythm, which is beneficial for biological and medical applications. Circadian-controlled multi-color white DC-LEDs can function as human-centered lighting system in the near future.

Overall, the development of phosphor architecture plays an important role in the evolution for five generations of white-light DC-LEDs. Although considerable phosphors have been investigated in the past decades, the phosphor material pool that is suitable for LEDs is still restricted. Nowadays, the pursuit for novel phosphors in general lighting system is one of the most significant and urgent tasks to be undertaken by advanced science and technology. Perovskite quantum dots (PQDs) with unique optical properties, such as tunable wavelength spanning the entire visible spectrum, narrow emission wavelength, and high PLQY approaching unity, are promising candidates for novel phosphors applied in solid lighting sources. Notably, the wide color selective range of PQDs supplies the chance to obtain high CRI and suitable CCT values containing less blue and ultraviolet emission, which is benefit for the achievement of healthy lighting. Wang and co-workers first applied the all-inorganic narrow-bandwidth $CsPbBr_{3-x}I_x$ (x = 0–3) QDs to modify the $Ce^{3+}$:YAG phosphors in the single down-conversion layer to fabricate the white-light DC-LEDs [45]. These optimized remote-type PQDs were chosen as red-emitting phosphors, which presented excellent warm white light emission with LE of 58 lm/W at an operation current of 20 mA, CIE coordinates of (x = 0.3248, y = 0.3162), CCT of 5907 K, and CRI of 90. Significantly, high operational current up to 350 mA did not induce serious degeneration of the electroluminescent property in the PQD-based white-light DC-LEDs, which indicated $CsPbBrI_2$ QD suitable for applications in high-power lightings. Zhang and Jiang reported the highly fluorescent $CsPbX_3$ PQDs combined with red-emissive phosphors of $(Ba, Ca, Sr)_3SiO_5$:Eu (BCS) uniformly dispersed in epoxy silicon to construct the white-light DC-LEDs [46]. The emitted light demonstrated promising performance with a high CRI of 93.2 and CCT of 5447 K at 20 mA current. Moreover, the spectral profiles of the devices preserved a fairly high level, such as high CRI values of 91.4–93.2 at different currents, even after storage in open air for 1 month. Do et al. proposed a six-color $CsPbX_3$ QDs-based white-light DC-LEDs system, including red, green, blue, cyan, yellowish green, and orange emitting monocolors [47]. Unlike previous works directly mixing the PQDs into the single down-conversion layer, this work demonstrated multilayer structures where emitters were located in separate domains. The multipackage LEDs emitted true-to-life spectra distributions with high color quality, and wide circadian tunable range, closing the gap between yellow and cyan in traditional three-primary-color (RGB) lighting. The optimized six-color multi-package device exhibited superior performance of LE of 62 lm/W, CRI of 96, CCT of 6500 K, and a wide color gamut covering the NTSC up to 145%. Later on, the same group fabricated $CsPb(Br_{0.35}I_{0.65})_3$, $CsPb(Br_{0.5}I_{0.5})_3$, $CsPb(Br_{0.6}I_{0.4})_3$, $CsPb(Br_{0.65}I_{0.35})_3$, $CsPbBr_3$, and $CsPb(Br_{0.75}Cl_{0.25})_3$ PQDs for red, orange, amber, yellowish green, green, and cyan emission with wavelengths of 644, 585, 567, 553, 517, and 488 nm, FWHMs of 31, 27, 24, 22, 17, and 16 nm, and PLQYs of 0.72, 0.53, 0.50, 0.48, 0.80, and 0.64, respectively [39]. Monochromatic PQDs-based LEDs were produced by combination of the NOA 63 binder capped with a blue-mirror-yellow-window long-wavelength pass-dichroic filter (LPDF). The PQD-containing multilayered white-light LEDs presented optimized visual energy efficiency levels and wide color gamut, with superior performance of LE of 58.8 lm/W, CRI of 95, CCT of 6459 K, CFI of 91.4, and CGI of 102. Rogach and co-workers demonstrated a white light DC-LEDs with the configuration of $CsPbBr_3$@POSS, $CsPb_{1.2}Br_{1.8}$ (silicone resin)/InGaN chip [48]. The device showed a LE of 14.1 lm/W, CIE coordinate of (0.349, 0.383), and CRI of 81. Later on, the same group reported a white light DC-LEDs with the structure of $MAPbBr_3$ QDs@POSS, $K_2SiF_6:Mn^{4+}$ (silicone resin)/InGaN chip [49]. The device presented a LE of 38 lm/W, which was significantly higher than that of incandescent lamp (17 lm/W). Besides, the CIE coordinate was (0.30, 0.33) with enhanced stability of the perovskite QDs-based DC-LEDs. These multi-package WLEDs were considered as human-oriented lighting devices, due to its vivid spectral distributions with high-fidelity, and health-promoting effects.

### *4.2.2 Backlight Displays*

More recently, the application of down-converted perovskite QLEDs has become popular in the area of modern display industry as backlighting components [50]. Figure 4.2a demonstrated the schematic diagram of a LCD module. Specifically, the whit-light emitted from the LED subsequently passed through a polarization filter, a TFT matrix with liquid crystals, a color filter, and a second polarization filter. The orientation of the exited light can be easily adjusted by electricity. In order to achieve the best-performance displays, the parameter of color gamut should be highlighted and addressed. Notably, the color gamut value depended on the position in the CIE diagram, which was regulated by National Television Standard Committee.

Figure 4.2b demonstrated the CIE diagram with different color gamut. Actually, the color gamut value can be enhanced by employment of green emitter with narrow-bandwidth, since human eyes are more sensitive in this region. Narrow-band β-SiAlON:$Eu^{2+}$ with green-emission combined with red-emissive $K_2SiF_6:Mn^{4+}$ has been reported as an effective approach to fulfil the LEDs with high-gamut value [51]. Further investigation of narrow-band solid-state materials could benefit for the enhancement of color gamut to obtain more efficient LED backlighting. Perovskite quantum dots are currently come up as a novel category fitted in the area of backlighting due to their high PLQY, extra narrow bandwidth, and tunable emission wavelength. Specifically, the narrow PL bandwidth PQDs exhibit superior color

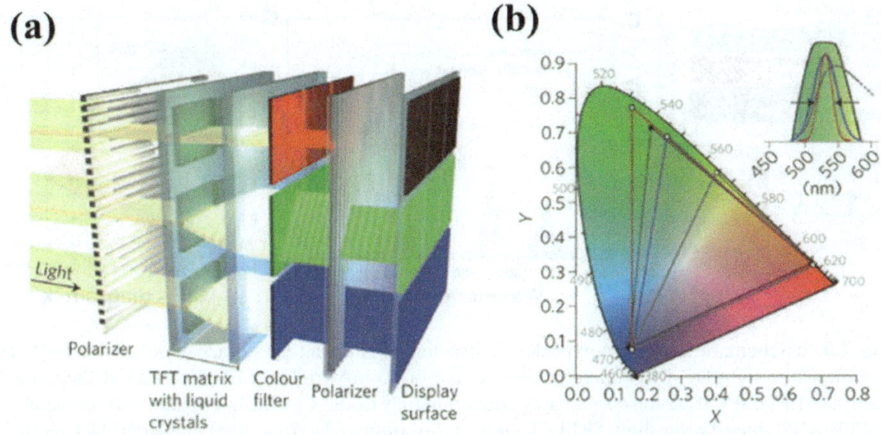


**Fig. 4.2** **a** Schematic diagram of LCD display. **b** CIE (1931) diagram with various color gamuts. Reproduced with permission from Ref. [50] Copyright 2015, Nature Publishing Group

gamut and purity than the traditional Cd quantum dots, which can improve the definition of display prototype. Moreover, the NTSC of organic-inorganic hybrid and all-inorganic perovskite quantum dots can reach up to 130% and 150% before the color filter, respectively [52].

Zhong and co-workers developed wide-color gamut white DC-LEDs by green-emitting organic-inorganic hybrid $MAPbBr_3$ PQDs and red-emitting rare-earth $K_2SiF_6$:$Mn^{4+}$ phosphors as color down-converters, providing improved color performance for display technology [24]. The corresponding optimized device showed a superior white light characteristics with LE of 48 lm/W at 4.9 mA, a CIE coordinates of (0.33, 0.27), reaching 130% of NTSC. Liu et al. first reported the all-inorganic PQDs in on-chip white-light DC-LEDs for backlight displays [53]. They solved the anion exchange issue by introduction of mesoporous silica nanocomposites to enhance thermal and photo stability. The green-emissive $CsPbBr_3$ QDs were mixed with red PQDs in silicon resin, which were excited by InGaN chips. The white-light device demonstrated a color gamut of 113% NTSC and 85% of Rec. 2020 after passing through a color filter, which surpassed the Cd QDs-based devices. Besides, the LE of the device is 30 lm/W with color coordinates of (0.24, 0.28). Chen and co-workers employed microfluidic spinning microreactors in a green way to fabricate the in situ $CsPbBr_3$/poly(methyl)methacrylate (PMMA) composites [54]. The $CsPbBr_3$/PMMA composites combined with CdSe QDs to function as color converters in white light DC-LEDs with color coordinates of (0.34, 0.33), CRI of 89.2, and 105% of color gamut.

Zhong and co-workers presented a white LED with configuration of $MAPbBr_3$/PVDF/$K_2SiF_6$:$Mn^{4+}$/InGaN chip [55]. The optimized device exhibited an excellent performance with a high LE of 109 lm/W, color coordinates of (0.272, 0.278), and a wide color gamut of 121% NTSC with 100% matching rate. In this report, the device performance surpassed the commercial CdSe QDs-based

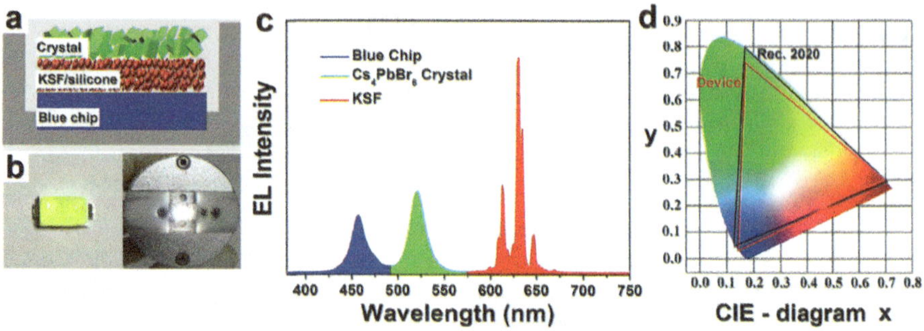


**Fig. 4.3** **a** Schematic of the prototype device structure. **b** An image of surface mounted device (left) and the white-light DC-LED operated at a current of 20 mA (right). **c** EL spectrum of the white-light DC-LED. **d** CIE color coordinates corresponding to the $Cs_4PbBr_6$ crystals with embedded $CsPbBr_3$ NCs based white-light DC-LED (red circle) compared to Rec. 2020 standard (black circle). Reproduced with permission from Ref. [57]. Copyright 2018, Wiley-VCH

white light DC-LEDs, demonstrating the bright future for PQD-based display backlights [56]. Later on, the same group improved the LE of the white light DC-LEDs to 151 lm/W and color gamut of 90.6% Rec. 2020 with matching rate of 89.5% at 20 mA by employing the nonstoichiometry centimeter-sized $Cs_4PbBr_6$ crystals with doped $CsPbBr_3$ QDs [57]. The device structure is consisted of $Cs_4PbBr_6/CsPbBr_3/K_2SiF_6:Mn^{4+}$/GaN chips as shown in Fig. 4.3a. Figure 4.3b, c presented the as-fabricated device picture and the strong EL spectrum of the white light DC-LEDs. Moreover, the coordinate value of the device is (0.331, 0.331) (Fig. 4.3d). These results demonstrated the huge opportunities of PQDs for their applications in display backlights of LCDs. Sun and co-workers also reported $Cs_4PbBr_6/CsPbBr_3$ composites combined with $K_2SiF_6:Mn^{4+}$ phosphors to construct the white light DC-LEDs. The device exhibited a high LE at the driving current of 5 mA [58]. All the device architectures and performances of down-converted perovskite QLEDs are elaborated and summarized in Table 4.1.

### 4.2.3 *Stability and Lifetime*

Although perovskite QDs have been developed rapidly in the past few years, the intrinsic poor stability still limited their commercial applications in optoelectronic lightings and displays. Actually, ground-state perovskite QDs are chemically stable, and only slight QDs are hydrated in ambient environment. The fluorescence quenching in down-converted perovskite QLEDs is chiefly aroused from the increased surface traps and QD agglomerations. In order to improve the lifetime of down-converted perovskite QLEDs, strategies including compositional engineering, surface engineering, matrix encapsulation and device encapsulation have

**Table 4.1** Recent advances of down-converted perovskite QLEDs

| Composites | Device structure | LE (lm/W) | CRI | CCT (K) | CIE coordinates | References |
|---|---|---|---|---|---|---|
| $CsPbBrI_2$ | InGaN chip/YAG:Ce PiG/$CsPbBrI_2$ (PMMA) | 58@20 mA | 90 | 5907 | (0.3248, 0.3162) | [45] |
| $CsPbBr_3$ | Blue chip/$CsPbBr_3$: BSC (epoxy silicon) | – | 93.2 | 5447 | (0.3339, 0.3617) | [46] |
| $CsPbX_3$ | UV chip/$CsPbX_3$ (NOA 63)/LPDF | 62@120 mA | 96 | 6500 | 145% NTSC | [47] |
| $CsPbX_3$ | InGaN chip/$CsPbX_3$ (NOA 63)/LPDF | 58.8 | 95 | 6459 | (0.313, 0.334) | [39] |
| $CsPbBr_3$@POSS | InGaN chip/$CsPbBr_3$@POSS, $CsPb_{1.2}Br_{1.8}$ (silicone resin) | 14.1 | 81 | – | (0.349, 0.383) | [48] |
| $MAPbBr_3$@POSS | InGaN chip/$MAPbBr_3$@POSS, KSF (silicone resin)/ | 38 | – | – | (0.30, 0.30) | [49] |
| $MAPbBr_3$ | GaN chip/KSF (silicone)/$MAPbBr_3$ (PMMA) | 48@4.9 mA | – | – | (0.33, 0.27), 130% NTSC | [24] |
| $CsPbBr_3$–$SiO_2$ | InGaN chip/$CsPbBr_3$–$SiO_2$, $CsPb(Br_{0.4}I_{0.6})_3$ (silicone resin) | 30 | – | – | (0.24, 0.28), 130% NTSC | [53] |
| $CsPbBr_3$/PMMA | GaN chip/$CsPbBr_3$/PMMA, CdSe (silicone resin) | – | 89.2 | – | (0.34, 0.33), 105% NTSC | [54] |
| $MAPbBr_3$/PVDF | InGaN chip/KSF (adhesive)/$MAPbBr_s$/PVDF | 109@20 mA | – | – | (0.272, 0.278), 121% NTSC | [55] |
| $CsPbBr_3$/$Cs_4PbBr_6$ | GaN chip/KSF (silicone resin)/$CsPbBr_3$/$Cs_4PbBr_6$ | 151@20 mA | – | – | 95% rec. 2020 | [56] |
| $CsPbBr_3$/$Cs_4PbBr_6$ | InGaN chip/$CsPbBr_3$/$Cs_4PbBr_6$, KSF | 73.8 @5 mA | – | – | – | [58] |

**Table 4.2** Summary of the strategies to improve stability of down-converted perovskite QLEDs

| Composites | Device structure | LE | Stability | Other index | References |
|---|---|---|---|---|---|
| $MAPbBr_3/NaNO_3$ | Blue chip/$MAPbBr_3/NaNO_3$, KSF (siliconeresin) | 22.39 lm/W (20 mA) | EL intensity remains 50% (bias current of 20 mA for 4 h) | (0.28, 0.32), 127% of NTSC CRI: 86.6 | [59] |
| $CsPbBr_s$-TDPA | Blue chip/$CsPbBr_3$-TDPA, KSF (silicone resin) | 63 lm/W (20 mA) | 57 lm/W (relative humidity of 60% for 15 h) | (0.31,0.29), 122% of NTSC CRI: 83, CCT: 7072 K | [60] |
| $MAPbX_3/SiO_2$ | InGaN chip/$MAPbBr_3/SiO_2$, $MAPbBr_{1.05}I_{1.95}/SiO_2$ (polystyrene) | 54 lm/W | PL intensity remains 75% (15 min-heating at 60 °Cl | (0.31, 0.34) CRI: 85, CCT: 6581 K | [61] |
| $CsPbBr_3/SiO_2$-$Al_2O_3$ | Blue chip/$CsPbBr_3$, $CsPbBr_s/SiO_2$-$Al_2O_3$ (POMS) | 80.77 (5 mA) | PL intensity remains 90%(illumination for 96 h) | – | [62] |
| $CsPbBr_3/SiO_2$ | GaN chip/CdSe NCs (PMMA)/CsPbBr3/$SiO_2$ (PMMA) | 56 lm/W (5 mA) | Unchanged after working for 1 h (1 mA) | (0.30, 0.32), 138% NTSC | [63] |
| $CsPbBr_3$/PSZ | Blue chip/$CsPbBr_3$/PSZ, KSF (PMMA) | 138.6 (60 mA) | unchanged PL intensity (1 month) | (0.31, 0.33), 111% NTSC CCT: 6762 | [64] |
| $CsPbBr_3$-PLA | Bule chip/$CsPbBr_r$PLA (PMMA)/BCS | 62.93 lm/W (20 mA) | 50 lm/W (Working for 50 h) | (0.38, 0.37) CCT: 4000 | [65] |

been reported and summarized in Table 4.2. Among the different protection strategies, compositional or surface engineering can keep or even improve the photoluminescence characteristics of perovskite QDs, showing significant potential in the optoelectronic devices. Matrix encapsulation with compact protective polymer film covering perovskite QDs is the most widespread approach to improve the water resistance of QDs. However, the conductivity of perovskite QDs may be sacrificed after the coatings of compact polymers or inorganics materials. Moreover, additional device encapsulation over perovskite QDs-polymer layers can further prevent the agglomeration and improve the duration. Effective combination of the protective strategies can lead to a tremendous advance for the stability of the down-converted perovskite QLEDs.

Zhong and co-workers employed organic-inorganic hybrid PQDs into ionic matrices to fabricate the stable $MAPbBr_3/NaNO_3$ nanocomposites [59]. After mixing the nanocomposites and red-emissive $K_2SiF_6:Mn^{4+}$ phosphors in silicone resin, wide color gamut DC-LEDs reaching up to 127% of the North American National Television Standard Committee (NTSC) specification were achieved, transcending the typical Cd-based LEDs (NTSC 104%) and commonly used phosphor LEDs (NTSC 86%). The corresponding devices displayed a LE of 22.39 lm/W, CIE coordinates of (0.28, 0.32), and a CRI value of 86.6. The stability of the LEDs was further demonstrated, and the EL spectra exhibited nearly half degradation after 4 h at the operational bias current of 20 mA. Incorporation with inorganic-salt provides an effective strategy to improve the thermal and photo stability of organic-inorganic hybrid PQDs and paves the way for application in solid-state lighting. Wang and co-workers developed alkyl phosphate (TDPA) coated $CsPbBr_3$ QDs to improve the water and thermal stability [60]. The $CsPbBr_3$-TDPA composite were combined with $K_2SiF_6:Mn^{4+}$ phosphors to function as color converters in white light DC-LEDs. The devices demonstrated a high LE of 63 lm/W at an operational current of 20 mA, a wide color gamut of 122% of NTSC, CRI of 83, and CCT of 7072 K, which showed potential in backlight for LCDs. Besides, the corresponding device presented enhanced stability with maintaining 90% LE under a relative humidity of 60% for 15 h. Zhang, Fan and Liu demonstrated a structure of $MAPbBr_3/SiO_2$, $MAPbBr_{1.05}I_{1.95}/SiO_2$ (polystyrene)/InGaN chip [61]. The organic-inorganic hybrid PQDs exhibited significantly enhanced environmental and thermal stability after the novel silica encapsulation process. Besides, anion exchange reactions were not happened between PQDs, which maintained their PL spectra and benefit for the white-light DC-LEDs fabrication. The corresponding DC-LEDs showed a LE of 54 lm/W, CIE coordinates of (0.31, 0.34), and CRI of 85, which is suitable for applications in lighting and displays. Li and co-workers proposed the silica/alumina monolith (SAM) coated $CsPbBr_3$ QDs by sol-gel method [62]. Figure 4.4a, b showed that the emission peak intensity of $CsPbBr_3$ QDs-SAM powders was much stronger than the commercial $Sr_2SiO_4:Eu^{2+}$ green phosphor, indicating the high PLQYs of these QDs. The composite displayed an enhanced photostability for 300 h under the strong blue light illumination, surpassing the reported bare $CsPbBr_3$, DDAB-treated $CsPbBr_3$, $SiO_2$-coated $CsPbBr_3$ powders as shown in Fig. 4.4c. The $CsPbBr_3$ QDs-SAM were

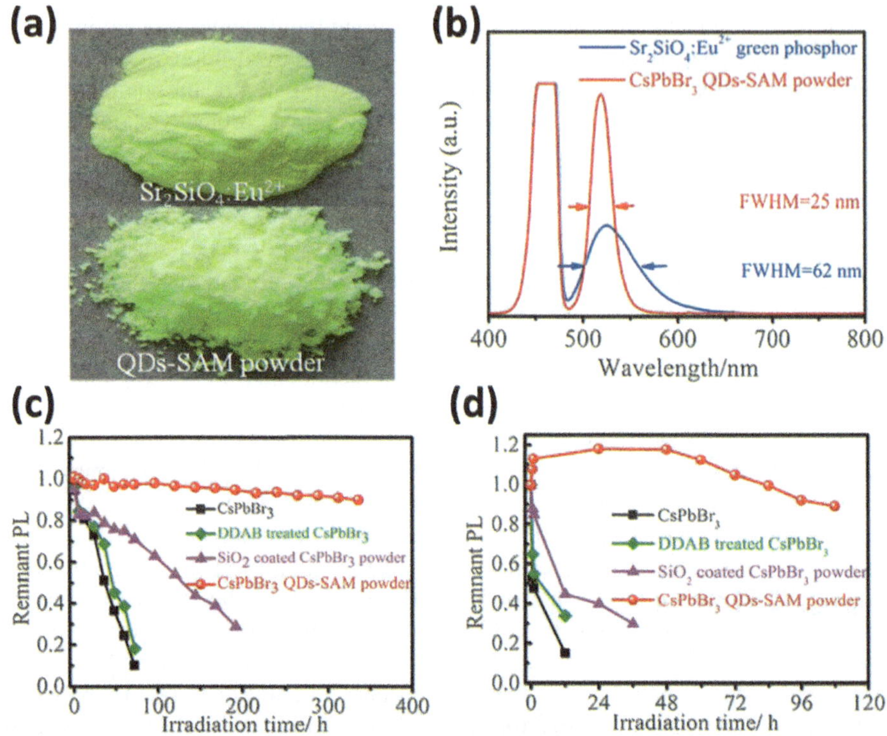


**Fig. 4.4** **a** Optical images and **b** PL spectra of $Sr_2SiO_4$:$Eu^{2+}$ green phosphor and $CsPbBr_3$ QDs-SAM powder. Stability performance of the devices with different powders **c** under illumination with a 470 nm LED light and **d** sealed with PDMS on the LED chip at 5 mA and 2.7 V. Reproduced with permission from Ref. [62]. Copyright 2017, Wiley-VCH

facially incorporated with the PDMS resins and functioned as color-converting materials with curing on blue-emitting LED chips. The device showed a high luminous efficacy of 80.77 lm/W at the operational current of 5 mA and improved stability, which remained 90% of the initial PL emission under illumination for 96 h.

Sun et al. developed monodisperse $CsPbX_3/SiO_2$ and $CsPbBr_3/Ta_2O_5$ Janus nanoparticles to improve the stability of PQDs against the damage induced by water, oxygen and light irradiation at a single particle level [63]. Besides, these Janus nanoparticles could form thin film with high-quality because of their small size and superior dispersibility. The $CsPbBr_3/Ta_2O_5$ film exhibited almost no degradation of the PL intensity (only 2% drop) after continuous intense ultraviolet irradiation for 10 h, in contrast to the degradation of 35% with bare $CsPbBr_3$ QDs. Furthermore, the green-emitting $CsPbBr_3/SiO_2$ was combined with red-emitting CdSe QDs and GaN chip to fabricate the prototype of white light DC-LEDs. The devices demonstrated a LE of 56 lm/W at 5 mA, an EQE of 16.14%, a color gamut of 138% of the NTSC, CRI of 63, and color coordinates of (0.30, 0.32). Notably, the color coordinates were still in the area of white emission under continuous current of 1 mA forl h. Do and co-workers tangled the instability issues of PQDs by introduction of the dual-silicon nitride ($SiN_x/SiN_xO_y$) and silicon oxide ($SiO_y$) ligands of the polysilazane (PSZ) to form a protective layer coated on $CsPbBr_3$ QDs and passivate the surface defects with enhanced thermal, photo, air, and humidity stability even in coarse environment [64]. The $CsPbBr_3$-PSZ composite powders showed unchanged PL intensity for one month at room temperature and RH 60%; while the intensity levels were degraded by 9.8%, 47.9%, and 66.3% when heated to 60 °C, 100 °C, and 140 °C, respectively. Furthermore, the $CsPbBr_3$-PSZ powders were utilized to fabricate white light DC-LEDs with the structure of $CsPbBr_3$/PSZ/$K_2SiF_6$:$Mn^{4+}$ (PMMA)/blue chip, which demonstrated a high LE of 138.6 lm/W at 60 mA, an EQE of 51.4%, a wide color gamut of 111% and 128% with and without color filters, color coordinates of (0.308, 0.328), and CCT value of 6762 K.

Li and Zhang reported $CsPbX_3$-PLA composites to improve the long-term stability under ambient environment of $CsPbX_3$ QDs, especially for the $CsPbCl_3$ QDs [65]. The $CsPbX_3$ QDs were exploited into white light DC-LEDs with the configuration of $CsPbBr_3$-PLA (PMMA)/(Ba, Ca, $Sr)_3SiO_5$:Eu (silicone gel)/blue LED chip. Figure 4.5a presented that the $CsPbBr_3$-PLA had a narrow EL bandwidth, coinciding with the requirement of backlight displays. The intensity of EL spectra enhanced as the driving current increased as shown in Fig. 4.5b. However, the spectra shape was still unchanged, showing that the $CsPbBr_3$ QDs were not saturated under the irradiation of blue light. The device achieved a white emission with a LE of 62.93 lm/W under 20 mA, CCT about of 4000 K, CIE chromaticity coordinates of (0.38, 0.37) (Fig. 4.5c, d). Moreover, the device would exhibit a warmer white color as the driving current increased, consistent with the Planckian locus line. Figure 4.5e showed the durability of the white light DC-LEDs for different operational times. No obvious changes of the EL spectra had been observed after working for 1 h; while the intensity of the green band corresponding to $CsPbX_3$ QDs was slightly dropped after working for 5 hs. Even the working time lasted for 50 hs, the EL intensity of the white light DC-LEDs remained 80% of the initial value.

For the commercial applications, it is meaningful to in-depth understand the quenching mechanism of perovskite QDs-based LEDs. Oxygen and moisture promoted ion migration may influence crystal growth and etch unstable perovskite QDs, which reduces the PLQYs. So far, although surface treatment, and compact and monodispersed shell shielding have been developed to improve the thermal stability, the lifetime of down-converted perovskite QLEDs is still far from 1000 hs under 20 W/cm$^2$ of incident light and 110 °C. In this respect, the thermal-induced degradation of perovskite QDs-based phosphors is the most difficult issue for white light DC-LEDs, which is similar with the CdSe QDs-based LEDs. Although several approaches have been explored to improve the thermal stability, such as the remote-type device configuration or a PiG technique, the thermal stability should be further improved in future.

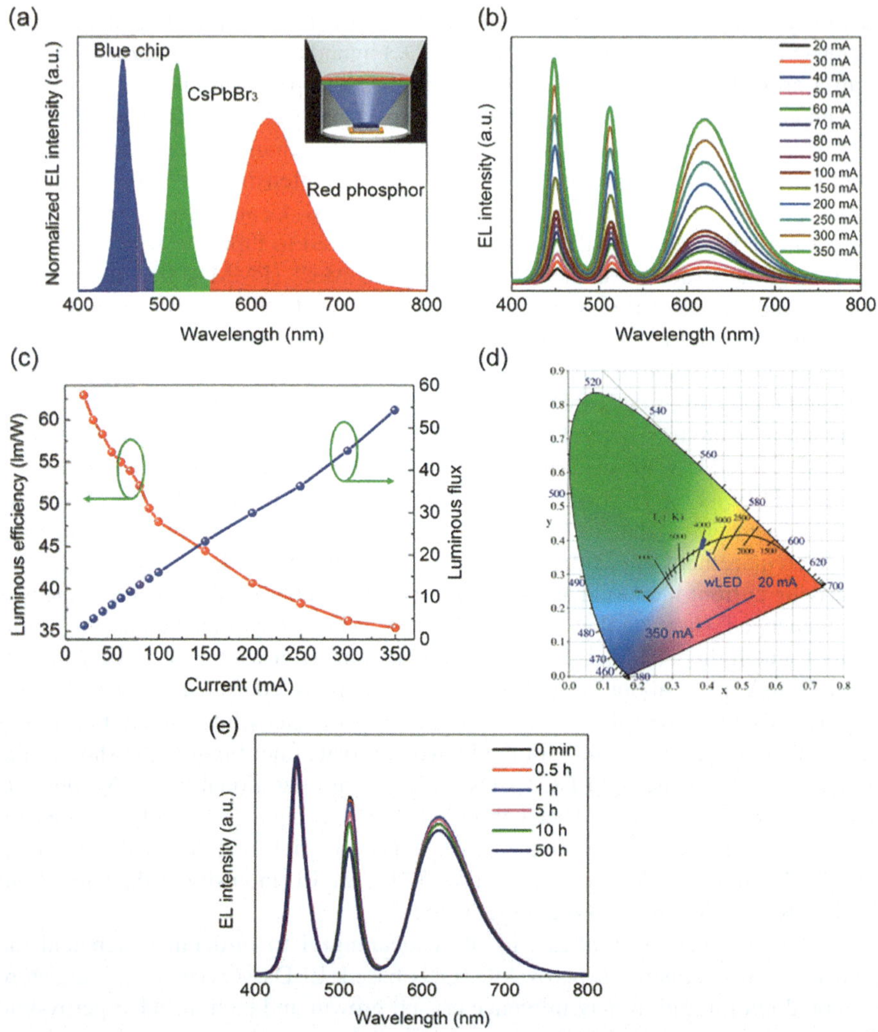


**Fig. 4.5** **a** Schematic illustration and EL spectra of the white-light DC-LEDs. **b** The EL spectra of the device operated under different forward-bias currents. **c** Luminance efficiency and luminance flux of the device. **d** The corresponding color coordinates of the device in a CIE diagram. **e** The EL spectra of the device measured at different working times. Reproduced with permission from Ref. [65]. Copyright 2018, The Royal Society of Chemistry

## 4.3 Perovskite AM-QLEDs

For backlight displays and lighting sources, self-emissive device which uses active matrix (AM) mode is the optimal selection due to the large-area, energy saving, wide color gamut and flexible characteristics. Active matrix CdSe or InP-based QD technologies are favored by the display manufacturers, such as Samsung, LG, and Sony, which are fulfilled by array pixels of RGB QLEDs combined with thin film transistor (TFT) backplanes [6, 66]. In contrast with the traditional white-light DC-LEDs, AM-QLEDs exhibit simplified structures without liquid crystals and color filters. The AM-QLEDs are composed of high-resolution RGB pixels, so the efficiency and lifetime of each color are crucial. Besides, each pixel is controlled individually by current driving mode in AM-QLEDs, which results in decreased power consumption. Moreover, AM-QLEDs also possess higher contrast ratio, wider viewing angle, and shorter response time [67].

Compared to these traditional CdSe or InP QDs, perovskite QDs show superior semiconductor characteristics, namely high photoluminescence quantum yields (PLQYs), narrow emission bands, wide gamut of highly saturated colors, and broadly tunable color range, rendering them potential candidates for next-generation AM-QLEDs [68, 69]. Furthermore, the defect tolerance and solution processability of perovskite QDs make them possible to apply in flexible and stretchable devices [70–72]. A slurry of research efforts have been poured in improving the performance of perovskite QLEDs since it has been discovered, and their efficiency has been enhanced significantly with state-of-the-art EQE value of 21.3% [29]. However, there are still some challenges for their commercial applications, namely, the instability and toxicity (lead-based) of perovskite QDs based AM-QLEDs.

### *4.3.1 Evolution of Device Performance*

So far, a plenty of research efforts have been devoted to improve morphology control of PQD films, carrier injection and transport efficiency, and band match of energy level, which results in enhanced performances of perovskite QLEDs. The efficiency of perovskite QLEDs has been significantly enhanced during the past few years, which can be ascribed to the development of perovskite materials, device structure optimizations, and in-depth comprehension of fundamental device physics [73]. The representative evolution of all-type efficient perovskite QLEDs are demonstrated in this section.

**$MAPbX_3$ QLEDs**. The first perovskite QLEDs were fabricated with $MAPbX_3$ QDs and proposed in 2014 by Galian and Pérez-Prieto [20]. However, the demonstrated luminance of this device was very low, which was smaller than 1 cd/m. Later on, Pei and Zhong employed size-tunable $MAPbBr_3$ QDs to fabricate QLEDs, which exhibited an improved performance with maximum brightness of 2503 cd/m$^2$, current efficiency (CE) of 4.5 cd/A, power efficiency (PE) of 3.5 lm/W, and EQE of 1.1% [74]. Xiong et al. adopted the device structure of ITO/PEDOT:PSS/$MAPbX_3$ QDs/TPBi/$Cs_2CO_3$/Al and obtained the superior performance with CE of 11.49 cd/A, PE of 7.84 lm/W, and an EQE of 3.8% (Fig. 4.6) [75]. Sargent, Xiong, and Demir presented an efficient QLED based on $MAPbBr_3$ QDs by achieving charge balance

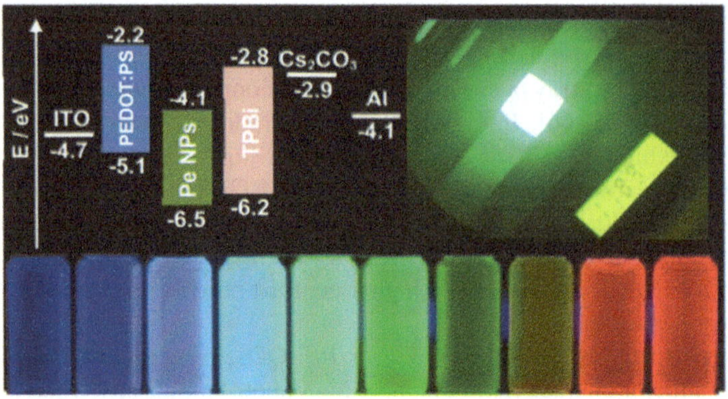


**Fig. 4.6** Schematic illustrations of $MAPbX_3$ QDs-based devices. Reproduced with permission from Ref. [75]. Copyright 2016, American Chemical Society

and suppressing the Auger recombination in low driving voltage [26]. The devices showed a maximum luminance of 43440 cd/m$^2$, PE of 30.3 lm/W, and EQE of 12.9%. Wu, Zheng, Yin optimized the device structure of perovskite QLEDs with the employment of poly[bis(4-phenyl) (2,4,6-trimethylphenyl)amine] (PTAA) doped poly-(Nvinylcarbazole) (PVK) as hole transport layer [76]. The optimal device demonstrated a maximum luminance of 7352 cd/m$^2$, a CE of 11.1 cd/A, and a turn-on voltage of 3.2 V.

**$FAPbX_3$ QLEDs**. Compared to $MAPbX_3$ QDs, organic-inorganic hybrid formamidinium lead halide ($FAPbX_3$) QDs exhibit higher stability both in the solution and film conditions, which are wise emitter choices for QLEDs. Lee and co-workers utilized short length amine ligand to enhance efficiency of $FAPbBr_3$-based QLEDs [77]. Actually, surface ligand engineering plays an important role in determining the performance of QLEDs. On one hand, surface capping ligands can remove surface trap states and improve stability of PQDs. However, they also form an insulating film and hinder the charge injection and transport inside QLEDs. Thus, appropriate controlling of surface ligands benefits for the performance enhancement of perovskite QLEDs. The reported $FAPbBr_3$-n-butylamine-based QLED yielded a high performance with CE of 9.16 cd/A, PE of 6.4 lm/W, and EQE of 2.5%. Wang, Zou and Zhong exploited 3,3-diphenylpropylamine bromide (DPPA-Br) as surface capping ligands to fabricate uniform $FAPbBr_3$ QDs [27]. The resulting QLED demonstrated a high performance with maximum luminance of 13970 cd/m$^2$, CE of 66.3 cd/A, and EQE of 16.3% (Fig. 4.7). Yang, Shen, and Li demonstrated a highly luminescent and color-tunable formamidinium lead halide ($FAPbX_3$, X = Cl, Br, I or mixed) QDs synthesized at room-temperature for efficient light emitting [78]. The $FAPbX_3$ QDs showed single decay dynamics and nearly non-blinking at single dot level. The resulting $FAPbBr_3$ QDs-based QLED presented a high luminance of 403 cd/m$^2$ and EQE of 2.8%.

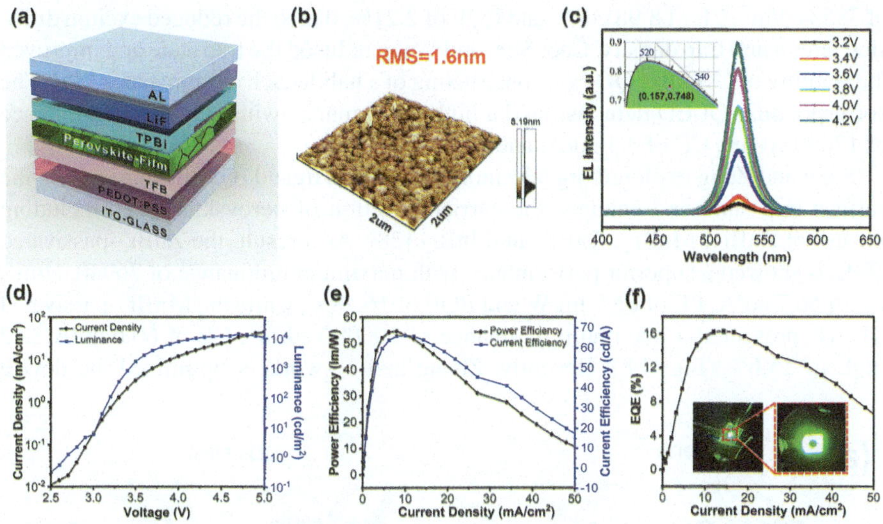


**Fig. 4.7** **a** Device architecture of $FAPbBr_3$-based LEDs. **b** AFM image of the surface of $FAPbBr_3$ PNC film. **c** EL spectra and CIE coordinates of $FAPbBr_3$-based LEDs under different voltages. **d** Current density and luminance of the devices as a function of voltage. **e–f** CE, PE, and EQE as a function of current density and lit pictures of the LEDs. Reproduced with permission from Ref. [27]. Copyright 2018, American Chemical Society

**$CsPbX_3$ QLEDs**. Kovalenko and co-workers developed a novel approach for mixed halide perovskites (MHPs) by designing all-inorganic $CsPbX_3$ QDs [21]. Subsequently, Zeng et al. first applied the all-inorganic $CsPbX_3$ QDs into LEDs with luminances of 742 cd/m$^2$, 946 cd/m$^2$, and 528 cd/m$^2$, and EQEs of 0.07%, 0.12%, and 0.09% for blue, green, and red emission, respectively [79]. The electrically insulating intrinsic of oleic acid (OA) and oleylamine (OAm) ligands on the surface of $CsPbX_3$ QDs hindered he charge carrier injection and transport in QLEDs, which led to a modest device performance. The same group solved this issue through a wash technique to control the ligand density, which resulted in an enhanced performance of QLEDs. The resulting device achieved a maximum luminance of 15185 cd/m$^2$ and EQE of 6.27% [80]. Choy and Rogach adopted an intercalated perfluorinated ionomer (PFI) layer between HTL and active layer to improve the hole injection efficiency [81]. The device demonstrated a brightness of 1377 cd/m$^2$, CE of 0.19 cd/A and EQE of 0.06%. Greenham and Tan proposed a trimethylaluminum (TMA) vapor-based crosslink to protect QD films; therefore, orthogonal solvents were not essential for the charge transport layer deposition in QLEDs [82]. The near-complete coverage of $CsPbBr_3$ QDs film promoted charge capture and induced high luminances of 8.7, 2335, 1559, and 206 cd/m$^2$ and EQEs of 0.0074%, 0.19%, 1.4% and 5.7% for blue, green, orange, and red emission, respectively. Wang and Sun reported an innovative dual-phase $CsPbBr_3$-$CsPb_2Br_5$ QD composite to apply in QLEDs [83]. The resulting devices exhibited a significantly increased performance with maximum luminance of 3853 cd/m$^2$, CE of 8.98 cd/A, and EQE of 2.21%, due to the reduced exciton diffusion length and trap density. Cao, Sun, and Wang reduced the trap state and improved the stability of $CsPbBr_3$ QDs via constructing of a halide-rich circumstance [84]. The corresponding QLED demonstrated a high performance with maximum luminance of 12090 cd/m$^2$, CE of 3.1 cd/A, and EQE of 1.194%.

Song and Zeng explored organic-inorganic hybrid ligand (OIHL) to passivate the surface trap state and enhance the carrier injection of perovskite QDs, including the use of $ZnBr_2$, $MnBr_2$, $GaBr_3$, and $InBr_3$ [28]. As a result, the $ZnBr_2$-passivated QLEDs showed a superior performance with maximum luminance of 76940 cd/m$^2$, CE of 66.7 cd/A, PE of 65.9 lm/W and EQE of 16.48%; while the $MnBr_2$-passivated QLEDs presented a maximum luminance of 100080 cd/m$^2$, CE of 60.6 cd/A, and EQE of 15.6% (Fig. 4.8). Recently, Zhang and co-workers optimized the device

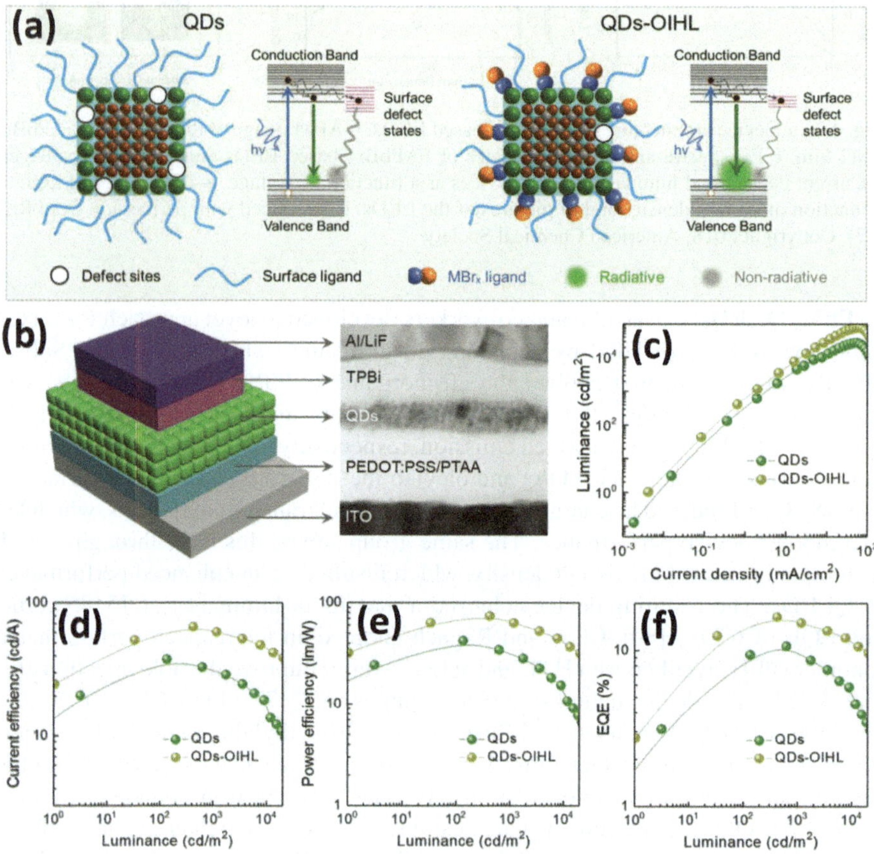


**Fig. 4.8** **a** Schematic demonstration of radiative/nonradiative recombination of perovskite QDs with and without OIHL. **b** Illustration of multilayer perovskite QLED device and cross-sectional TEM image. **c–f** Luminance, CE, PE, and EQE of the devices. Reproduced with permission from Ref. [28]. Copyright 2018, Wiley-VCH

architecture and energy level alignment of $CsPbBr_3$-based QLEDs through the introduction of Li-doped $TiO_2$ nanoparticles as the electron transport layer (ETL) [85]. The resulting device exhibited a reduced turn-on voltage of 2 V and an enhanced CE of 15.2 cd/A for inverted QLEDs with Li-doped $TiO_2$ as ETL, in contrast with the turn-on voltage of 3 V and CE of 5.6 cd/A.

For red emission, Ning and Bakr exploited bidentate ligands to passivate $CsPbI_3$ QDs through a postsynthesis passivation approach [86]. The passivated $CsPbI_3$-based QLEDs allowed a red emission with a maximum luminance of 748 cd/m$^2$ and an EQE of 5.02%. Later on, Chiba and Kido made a significant breakthrough for red-emission perovskite QDs via an anion-exchange approach with long alkyl based oleylammonium iodide (OAM-I) and aryl-based aniline hydroiodide (An-HI) ligands (Fig. 4.9) [29]. This enabled efficient red-emissive QLEDs with EL peaks of 653 and 645 nm, turn on voltages of 2.8 and 2.7 V, maximum luminances of 500 cd/m$^2$ and 749 cd/m$^2$, CEs of 10.6 cd/A, 11.6 cd/A, PEs of 11.9 lm/W, 12.9 lm/W, and EQEs of 21.3%, 14.1% for OAM-I and An-HI-passivated perovskite QLEDs, respectively.

In the past decade, red and green emissive perovskite-based QLEDs have made rapid advances; however, the performances of blue-emissive perovskite QLEDs remained a challenge. In this aspect, Liao and Bakr developed nonpolar solvent-soluble organic pseudohalide of n-dodecylammonium thiocyanate (DAT) to passivate the Cl vacancies and remove defects in Cs-based MHP QDs, which enabled the QLEDs with blue emission and improved device efficiency [87]. The DAT-treated blue-emission device showed a high luminance of 465 cd/cm$^2$, and a maximum EQE of 6.3%, which was a record for perovskite QLEDs emitting in the range of 460–480 nm according to Rec. 2020 display standards, and a half lifetime of ~99 s. Bai and Song incorporated $Ni^{2+}$ ions into $CsPbCl_xBr_{3-x}$ QDs by supersaturated recrystallization synthetic method and modulated the Cl/Br element ratios to obtain efficient blue-emissive PQDs [88]. The corresponding device presented a maximum luminance of 612 cd/m$^2$ and EQE of 2.4% (Fig. 4.10).

**Mixed cation perovskite QLEDs**. Mixing cations had been investigated as an effective strategy to construct high efficiency perovskite QLEDs. Chen and Sun first reported the synthesis of $MA_{1-x}Cs_xPbBr_3$ QDs and the corresponding $MA_{0.7}Cs_{0.3}PbBr_3$-based QLEDs illustrated a maximum luminance of 24510 cd/m$^2$, CE of 4.1 cd/A, and EQE of 1.3% [10]. Zhang, Chen and Sun exploited the mixed cation $FA_{0.8}Cs_{0.2}PbBr_3$ to fabricate QLED, which achieved a high luminance of 55005 cd/m$^2$, CE of 10.09 cd/A and EQE of 2.80% [89]. Recently, Zhong and co-workers utilized red emissive $FA_{0.87}Cs_{0.13}PbI_3$ NCs with GBL as a solvent to fabricate the QLEDs [90]. The resulting device showed a maximum luminance of 218 cd/m$^2$ and a peak EQE of 15.8%.

Table 4.3 summarized the device architecture and performance of the perovskite QLEDs in detail. The PLQYs of perovskite QD films related to the QLED performance differed with the QD solutions, which show PLQYs approaching unity. Although surface passivation ligands can improve the PLQY of QD solution, their insulating nature impedes the charge injection in QD film and further negatively impacts the QLED performance. Hence, appropriate ligand choice and effective

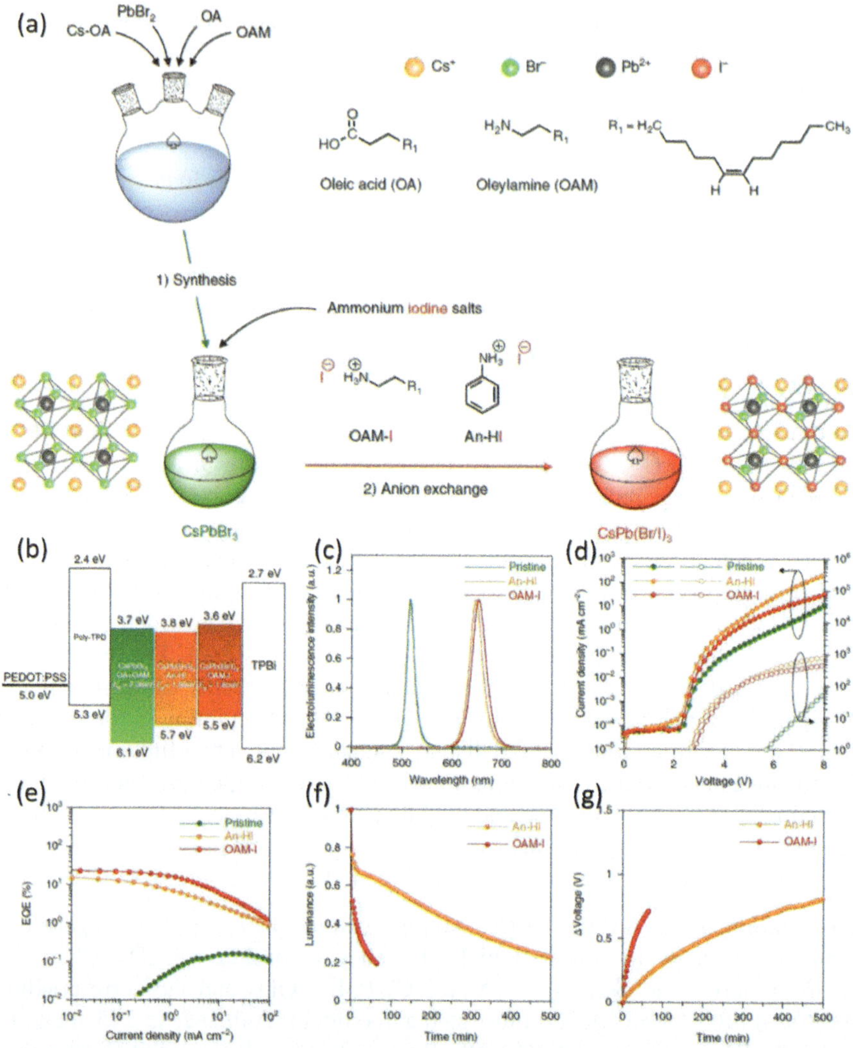


**Fig. 4.9** **a** Illustration of the anion-exchange method to fabricate red-emissive perovskite QDs from $CsPbBr_3$ QDs. **b** Energy level diagram of the perovskite QLEDs. **c** Evolution of EL performance with different anions. **d–e** J-V-L and EQE of the QLEDs. **f–g** Luminances and voltage variations of perovskite QLEDs versus running time. Reprinted with permission from Ref. [29]. Copyright 2018, Nature Publishing Group

ligand exchange play important roles in building high-efficiency perovskite QLEDs. Moreover, modifying device structures with matched energy levels to increase radiative recombination efficiency is another major method in the achievement of efficient perovskite QLEDs.

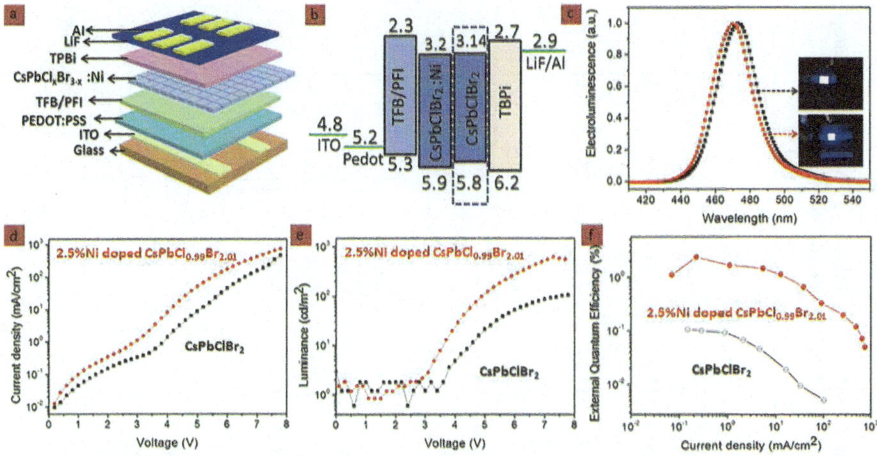


**Fig. 4.10** **a**, **b** Schematic diagram and band level of the LED based on 2.5% $Ni^{2+}$ ions doped $CsPbCl_{0.99}Br_{2.01}$ QDs. **c** The EL spectra and photos of the LEDs based on $CsPbClBr_2$ QDs (black) and 2.5% $Ni^{2+}$ doped $CsPbCl_{0.99}Br_{2.01}$ QDs (red), respectively. **d–f** current density, luminance, and EQE of the corresponding devices. Reprinted with permission from Ref. [88]. Copyright 2020, American Chemical Society

## *4.3.2 Stability and Lifetime*

In recent decades, the rapid development of perovskite QLEDs with significantly enhanced efficiency has made them draw great interest in active mode devices. However, the foremost hindrance to the commercialization of perovskite QLEDs originates from their instability in air. So far, the reported operational lifetime of perovskite QLEDs is still far below the commercial standard of more than 10,000 hs; hence, improving the long-term stability is the primary aim for their commercial applications. In the initial development stage of perovskite QLEDs, the reported lifetime was rather short. Docampo and Costa demonstrated that $MAPbBr_3$ and $FAPbBr_3$ QDs based light-emitting electrochemical cells (LECs) underwent a sharply reduce in luminance over the first minutes and then stayed constant with the luminances of 1.8 and 0.8 cd/m$^2$, and CEs of 0.013 and 0.005 cd/A, respectively [91]. It has been investigated that the lifetime was only few hours for these $MAPbBr_3$ and $FAPbBr_3$ QDs-based LECs. Indeed, anion exchange, particle aggregation, and ligand detachment may happen during the fabrication of perovskite QLEDs, which has a negative effect on their long-term stability [68]. To circumvent these issues, mixed cations and postsynthesis passivation have been reported to improve the stability of perovskite QDs; while the employment of inorganic transport layers and device encapsulation could further enhance the lifetime of perovskite QLEDs.

**Compositional engineering**. Once, $CsPbI_3$ QDs with stable emission have been problematic to acquire owing to the phase transitions. Fundamentally, ionic radius has a significant effect on the structural stability of perovskite QDs. Rationally doping

**Table 4.3** Recent advances of perovskite AM-QLEDs

| QDs | Device structure | EL peak (nm) | Von (V) | Max. L (cd/m$^2$) | Max. CE (cd/A) | Max. PE (lm/W) | Max. EQE (%) | References |
|---|---|---|---|---|---|---|---|---|
| *$MAPbX_3$ QLEDs* | | | | | | | | |
| $MAPbBr_3$ | ITO/PEDOT:PSS/pTPD/QDs/Ba/Ag | 533 | – | <1 | – | – | – | [20] |
| $MAPbBr_3$ | ITO/PEDOT:PSS/QDs/TPBi/CsF/Al | 510 | 2.9 | 2503 | 4.5 | 3.5 | 1.1 | [74] |
| $MAPbBr_3$ | ITO/PEDOT:PSS/QDs/TPBi/$Cs_2CO_3$/Al | 512 | 2.8 | 11,830 | 11.49 | 7.84 | 3.8 | [75] |
| $MAPbBr_3$ | IT0/PED0T:PSS/QDs/B3PYMPM:TPBi/B3PYMPM:$CS_2CO_3$/Al | 524 | 2.9 | 43,400 | – | 30.3 | 12.9 | [26] |
| $MAPbBr_3$ | ITO/PEDOT:PSS/PVK:PTAA/QDs/LiF/Al | 524 | 3.2 | 7532 | 11.1 | – | 2.91 | [76] |
| *$FAPbX_3$ QLEDs* | | | | | | | | |
| $FAPbBr_3$ | ITO/Buf–H IL/QDs/TPBi/LiF/Al | 530 | – | – | 9.16 | 6.4 | 2.05 | [77] |
| $FAPbBr_3$ | ITO/PEDOT:PSS/TFB/QDs/TPBi/LiF/Al | 526 | 2.7 | 13,970 | 66.3 | – | 16.3 | [27] |
| $FAPbBr_3$ | ITO/PEDOT:PSS/TFB/QDs/LiF/Al | – | 3.1 | 403 | – | – | 2.8 | [78] |
| *$CsPbX_3$ QLEDs* | | | | | | | | |
| $CsPb(BrCl)_3$ | ITO/PEDOT:PSS/PVK/QDs/TPBi/LiF/Al | 455 | 5.1 | 742 | 0.14 | 0.07 | 0.07 | [79] |
| $CsPbBr_3$ | ITO/PEDOT:PSS/PVK/QDs/TPBi/LiF/Al | 516 | 4.2 | 946 | 0.43 | 0.18 | 0.12 | [79] |
| $CsPb(BrI)_3$ | ITO/PEDOT:PSS/PVK/QDs/TPBi/LiF/Al | 586 | 4.6 | 528 | 0.08 | 0.06 | 0.09 | [79] |
| $CsPbBr_3$ | ITO/PEDOT:PS5/PolyTPD/QDs/TPBi/LiF/Al | 512 | 3.4 | 15,185 | 13.3 | 5.24 | 6.27 | [80] |
| $CsPbBr_3$ | ITO/PEDOT:PSS/Poly-TPD/PFI/QDs/TPBi/LIF/Al | 516 | 3.5 | 1377 | 0.19 | – | 0.06 | [81] |
| $CsPbBr_{1.5}Cl_{1.5}$ | ITO/ZnO/QDsAFB/$MoO_3$/Ag | 480 | – | 8.7 | – | – | 0.0074 | [82] |
| $CsPbBr_3$ | ITO/ZnO/QDs/TFB/$MoO_3$/Ag | 523 | – | 2335 | – | – | 0.19 | [82] |
| $CsPbI_{2\cdot25}Br_{1.5}$ | ITO/ZnO/QDsAFB/$MoO_3$/Ag | 619 | – | 1559 | – | – | 1.4 | [82] |
| $CsPbI_3$ | ITO/ZnO/QDs/TFB/$MoO_3$/Ag | 698 | – | 206 | – | – | 5.7 | [82] |

(continued)

**Table 4.3** (continued)

| QDs | Device structure | EL peak (nm) | Von (V) | Max. L ($cd/m^2$) | Max. CE (cd/A) | Max. PE (Im/W) | Max. EQE (%) | References |
|---|---|---|---|---|---|---|---|---|
| $CsPbBr_3$–$CsPb_2Br_5$ | ITO/PEDOT:PSS/QDs/TPBi/LiF/Al | 527 | 4.6 | 3853 | 8.98 | 3.4 | 2.21 | [83] |
| $CsPbBr_3$ | ITO/PEDOT: PSS/Poly-TPD/QDs/TPBi/LiF/AI | 515 | 4.6 | 12,090 | 3.106 | 1.38 | 1.194 | [84] |
| $CsPbBr_3$–$MnBr_2$ | ITO/PEDOT:PSS/PTAA/QDs/TPBi/LiF/Al | – | – | 100,080 | 60.6 | – | 15.6 | [28] |
| $CsPbBr_3$ | FTO/$TiO_2$/$LiTiO_2$/QDs/poly-TPD/PEDOT:PSS/Au | – | 2 | 22,057 | 15.2 | – | – | [85] |
| $CsPbI_3$ | ITO/PEDOT:PSS/Poly-TPD/QDs/TPBi/LiF/Al | 688 | 4.5 | 748 | – | 0.47 | 5.02 | [86] |
| $CsPbBr_3$–OAM–I | ITO/PEDOT:PSS/Poly-TPD/QDs/TPBi/Liq/Al | 653 | 2.8 | 500 | 10.6 | 11.9 | 21.3 | [29] |
| $CsPbBr_3$–An–HI | ITO/PEDOT:PSS/Poly-TPD/QDs/TPBi/Liq/Al | 645 | 2.7 | 794 | 11.6 | 12.9 | 14.1 | [29] |
| $CsPb(Br_xCl_{1-x})_3$-DAT | ITO/TFB/PFI/QDs/3TPVMB/Liq/Al | 471 | – | 465 | – | – | 6.3 | [87] |
| $CsPbCl_{0.99}Br_{2.01}$–$Ni^{2+}$ | ITO/PEDOT:PSS/TFB/PFI/QDs/TPBi/LiF/Al | 470 | – | 612 | – | – | 2.4 | [88] |
| *Mixed cation perovskite QLEDs* | | | | | | | | |
| $MA_{0.7}Cs_{0.3}PbBr_3$ | ITO/PEDOT:PSS/TFB/QDs/TPBi/LiF/Al | – | 4 | 24,510 | 4.1 | – | 1.3 | [10] |
| $FA_{0.8}Cs_{0.2}PbBr_3$ | ITO/PEDOT:PSS/TFB/QD5/TPBi/LiF/Al | – | 3.5 | 55,005 | 10.09 | – | 2.8 | [89] |
| $FA_{0.87}Cs_{0.13}PbI_3$ | ITO/ZnO/PEI/QDs/CBP/MoOx/Au | 695 | 1.9 | 218 | 1.2 | – | 15.8 | [90] |

the A-site with a suitable larger cation is an effective way to stabilize the $CsPbI_3$ QDs structure, arising from the higher tolerance factor. Kovalenko and co-workers first realized the stable perovskite QLEDs with red-emission through doping larger size FA cations into $CsPbI_3$ QDs [92]. The corresponding $FA_{0.1}Cs_{0.9}PbI_3$-based QLEDs showed an EQE of 0.12%.

**Surface engineering**. Postsynthesis passivation is another adoptable method to address this stability issue of $CsPbI_3$ QDs. Traditional OLA and OA ligands are prone to be lost during the purification process. The ill-passivated surface results in the non-radiative recombination of carrier via surface trap states, which makes the perovskite QDs susceptible to oxygen and water in air. Hence, the employment of suitable branched surface ligands with large steric hindrance to substitute OA/OLA and strengthening the interactions between the surface and ligands plays a crucial role in enhancing the stability of perovskite QDs. Ning and Bakr utilized a bidentate ligand of 2,2′-iminodibenzoic acid to synthesize stable $CsPbI_3$ QDs via the postsynthesis passivation approach [86]. The passivated-$CsPbI_3$ based QLEDs achieved a maximum luminance of 748 cd/m$^2$ and an EQE of 5.02%.

**Inorganic CTLs**. Inorganic charge transport layers (CTLs) usually have better chemical stability than the organic materials, which can effectively function as barrier layers to prevent the penetration of moisture and oxygen into active layer. Li and Shan adopted inorganic ZnO and NiO nanoparticles as CTLs in QLEDs to improve the device performance and enhance stability [93]. The device reached a luminance of 6093.2 cd/m$^2$, CE of 7.96 cd/A and EQE of 3.79%. Moreover, the $CsPbBr_3$ based QLEDs investigated good operational stability, which could tolerate a high humidity (75%, 12 h) and a high working temperature (393 K, three heating/cooling cycles) even without encapsulation (Fig. 4.11). Song and Zeng also applied NiO and ZnO into $CsPbBr_3$-based QLEDs with a device structure of ITO/NiO/$CsPbBr_3$ QDs/ZnO/Al [94]. As a result, the device demonstrated a lifetime 3.5 times longer than the conventional organic CTL-based device. Furthermore, the devices could endure high moisture when soaked in water for about 20 s without encapsulation. These works present big strides for stable perovskite QLEDs in ambient or even harsh conditions.

**Device encapsulation**. Device encapsulation is the most common approach to enhance the stability of commercial optoelectronics, owing to its effective protection of devices from oxygen and water. Zhang and Jie had proved that this method was also suitable for perovskite QLEDs [95]. A cover glass slide and UV-curing adhesive were utilized to package a $MAPbBr_3$-based QLED in this work. As a consequence, the encapsulated device demonstrated small fluctuation in open air within 7 days (Fig. 4.12). In contrast, the bare QLED showed full degradation owing to the high humid sensitivity of $MAPbX_3$ QDs.

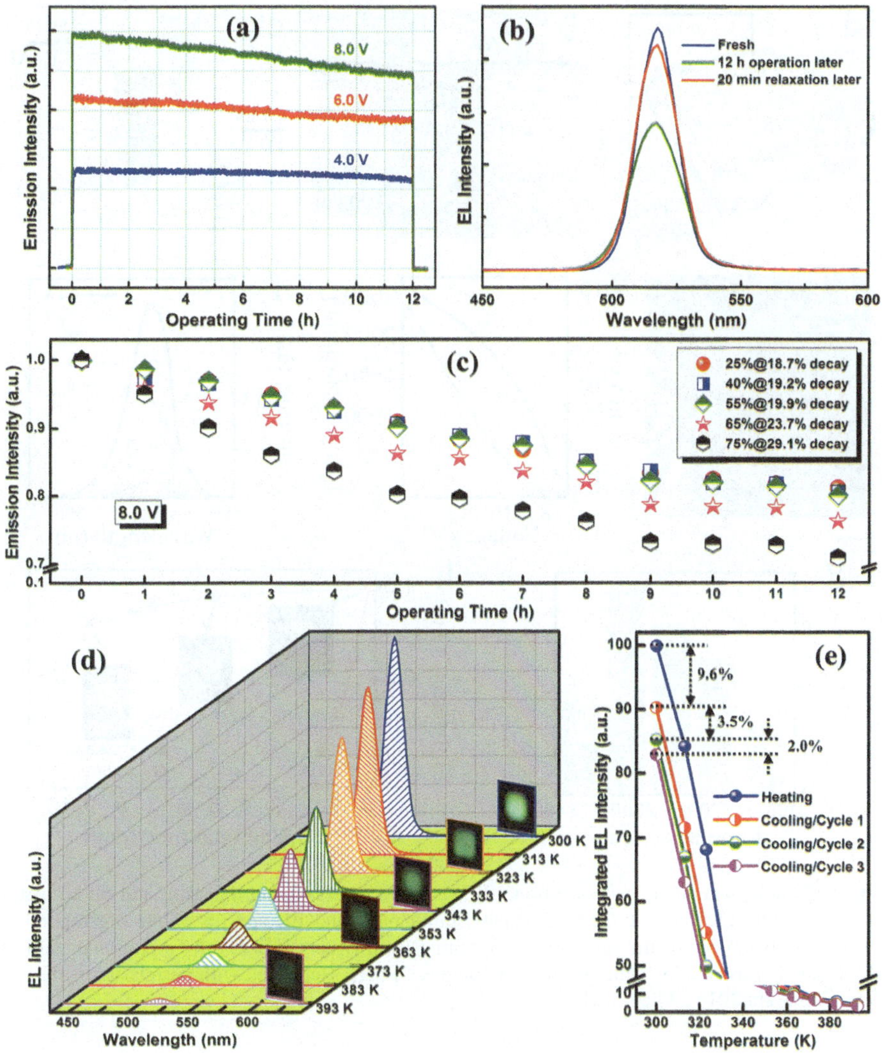


**Fig. 4.11** **a** Emission intensity of the perovskite LED with different voltages versus running time. **b** EL spectra measured before aging, after operation for 12 hs, and after a relaxation time of 20 mins. **c** Normalized emission intensity of the perovskite LED after different running periods under various humidity conditions. **d** Temperature-dependent EL spectra of the device at a fixed driving current density of 50 mA/cm$^2$. The insets demonstrate the photographs of an emitting unit of the LED captured at varying working temperature. **e** Integrated EL intensity of of the perovskite LEDs under three heating/cooling cycles at a fixed driving current density of 50 mA/cm$^2$. Reprinted with permission from Ref. [93]. Copyright 2018, American Chemical Society

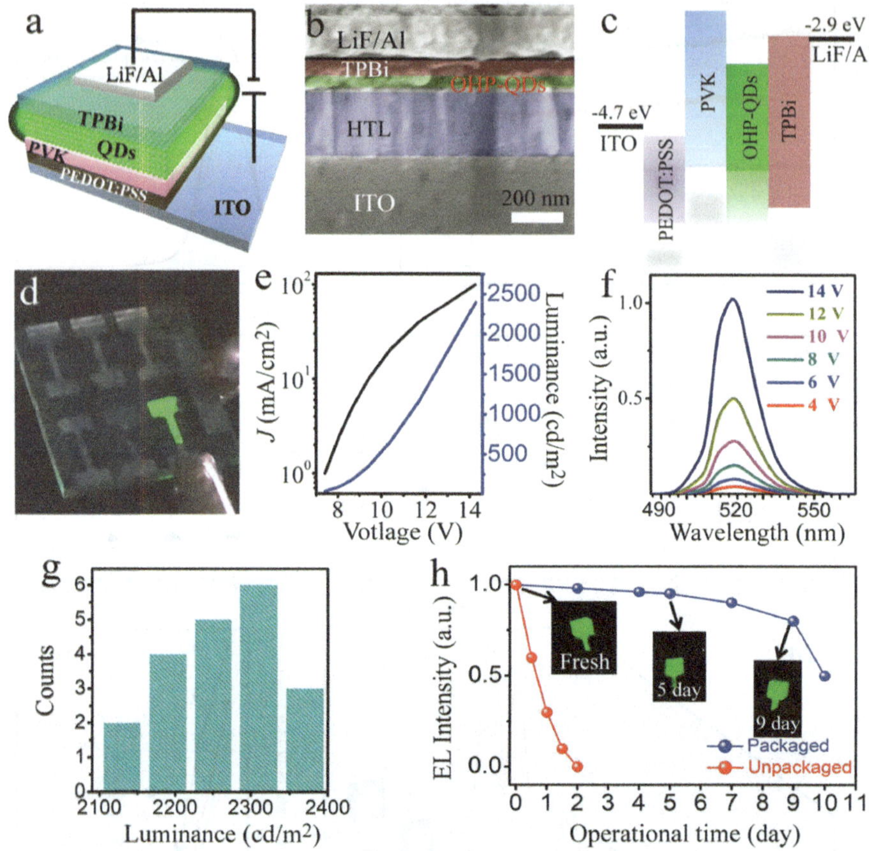


**Fig. 4.12** **a–d** Schematic, cross-sectional SEM image, energy band diagram, and photo image of a perovskite QLED. **e–f** Current density, luminance, and EL spectra of the corresponding devices. **g** Histogram of the luminances measured from twenty QLEDs. **h** Operational stability of QLEDs versus running time under a continuous voltage. Reprinted with permission from Ref. [95]. Copyright 2016, Wiley-VCH

## 4.3.3 Lead-Free Devices

Perovskite QDs are promising emitters for next-generation lighting sources and displays. To date, $Pb^{2+}$ is still the most frequent B-site cation of perovskite QDs with the $ABX_3$ format because of the higher stability and PLQYs. However, the inherent toxicity of $Pb^{2+}$ is unfriendly to the environment, which remains a tough issue. According to Restriction of Hazardous Substances (RoHS), total or partial replacement of $Pb^{2+}$ is required to control the heavy metal dosage in QLEDs [68]. Optical properties and intrinsic stability of the perovskite QDs will substantially changed after the replacement of other non-lead analogous metals ions, since the imbalance in the electronic dimensionality [96]. Here, a thorough comprehension of structural reorganization after component engineering at the atom level is strongly expected. Density functional theory (DFT) has been reported to provide a strong support in screening potential candidates for reasonable molecular designs, depending on the estimates of bandgap and thermodynamic stability [97–99]. At present, $Sn^{2+}$ ions are most appropriate substitutes for less toxic elements, which have been widely applied in solar cells [100, 101]. However, only few studies have been reported on Sn-based perovskite QLEDs. Wang and Sun demonstrated partial $Sn^{2+}$ ions-substituted $CsPb_{1-x}Sn_xBr_3$ QDs to alleviate toxicity, and applied them into QLEDs for the first time [102]. As the Sn content enhanced, the perovskite QDs showed a blue-shift of bandgap and emission peak. The corresponding device achieved a relatively good performance with luminance of 5495 cd/m$^2$ and CE of 3.60 cd/A. It is noteworthy that excess Sn addition could cause fast decay of luminance, leading to the inferior device performance.

### 4.3.4 Flexible and Stretchable Devices

Colloidal perovskite QDs can be dispersed in nonpolar solvents and maintain functionality under tensile strains, enabling their applications in flexible/stretchable devices and satisfying the future development trend of QLEDs. Zhong and Pei applied $MAPbBr_3$ QDs into flexible light-emitting devices with the configuration of silver nanowires–polymer composite substrate/PEDOT:PSS/PVK:TAPC/QDs/TmPyPB/CsF/Al [103]. The resulting device reached a high performance with CE of 10.4 cd/A, LE of 8.1 lm/W, and EQE of 2.6% at the luminance of 1000 cd/m$^2$. Besides, the device could tolerate high bending radius up to 2.5 mm and survive 1000 bending cycles under 4 mm radius, exhibiting high mechanical stability. Later on, Sun and Pei further exploited the $MAPbBr_3$ QDs into the stretchable perovskite QLEDs for the first time, which was realized by conforming an ultrathin flexible perovskite QLED stack on a buckled elastomer substrate (Fig. 4.13) [104]. The device adopted an architecture of polyimide-silver nanowires composite substrate/PEDOT:PSS/PVK:TAPC/$MAPbBr_3$ QDs/TPBi/CsF/Al and achieved high luminance of 3187 cd/m$^2$ and CE of 9.2 cd/A. Moreover, the stretchable perovskite QLEDs maintained high performance under mechanical deformation up to 50% and tolerated repeated 1000 stretch-release cycles under 20% tensile strain.

## 4.4 Summary and Outlook

Perovskite QDs with intriguing optoelectronic characteristics, such as low cost, high color-rendering index, wide color gamut, energy saving, and compatibility with flexible/stretchable devices have drawn worldwide tractions in a wide range of photonic

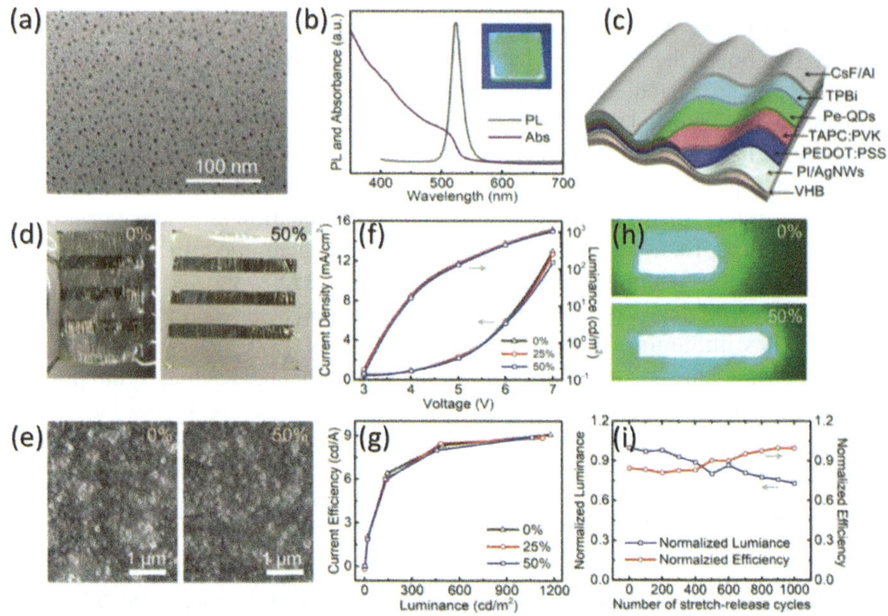


**Fig. 4.13** **a**, **b** TEM image and absorption/PL spectra of the $MAPbBr_3$ QDs. **c** Device structure of the stretchable QLEDs. **d** Optical photographs of the devices with 0% and 50% strains. **e** SEM images of $MAPbBr_3$ QDs under 0% and 50% strains. **f–g** J-V-L and CE characteristics of the QLEDs under 0, 25 and 50% strains. **h** Optical photographs of the perovskite QLEDs under strains of 0 and 50%. **i** Device robustness measured under stretch-release cycles between 0 and 20% strains. Reprinted with permission from Ref. [104]. Copyright 2019, Wiley-VCH

and optoelectronic applications. Applying perovskite QDs as phosphors in white-light down-converted LEDs or emitter materials in AM-QLEDs to obtain high efficiency and wide color gamut is significantly promising for next-generation high-definition information displays and high-quality lighting sources applications. In this chapter, the evolution of device performance of perovskite QDs-based LEDs has been demonstrated; while significant enhancement of device performance has been realized through appropriate ligand choice and effective ligand exchange together with modifying device architectures. Particularly, blue-emission perovskite QLEDs are also investigated with improved efficiency and stability in the past few years. For the commercial applications, further efforts are required to elucidate the physical mechanisms and improve the device performance of perovskite QLEDs. Besides, the poor stability of perovskite QLEDs in ambient and harsh conditions has attracted a wide concern, and simultaneously, several strategies have been explored to circumvent this issue, including compositional engineering, surface engineering, matrix encapsulation and device encapsulation. Although significant advances have been made in this regard, the lifetime of perovskite QLEDs remains far away from the commercial requirements. Further strategies to stabilize perovskite QDs and prevent their decomposition are in demand. Moreover, to avoid the toxicity problem, stable lead-free metal halide analogues with high optoelectronic performance are also needed. Furthermore, colloidal perovskite QDs can maintain functionality under tensile strains, opening new possibilities for their applications in flexible/stretchable electronics to satisfy the future development trend of QLEDs. Encapsulation technology of such kind of devices should be further investigated. We hope this book could accelerate new insights and further research studies to boost the development of perovskite QDs-based DC-LEDs and AM-QLEDs for vivid color displays and high-quality solid lighting sources. And we are convinced that a highly efficient and stable perovskite QDs-based LEDs will fulfil in the near future.

**Acknowledgements** This work was supported by the National Key Research and Development Program of China and the National Natural Science Foundation of China (NSFC) under Grants 2017YFB0404500, 61590930, 61825402, 61675085, 61805096, and 61705075.

## References

1. A. Kojima, K. Teshima, Y. Shirai, T. Miyasaka, J. Am. Chem. Soc. **131**, 6050–6051 (2009)
2. D.W. de Quilettes, S.M. Vorpahl, S.D. Stranks, H. Nagaoka, G.E. Eperon, M.E. Ziffer, H.J. Snaith, D.S. Ginger, Science **348**, 683–686 (2015)
3. N.J. Jeon, J.H. Noh, W.S. Yang, Y.C. Kim, S. Ryu, J. Seo, S.I. Seok, Nature **517**, 476–480 (2015)
4. H. Tsai, W. Nie, J.-C. Blancon, C.C. Stoumpos, R. Asadpour, B. Harutyunyan, A.J. Neukirch, R. Verduzco, J.J. Crochet, S. Tretiak, L. Pedesseau, J. Even, M.A. Alam, G. Gupta, J. Lou, P.M. Ajayan, M.J. Bedzyk, M.G. Kanatzidis, A.D. Mohite, Nature **536**, 312–316 (2016)
5. N. Li, Z. Zhu, C.-C. Chueh, H. Liu, B. Peng, A. Petrone, X. Li, L. Wang, A.K.Y. Jen, Adv. Energy Mater. **7**, 1601307 (2017)
6. T. Senden, R.J.A. van Dijk-Moes, A. Meijerink, Light: Sci. Appl. **7**, 8 (2018)
7. C.-C. Hou, H.-M. Chen, J.-C. Zhang, N. Zhuo, Y.-Q. Huang, R.A. Hogg, D.T. Childs, J.-Q. Ning, Z.-G. Wang, F.-Q. Liu, Z.-Y. Zhang, Light: Sci. Appl. **7**, 17170 (2018)
8. F. Ye, H. Chen, F. Xie, W. Tang, M. Yin, J. He, E. Bi, Y. Wang, X. Yang, L. Han, Energy Environ. Sci. **9**, 2295–2301 (2016)
9. Z.-K. Tan, R.S. Moghaddam, M.L. Lai, P. Docampo, R. Higler, F. Deschler, M. Price, A. Sadhanala, L.M. Pazos, D. Credgington, F. Hanusch, T. Bein, H.J. Snaith, R.H. Friend, Nat. Nanotechnol. **9**, 687–692 (2014)
10. H. Cho, S.H. Jeong, M.H. Park, Y.H. Kim, C. Wolf, C.L. Lee, J.H. Heo, A. Sadhanala, N. Myoung, S. Yoo, S.H. Im, R.H. Friend, T.-W. Lee, Science **350**, 1222–1225 (2015)
11. M. Liu, M.B. Johnston, H.J. Snaith, Nature **501**, 395 (2013)
12. W.S. Yang, B.-W. Park, E.H. Jung, N.J. Jeon, Y.C. Kim, D.U. Lee, S.S. Shin, J. Seo, E.K. Kim, J.H. Noh, S.I. Seok, Science **356**, 1376–1379 (2017)
13. T.M. Brenner, D.A. Egger, L. Kronik, G. Hodes, D. Cahen, Nat. Rev. Mater. **1**, 15007 (2016)
14. N.J. Jeon, H. Na, E.H. Jung, T.-Y. Yang, Y.G. Lee, G. Kim, H.-W. Shin, S.Il Seok, J. Lee, J. Seo, Nat. Energy **3**, 682–689 (2018)
15. W. Wei, Y. Zhang, Q. Xu, H. Wei, Y. Fang, Q. Wang, Y. Deng, T. Li, A. Gruverman, L. Cao, J. Huang, Nat. Photonics **11**, 315 (2017)
16. S. Shrestha, R. Fischer, G.J. Matt, P. Feldner, T. Michel, A. Osvet, I. Levchuk, B. Merle, S. Golkar, H. Chen, S.F. Tedde, O. Schmidt, R. Hock, M. Ruhrig, M. Goken, W. Heiss, G. Anton, C.J. Brabec, Nat. Photonics **11**, 436 (2017)

17. Q. Chen, J. Wu, X. Ou, B. Huang, J. Almutlaq, A.A. Zhumekenov, X. Guan, S. Han, L. Liang, Z. Yi, J. Li, X. Xie, Y. Wang, Y. Li, D. Fan, D.B.L. Teh, A.H. All, O.F. Mohammed, O.M. Bakr, T. Wu, M. Bettinelli, H. Yang, W. Huang, X. Liu, Nature **561**, 88–93 (2018)
18. X. Dai, Z. Zhang, Y. Jin, Y. Niu, H. Cao, X. Liang, L. Chen, J. Wang, X. Peng, Nature **515**, 96 (2014)
19. L.E. Brus, J. Chem. Phys. **80**, 4403 (1984)
20. L.C. Schmidt, A. Pertegás, S. González-Carrero, O. Malinkiewicz, S. Agouram, G. Mínguez Espallargas, H.J. Bolink, R.E. Galian, J. Pérez-Prieto, J. Am. Chem. Soc. **136**, 850 (2014)
21. L. Protesescu, S. Yakunin, M.I. Bodnarchuk, F. Krieg, R. Caputo, C.H. Hendon, R.X. Yang, A. Walsh, M.V. Kovalenko, Nano Lett. **15**, 3692 (2015)
22. S. Gonzalez-Carrero, R.E. Galian, J. Pérez-Prieto, J. Mater. Chem. A **3**, 9187 (2015)
23. J. Pan, L.N. Quan, Y. Zhao, W. Peng, B. Murali, S.P. Sarmah, M. Yuan, L. Sinatra, N.M. Alyami, J. Liu, E. Yassitepe, Z. Yang, O. Voznyy, R. Comin, M.N. Hedhili, O.F. Mohammed, Z.H. Lu, D.H. Kim, E.H. Sargent, O.M. Bakr, Adv. Mater. **28**, 8718 (2016)
24. F. Zhang, H. Zhong, C. Chen, X.-G. Wu, X. Hu, H. Huang, J. Han, B. Zou, Y. Dong, ACS Nano **9**, 4533 (2015)
25. Q. Shan, J. Song, Y. Zou, J. Li, L. Xu, J. Xue, Y. Dong, B. Han, J. Chen, H. Zeng, Small **13**, 1701770 (2017)
26. F. Yan, J. Xing, G. Xing, L. Quan, S.T. Tan, J. Zhao, R. Su, L. Zhang, S. Chen, Y. Zhao, A. Huan, E.H. Sargent, Q. Xiong, H.V. Demir, Nano Lett. **18**, 3157 (2018)
27. D. Han, M. Imran, M. Zhang, S. Chang, X.-G. Wu, Xi Zhang, J. Tang, M. Wang, S. Ali, X. Li, G. Yu, J. Han, L. Wang, B. Zou, H. Zhong, ACS Nano **12**, 8808 (2018)
28. J. Song, T. Fang, J. Li, L. Xu, F. Zhang, B. Han, Q. Shan, H. Zeng, Adv. Mater. **30**, 1805409 (2018)
29. T. Chiba, Y. Hayashi, H. Ebe, K. Hoshi, J. Sato, S. Sato, Y.-J. Pu, S. Ohisa, J. Kido, Nat. Photonics **12**, 681 (2018)
30. H.-C. Wang, Z. Bao, H.-Y. Tsai, A.-C. Tang, R.-S. Liu, Small **14**, 1702433 (2018)
31. N.C. George, K.A. Denault, R. Seshadri, Annu. Rev. Mater. Res. **43**, 481–501 (2013)
32. G. Li, Y. Tian, Y. Zhao, J. Lin, Chem. Soc. Rev. **44**, 8688–8713 (2015)
33. Z. Xia, A. Meijerink, Chem. Soc. Rev. **46**, 275–299 (2017)
34. Y. Wei, Z. Cheng, J. Lin, Chem. Soc. Rev. **48**, 310–350 (2019)
35. Y. Ohno, Fourth international conference on solid state lighting (2004)
36. Y. Cui, T. Song, J. Yu, Y. Yang, Z. Wang, G. Qian, Adv. Funct. Mater. **25**, 4796 (2015)
37. B.W. D'Andrade, S.R. Forrest, Adv. Mater. **16**, 1585 (2004)
38. J. Zhang, R. Hu, X. Yu, B. Xie, X. Luo, Opt. Laser Technol. **88**, 161–165 (2017)
39. H.C. Yoon, J.H. Oh, S. Lee, J.B. Park, Y.R. Do, Sci. Rep. **7**, 2808 (2017)
40. C.C. Lin, R.S. Liu, J. Phys. Chem. Lett. **2**, 1268 (2011)
41. B. Wang, H. Lin, J. Xu, H. Chen, Y. Wang, A.C.S. Appl, Mater. Interfaces **6**, 22905 (2014)
42. Y.Q. Li, J.E.J. van Steen, J.W.H. van Krevel, G. Botty, A.C.A. Delsing, F.J. DiSalvo, G. de With, H.T. Hintzen, J. Alloys Compd. **417**, 273–279 (2006)
43. K. Uheda, N. Hirosaki, Y. Yamamoto, A. Naito, T. Nakajima, H. Yamamoto, Electrochem. Solid-State Lett. **9**, H22–H25 (2006)
44. H. Zhu, C.C. Lin, W. Luo, S. Shu, Z. Liu, Y. Liu, J. Kong, E. Ma, Y. Cao, R.-S. Liu, X. Chen, Nat. Commun. **5**, 4312 (2014)
45. J. Zhou, F. Huang, H. Lin, Z. Lin, J. Xu, Y. Wang, J. Mater. Chem. C **4**, 7601 (2016)
46. G. Li, H. Wang, T. Zhang, L. Mi, Y. Zhang, Z. Zhang, W. Zhang, Y. Jiang, Adv. Funct. Mater. **26**, 8478 (2016)
47. H.C. Yoon, H. Kang, S. Lee, J.H. Oh, H. Yang, Y.R. Do, A.C.S. Appl, Mater. Interfaces **8**, 18189 (2016)
48. H. Huang, B. Chen, Z. Wang, T.F. Hung, A.S. Susha, H. Zhong, A.L. Rogach, Chem. Sci. **7**, 5699–5703 (2016)
49. H. Huang, Q. Xue, B. Chen, Y. Xiong, J. Schneider, C. Zhi, H. Zhong, A.L. Rogach, Angew. Chem. Int. Ed. **56**, 9571–9576 (2017)
50. P. Pust, P.J. Schmidt, W. Schnick, Nat. Mater. **14**, 454 (2015)

51. D.-H. Kim, J. Ryu, S.-Y. Cho, Appl. Phys. A **102**, 79–83 (2011)
52. J.R. Haynes, Phys. Rev. **98**, 1866–1868 (1955)
53. H.C. Wang, S.Y. Lin, A.C. Tang, B.P. Singh, H.C. Tong, C.Y. Chen, Y.C. Lee, T.L. Tsai, R.S. Liu, Angew. Chem. Int. Ed. **55**, 7924–7929 (2016)
54. K. Ma, X.-Y. Du, Y.-W. Zhang, S. Chen, J. Mater. Chem. C **5**, 9398–9404 (2017)
55. Q. Zhou, Z. Bai, W.G. Lu, Y. Wang, B. Zou, H. Zhong, Adv. Mater. **28**, 9163–9168 (2016)
56. E. Jang, S. Jun, H. Jang, J. Lim, B. Kim, Y. Kim, Adv. Mater. **22**, 3076 (2010)
57. X. Chen, F. Zhang, Y. Ge, L. Shi, S. Huang, J. Tang, Z. Lv, L. Zhang, B. Zou, H. Zhong, Adv. Funct. Mater. **28**, 1706567 (2018)
58. Y.-M. Chen, Y. Zhou, Q. Zhao, J.-Y. Zhang, J.-P. Ma, T.-T. Xuan, S.-Q. Guo, Z.-J. Yong, J. Wang, Y. Kuroiwa, C. Moriyoshi, H.-T. Sun, A.C.S. Appl, Mater. Interfaces **10**, 15905–15912 (2018)
59. G. Yang, Q. Fan, B. Chen, Q. Zhou, H. Zhong, J. Mater. Chem. C **4**, 11387 (2016)
60. T. Xuan, X. Yang, S. Lou, J. Huang, Y. Liu, J. Yu, H. Li, K.-L. Wong, C. Wang, J. Wang, Nanoscale **9**, 15286–15290 (2017)
61. Z. Liu, Y. Zhang, Y. Fan, Z. Chen, Z. Tang, J. Zhao, Y. Lv, J. Lin, X. Guo, J. Zhang, X. Liu, ACS Appl. Mater. Interfaces **10**, 13053–13061 (2018)
62. Z. Li, L. Kong, S. Huang, L. Li, Angew. Chem. Int. Ed. **129**, 8246–8250 (2017)
63. H. Hu, L. Wu, Y. Tan, Q. Zhong, M. Chen, Y. Qiu, D. Yang, B. Sun, Q. Zhang, Y. Yin, J. Am. Chem. Soc. **140**, 406–412 (2017)
64. H.C. Yoon, S. Lee, J.K. Song, H. Yang, Y.R. Do, A.C.S. Appl, Mater. Interfaces **10**, 11756–11767 (2018)
65. L. Rao, Y. Tang, C. Yan, J. Li, G. Zhong, K. Tang, B. Yu, Z. Li, J.Z. Zhang, J. Mater. Chem. C **6**, 5375–5383 (2018)
66. Y.E. Panfil, M. Oded, U. Banin, Angew. Chem. Int. Ed. **57**, 2 (2018)
67. X. Dai, Y. Deng, X. Peng, Y. Jin, Adv. Mater. **29**, 1607022 (2017)
68. M.V. Kovalenko, L. Protesescu, M.I. Bodnarchuk, Science **358**, 745 (2017)
69. M. Liu, H. Zhang, D. Gedamu, P. Fourmont, H. Rekola, A. Hiltunen, S.G. Cloutier, R. Nechache, A. Priimagi, P. Vivo, Small **15**, 1900801 (2019)
70. D.N. Dirin, L. Protesescu, D. Trummer, I.V. Kochetygov, S. Yakunin, F. Krumeich, N.P. Stadie, M.V. Kovalenko, Nano Lett. **16**, 5866 (2016)
71. A. Swarnkar, V.K. Ravi, A. Nag, ACS Energy Lett. **2**, 1089 (2017)
72. Y.-F. Li, J. Feng, H.-B. Sun, Nanoscale **11**, 19119–19139 (2019)
73. L.N. Quan, F.P.G. de Arquer, R.P. Sabatini, E.H. Sargent, Adv. Mater. **2018**, 180 (1996)
74. H. Huang, F. Zhao, L. Liu, F. Zhang, X.-G. Wu, L. Shi, B. Zou, Q. Pei, H. Zhong, A.C.S. Appl, Mater. Interfaces **7**, 28128 (2015)
75. J. Xing, F. Yan, Y. Zhao, S. Chen, H. Yu, Q. Zhang, R. Zeng, H.V. Demir, X. Sun, A. Huan, Q. Xiong, ACS Nano **10**, 6623 (2016)
76. X. Lin, X. Wu, J. Zheng, H. Rui, Z. Zhang, Y. Hua, S. Yin, IEEE Electron. Device Lett. **40**, 1479–1482 (2019)
77. Y.-H. Kim, G.-H. Leed, Y.-T. Kim, C. Wolf, H.J. Yun, W. Kwon, C.G. Park, T.-W. Lee, Nano Energy **38**, 51 (2017)
78. T. Zhang, H. Li, P. Yang, J. Wei, F. Wang, H. Shen, D. Li, F. Li, Organ. Electron. **68**, 76–84 (2019)
79. J. Song, J. Li, X. Li, L. Xu, Y. Dong, H. Zeng, Adv. Mater. **27**, 7162 (2015)
80. J. Li, L. Xu, T. Wang, J. Song, J. Chen, J. Xue, Y. Dong, B. Cai, Q. Shan, B. Han, H. Zeng, Adv. Mater. **29**, 1603885 (2017)
81. X. Zhang, H. Lin, H. Huang, C. Reckmeier, Y. Zhang, W.C. Choy, A.L. Rogach, Nano Lett. **16**, 1415 (2016)
82. G. Li, F.W.R. Rivarola, N.J.L.K. Davis, S. Bai, T.C. Jellicoe, F. de la Peña, S. Hou, C. Ducati, F. Gao, R.H. Friend, N.C. Greenham, Z.-K. Tan, Adv. Mater. **28**, 3528 (2016)
83. X. Zhang, B. Xu, J. Zhang, Y. Gao, Y. Zheng, K. Wang, X.W. Sun, Adv. Funct. Mater. **26**, 4595 (2016)
84. Y. Li, Y. Zeng, R. Pan, S. Chen, W. Cao, X.W. Sun, K. Wang, Chem. Mater. **29**, 5168 (2017)

85. Q. Khan, A. Subramanian, G. Yu, K. Maaz, D. Li, R. Ur, R. Sagar, K. Chen, W. Lei, B. Shabbir, Y. Zhang, Nanoscale **11**, 5021–5029 (2019)
86. J. Pan, Y. Shang, J. Yin, M.D. Bastiani, W. Peng, I. Dursun, L. Sinatra, A.M. El-Zohry, M.N. Hedhili, A.-H. Emwas, O.F. Mohammed, Z. Ning, O.M. Bakr, J. Am. Chem. Soc. **140**, 562 (2018)
87. X. Zheng, S. Yuan, J. Liu, J. Yin, F. Yuan, W.-S. Shen, K. Yao, M. Wei, C. Zhou, K. Song, B.-B. Zhang, Y. Lin, M.N. Hedhili, N. Wehbe, Y. Han, H.-T. Sun, Z.-H. Lu, T.D. Anthopoulos, O.F. Mohammed, E.H. Sargent, L.-S. Liao, O.M. Bakr, ACS Energy Lett. (2020)
88. G. Pan, X. Bai, W. Xu, X. Chen, Y. Zhai, J. Zhu, H. Shao, N. Ding, L. Xu, B. Dong, Y. Mao, H. Song, ACS Appl. Mater. Interfaces (2020)
89. X. Zhang, H. Liu, W. Wang, J. Zhang, B. Xu, K.L. Karen, Y. Zheng, S. Liu, S. Chen, K. Wang, X.W. Sun, Adv. Mater. **29**, 1606405 (2017)
90. X. Zhang, D. Han, C. Wang, I. Muhammad, F. Zhang, A. Shmshad, X. Xue, W. Ji, S. Chang, H. Zhong, Adv. Opt. Mater. **7**, 1900774 (2019)
91. M.F. Aygüler, M.D. Weber, B.M.D. Puscher, D.D. Medina, P. Docampo, R.D. Costa, J. Phys. Chem. C **119**, 12047 (2015)
92. L. Protesescu, S. Yakunin, S. Kumar, J. Bär, F. Bertolotti, N. Masciocchi, A. Guagliardi, M. Grotevent, I. Shorubalko, M.I. Bodnarchuk, C.-J. Shih, M.V. Kovalenko, ACS Nano **11**, 3119 (2017)
93. Z. Shi, S. Li, Y. Li, H. Ji, X. Li, D. Wu, T. Xu, Y. Chen, Y. Tian, Y. Zhang, C. Shan, G. Du, ACS Nano **12**, 1462 (2018)
94. Q. Shan, J. Li, J. Song, Y. Zou, L. Xu, J. Xue, Y. Dong, C. Huo, J. Chen, B. Han, H. Zeng, J. Mater. Chem. C **5**, 4565 (2017)
95. W. Deng, X. Xu, X. Zhang, Y. Zhang, X. Jin, L. Wang, S.-T. Lee, J. Jie, Adv. Funct. Mater. **26**, 4797 (2016)
96. Z. Xiao, W. Meng, J. Wang, D.B. Mitzi, Y. Yan, Mater. Horiz. **4**, 206 (2017)
97. F. Giustino, H.J. Snaith, ACS Energy Lett. **1**, 1233 (2016)
98. N. Rajeev Kumar, R. Radhakrishnan, Mater. Lett. **227**, 289 (2018)
99. A. Jain, O. Voznyy, E.H. Sargent, J. Phys. Chem. C **18**, 28 (2017)
100. T.C. Jellicoe, J.M. Richter, H.F.J. Glass, M. Tabachnyk, R. Brady, S.E. Dutton, A. Rao, R.H. Friend, D. Credgington, N.C. Greenham, M.L. Böhm, J. Am. Chem. Soc. **138**, 2941 (2016)
101. W. van der Stam, J. J. Geuchies, T. Altantzis, K.H.W. van den Bos, J. D. Meeldijk, S.V. Aert, S. Bals, D. Vanmaekelbergh and C. de Mello Donega, J. Am. Chem. Soc. **139**, 4087 (2017)
102. X. Zhang, W. Cao, W. Wang, B. Xu, S. Liu, H. Dai, S. Chen, K. Wang, X.W. Sun, Nano Energy **30**, 511 (2016)
103. F. Zhao, D. Chen, S. Chang, H. Huang, K. Tong, C. Xiao, S. Chou, H. Zhong, Q. Pei, J. Mater. Chem. C **5**, 531 (2017)
104. Y.-F. Li, S.-Y. Chou, P. Huang, C. Xiao, X. Liu, Y. Xie, F. Zhao, Y. Huang, J. Feng, H. Zhong, H.-B. Sun, Q. Pei, Adv. Mater. **31**, 1807516 (2019)

# Chapter 5
# Polarized Emission from Perovskite Nanocrystals

**Qiang Jing and Kai Wang**

**Abstract** Polarization is a fundamental property of light and polarized light, including linearly polarized light, circularly polarized light and elliptically polarized light, plays a very important role in optoelectronic devices and systems, such as liquid crystal display (LCD), labeled biological systems, optical quantum computers, 3D display system, polarization photodetection, etc. Halide perovskites nanocrystals (PNCs), in terms of perovskite quantum dots, nanowires, nanorods, or nanoplates, have been found to demonstrate linearly and circularly polarized luminescence. This chapter provides an overview of the polarized emission mechanism of PNCs and linearly polarized luminescence of PNCs, including several techniques have been developed to align anisotropic PNCs into ordered patterns, for example, electrospinning, mechanical stretching, template assisted growth, solution-phase growth and chemical vapor deposition growth. Finally, the circularly polarized luminescence of PNCs and its potential applications also have been introduced.

Polarization is a fundamental property of light and it means the property of certain electromagnetic radiations in which the direction and magnitude of the vibrating electric field are related in a specified way. For a light wave, its vibrating electric vector associated with each wave is perpendicular to the direction of propagation. According to its property, polarized light usually can be divided into linearly polarized light, circularly polarized light and elliptically polarized light. As shown in Fig. 5.1, a beam of unpolarized light consists of waves moving in the same direction with their electric vectors pointed in random orientations about the axis of propagation. However, plane polarized light, also known as linearly polarized light, consists of waves in which the direction of vibration is the same for all waves. In circular polarization the electric vector rotates about the direction of propagation as the wave progresses. Circularly polarized light includes left-handed circularly polarized light and right-handed circularly polarized light. In elliptically polarized light, the vector

Q. Jing · K. Wang (✉)
Department of Electrical and Electronic Engineering, Southern University of Science and Technology, Shenzhen, China
e-mail: wangk@sustc.edu.cn



Y. Zhou and Y. Wang (eds.), *Perovskite Quantum Dots*, Springer Series in Materials Science 303, https://doi.org/10.1007/978-981-15-6637-0_5

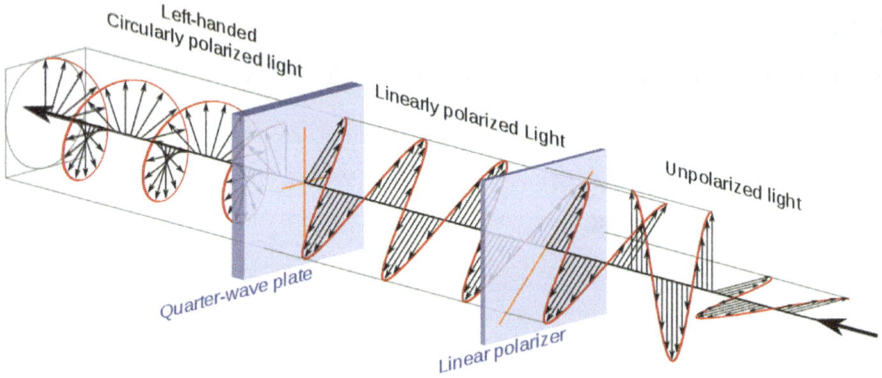


**Fig. 5.1** Schematic of kinds of polarized light. (www.miniphysics.com)

also rotates about the propagation direction, but the amplitude of the vector changes so that the projection of the vector on a plane at right angles to the propagation direction describes an ellipse. Moreover, the polarization ratio is a parameter that can measure the degree of polarization of linearly polarized light, which can be defined as:

$$P = \frac{I_{\max} - I_{\min}}{I_{\max} + I_{\min}}$$

where $\boldsymbol{P}$ is the polarization ratio, $\boldsymbol{I_{max}}$ and $\boldsymbol{I_{min}}$ are respectively the maximum and minimum photoluminescence (PL) intensities of the light that are collected after passing through the optical media.

Polarized light plays a very important role in optoelectronic devices and systems, such as liquid crystal display (LCD), labeled biological systems, optical quantum computers, 3D display system, polarization photodetection, etc. Especially, luminescent materials with polarized emission are essential for the polarization applications. For example, as shown in Fig. 5.2, there are two orthorhombic polarizers exist at two surfaces of an LCD cell. However, light emitted from LED backlighting is unpolarized, and at most only half of the light from LEDs can pass through the LCD cell, limiting the enhancement of brightness. If we turn the unpolarized light from LEDs into linearly polarized light through aligned quantum rods with very low absorption loss, more than half light is able to pass through the LCD cell, enhancing the brightness and decreasing the power consumption of the display, which is very important for extending endurance time of kinds of mobile electronics and wearable electronics, such as watch, cell phone, pad, laptop, etc. Therefore, the investigation of polarization properties of luminescent materials and ways to enhance the polarized emission is necessary and important.

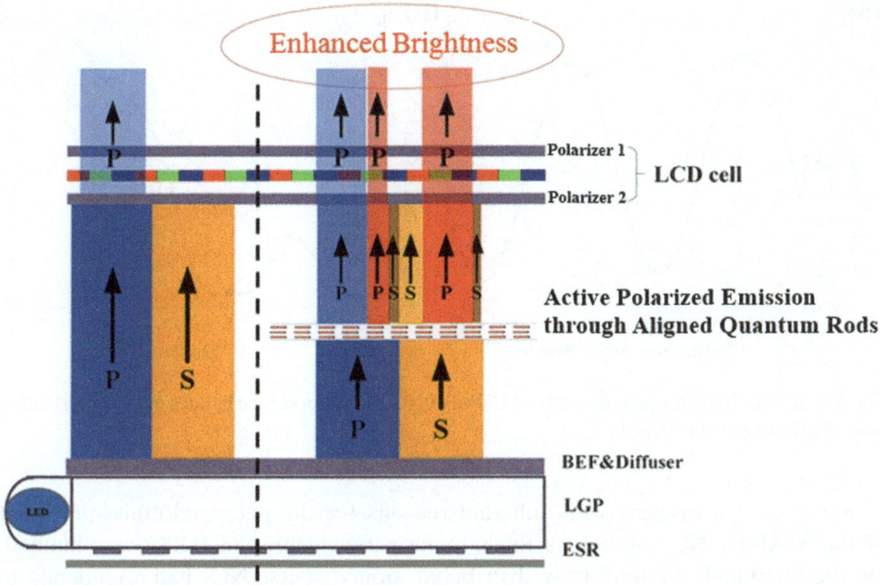


**Fig. 5.2** Schematic of enhancing the brightness of LCD module by aligned quantum rods with polarized emission

Halide perovskites nanocrystals (PNCs), in terms of perovskite quantum dots, nanowires, nanorods, or nanoplates, have been found to demonstrate linearly and circularly polarized luminescence. This chapter will introduce the polarization of PNCs.

## 5.1 Polarized Emission Mechanism of PNCs

Polarized emission of PNCs is mainly induced by three mechanisms: (1) intrinsic and extrinsic transition dipole moment anisotropy, (2) optical matrix element anisotropy induced by quantum size effect, and (3) excitation field anisotropy due to dielectric confinement of optical electric field.

Wang et al. [1] first investigated the polarization characteristics of all-inorganic $CsPbX_3$ NCs (X = Cl, Br, or I) (Fig. 5.3). They found that $CsPbI_3$ NCs were distorted from the cubic structure caused by bigger iodine atom, the distorted cubic structure broke the space inversion symmetry and the resulted asymmetrical structure was responsible for the polarization property of $CsPbI_3$ NCs. Another possible reason for the polarized emission they found was that $CsPbX_3$ NCs were highly ionized, which facilitated self-organization forming ordered packing structure in hexane. As a result, the polarization of $CsPbI_3$ reached 0.36 in hexane and 0.40 in film.

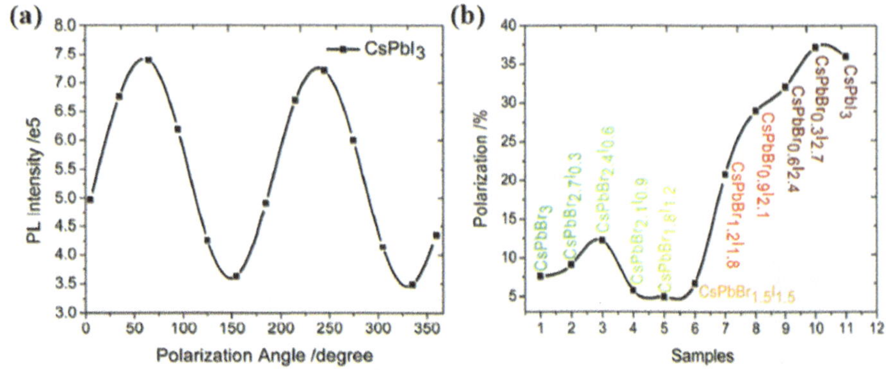


**Fig. 5.3** **a** Polarized emission property of $CsPbI_3$ PNCs. **b** Effects of the halides on the polarization ratio of all-inorganic PNCs [1]

Shi et al. [2] discussed the inherent reasons for the polarized emission effect of the $MAPbI_3$ NC solution, a linear polarization degree of 0.28 was obtained. On the one hand, the uniformly distributed monodisperse NCs had a tendency to show self-assembled agglomerates in a large area, which wound favor polarization without further alignment processes. On the other hand, the special crystal structure of $MAPbI_3$ NCs differed from the relatively stable cubic structure, such as $MAPbBr_3$, which would result in the migration of ionic species in the crystal because the $MAPbI_3$ perovskites were distorted from the iodine atoms. In theory, it would break the space inversion symmetry under the "built-in" potential of perovskites and produced the dipole moment, so the resulting asymmetrical structure was another important reason for the polarization effect of $MAPbI_3$ NCs.

Liu et al. [3] reported a systematic study on the optical properties of single $FAPbBr_3$ NCs, they observed a polarization ratio of 0.21 in the optical emission of single $FAPbBr_3$ NCs at the room temperature, which then enhanced to ~0.92 at the cryogenic temperature of 4 K. In the single $FAPbBr_3$ NC, the crystal structure would be distorted by the large organic cation of $FA^+$, causing transitions from the cubic to the tetragonal and orthorhombic phases at the room and cryogenic temperatures, respectively. The linearly-polarized optical emission of a single $FAPbBr_3$ NC originated from the 1D dipole moment of a bright-exciton state. When this dipole moment was perpendicular (parallel) to the objective optical axis, a maximum (minimum) value of degree of linear polarization would be expected. Compared to the room-temperature value, the significantly enhanced degree of linear polarization reflected the reduced thermal mixing between the exciton states carrying 2D and 1D dipole moments at the cryogenic temperature.

Anisotropic PNCs should induce polarization of both the absorption and emission due to the unequal electric field strength along the different particle dimensions. By adapting the strategies from the synthesis of conventional II–VI semiconductor NCs, anisotropic PNCs, such as nanoplatelets, nanowires, and nanorods have been successfully fabricated and their polarized emission properties also have been studied.

## 5.2 Linearly Polarized Luminescence of PNCs

Although single NC exhibit inherently polarized emissions, the polarization effects can be reduced when they are randomly distributed. Fabrication of well-aligned PNCs is essential to make use of their polarization properties. Several techniques have been developed to align anisotropic PNCs into ordered patterns, including electrospinning, mechanical stretching, template assisted growth, solution-phase growth and chemical vapor deposition growth.

### *5.2.1 Electrospinning Technique*

Electrospinning is a process that can produce nano-to micro-scaled fibers in diameter via the high potential difference between the nozzle of a solution container and a grounded and conductive collector. It is a facile and inexpensive process to obtain nanosized polymeric and composite fibers. Moreover, it provides full control of the morphology of individual fibers (fiber diameter, shape, and surface chemistry). Electrospinning provides not only a way to obtain uniaxial fibers but also can be used to align nanostructures. The alignment of the fibers provides a 1D hierarchical structure that can potentially be used in a polarization filter.

For example, as shown in Fig. 5.4a, Demir et al. [4] first synthesized $CsPbBr_3$ nanowires (NWs) with a width of ~4 nm and length of hundreds of nanometers, then the NWs were integrated into electrospun polyurethane (PU) fibers for electrospinning. As shown in Fig. 5.4b, the resulting composite fibers are almost perfectly aligned with a polarization ratio of 0.3. This method can be applied to different all inorganic halide perovskites such as $CsPbI_3$ to further enhance the polarization of the composite system.

In addition, as shown in Fig. 5.4c, Zhong et al. [5] reported an in-situ fabrication of PNCs in polyvinylalcohol (PVA) nanofibers ($MAPbX_3$@PVA nanofibers) through electrospinning a perovskite precursor solution. With the content of the precursors increased, the resulting $MAPbBr_3$ nanocrystals in PVA matrix changed the shape from ellipsoidal to pearl-like, and finally into rods-like. By changing composition of halides, the emission spectra of $MAPbX_3$ PNCs can be tuned from 400 to 700 nm. From single nanofiber PL measurements, they demonstrated correlations between polarization anisotropy and shape evolution of the $MAPbBr_3$ NCs in PVA nanofiber. The emission polarization ratio increased from $0.36 \pm 0.03$ to $0.53 \pm 0.06$ when their aspect ratios increased from 1.5 to 6. By adapting parallel positioned conductor strips with an air gap as a collection electrode, an emission polarization ratio of $0.42 \pm 0.03$ was achieved for the $MAPbX_3$@PVA film of 0.5 cm × 2 cm in size.

Moreover, as shown in Fig. 5.4d, by adapting the electrospinning technique, Zhong et al. [6] further fabricated inch-size aligned polymer nanofibers films with embedded $MAPbBr_3$ using a folded aluminium foil as collector. They found the $MAPbBr_3$ gradually crystallized and had a high-density arrangement along the nanofibers,

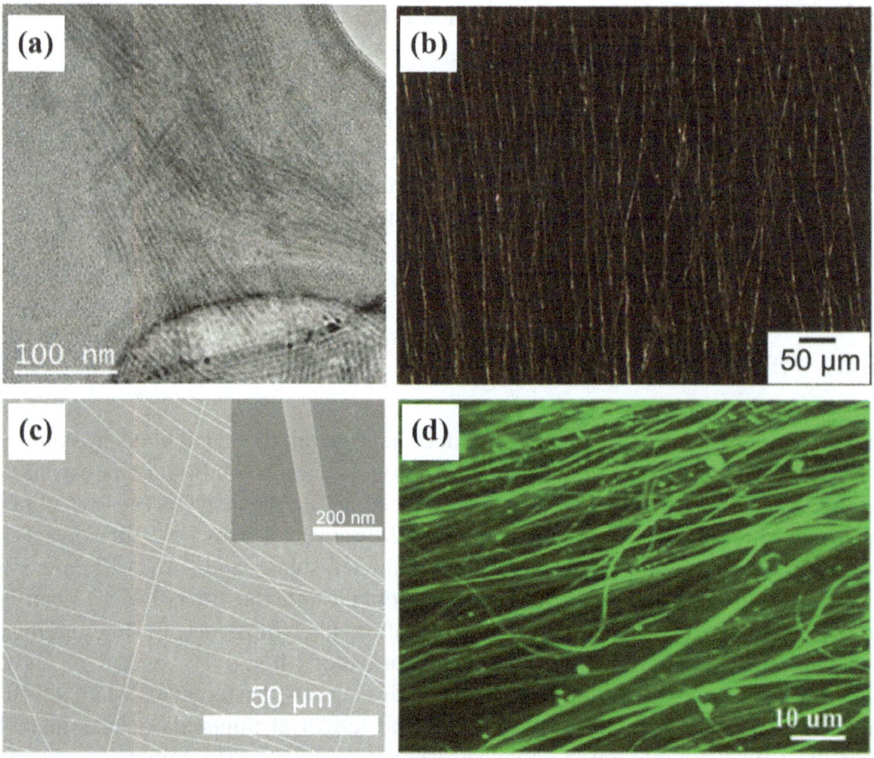


**Fig. 5.4** **a** TEM images of the NWs cast from dichloromethane dispersion showing a uniform thickness of 4 nm. **b** Optical microscopy image of aligned fibers [4]. **c** SEM image of $MAPbBr_3$@PVA nanofibers, inset shows an enlarged view of a single nanofiber [5]. **d** Laser scanning confocal microscopy image of nanofibers with $MAPbBr_3$ of 10% [6]

at the same time, adding appropriate amount of cyanoethyl cellulose can improve the dispersion and PLQYs of embedded $MAPbBr_3$ nanofibers. Finally, an emission polarization ratio was achieved 0.43 for inch-sized aligned nanofiber films. The combination of easy fabrication and high polarization ration in these aligned nanofiber film make them potential candidates for polarization enhanced photonic and optoelectronic devices.

### 5.2.2 *Mechanical Stretching Technique*

Mechanical stretching of polymer matrix is another mostly used method to align nanostructures. Lu et al. [7] fabricated $MAPbX_3$/PVDF polymer composite films with strong polarized PL by combining the in-situ fabrication with controllable mechanical stretching. The fabrication of $MAPbX_3$/PVDF-based composite polymer film can be divided into two stages, PVDF first crystallized into colorless films along with the evaporation of N, N-dimethylformamide (DMF) (time node 1). With the evaporation of residual DMF, the nucleation of perovskite initialized and small-sized nanoparticles formed in the polymeric matrix due to the spatial confinement effects (time node 2). The stretched films at time node 1 and 2 can be seen in Fig. 5.5a. As shown in Fig. 5.5b, the polarization ratio is about 0.16~0.22 for the composite films stretched at time node 1 and 0.24~0.28 for the composite films stretched at the time node 2.

In addition, as shown in Fig. 5.6a, Raja et al. [8] first synthesized $CsPbBr_3$ NWs, then they mixed solutions of NWs and SEBS (poly(styrene-ethylene-butylene-styrene)) in toluene to form composites. SEBS is a widely used low-cost structural tri-block copolymer which is a thermoplastic elastomer, it combines high formability and recyclability with high toughness. SEBS is highly ductile and can be drop-casted into thick samples that are optically clear. As shown in Fig. 5.6b, after mechanical stretching of the drop-casted NWs-SEBS composite film, a polarization ratio of $0.44 \pm 0.05$ in the aligned nanowire-polymer composite emission was achieved, as compared to $0.08 \pm 0.01$ in the case of nanocube-polymer composites, where the two samples are prepared and measured under the same conditions.

Moreover, He et al. [9] reported an in-situ catalyst-free strategy to synthesize $CsPbBr_3$ nanorods in polymer matrix (NRs-PS) with high stability. As shown in Fig. 5.7a, the NRs-PS film was heated with a heat gun in air and steadily stretched to a final length that is 8 times longer. Well aligned NRs were found with 37% of them lying in the $\pm 1.5°$ range and 80% lying in $\pm 4.5°$ range (Fig. 5.7b) with respect to the stretching direction. As shown in Fig. 5.7c, the minor polarization of the NRs-PS film before stretching was attributed to uneven distribution of NR orientations. The PL angular dependence of the stretch-aligned NRs-PS complies well with a sinusoidal shape, with the maximum along the rod axis and the obtained highest polarization ratio value around 0.23.

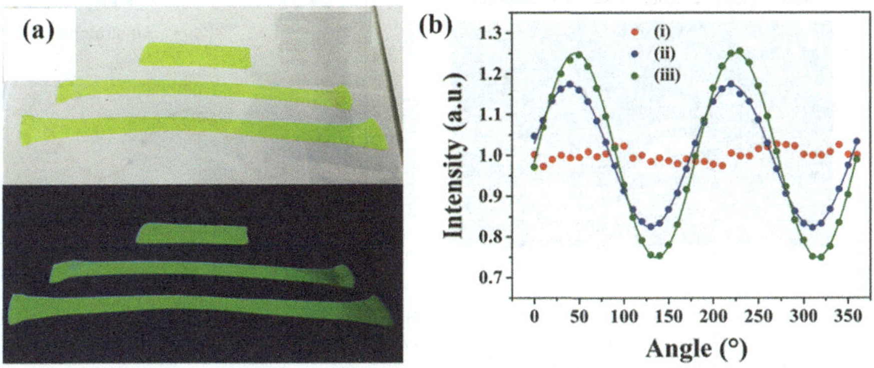


**Fig. 5.5** **a** Photographs of films under daylight and UV irradiation. Both from top to bottom: films without stretching, films stretched at time node 1, and stretched at time node 2. **b** The polarization measurements of the composite films (the mass fraction of perovskite precursors is 8%). (i) Films without stretching; (ii) films stretched at time node 1; (iii) films stretched at time node 2 [7].

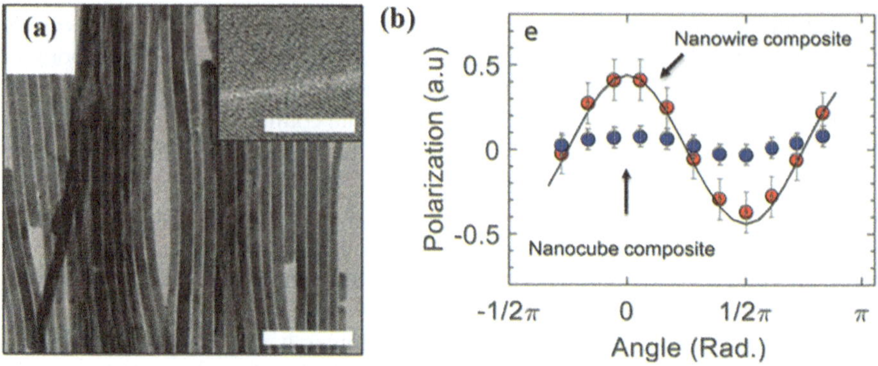


**Fig. 5.6** **a** TEM micrograph of one-dimensional single nanowires. Scale bar is 100 and 10 nm for inset. **b** Calculated polarization as a function of polarized analyzer angle of nanowire-polymer fiber (red) and nanocube-polymer fiber (blue) demonstrating significant polarized emission (P = 0.44) in the nanowire composite parallel to the fiber axis, as compared to the nanocube composite (P = 0.08) [8].

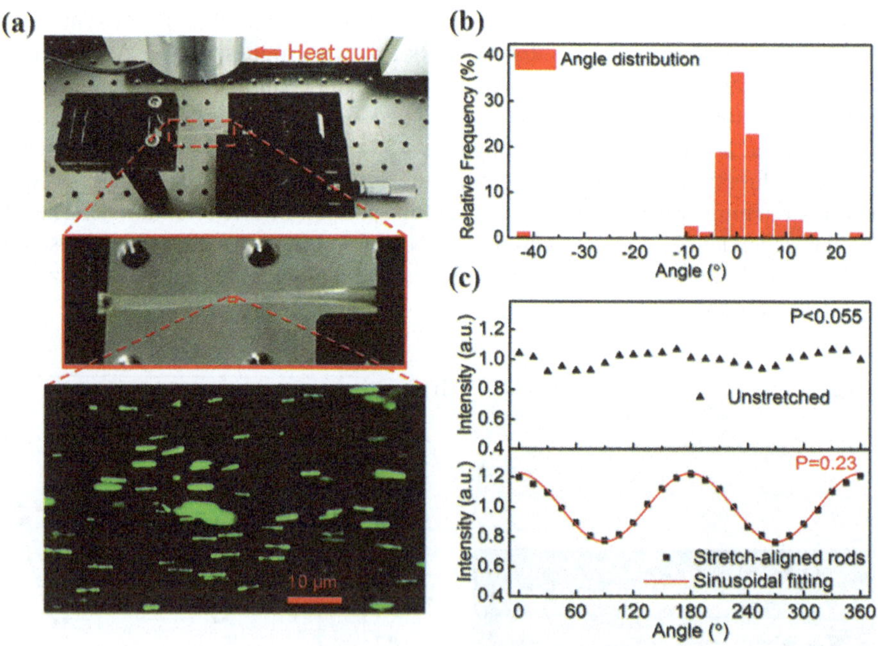


**Fig. 5.7** Stretch-alignment of CsPbBr3 NRs-PS [9]. **a** Up: setup for stretching the NRs-PS film, Middle: enlarged NRs-PS sample illustration, Down: Fluorescent microscopic image of stretch-aligned NRs. **b** Angle distribution of stretch-aligned NRs (in image a down). Elongation ratio is 8 to 1. **c** PL polarization of unstretched (up) and stretchaligned (down) NRs-PS

### 5.2.3 Template Assisted Growth Technique

Template-assisted growth is an efficient way to grow anisotropic perovskite nanorods or nanowires. In the process of growth, the confined effect of the template can lead the one-dimensional growth of crystals along the arrays, forming high-quantity nanowires.

Deng et al. [10] reported a method to produce highly aligned single-crystalline $MAPb(I_{1-x}Br_x)_3$ (x = 0, 0.1, 0.2, 0.3, 0.4) NWs arrays with continuously tunable absorption edges from 680 to 780 nm. Periodically aligned SU-8 photoresist stripes on the $SiO_2$/Si substrate was used as the template, and the perovskite NW arrays were fabricated by a fluid-guided antisolvent vapor assisted crystallization (FGAVC) method. The template was first dipped into $MAPbI_3$/DMF solution and placed in the saturated antisolvent ($CH_2Cl_2$) vapor. The diffusion of antisolvent lead to the precipitation of $MAPbI_3$ nanocrystals along the sides of SU-8 photoresist strips, the schematic diagram can be seen in Fig. 5.8a.

In addition, Liu et al. [11] prepared NWs arrays using imprinted PDMS as the template, as the evaporation of DMF of MAX and $PbX_2$ stock solution, the template confinement effect led to the directional growth of $MAPbX_3$ along the arrays. Well defined dimensions and uniform geometries enabled individual NWs to function as high-quality FP lasers with almost identical optical modes and similarly low lasing

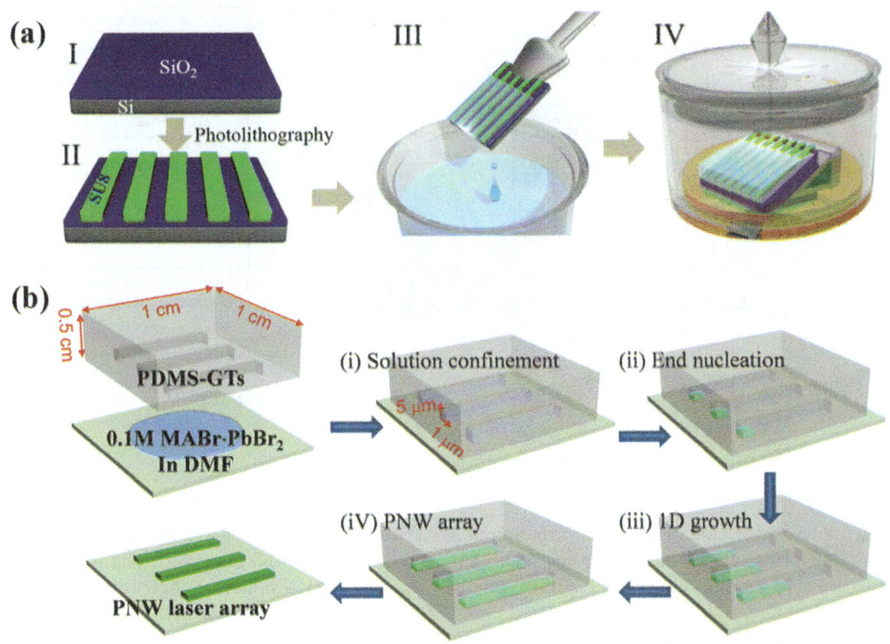


**Fig. 5.8** **a** Schematic illustration of FGAVC method for the fabrication of $MAPbI_3$ NW array [10]. **b** schematic diagram of the preparation procedures [11]

thresholds (220 $nJ/cm^2$), allowing igniting them simultaneously as a laser array, the schematic diagram can be seen in Fig. 5.8b. The uniform and regular alignment of the anisotropic PNCs fabricated by template-assisted method appropriate for electronic and photonic devices like laser in on-chip system or detector, but for a display system, the difficulty of large scale alignment of this method may limit its application.

### *5.2.4 Solution-Phase Growth Technique*

Solution-phase growth process affords low-temperature solution processing, which can be adjusted by the perovskite precursors. Liu et al. [12] reported the synthesis of uniform and luminescent $MAPbBr_3$ nanoplatelets (NPLs) through self-organization of pre-formed colloidal $MAPbBr_3$ nanodots (NDs). As shown in Fig. 5.9a, shape transformation from spherical NDs to square or rectangular NPLs can be achieved by keeping the preformed colloidal nanocrystals at a high concentration (3.5 mg/mL) for 3 days. Theoretical simulations and analysis reveal that the dipole–dipole interactions with a realignment of dipolar vectors are the main factors driving the self-organization process, which is a new mechanism for the assembly of NCs. A linearly polarization degree of 0.11 was obtained from the stretched $MAPbBr_3$ NPL-embedded polymer

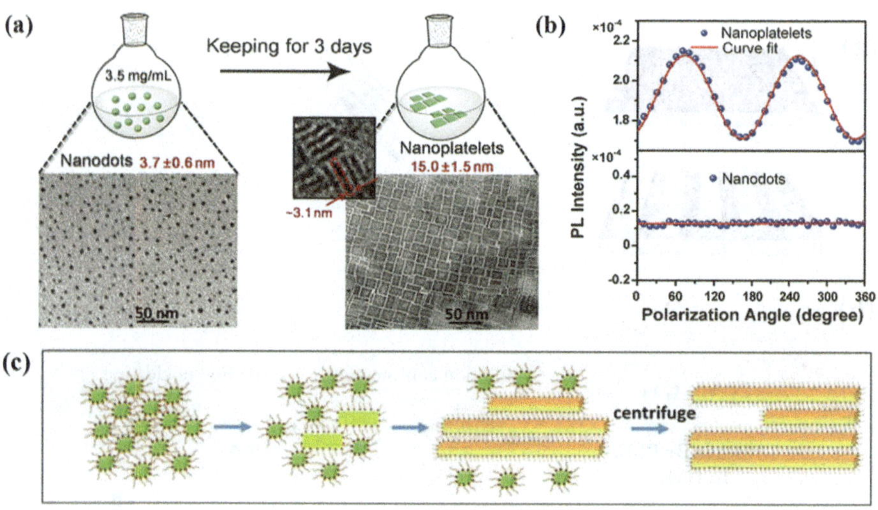


**Fig. 5.9** **a** Illustration of the transformation of $MAPbBr_3$ NCs from NDs to NPLs, and the corresponding TEM images of the original NDs (left) and the resultant NPLs (right). The inset is the enlarged side view of the NPLs [12]. **b** PL intensity as a function of polarizer angle (blue spheres), and can be fitted to sinusoidal function (red line) [12]. **c** Schematic illustration of the transformation of precursor powders into NWs by ultrasonication, in which initially formed $CsPbBr_3$ nanocubes are transformed into NWs [13]

films (Fig. 5.9b). This work opens up an alternative way to realize shape control of halide perovskite NCs and provides potential polarized emissive materials.

In addition, as shown in Fig. 5.9c, Tong et al. [13] demonstrated a single-step ligand-assisted synthesis of precursor powders into single-crystalline colloidal $CsPbBr_3$ NWs of uniform thickness. The optical properties of the NWs can be tuned across the entire visible range by varying the halide (Cl, Br and I) composition through a subsequent halide ion exchange. Single particle studies show that these NWs exhibit strong polarized emission with a polarization anisotropy of 0.36.

Bekenstein et al. [14] synthesized $CsPbBr_3$ perovskite NPLs by a colloidal method. $PbBr_2$ was solubilized in octadecene (ODE) with oleic acid and oleylamine, and then Cs-oleate was injected at elevated temperatures (90~130 °C) to form NPLs. By adjusting the reaction temperature, a different shape can be obtained. Reactions conducted at 150 °C produce mostly symmetrical nanocubes with green-color PL emission, at 130 °C, cyan emitting lower symmetry nanoplates can be obtained, at 90–100 °C, very thin NPLs were detected along with lamellar structures ranging 200~300 nm in length. The HRTEM and XRD characteristics are shown in Fig. 5.10a. The NPLs show high photoluminescence (PLQYs up to 84%).

Moreover, Li et al. [15] realized the direct synthesis of $CsPbBr_3$ NRs with an average width of around 5 nm and average lengths of 10.8 and 23.2 nm (Fig. 5.10b),

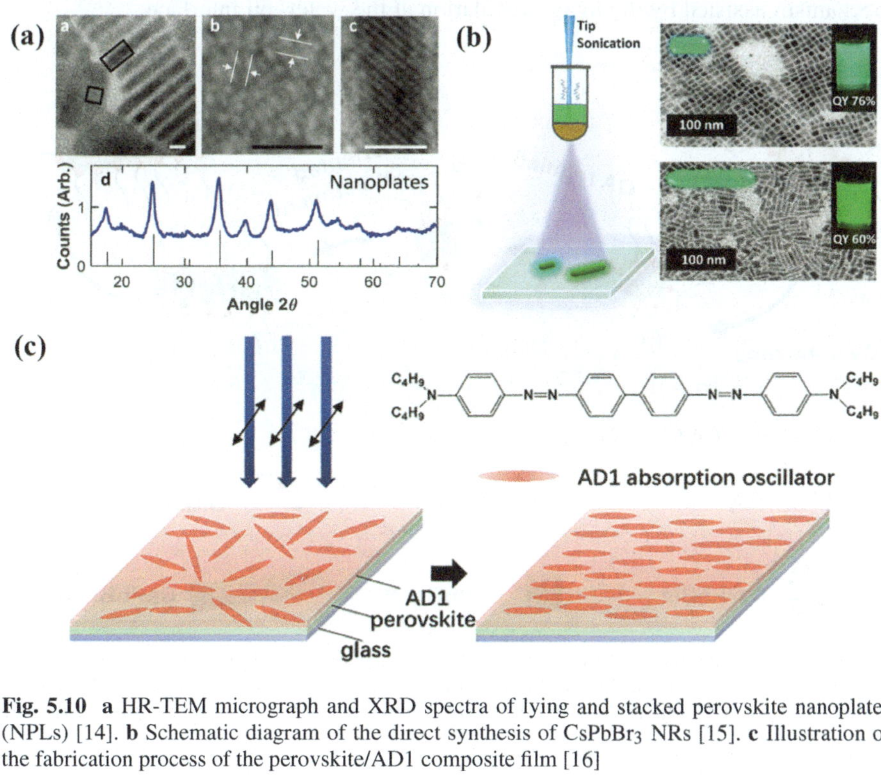


**Fig. 5.10** **a** HR-TEM micrograph and XRD spectra of lying and stacked perovskite nanoplates (NPLs) [14]. **b** Schematic diagram of the direct synthesis of $CsPbBr_3$ NRs [15]. **c** Illustration of the fabrication process of the perovskite/AD1 composite film [16]

respectively, in two samples, which show a high photoluminescence quantum yield of 60–76% and reasonably high emission anisotropy of about 0.2 for longer rods. Both $CsPbCl_3$ and $CsPbI_3$ NRs with similar dimensions have then been derived from the $CsPbBr_3$ NRs by anion-exchange reactions. Remarkably, the synthesis of the NRs has been achieved in polar alcohols, a class of solvents not usually found to be beneficial in classical perovskite nanoparticle synthesis. This work not only offers the possibility to control the shape of chemically synthesized perovskite nanocrystals but also constitutes the hitherto less common strategy of synthesizing perovskite nanoparticles in polar rather than nonpolar or only weakly polar solvents.

In addition, as shown in Fig. 5.10c, Zhang et al. [16] demonstrated a strong linearly polarized PL from an azobenzene dye (AD1) functionalized full-color perovskites. The PL of the red (R), green (G), and blue (B) perovskite/AD1 composite films shows an exceptional high degree of polarization ration of 0.86~0.93. In addition, the AD1 functionalized strategy is successfully applied to generate linearly polarized electroluminescence (EL) from an inverted perovskite light emitting diode, showing a polarization ration of 0.47.

As shown in Fig. 5.11, Jing et al. [17] reported highly luminescent $CsPbBr_3$ NRs with aspect ratios around 2~10 through chemical transformation from $Cs_4PbBr_6$ nanocrystals. Detailed structural characterization shows that the NRs are cubic phase with uniform growth direction. The NRs are formed through an oriented-attachment mechanism assisted by the ligand-regulation at the water–oil interface.

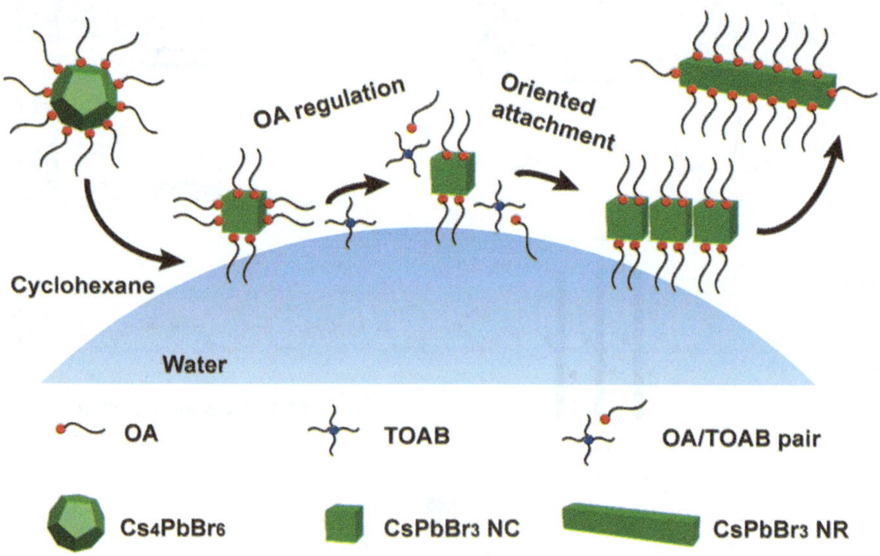


**Fig. 5.11** Schematic illustration of the formation mechanism of the $CsPbBr_3$ NRs. The NRs are assembled by the oriented attachment of intermediately formed nanocubes at the water–oil interface. Cation $(C_8H_{17})_4\ N^+$ from TOAB can effectively combine with the OA anion to form an ion pair [17]

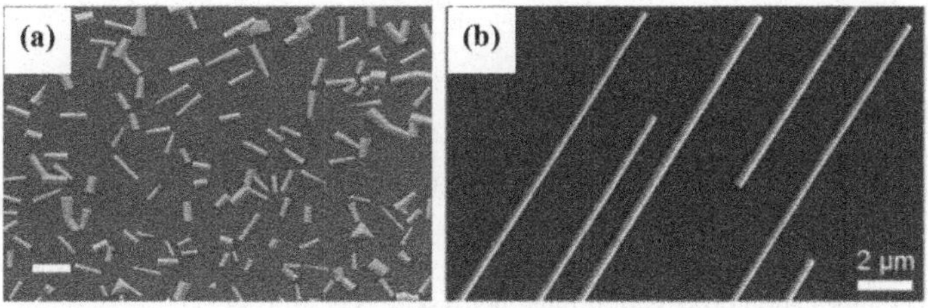


**Fig. 5.12** **a** SEM image of as-grown $CsPbBr_3$ perovskite nanostructures (scale bar, 10 μm) [18]. **b** SEM images of the directional $CsPbBr_3$ NWs growth along faceted M-plane sapphire [19]

### *5.2.5 Chemical Vapor Deposition Growth Technique*

Contrary to solution-phase growth technique, the nanostructure prepared by chemical vapor deposition (CVD) technique has higher crystallization quality, lower defect density and better repeatability. As shown in Fig. 5.12a, Zhou et al. [18] reported the growth of high-quality all-inorganic cesium lead halide alloy perovskite micro/nanorod with complete composition tuning by vapor-phase deposition. The as-grown micro/nanorods are single-crystalline with a triangular cross section and show strong photoluminescence which can be tuned from 415 to 673 nm by varying the halide composition. Furthermore, these single-crystalline perovskite micro/nanorods themselves function as effective Fabry–Perot cavities for nanoscale lasers.

In addition, as shown in Fig. 5.12b, Shoaib et al. [19] reported the successful vapor growth of ultralong directional $CsPbBr_3$ perovskite NWs using the graphoepitaxial effect on annealed M-plane sapphire substrates. The wires have a diameter of several hundred nanometers, with lengths up to several millimeters. Microstructure characterization shows that $CsPbBr_3$ NWs are high-quality single crystals, with smooth surfaces and well-defined cross section.

Gao et al. [20] synthesized well-defined $CsPbX_3$ ($X = Br$, Cl) NW arrays with average thickness (T) below 15 nm (7 nm) on muscovite mica substrate by a facile CVD method. Systematical size-dependence of emission anisotropy is explored and the results show that the polarization ratio undergoes invariant ($T < 40$ nm), decrease ($T \approx 50 \sim 140$ nm), and oscillation regions ($T > 140$ nm), respectively. Strong emission anisotropy with polarization ratio up to $\approx 0.78$ is demonstrated in the nanowires with $T < 40$ nm due to the electrostatic dielectric confinement. With the increasing of thickness, the polarization ratio remarkably reduces monotonously to $\approx 0.17$ until $T \approx 140$ nm.

### *5.2.6 Other Techniques*

Inkjet printing is regarded as the forefront of microelectronics fabrication technology due to its simple, flexible, rapid and cost-effective manufacturing approach, which

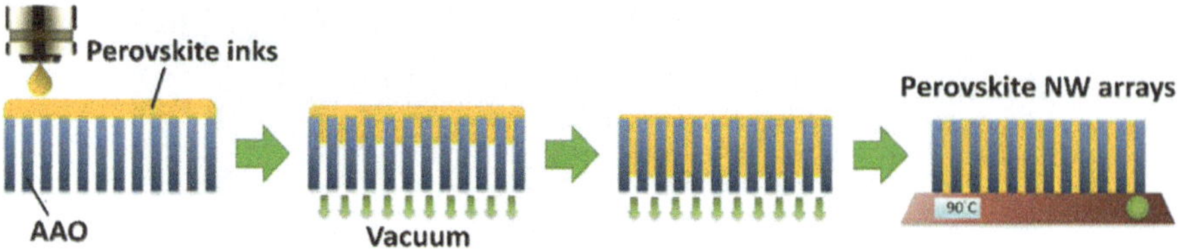


**Fig. 5.13** The fabrication process of printed perovskite NW arrays [21]

provides an efficient way for mass production. As shown in Fig. 5.13, Lin et al. [21] realized the anisotropic growth of perovskite and successfully fabricated high quality and orientationally aligned perovskite NW array films for extremely polarized emission by incorporating inkjet printing technique and nanoporous anodic aluminum oxide (AAO) confinement strategy. The AAO displays high transmittance > 70%, thereby allowing optical excitation and light emission of perovskite NWs. Using blue GaN light-emitting diodes (LEDs) as optical pumping source, the printed $MAPbBr_3$ NW arrays demonstrate intense emission featuring polarization ratio up to 0.84. In addition, Zhou et al. [22] developed a nanocomposite ink composed of perovskite nanowire bundles suspended in a cylindrical microphase-forming polystyrene-polyisoprene-polystyrene (SIS) block copolymer matrix. Then they fabricated macroscopically polarized films with unidirectionally aligned $CsPbBr_3$ perovskite NWs with a polarization anisotropy of 0.36 by direct ink writing (DIW), an extrusion-based 3D-printing method.

## 5.3 Circularly Polarized Luminescence of PNCs

Chiral functional materials with circularly polarized luminescence (CPL) have attracted increasing attention in recent years for their wide potential applications in 3D optical displays, encrypted transmission, storage of information, bio-encoding, photoelectric devices, chiroptical materials and even catalysts for asymmetric photochemical synthesis. Currently, CPL is often produced from nonpolarized light by the use of filters, which leads to a loss of light intensity, and in the case of light emitting devices, to more complex and bulky device architectures. Therefore, there is a need for developing efficient devices that can emit CPL directly, which would be more efficient in terms of device simplicity (thickness), production costs, and energy consumption, compared with conventional circular polarized devices employing polarizing filters.

While the palette of CPL-active materials obtained from organic nanostructures is broad, the variety of inorganic materials used for chiral emissive material is limited. To date, only a few CPL-active chiral inorganic nanomaterials were developed either through the tedious synthesis or with unregulated chiroptical properties. It has been demonstrated that chalcogenide semiconductor nanomaterials such as CdS or CdTe quantum dots (QDs) or quantum rods (QRs), capped by chiral amino acids or protein, would adopt a chiral crystal structure, resulting in CPL activity. Other emissive inorganic nanomaterials, such as nanostructured ZnO films and silver nanoclusters, have been reported to exhibit CPL after modification with chiral capping reagent. Recently, circularly polarized luminescence of PNCs also has been widely researched.

As shown in Fig. 5.14, Shi et al. [23] successfully endowed the PNCs with a novel CPL through supramolecular coassembly of achiral PNCs and chiral lipid gelator. Various colorful achiral PNCs could coassembled with chiral gelator to form cogels, exhibiting colorful circularly polarized emission. The CPL sign could be regulated by the supramolecular chirality of the chiral gels. This work provides a new perspective and general approach for fabricating CPL active emissive nanomaterials.

In addition, as shown in Fig. 5.15, Ma et al. [24] reported the synthesis of chiral 2D perovskite (S- and R-MBA)$_2$PbI$_4$ crystals (MBA = $C_6H_5C_2H_4NH_3$) with a high degree of circularly polarized PL and sensitive CPL detection. The as-synthesized (S- and R-MBA)$_2$PbI$_4$ crystals both have a needle shape and exhibit strong oppositely

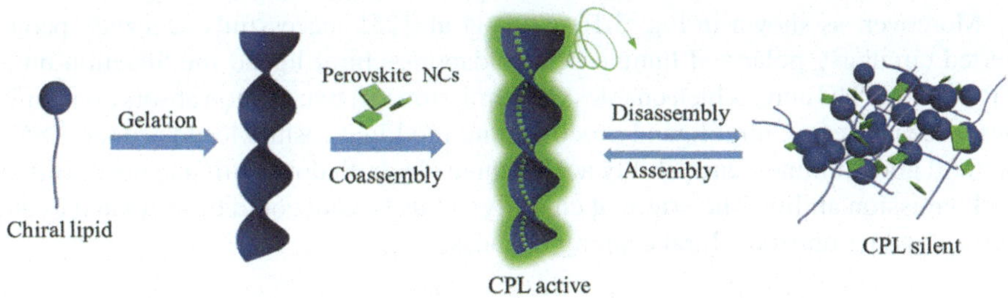


**Fig. 5.14** Illustration of the possible coassembly induced chirality of perovskite NCs in chiral gels [23]

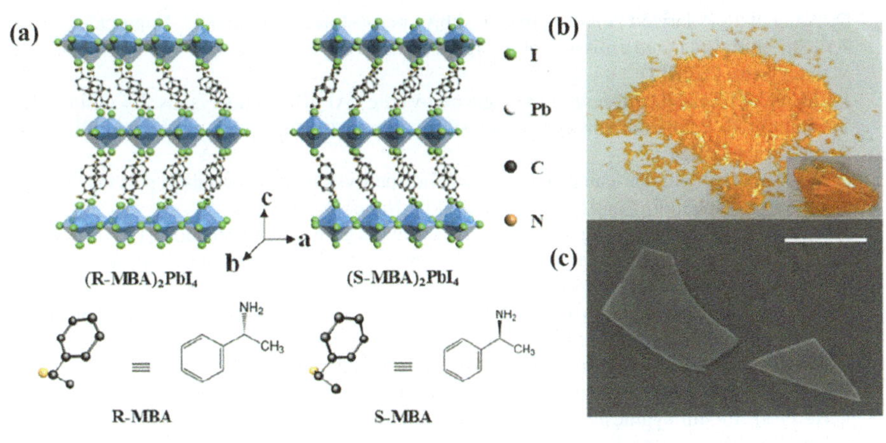


**Fig. 5.15** **a** Schematic illustrations of crystal structures of (R- and S-MBA)$_2$PbI$_4$. **b**, **c** optical microscope image **b** and SEM image **c** of (R-MBA)$_2$PbI$_4$ [24]

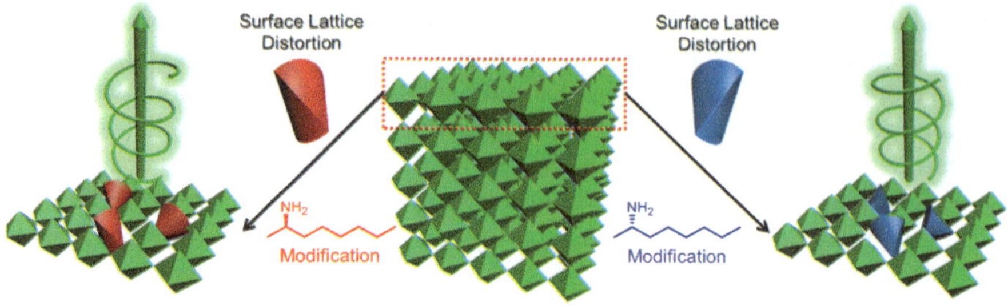


**Fig. 5.16** Schematic illustration of the origin of chirality in chiral $CsPbBr_3$ perovskite [25]

signed CD signals. The average degree of circularly polarized PL is around 10% with a maximum value of 17.6% at 77 K, nearly six times larger than that in the reduced dimensional chiral perovskites at 2 K, which is 3% at zero magnetic field. The degree of the circularly polarized PL reduces with temperature, implying the reduction of the chirality transfer or the increase of the spin flip as the temperature increases.

Moreover, as shown in Fig. 5.16, Chen et al. [25] successfully realized upconverted circularly polarized luminescence adapting chiral ligand modification on a TPA-active PNC core, which enables CPL emission in a two-photon absorption luminescence process. Chiral ligand α-octylamine-modified cesium lead bromide PNCs showed good luminescence ability accompanied by both downshift and upconverted CPL emission ability. The origin of chirality in this system could be attributed to the surface induction from chiral capping ligandss.

## References


1. D. Wang et al. Polarized emission from $CsPbX_3$ perovskite quantum dots. Nanoscale (2016)
2. Z.F. Shi et al. Polarized emission effect realized in $CH_3NH3PbI_3$ perovskite nanocrystals. J. Mater. Chem. C (2017)
3. J. Liu et al. Polarized emission from single perovskite $FAPbBr_3$ nanocrystals. J. Lumin (2020)
4. T. Güner et al. Polarized emission from $CsPbBr_3$ nanowire embedded-electrospun PU fibers. Nanotechnol (2018)
5. L. Meng et al. In-situ fabricated anisotropic halide perovskite nanocrystals in polyvinylalcohol nanofibers: shape tuning and polarized emission. Nano. Res. (2019)
6. Y. Wang et al. Inch-size aligned polymer nanofibers films with embedded $CH_3NH_3PbBr_3$ nanocrystals: electrospinning fabrication using a folded aluminium foil as collector. Nanotechnol. (2019)
7. W.G. Lu et al. Strong polarized photoluminescence from stretched perovskite nanocrystal embedded polymer composite films. Adv. Opt. Mater. (2017)
8. S.N. Raja et al. Encapsulation of perovskite nanocrystals into macroscale polymer matrices: enhanced stability and polarization. ACS Appl. Mater. Interfaces (2016)
9. J. He et al. In-situ synthesis and macroscale alignment of $CsPbBr_3$ perovskite nanorods in polymer matrix. Nanoscale (2018)
10. W. Deng et al. Ultrahigh-responsivity photodetectors from perovskite nanowire arrays for sequentially tunable spectral measurement. Nano. Lett. (2017)

11. P. Liu et al. Organic-inorganic hybrid perovskite nanowire laser arrays. ACS Nano (2017)
12. L. Liu et al. Colloidal synthesis of $CH_3NH_3PbBr_3$ nanoplatelets with polarized emission through self-organization. Angew. Chem. Int. Ed (2017)
13. Y. Tong et al. Precursor powders-to-$CsPbX_3$ perovskite nanowires: one-pot synthesis, growth mechanism and oriented self-assemblies. Angew. Chem. Int. Ed (2017)
14. Y. Bekenstein et al. Highly luminescent colloidal nanoplates of perovskite cesium lead halide and their oriented assemblies. J. Am. Chem. Soc (2015)
15. Y. Li et al. Using polar alcohols for the direct synthesis of cesium lead halide perovskite nanorods with anisotropic emission. ACS Nano (2019)
16. J. Zhang et al. Strong linearly polarized photoluminescence and electroluminescence from halide perovskite/azobenzene dye composite film for display applications. Adv. Opt. Mater (2020)
17. Q. Jing et al. Highly luminescent $CsPbBr_3$ nanorods synthesized by a ligand-regulated reaction at the water–oil interface. J. Mater. Chem. C (2019)
18. H. Zhou et al. Vapor growth and tunable lasing of band gap engineered cesium lead halide perovskite micro/nanorods with triangular cross section. ACS Nano (2016)
19. M. Shoaib et al. Directional growth of ultralong $cspbbr_3$ perovskite nanowires for high-performance photodetectors. J. Am. Chem. Soc (2017)
20. Y. Gao et al. Ultrathin $CsPbX_3$ nanowire arrays with strong emission anisotropy. Adv. Mater (2018)
21. C.H. Lin et al. Giant optical anisotropy of perovskite nanowire array films. Adv. Funct. Mater (2020)
22. N. Zhou et al. Perovskite nanowire–block copolymer composites with digitally programmable polarization anisotropy. Sci. Adv (2019)
23. Y. Shi et al. Endowing perovskite nanocrystals with circularly polarized luminescence. Adv. Mater (2018)
24. J. Ma et al. Chiral 2D perovskites with a high degree of circularly polarized photoluminescence. ACS Nano (2019)
25. W. Chen et al. Two-photon absorption-based upconverted circularly polarized luminescence generated in chiral perovskite nanocrystals. J. Phys. Chem. Lett (2019)

# Chapter 6
# Characterization of Lead Halide Perovskites Using Synchrotron X-ray Techniques

**Lijia Liu and Zhaohui Dong**

**Abstract** In this chapter, we are going to introduce several characterization methods that utilize synchrotron X-rays as the probing source for understanding electronic structure, crystal structure, and optical properties of lead halide perovskite materials. We will show how X-ray absorption fine structure (XAFS) can be used to understand the dissociation mechanism, to identify the presence of defect, and to analyze the influence of dopant introduction on structural modification. We will demonstrate that an X-ray excited optical luminescence (XEOL) in combination with XAFS analysis helps to understand the luminescence mechanism of doped perovskite. We will also present the use of synchrotron X-ray diffraction (XRD) to study the creation of new crystal phases induced by high pressure and the phase transformation of perovskite in situ. The working principle of each technique will be introduced, followed by examples from recently published research articles.

## 6.1 Introduction

Lead halide perovskites (LHPs) has attracted great attention since its first successful application in solid-state dye-sensitized solar cells in 2012 [1]. Potential applications of LHPs have been extended to many fields, such as solar cells, light-emitting devices, catalysis, and lasers [2, 3]. LHP has a perovskite-type structure of $ABX_3$, in which A is a small organic alkylamine cation (typically $CH_3NH_3^+$ or $NH_2CH = NH_2^+$) or a heavy alkaline cation (i.e. $Cs^+$), B is $Pb^{2+}$, and X is a halide ($Cl^-$, $Br^-$, or $I^-$). The band gap of LHPs can be easily tuned throughout the entire visible wavelength by adjusting the composition of halide ion species and/or the nanocrystal size through

---

L. Liu (✉)
Department of Chemistry, University of Western Ontario, 1151 Richmond Street, London, ON N6A 5B7, Canada
e-mail: lijia.liu@uwo.ca

Z. Dong
Shanghai Synchrotron Radiation Facility (SSRF), Shanghai Advanced Research Institute, Chinese Academy of Science, Shanghai 201204, People's Republic of China
e-mail: dongzhaohui@zjlab.org.cn



Y. Zhou and Y. Wang (eds.), *Perovskite Quantum Dots*, Springer Series in Materials Science 303, https://doi.org/10.1007/978-981-15-6637-0_6

quantum size-effect [4]. The I-based LHPs ($ABI_3$) absorbs red light, making them suitable for solar cell applications. The power conversion efficiency of LHP solar cell has significantly improved since it was first used in dye-sensitized solar cells in 2009 [5]. As the light absorbing material in solar cell, LHPs have demonstrated suitable band gap and trap-state density, high absorption coefficient, and long carrier diffusion lengths [6, 7]. Br- and Cl- based LHPs, on the other hand, have slightly larger band gap, and are more widely used for light-emitting applications. As a light-emitter, they exhibit single-color emission, high photoluminescence quantum yield (PLQY) [3]. By introducing additional divalent or trivalent ions as dopants, the PLQY of intrinsic LHP can be further enhanced [8, 9], and some dopant ions introduce additional radiative recombination paths, making LHP a dual-color light emitter [10, 11].

Although LHPs have brought exciting opportunities for the optoelectronics market, commercial applications are still challenging. Stability and toxicity are two major concerns LHP-based devices. The organometal LHPs are unstable under light, heat, and oxygen [12, 13], and almost all LHPs are vulnerable to moisture [14]. For example, bare methylammonium lead iodide ($MAPbI_3$) thin films are only stable in an oxygen- and water-free environment. Bare $CsPbBr_3$ suffers serious photoluminescence loss under sustained operational conditions [15, 16]. The presence of Pb in LHP also raises environmental safety concern if LHP-based devices are produced in large quantity. Although attempt has been made to synthesis Pb-free perovskites for optoelectronic applications, the device performance is not satisfactory [17, 18]. To address these issues, a deep understanding on the fundamental properties of LHPs is essential. In this chapter, we will provide a summary on a series of characterization techniques for LHP research that utilize X-rays generated from synchrotron radiation.

Synchrotron radiation is produced by accelerating electrons approaching the speed of light. When the electrons are directed in a curved orbit under magnetic field, electromagnetic radiation is produced tangentially to the orbit. The energy of the electromagnetic radiation covers a wide range, from near-infrared to hard X-ray, and a monochromator is used for energy selection to meet specific research needs. For research that involves the use of X-rays, synchrotron X-ray has several advantages over the traditional X-ray sources: (1) the photon energy is tunable from soft X-ray (a few tens of eV) to hard X-rays (several tens of keV); (2) the photon flux is more than 10 orders of magnitude than traditional X-rays; (3) it can provide highly collimated beam with spatial resolution down to several microns or in some cases, several tens of nanometers; (4) the photons can be either linearly or circularly polarized which allows polarization-dependent studies; (5) electrons can be produced in bunches with the use of radio frequency cavities inserted in their travelling orbit. The emitted light in this case is pulsed with a few nanoseconds between pulses, which allows time-resolved spectroscopy.

Several synchrotron X-ray spectroscopic and microscopic techniques have been successfully applied on studying the fundamental properties of LHP materials and LHP-based optoelectronic devices. For example, X-ray absorption fine structure

(XAFS) is used to study the degradation mechanism and local structural defect [19–21]; X-ray excited optical luminescence (XEOL) spectroscopy provides evidence on the luminescence mechanism of ion-doped LHP [22]; X-ray diffraction (XRD) acquired under high pressure can track the crystal structure evolution during phase transformation and allows identification of new crystal phases [23, 24]; X-ray photoelectron spectroscopy (XPS) studies the surface composition and perovskite-ligand charge transfer [25, 26]; X-ray scattering helps to understand formation and orientation of crystalline grains in perovskite films [27, 28]. In the following, we will focus on three X-ray probing techniques, which are XAFS, XEOL and XRD, and demonstrate how synchrotron X-ray helps in materials analysis with particular emphasis on LHP materials. The goal of this chapter is to provide insight on using these spectroscopic techniques to guide future material synthesis and device design.

## 6.2 Techniques and Applications

### *6.2.1 X-ray Absorption Spectroscopy*

X-ray absorption spectroscopy (XAS) measures the absorption behavior of materials as a function of incident photon energy. XAS can only be performed using a synchrotron source, where tunable X-ray source is provided. There are several excellent resources that provide detailed theory background as well as data analysis protocols [29, 30]. The readers are welcome to refer to them for an in-depth learning on such technique. In the following, we outline the basic working principle of XAS and the key knowledge required to understand the results discussed in literatures.

#### 6.2.1.1 Working Principle

XAS measures the absorption coefficient of a material upon interacting with incident X-rays. The general absorption principle follows the Beer's law as:

$$I_t = I_0 e^{-\mu l} \tag{6.1}$$

in which $I_0$ is the intensity of incident photons, $I_t$ is the intensity of the transmitted photons, $l$ is sample thickness and $\mu$ is the absorption coefficient. The absorption coefficient $\mu$ depends on (1) the type of atoms in the materials, (2) how they are distributed and their bonding environment, and (3) the polarization and energy of the X-rays. Experimentally it is determined by comparing the intensities of incident light and the ones passing the material with known thickness.

For a monatomic sample, once the incident energy ($E_{ex}$) is sufficient to promote an electron from a core level to a previously occupied energy level, there will be an abrupt increase in the absorption coefficient. The absorption spectrum (Fig. 6.1a)

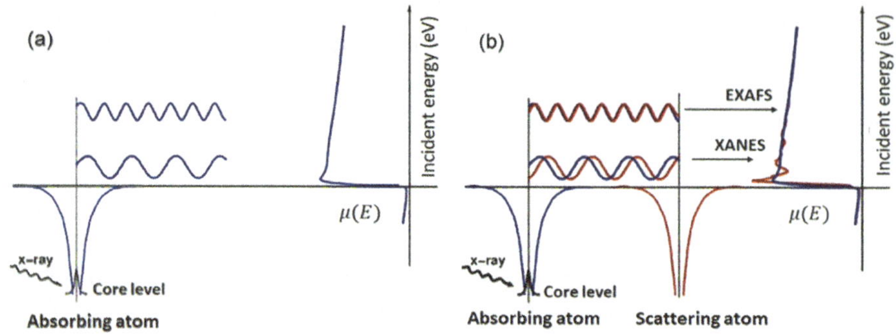


**Fig. 6.1** Schematic illustration of the outgoing photoelectron wave produced by **a** a free atom and **b** a diatomic system upon X-ray absorption and the corresponding XAFS spectrum

exhibits a flat line at energy lower than the absorption threshold, followed by a sharp rise in intensity, then a monotonic decay. The energy ($E_0$) where the intensity rise occurs is referred to as an absorption edge (or simply, an edge). Above $E_0$, photoelectrons are created which travel as a wave with wavenumber proportional to $\sqrt{E_{ex} - E_0}$. Table 6.1 lists the letter symbols that label the absorption edges and the corresponding electronic transitions. The edge energy $E_0$ is dependent on the type of element and the energy level of the electron being excited. The detailed values are available in the X-ray data booklet [31]. Please note that these tabulated values apply to monatomic elements. For compounds that contain multiple atoms, some deviation occurs, which is used to identify the change in oxidation states of the element and their surrounding chemical environment.

If the excited electrons are from an element in a polyatomic compound, the outgoing photoelectron wave will be interfered with the neighboring atoms. The construction and destruction waves between outgoing and backscattered photons modulate absorption coefficient $\mu(E)$ and create oscillations in absorption intensities above $E_0$, producing X-ray absorption fine structure (XAFS), shown in Fig. 6.1b.

**Table 6.1** Symbols for absorption edges and the corresponding electronic transition

| Principal quantum number | Absorption edge | Electronic transition |
|---|---|---|
| **1** | **K** | $1s \rightarrow \boldsymbol{np}$ |
| 2 | $L_1$ | $2s \rightarrow np$ |
| | $\mathbf{L_{2,3}}$ | $\boldsymbol{2p_{1/2,3/2} \rightarrow ns, nd}$ |
| 3 | $M_1$ | $3s \rightarrow np$ |
| | $M_{2,3}$ | $3p_{1/2,3/2} \rightarrow ns, nd$ |
| | $M_{4,5}$ | $3d_{3/2,5/2} \rightarrow np, nf$ |

Content in bold is the ones that are most frequently measured

In terms of data interpretation, XAFS is usually divided into two energy regions. From a few eV below $E_0$ to 50 eV above is the region of X-ray absorption near-edge structure (XANES). In XANES, the modulation of $\mu(E)$ is mainly caused by multiple scattering from surrounding atoms, because the kinetic energy of outgoing photoelectrons is low. It is highly sensitive to the local chemical environment, such as oxidation states, unoccupied electronic states, and local symmetry, so it can be used as "fingerprints" of the materials. At this region, the outgoing signal is dominated by single scattering, and the modulation of the absorption coefficient is be.

From 50 eV up to 1000 eV above the edge, the strong oscillations start to disappear. This is the region of extended X-ray absorption fine structure (EXAFS). EXAFS is dominated by single scattering (the outgoing electron has a high kinetic energy). Local structures such as bond length within first 2 or 3 atomic shells can be obtained by analyzing the EXAFS data [32, 33]. The application of EXAFS is not limited to solid powders, thin films, liquids, and devices in operando can also be directly measured at the experimental stations [34].

Data analysis of EXAFS is less intuitive. As seen in Fig. 6.1b, at the EXAFS region, the intensity oscillation in much weaker compared to the one at the XANES region, therefore a "fine-structure" function $\chi(E)$ is introduced to define the change in absorption amplitude as a function of X-ray energy:

$$\chi(E) = \frac{\mu(E) - \mu_0(E)}{\Delta\mu_0(E)} \tag{2.2}$$

where $\mu(E)$ is the measured absorption coefficient, $\mu_0(E)$ represent the absorption of an isolated atom (i.e. the background), and $\Delta\mu_0(E)$ is the amount of jump in intensity at the absorption threshold. Since the spectral oscillation diminishes quickly above the absorption edge, the spectrum is also converted from the X-ray energy space (E-space) to the wavenumber space (k-space). The conversion is shown in the equation:

$$k = \sqrt{\frac{2m(E - E_0)}{\hbar^2}} \tag{2.3}$$

The function, $\chi(k)$, is referred to as the "EXAFS". It is a summary of scattering waves, which is from the interference with different atoms sitting in the near-neighbor coordination shells. A Fourier transform of the EXAFS spectrum provides photo-electron scattering profile as a function of the radial distance from the absorbing atom. EXAFS displayed in this way is referred to as the "R-space" profile.

The full EXAFS equation has the expression:

$$\chi(k) = \sum_j \frac{N_j S_0^2 f_j(k) e^{-2k^2\sigma_j^2} e^{-2R_j/\lambda_j(k)}}{kR_j^2} \sin\left[2kR_j + \delta_j(k) + 2\varphi_C(k)\right] \tag{2.4}$$

where $N_j$ is the number of the neighboring atoms in the jth coordination shell, $S_0^2$ is the amplitude reduction factor, $f(k)$ and $\delta(k)$ are scattering properties of the neighboring atoms, $\varphi_C(k)$ is the phase-shift of the absorbing atom, $R$ is the distance to the neighboring atom, and $\sigma^2$ is the disorder in the interatomic distance. Quantitative analysis of EXAFS helps to determine the local structure of the absorbing atoms, including parameters such as coordination numbers, interatomic distances, and degree of structural disorder. This is performed by first establishing a structure model, and use this model to generate the scattering parameters in the EXAFS equation to fit the measured EXAFS spectrum. These calculated parameters will then be used to predict and modify the guessed structural parameters until a best fit is obtained.

The choice of XAFS techniques depends on specific scientific questions to be addressed. In the next two sections, we are going to demonstrate how XANES and EXAFS are used in advancing the knowledge on perovskite materials.

#### 6.2.1.2 XANES using Soft X-rays: The Dissociation Mechanism

Soft X-rays usually refer to X-rays with energy lower than 2000 eV, which is suitable for studying low Z elements (first two rows of the main group elements in the periodical table and some transition metals). At this range, the energy separation between adjacent absorption edges are small, so EXAFS is often not feasible. Instead, XANES is the main focus, and the rich spectral features at this region is useful for identifying chemical species and the oxidation state of the element of interests. For characterization of perovskites, XANES has been used to study the alignment of the cation groups [35, 36] and the band structure of formamidinium lead halide [37]. In the following, we will focus on the research dissociation mechanism study of methylammonium lead halide ($MAPbI_3$).

The first generation of LHP for optoelectronic application has a hybrid structure, which contains inorganic framework $PbI_6^{2-}$ and organic methylammonium cations $CH_3NH_3^+$ [38]. This hybrid perovskite has a chemical formula of $CH_3NH_3PbI_3$ ($MAPbI_3$), and can be easily synthesized by mixing two precursors $CH_3NH_2$ and $PbI_2$ in an organic solvent such as $\gamma$-butyrolactone [5]. $MAPbI_3$ has a black color, which is ideal for light harvesting. As the light absorbing layer in photovoltaic cells, $MAPbI_3$ (or $MAPbI_{3-x}Cl_x$) has long carrier diffusion lengths and low charge recombination rates [6, 39]. However, the greatest challenge for commercialization of $MAPbI_3$-based solar cell is the insufficient lifetime. $MAPbI_3$ is vulnerable to various environmental factors, such as water, temperature, and UV-exposure [12–14]. There are several decomposition mechanisms proposed by several groups and there is still controversy among them [40, 41]. Aside from decomposition, there is also possible by-product that co-exist with $MAPbI_3$, and might play a negative role in the photovoltaic performance. It is therefore crucial to identify the chemical species present in thin film $MAPbI_3$.

The corner-sharing octahedral $PbI_6^{2-}$ framework is relatively stable. The $CH_3NH_3^+$ cations sitting between those octahedra are vulnerable to decomposition.

The presence of multiple carbon species in $MAPbX_3$ films have been observed using X-ray photoelectron spectroscopy by various researchers [42–44]. It is however, difficult to identify the origin of the carbon species. Some might be related to carbon impurities in the environment, but for $MAPbI_3$ prepared under vacuum, dual carbon species were also observed [45].

XANES at the C K-edge has been successfully used as a probe for the carbon species in $MAPbI_3$ [19]. Two $MAPbI_3$ films were investigated: one is freshly prepared $MAPbI_3$ film, and the other is a film prepared with identical method but under prolonged X-ray exposure prior to XANES measurement. Figure 6.2, XANES of both films contain a sharp peak at ~282 eV, followed by a doublet at higher energy side. The ratio between the low energy and high energy feature differs between the two samples. To better interpret the spectra, simulated XANES based on a perfect cubic $MAPbI_3$ was included for comparison. From the simulated result, we can see only features at high energy were reproduced. The low energy peak does not belong to a perfect $MAPbI_3$ structure. Instead, if the C–N bond length in the $CH_3NH_3^+$ unit is stretched, a new peak appears at the low energy, and the separation between the

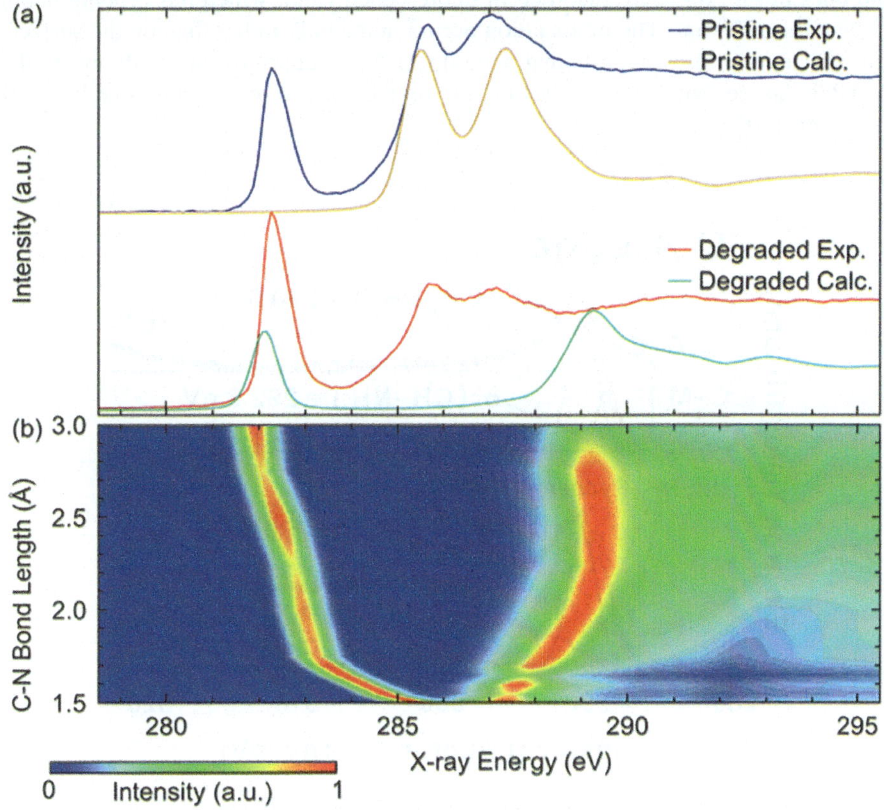


**Fig. 6.2** XANES at the C K-edge of a $MAPbI_3$ thin film. Adapted from Ref. [19]

new peak and the main peak increases as the C–N bond length increases. Therefore, a possible dissociation mechanism of $MAPbI_3$ was proposed: $CH_3NH_3^+$ decomposes into $CH_3^+$ which further bonds to iodide of the main $PbI_6^{2-}$ framework, and $NH_3$ migrates out of the perovskite lattice as a neutral molecule. This hypothesis was also supported by other characterization results such as X-ray photoelectron spectroscopy and mass spectroscopy [45–48].

It is worth mentioning that another study also used the C K-edge XANES to identify the organic species in $MAPbI_3$ thin films, shown in Fig. 6.3 [20]. In this study, the XANES spectra were analyzed by peak fitting, which is done by first defining a step function as the background (single-atom absorption), and fit the main resonance with several Gaussian peaks. The PEY and TEY in the figure refer to partial electron yield and total electron yield, respectively. In this analysis, a neutral $CH_3NH_2$ species was identified. Comparing with the C K-edge XANES of $MAPbI_3$ in Fig. 6.2, the apparent difference is that the low energy feature is absent in this study. One possible explanation is that the spectra acquired in the two studies probe samples at different depths. XANES in Fig. 6.3 were recorded with electron yield, which has a shallow probing depth (25 nm for TEY and 6 nm for PEY), while the ones in Fig. 6.2 were recorded in fluorescence yield, which has probing depth greater than 200 nm. The dissociation occurs in the bulk rather than on the surface. The calculation also suggests that neutral $CH_3NH_2$ cannot reside stably inside the $MAPbI_3$ lattice, but rather resides mainly on the surface or at grain boundaries of thin films [20].

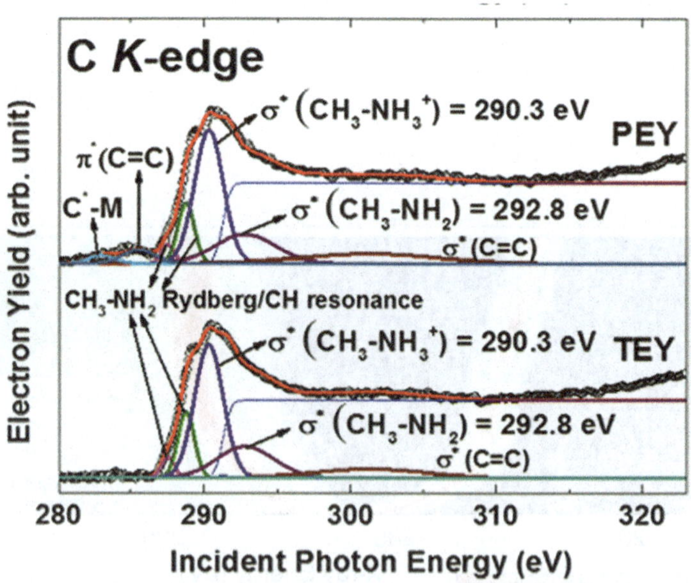


**Fig. 6.3** The C K-edge XANES of $MAPbI_3$ film acquired with PEY and TEY. Curve fitting and the assignment of the composition were indicated in the figure. Adapted from Ref. [20]

### 6.2.1.3 EXAFS using Hard X-rays: The Local Structure Probe

As a structure probe, EXAFS has been used to analyze perovskite thin films and nanocrystals. The coordination environment of Pb extracted from the Pb $L_3$-edge EXAFS can be correlate with the stability of LHP and the performance of LHP-based optoelectronic devices. There are reports on using the Pb $L_3$-edge EXAFS to study the temperature and light-induced degradation product of $MAPbI_3$, and the non-LHP species such as metallic Pb and $PbI_2$ can be identified by their characteristic bonding environment [49, 50].

Although an ideal perovskite structure contains $PbX_6^{2-}$ octahedral units in which Pb is 6-coordinated with halide ions, in practical case, as-synthesized $MAPbI_3$ contains various intrinsic defects [51, 52]. The presence of halide vacancies are beneficial for charge transport, but too much vacancies will act as traps which cause negative impact on the device performance [53]. It was found that adding a small amount of Cl in $MAPbI_3$ drastically improves the power conversion efficiency of $MAPbI_3$-based solar cells [54]. However, the ratio of Cl and I need to be precisely tuned to achieve the best performance. The "magic ratio" of the two halide precursors, $CH_3NH_3I$ and $PbCl_2$ is optimized to be 3:1 and confirmed by multiple studies. An EXAFS study was conducted to examine how the perovskite structure is affected by the precursor concentrations [21]. Figure 6.4, EXAFS was displayed at R-space and k-space, respectively, and a multiple-shell fitting model was used including Pb-Cl, Pb-I, and Pb-C/N scattering paths. The results show that the halide vacancies vary when different ratios of $PbCl_2$ and $CH_3NH_3I$ were used. The power conversion

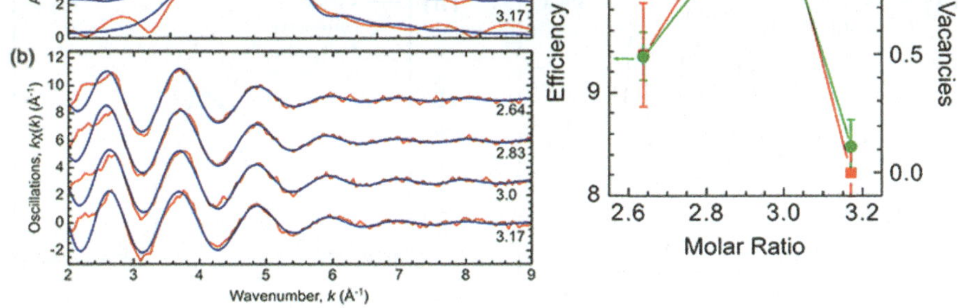


**Fig. 6.4** The Pb $L_3$-edge EXAFS of $MAPbI_{3-x}Cl_x$ prepared with different I: Cl ratios. The spectra are shown in **a** k-space and **b** R-space, respectively, with both experimental and fitted results. **c** The power conversion efficiencies and the halide vacancies plotted as a function of precursor ratio. Adapted from Ref. [21]

efficiencies of solar cells align perfectly with the amount of vacancies. $MAPbI_3$ with highest halide vacancies is achieved at the optimal 3:1 precursor ratio. Higher or lower ratio produces fewer vacancies and hinders the solar cell performance.

Doping ions into perovskite has demonstrated great potential in modifying the light-emitting properties of perovskites. For example, $Mn^{2+}$-doped $CsPbCl_3$ exhibits orange emission [10, 55]. $Yb^{3+}$-doping of $CsPbCl_3$ makes it a near-infrared light emitter [56]. $CsPbCl_3$ nanocrystals co-doped with $Bi^{3+}$ and $Mn^{2+}$ emit white light [57]. The intrinsic luminescence of $CsPbX_3$ can be greatly enhanced by introducing $Ni^{2+}$, $Cu^{2+}$, and $Y^{3+}$ [8, 58, 59]. We will show in Sect. 2.2 how synchrotron X-rays is used to study the luminescence mechanism of doped perovskite. In this section, we focus on the use of EXAFS technique in studying the local structure of dopant ions as well as how the main LHP framework is influenced upon dopant introduction.

EXAFS provides direct evidence that dopant introduction greatly impacts the pristine structure of the Pb-X framework. It was reported that $Ni^{2+}$ doping greatly enhances the intrinsic luminescence of $CsPbX_3$ regardless of the halide species. EXAFS at the Pb $L_3$-edge reveals that the luminescence enhancement is due to an improved short-range order of the $CsPbX_3$ lattice [58]. Similar phenomenon was also observed in Cu-doped $CsPbBr_3$. Doping significantly enhances the thermal stability and photoluminescence quantum yield of $CsPbBr_3$. Figure 6.5 shows the Pb $L_3$-edge EXAFS in k-space and R-space of Cu-doped $CsPbBr_3$ [8]. The detailed structural parameters derived from the fitting are summarized in Table 6.1. It was found that upon doping, the Pb–Br bond length decreases slightly due to lattice contraction. Such contraction increases the interaction between Pb and Br, and improves the formation energy of the $CsPbBr_3$ lattice. The coordination number of Pb is greatly increased, suggesting that doping reduces halide vacancies, and therefore enhance the luminescence intensity of $CsPbBr_3$.

EXAFS of the dopant ions can also be measured. In most of the cases, the R-space spectra at the dopant absorption edge show reduced dopant-halide bond length. This has been attributed to the smaller ionic radius of the substitute ions

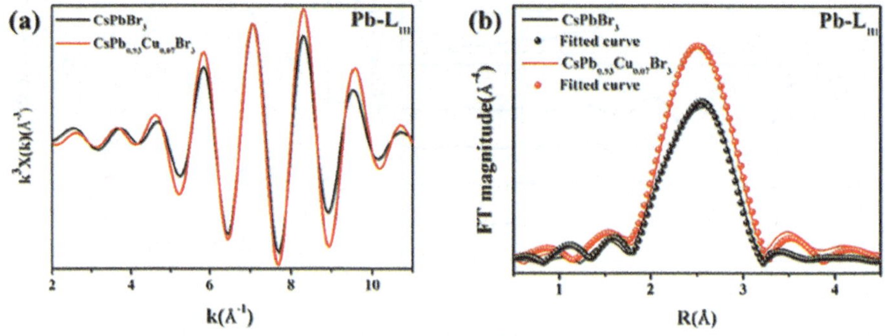


**Fig. 6.5** EXAFS spectra of $CuPbBr_3$ and Cu-doped $CsPbBr_3$ quantum dots. **a** $k_3$-weighted EXAFS in k-space, **b** Fourier transformed R-space EXAFS. Adapted from Ref. [8]

[8, 58, 60]. However, due to the low doping concentration, quantitative EXAFS analysis is limited, presumably due to the low signal to noise ratio at energy well-beyond the absorption edge.

## 6.2.2 *X-ray Excited Optical Luminescence (XEOL)*

### 6.2.2.1 The Working Principle

Upon photon absorption, electrons are promoted to the previously unoccupied electronic states, and settled at the bottom of the conduction band, while holes drift to the top of the valence band. Radiative recombination of electron-hole pairs produces luminescence. Luminescence can be produced with visible and UV light excitation, such as the light source used in conventional laboratory spectrofluorometer. X-ray on the other hand, can also be used as an excitation source, and the production of luminescence is illustrated in Fig. 6.6. During this process, core-electrons are excited, and the holes leaving behind undergoes a series of thermalization processes to reach to the top of the valance band of the materials. During thermalization, electrons at shallower levels fill the core hole, producing Auger electrons and a hole at an outer shell. The energetic electrons continue to travel in the solid, losing its energy and producing shallower core holes. This cascade process is repeated until holes are created at the top of the valence band, and recombine with electrons at the bottom of the conduction band.

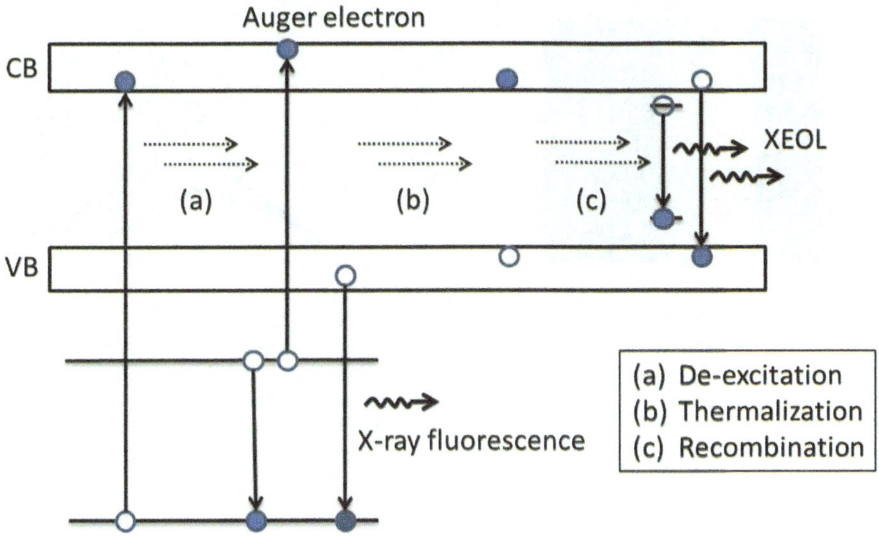


**Fig. 6.6** Schematic illustration of the production of luminescence following a core-electron excitation with X-ray

The biggest advantage of using X-ray as the excitation source is that the wavelength and intensity of the emitted light can be related to the de-excitation process of electrons at a particular core-level. X-ray excited optical luminescence (XEOL) is often collected at the same energy range as one chooses to acquire an XAFS spectrum of an element of interest. The light emission can then be related to the absorption behavior of the material. By tracking the XEOL profile as a function of excitation energy, the origin of the luminescence can be identified [61]. A combined XEOL-XANES analysis has been successfully used in understanding luminescence mechanism of heterostructures, nanostructured metal oxides, solid solutions and organic light-emitting materials [61–63].

For bright light-emitting materials, XEOL is collected simultaneously with XANES, generating a 2D contour plot, shown in Fig. 6.7a. The horizontal axis is the excitation energy, the vertical axis is the wavelength, and the color-coded z-axis is the intensity of the emitted light. In this plot, the excitation energy-dependency of luminescence can be visualized: the horizontal cut of the contour map shows an XEOL spectrum at a selected excitation energy, and the intensity variation of emission at certain wavelength can be obtained from a vertical cut from the map. For weak light-emitting materials, XEOL can be collected at selected excitation energies, where abrupt change in absorption coefficient occurs, shown in Fig. 6.7b. The intensity and emission profile change as a function of excitation energy can be interpolated.

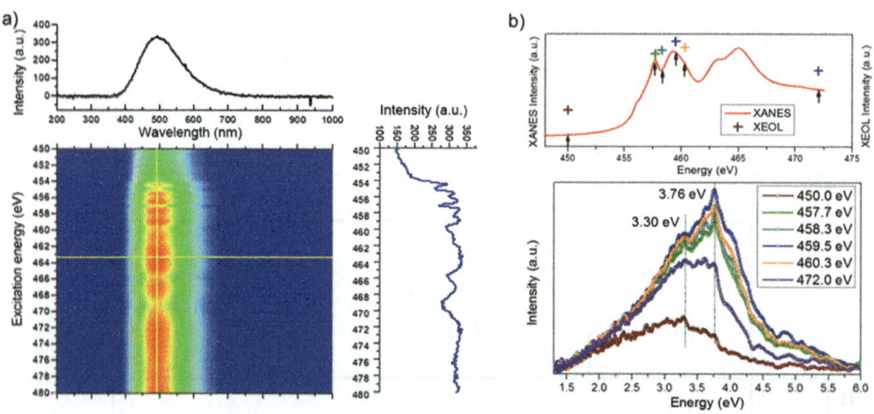


**Fig. 6.7** **a** a 2D XANES-XEOL contour plot of $TiO_2$ nanowires. A horizontal cut (XEOL spectrum) and a vertical cut (intensity variation at maximum XEOL as a function of energy) are shown on top and on the right of the contour plot, respectively. Adapted from Ref. [64] **b** top: integrated XEOL intensities (cross marks) of $TiO_2$ nanograss at selected excitation energies (indicated by arrows) in comparison with the Ti $L_{3,2}$-edge XANES. Bottom: XEOL spectra of nanograss under selected excitation energies. Adapted from Ref. [65]

#### 6.2.2.2 XEOL-XAFS Analysis: Luminescence Mechanism

Ion doping of $CsPbCl_3$ and $CsPbBr_3$ has been demonstrated an effective strategy of modifying their luminescence properties. In particular, $Mn^{2+}$ doping introduces an additional emission band at the orange–yellow region. Detailed charge transfer mechanism was investigated by an XEOL-XANES analysis on $Mn^{2+}$-doped $CsPb(Cl/Br)_3$ nanocrystals [22]. The nanocrystals exhibit dual emission bands, similar to most of the Mn-doped LHPs. The blue emission at ~430 nm is from the near band-gap recombination of perovskite host lattice, and the orange emission at ~600 nm is from the Mn d–d transition between the $^4T_1$ and $^6A_1$ configurations [10, 66].

Shown in Fig. 6.8, the XANES at the Mn $L_{3,2}$-edge of Mn-doped $CsPb(Br/Cl)_3$ has typical $Mn^{2+}$ features. Several XEOL spectra were recorded at selected excitation energies from below to above the Mn absorption edge. It is interesting to observe that not only the intensity of the luminescence is dependent on the excitation energy, a minor wavelength shift of the perovskite near-band-gap emission is also observed. Careful analysis reveals that this near-band-gap emission in fact contains two components: one is excitation energy-dependent, while the other is less sensitive to the excitation energy. When the Mn electrons absorb most of the incoming photons, a decrease in the Mn-related emission was observed. This suggests that

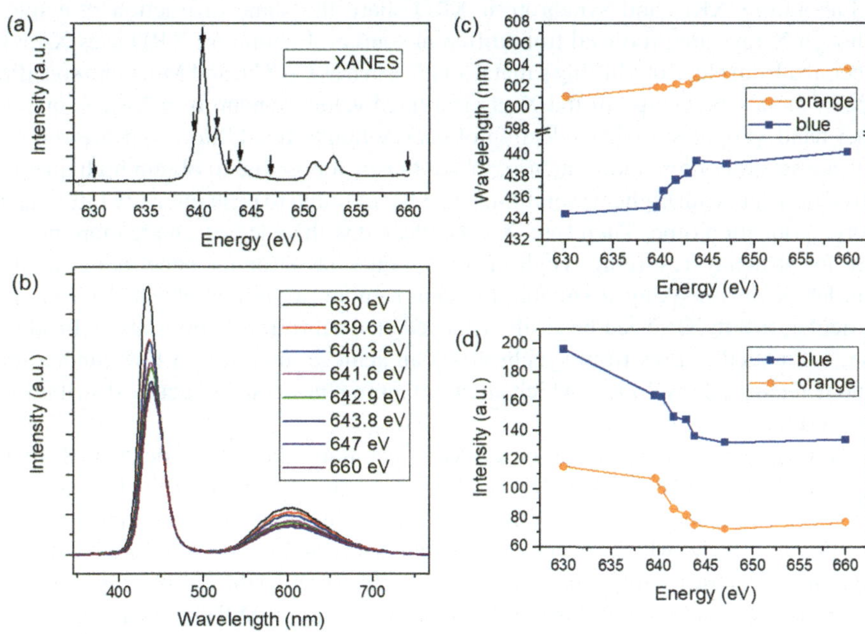


**Fig. 6.8** Excitation energy-dependent luminescence of Mn-doped $CsPb(Br/Cl)_3$. **a** Energy selection from the XANES spectrum. **b** XEOL at each excitation energy. **c** The trend of emission peak wavelength as a function of excitation energy for each emission band. **d** The trend of emission peak intensity as a function of excitation energy for each emission band. Adapted from Ref. [22]

directly promoting the electronic transition in Mn is not the effective radiative recombination channel. Instead, luminescence from Mn-doped $CsPb(Br/Cl)_3$ is facilitated by energy transfer from the excitons formed within the perovskite host lattice.

### *6.2.3 X-ray Diffraction (XRD)*

#### 6.2.3.1 XRD with Synchrotron Radiation

X-ray diffraction (XRD) is the most commonly used technique for determination of crystal structure at atomic resolution. The basic principle for XRD is given by the Bragg equation:

$$\lambda = 2d_{hkl}\sin\theta \tag{6.5}$$

where $\lambda$ is the wavelength of the incident X-ray, $d_{hkl}$ is the spacing of the lattice planes with the Miller indices *h, k, l*, and $\theta$ is the angle of specular reflection with respect to these planes. The diffraction angle, defined as the angle between the incident primary beam and the diffracted beam, is therefore equal to $2\theta$.

Laboratory XRD and synchrotron XRD share the same diffraction principle, although X-rays are produced from different sources. Laboratory XRD uses X-rays generated from electrons hitting a metal anode, such as Co, Cu, and Mo. For a specific type of metal, the energy of the X-ray is a fixed value. Synchrotron X-rays, on the other hand, provides a wide selection of energy range. In addition, synchrotron X-rays are several orders more intense, which makes it possible to obtain high quality diffraction data with higher resolution, lower background to peak intensity ratios, and shorter collection time. Therefore, XRD collected with synchrotron are more favorable for structure resolving. High energy X-rays has stronger penetrating ability than lab X-rays, making it suitable to adapt sample of various physical forms. In particular, in situ XRD can be easily achieved at synchrotron X-ray station. In addition, collimated X-rays from synchrotron can achieve micron to a few hundred of nanometer spatial resolution, which allows mapping the crystal structure distribution on a surface.

Various types of synchrotron-based XRD techniques have been developed, such as angle dispersive XRD (ADXRD) and energy dispersive XRD (EDXRD), shown in Fig. 6.9a,b, respectively. In both techniques, data acquisition is conducted in a transmission mode, which means the incident X-ray and the detector are located at different side of the sample. This requires either the energy of the X-ray is high enough or the sample is sufficiently thin for X-ray to penetrate. ADXRD uses a monochromatic beam as the source, and the diffraction pattern is collected from all solid angles. EDXRD, on the other hand, is almost exclusively performed using synchrotron X-rays. It uses polychromatic photons as the incident source and the detector is place

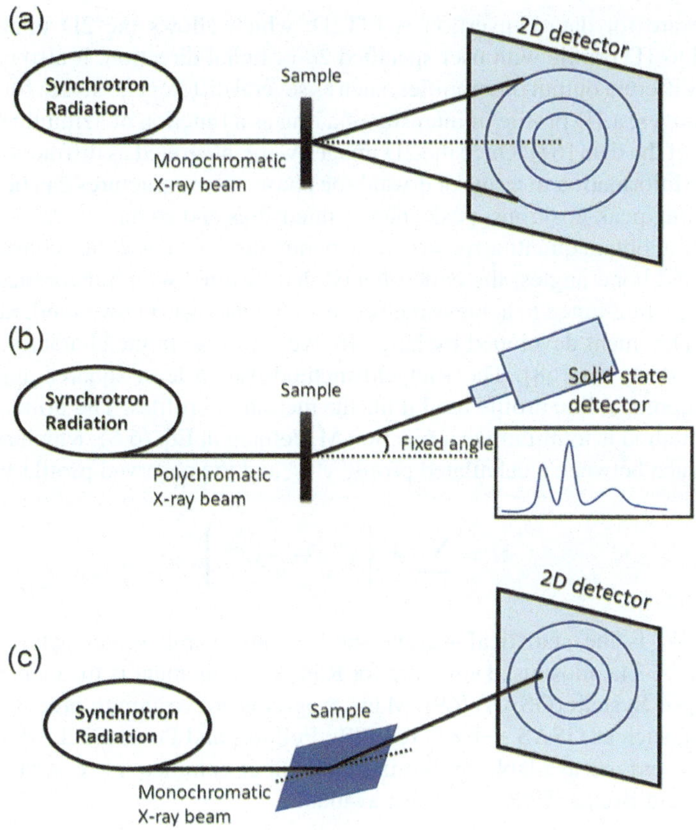


**Fig. 6.9** Schematic illustration of the experimental layout in **a** ADXRD, **b** EDXRD, and **c** GIXRD

at a fixed angle. Using the high penetrating X-ray from the synchrotron source, the change of crystal structure of materials can be tracked from deep within the sample. It allows study of deep strain profiling and material phase changes under extreme conditions. The fast data acquisition rate using EDXRD mode (because there is no need to move the detector mechanically during data collection) is ideal for materials that are unstable under X-ray, and for capturing structural change during chemical reactions in situ. However, for thick samples or research that focus on the material surfaces (e.g. thin films), transmission detection mode is no longer applicable. In this case, a grazing incidence XRD (GIXRD) setup is adapted as shown in Fig. 6.9c, where the incident beam and detector sitting on the same side of the sample (similar to laboratory XRD).

Different from laboratory XRD, the synchrotron XRD data is often collected in 2D as a series of Debye–Scherrer rings (except EDXRD). Such image needs to be converted into the form as intensity vs diffraction angle. The most widely used software for data convertion is FIT2D, which allows the 2D images to be integrated to 1D profile with user specified $2\theta$ or radial direction. It also provides a variety of different output possibilities, such as several different $2\theta$ scans, for different azimuth ranges; a 1D profile of intensity of a ring as a function of azimuth; or a polar transform of the data [67]. Once the 2D images were integrated as diffractions peaks, qualitative information in terms of crystalline phase, microstructures can be obtained by examining peak positions, peak shapes, intensities and so on.

In order to obtain quantitative structural parameters such as lattice constant, bond distance, and bond angles, the experimental data is fitted with a theoretical model, which can be fine-tuned to achieve the best match. One well-known method is called Rietveld refinement developed by Hugo Rietveld for use in the characterization of crystalline materials [68]. The Rietveld method uses a least squares approach to refine a theoretical line profile until it fits the measured profiles. The principle of the Rietveld method is to minimize a function M, defined in Eq. (6.6), which represents the difference between a calculated profile $y^{calc}$ and the observed profile $y^{obs}$.

$$M = \sum_i W_i \left\{ y_i^{obs} - \frac{1}{c} y_i^{cals} \right\}^2 \tag{6.6}$$

where $W_i$ is the statistical weight and c is an overall scale factor such that $y^{calc} = cy^{obs}$. The most used software for Rietveld refinement is the General Structure Analysis System (GSAS) [69]. Many programs are written to process Rietveld refinement, such as GSAS + EXPGUI [70], Fullprof and PowderCel, which can be used freely and are available on the internet. Other commercial programs such as MDI Jade and Bruker TOPAS are also available.

#### 6.2.3.2 In Situ XRD: Probing The Real-Time Structure Change

As mentioned above, the deeper penetration depth of high energy synchrotron X-rays and the fast data acquisition is easy to achieve due to the high X-ray intensity, in situ XRD can be conducted to capture the crystal structure change in real time. In situ XRD has been employed to investigate structure of hybrid and inorganic LHPs. So far, research has been focusing on one of the two topics: (1) the real-time structure change of LHP during a chemical reaction, which is often related to the study of degradation mechanism of LHPs; [24, 71, 72] (2) the pressure-induced structural change emphasizing on new phase identification, and to modify the physical and chemical property of LHPs using high pressure [23, 73–89].

For instance, the degradation process of the $CH_3NH_3PbI_3$ in moist air was investigated by Yang and co-works utilizing ss GIXRD [71]. In their study, a hydrated intermediate phase containing isolated $PbI_6^{4-}$ octahedral was identified as the first step of the degradation upon exposure of water vapor. Later, Zhao et al. designed an in situ cell to examine the stability of $CH_3NH_3PbI_3$ thin film in humid environment, shown in Fig. 6.10 [72]. They proposed a possible decomposition pathway

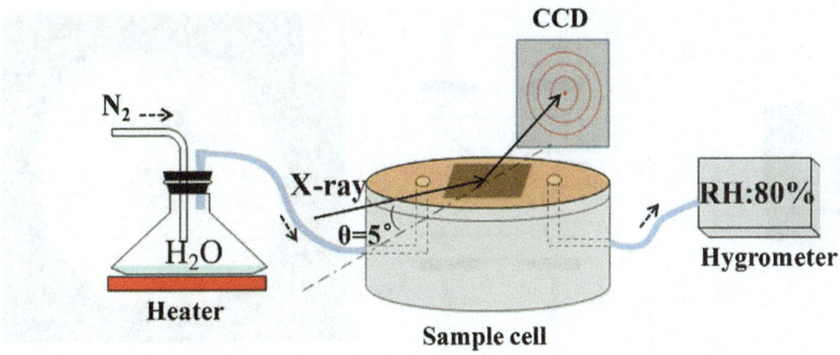


**Fig. 6.10** Schematic diagram of the experimental setup for an in situ XRD measurement. Adapted from Ref. [72]

$$CH_3NH_3PbI_3 + H_2O \leftrightarrow CH_3NH_3PbI_3 \cdot H_2O \quad (6.7)$$

$$CH_3NH_3PbI_3 \cdot H_2O \leftrightarrow PbI_2 + CH_3NH_3I + H_2O \quad (6.8)$$

which agrees with Yang's result. This is a direct evidence that the monohydrated species is formed during the decomposition. A similar in situ setup was recently applied to investigate the humid effect on stability of inorganic lead bromide perovskite [24]. They observed an on-site phase transformation from $Cs_4PbBr_6$ to $CsPbBr_3$ and $CsPb_2Br_5$ and also revealed that such transformation process is probably controlled by both the solubility of CsBr and the coordination number of Pb(II) in water.

The other major application of in situ XRD is to monitor the high-pressure behavior of LHPs, focusing on pressure-induced phase transformation and pressure-induced modifications in electronic and optical properties of LHPs. A diamond anvil cell (DAC) is used to generate high pressure. This is done by applying a static mechanical force perpendicularly to the surface of the sample placed in a vessel. Among the many types of DACs, the widely employed high-pressure vessel is symmetric DAC as shown in Fig. 6.11. The sample was placed between a pair of diamonds anvils. The tips of the diamond anvils are only hundreds of microns in diameter. Therefore, a moderate force can generate several millions of atmospheres (e.g., 1 N can generate 8 $\times 10^6$ Pa on a 400 μm culet.). The exact pressure values in the cell can be quantified using a pressure scale (e.g. ruby, Pt, Au). These materials have a known property that can be easily measured and has a direct correlation with applied pressure. For example, by exciting ruby with a green or blue laser, its fluorescence lines shift as a function of pressure [90]. Shown in Fig. 6.11c, during the measurement, a standard pressure scale (ruby ball) is loaded together with the sample of interest in the same DAC cell, and the real time structural change induced by pressure is then recorded by XRD.

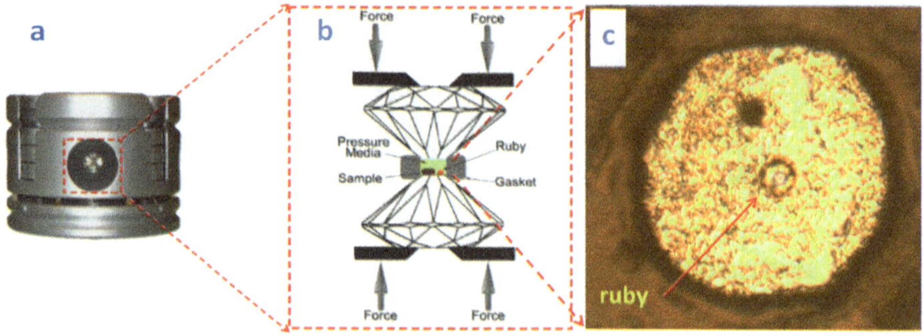


**Fig. 6.11** **a** A photograph of a symmetric DAC, **b** schematic layout of a DAC (side view), **c** top view of the sample hole in the DAC with a ruby ball loaded

### 6.2.3.3 Pressure-Induced Phase Transformation

A great number of high-pressure study have been carried out on LHPs and it was found the LHPs exhibit interesting pressure-induced structural change that leads to modifications in electronic and optical properties.

The pressure induced phase transformation of LHPs were summarized in Table 6.2. Rich polymorph crystal structures produced in LHPs at high pressure were detected by various groups of researchers. However, the pressure-induced phase transformation sequences are inconsistent. For example, in Wang's study, $MAPbCl_3$ was found undergoing a phase transformation sequence from cubic ($Pm\overline{3}m$) to isostructured cubic phase ($Pm\overline{3}m$) at 0.8 GPa and then to an orthorhombic phase ($Pnma$) at 2.0 GPa [80]. However, another study suggests that $MAPbCl_3$ turns into a cubic ($Im\overline{3}$) phase at 0.75 GPa and then amorphorizes directly at 2.0 GPa [81], while in Wang's study, amorphization was not observed until the pressure is above 5.6 GPa. The origin of these inconsistencies are not clear yet and more studies are needed.

Different from $MAPbCl_3$, the phase transformation of $MAPbBr_3$ from cubic ($Pm\overline{3}m$) to cubic ($Im\overline{3}$) was observed in several studies, although such transformation occurs at slightly difference pressures, which is likely due to the difference in hydrostatic conditions used in different experiments [23, 81, 84, 88, 89]. The mechanism of such phase transformation is attributed to pressure-induced shrinkage and titling of the $PbX_6^{2-}$ octahedra as well as the rotation of $MA^+$ cations. The isostructural phase transformation observed in $MAPbCl_3$ and $MAPbI_3$ more rely on rotation of the $MA^+$ cations [75, 79]. For $MAPbI_3$, a pressure-induced metallization was also observed when the applied pressure is increased up to 60 GPa, and $MAPbI_3$ exhibits new transport properties [78].

Along with the structural change, perhaps the most interesting property of LHP is its band gap narrowing when pressure increases. This leads to elongated carrier lifetime and such narrowed band gap is retained after pressure is released [77]. It should also be addressed that the first pressure-induced phase transformation in all

**Table 6.2** Summary of pressure induced phase transition in hybrid perovskites

| Materials original phase | | 2nd Phase | 3rd Phase | 4th Phase | References | *Note |
|---|---|---|---|---|---|---|
| $MAPbCl_3$ | Cubic $Pm\bar{3}m$ | Cubic ($Pm\bar{3}m$, 0.8GPa)* | Orthorhombic ($Pnma$, 2.0 GPa) | Amorphous (5.6 GPa) | [80] | Isostructural PT |
| | | Cubic ($Im\bar{3}$, 0.75 GPa) | N/A | Amorphous (2.1 GPa) | [81] | |
| $MAPbBr_3$ | Cubic $Pm\bar{3}m$ | Cubic ($Im\bar{3}$, 0.4 GPa) | Orthorhombic ($Pnma$, 1.8 GPa) | Amorphous (>2.0 GPa) | [23] | |
| | | Cubic ($Im\bar{3}$, 1.0 GPa) | N/A | Amorphous (2.8 GPa) | [89] | |
| | | Cubic ($Im\bar{3}$, 0.75 GPa) | N/A | Amorphous (2.7 GPa) | [81] | |
| | | Cubic (s, 0.8 GPa) | N/A | Amorphous (4.7 GPa) | [88] | |
| | | Cubic ($Im\bar{3}$, 0.9 GPa) | N/A | Amorphous (2.7 GPa) | [84] | |
| $MAPbI_3$ | Tetragonal $I4/mcm$ | Cubic ($Im\bar{3}$, 0.3 GPa) | Orthorhombic ($Immm$, 2.7 GPa) | Amorphous (4.7 GPa) | [83] | |
| | | Cubic ($Im\bar{3}$, 0.35 GPa) | N/A | Amorphous (2.5 GPa) | [81] | |
| | | Cubic ($Im\bar{3}$, 0.35 GPa) | Cubic ($Im\bar{3}$, 2.5 GPa)* | N/A | [75] | Isostructural PT |
| | Orthorhombic $Fmmm$ | Cubic ($Im\bar{3}$, 0.3 GPa) | N/A | Amorphous (3.4 GPa) | [84] | |
| $FAPbBr_3$ | Cubic $Pm\bar{3}m$ | Cubic ($Im\bar{3}$, 0.53 GPa) | Orthorhombic ($Pnma$, 2.2 GPa) | Amorphous (4.0 GPa) | [79] | |
| $FAPbI_3$ | Cubic $Pm\bar{3}m$ | Orthorhombic ($Imm2$, 0.34 GPa) | Orthorhombic ($Immm$, 2.7 GPa) | Amorphous (4.0 GPa) | [74] | |
| $CsPbBr_3$ | Orthogonal $Pbnm$ | Orthogonal ($Pbnm$, 1.2 GPa)* | N/A | Amorphous (2.4 GPa) | [73] | Isostructural PT |

*Note* PT stands for phase transformation. Phase transformations between structures with the same space group are transitions between two isostructural phases

LHP studied so far, occurs at a pressure below 1 GPa, especially for the cubic $MAPbI_3$ which can be obtained at a mild pressure of 0.35 GPa [75, 81, 83, 84]. This implies that modulation of LHP properties through applied pressure can be achieved without using DAC, and the phase transformation does not introduce additional chemical impurities, making this a practical and promising strategy of tuning the optical and electronic properties of LHPs.

## 6.3 Concluding Remarks

Synchrotron radiation technology has become an increasingly popular tool in materials characterization. It provides valuable information that cannot be obtained through laboratory techniques. This chapter summarizes three synchrotron X-ray techniques that have been used in studying LHP-based materials, which are X-ray absorption spectroscopy, X-ray excited optical luminescence, and X-ray diffraction. It should be noted that other synchrotron-based techniques, such as photoemission spectroscopy and X-ray scattering, are also employed in characterizing LHP-based devices, where LHPs are incorporated in the form of thin films. [26, 27, 91, 92]. These spectroscopic techniques allow detailed study on factors that influence charge transport in LHP films such as interface energetics and grain distribution. In addition, highly collimated X-rays from synchrotron source can also be produced with sub-micron spatial resolution. All the above-mentioned X-ray detection techniques can be conducted in the microscopy mode to obtain the local composition and structure of materials [93]. In summary, the fast advancement in synchrotron technology and analysis techniques provide an exciting future for characterization of LHPs. A deeper understanding on the fundamental properties of LHPs, the tunability of their structural and optical properties, and their stabilities in various environment will guide the development of LHP-based optoelectronic devices with outstanding performances.

## References


1. I. Chung, B. Lee, J. He, R.P. Chang, M.G. Kanatzidis, Nature **485**, 486–489 (2012)
2. H.S. Jung, N.G. Park, Small **11**, 10–25 (2015)
3. Q. Zhang, Y. Yin, ACS Central Sci. **4**, 668–679 (2018)
4. L. Protesescu, S. Yakunin, M.I. Bodnarchuk, F. Krieg, R. Caputo, C.H. Hendon, R.X. Yang, A. Walsh, M.V. Kovalenko, Nano Lett. **15**, 3692–3696 (2015)
5. A. Kojima, K. Teshima, Y. Shirai, T. Miyasaka, J. Am. Chem. Soc. **131**, 6050–6051 (2009)
6. D. Shi, V. Adinolfi, R. Comin, M. Yuan, E. Alarousu, A. Buin, Y. Chen, S. Hoogland, A. Rothenberger, K. Katsiev, Y. Losovyj, X. Zhang, P.A. Dowben, O.F. Mohammed, E.H. Sargent, O.M. Bakr, Science **347**, 519–522 (2015)
7. M.A. Green, A. Ho-Baillie, H.J. Snaith, Nat. Photonics **8**, 506–514 (2014)
8. C. Bi, S. Wang, Q. Li, S.V. Kershaw, J. Tian, A.L. Rogach, J. Phys. Chem. Lett. **10**, 943–952 (2019)
9. J.S. Yao, J. Ge, B.N. Han, K.H. Wang, H.B. Yao, H.L. Yu, J.H. Li, B.S. Zhu, J.Z. Song, C. Chen, Q. Zhang, H.B. Zeng, Y. Luo, S.H. Yu, J. Am. Chem. Soc. **140**, 3626–3634 (2018)
10. W. Liu, Q. Lin, H. Li, K. Wu, I. Robel, J.M. Pietryga, V.I. Klimov, J. Am. Chem. Soc. **138**, 14954–14961 (2016)
11. P.J.S. Rana, T. Swetha, H. Mandal, A. Saeki, P.R. Bangal, S.P. Singh, J. Phys. Chem. C **123**, 17026–17034 (2019)
12. D. Bryant, N. Aristidou, S. Pont, I. Sanchez-Molina, T. Chotchunangatchaval, S. Wheeler, J.R. Durrant, S.A. Haque, Energy Environ. Sci. **9**, 1655–1660 (2016)
13. B. Conings, J. Drijkoningen, N. Gauquelin, A. Babayigit, J. D'Haen, L. D'Olieslaeger, A. Ethirajan, J. Verbeeck, J. Manca, E. Mosconi, F. De Angelis, H.G. Boyen, Adv. Energy Mater. **5**, 1500477 (2015)

14. B. Salhi, Y.S. Wudil, M.K. Hossain, A. Al-Ahmed, F.A. Al-Sulaiman, Renew. Sustain. Energy Rev. **90**, 210–222 (2018)
15. S. Huang, Z. Li, B. Wang, N. Zhu, C. Zhang, L. Kong, Q. Zhang, A. Shan, L. Li, A.C.S. Appl, Mater. Interfaces **9**, 7249–7258 (2017)
16. J. Pan, S.P. Sarmah, B. Murali, I. Dursun, W. Peng, M.R. Parida, J. Liu, L. Sinatra, N. Alyami, C. Zhao, E. Alarousu, T.K. Ng, B.S. Ooi, O.M. Bakr, O.F. Mohammed, J. Phys. Chem. Lett. **6**, 5027–5033 (2015)
17. Z. Shi, J. Guo, Y. Chen, Q. Li, Y. Pan, H. Zhang, Y. Xia, W. Huang, Adv. Mater. **29**, 1605005 (2017)
18. L. Liang, P. Gao, Adv. Sci. **5**, 1700331 (2018)
19. W. Xu, L. Liu, L. Yang, P. Shen, B. Sun, J.A. McLeod, Nano Lett. **16**, 4720–4725 (2016)
20. M.-C. Jung, Y.M. Lee, H.-K. Lee, J. Park, S.R. Raga, L.K. Ono, S. Wang, M.R. Leyden, B.D. Yu, S. Hong, Y. Qi, Appl. Phys. Lett. **108**, 073901 (2016)
21. J.A. McLeod, Z. Wu, B. Sun, L. Liu, Nanoscale **8**, 6361–6368 (2016)
22. J. Ma, Q. Yao, J.A. McLeod, L.Y. Chang, C.W. Pao, J. Chen, T.K. Sham, L. Liu, Nanoscale **11**, 6182–6191 (2019)
23. Y. Wang, X. Lu, W. Yang, T. Wen, L. Yang, X. Ren, L. Wang, Z. Lin, Y. Zhao, J. Am. Chem. Soc. **137**, 11144–11149 (2015)
24. M. Liu, J. Zhao, Z. Luo, Z. Sun, N. Pan, H. Ding, X. Wang, Chem. Mater. **30**, 5846–5852 (2018)
25. B. Philippe, B.-W. Park, R. Lindblad, J. Oscarsson, S. Ahmadi, E.M.J. Johansson, H. Rensmo, Chem. Mater. **27**, 1720–1731 (2015)
26. V.K. Ravi, P.K. Santra, N. Joshi, J. Chugh, S.K. Singh, H. Rensmo, P. Ghosh, A. Nag, J. Phys. Chem. Lett. **8**, 4988–4994 (2017)
27. W. Huang, F. Huang, E. Gann, Y.-B. Cheng, C. McNeill, Adv. Funct. Mater. **25**, 5529–5536 (2015)
28. C.Y. Chang, Y.C. Huang, C.S. Tsao, W.F. Su, A.C.S. Appl, Mater. Interfaces **8**, 26712–26721 (2016)
29. G. Bunker, *Introduction to XAFS: A Practical Guide to X-Ray Absorption Fine Structure Spectroscopy* (Cambridge University Press, New York, 2010)
30. P. Willmott, *Spectroscopic Techniques, An Introduction to Synchrotron Radiation: Techniques And Applications* (John Wiley & Sons Ltd, UK, 2019)
31. An online version of the booklet is available at: https://xdb.lbl.gov/
32. E.A. Stern, D.E. Sayers, F.W. Lytle, Phys. Rev. B **11**, 4836–4846 (1975)
33. J.J. Rehr, R.C. Albers, Rev. Mod. Phys. **72** (2000)
34. Z. Sun, Q. Liu, T. Yao, W. Yan, S. Wei, Sci. Chin. Mater. **58**, 313–341 (2015)
35. J.A. McLeod, Z. Wu, P. Shen, B. Sun, L. Liu, J. Phys. Chem. Lett. **5**, 2863–2867 (2014)
36. Y. Wang, Z. Xia, L. Liu, W. Xu, Z. Yuan, Y. Zhang, H. Sirringhaus, Y. Lifshitz, S.T. Lee, Q. Bao, B. Sun, Adv. Mater. **29**, 1606370 (2017)
37. P. Shen, K. Nie, X. Sun, L. Liu, J.A. McLeod, Phys. Stat. Sol. RRL **10**, 677–681 (2016)
38. L. Etgar, P. Gao, Z. Xue, Q. Peng, A.K. Chandiran, B. Liu, M.K. Nazeeruddin, M. Gratzel, J. Am. Chem. Soc. **134**, 17396–17399 (2012)
39. S.D. Stranks, G.E. Eperon, G. Grancini, C. Menelaou, M.J. Alcocer, T. Leijtens, L.M. Herz, A. Petrozza, H.J. Snaith, Science **342**, 341–344 (2013)
40. J.M. Frost, K.T. Butler, F. Brivio, C.H. Hendon, M. van Schilfgaarde, A. Walsh, Nano Lett. **14**, 2584–2590 (2014)
41. G. Niu, X. Guo, L. Wang, J. Mater. Chem. A **3**, 8970–8980 (2015)
42. Q. Wang, R. Wang, P. Shen, C. Li, Y. Li, L. Liu, S. Duhm, J. Tang, Adv. Mater. Interfaces (2015)
43. S.R. Raga, M.-C. Jung, M.V. Lee, M.R. Leyden, Y. Kato, Y. Qi, Chem. Mater. **27**, 1597–1603 (2015)
44. T.-W. Ng, C.-Y. Chan, M.-F. Lo, Z.Q. Guan, C.-S. Lee, J. Mater. Chem. A (2015)
45. R. Wang, C. Wu, Y. Hu, J. Li, P. Shen, Q. Wang, L. Liao, L. Liu, S. Duhm, A.C.S. Appl, Mater. Interfaces **9**, 7859–7865 (2017)

46. N.K. Kim, Y.H. Min, S. Noh, E. Cho, G. Jeong, M. Joo, S.W. Ahn, J.S. Lee, S. Kim, K. Ihm, H. Ahn, Y. Kang, H.S. Lee, D. Kim, Sci. Rep. **7**, 4645 (2017)
47. G. Grancini, A.R. Srimath Kandada, J.M. Frost, A.J. Barker, M. De Bastiani, M. Gandini, S. Marras, G. Lanzani, A. Walsh, A. Petrozza, Nat. Photonics, **9**, 695–701 (2015)
48. E.J. Juarez-Perez, Z. Hawash, S.R. Raga, L.K. Ono, Y. Qi, Energy Environ. Sci. **9**, 3406–3410 (2016)
49. G. Abdelmageed, L. Jewell, K. Hellier, L. Seymour, b. Luo, F. Bridges, J.Z. Zhang, S. Carter, Appl. Phys. Lett. **109**, 233905 (2016)
50. G. Abdelmageed, C. Mackeen, K. Hellier, L. Jewell, L. Seymour, M. Tingwald, F. Bridges, J.Z. Zhang, S. Carter, Sol. Energy Mater. Sol. Cells **174**, 566–571 (2018)
51. J. Kim, S.-H. Lee, J.H. Lee, K.-H. Hong, J. Phys. Chem. Lett. **5**, 1312–1317 (2014)
52. W.-J. Yin, T. Shi, Y. Yan, Appl. Phys. Lett. **104**, 063903 (2014)
53. C.-C. Chueh, C.-Y. Liao, F. Zuo, S.T. Williams, P.-W. Liang, A.K.Y. Jen, J. Mater. Chem. A **3**, 9058–9062 (2015)
54. Q. Chen, H. Zhou, Y. Fang, A.Z. Stieg, T.B. Song, H.H. Wang, X. Xu, Y. Liu, S. Lu, J. You, P. Sun, J. McKay, M.S. Goorsky, Y. Yang, Nature Commun. **6**, 7269 (2015)
55. S. Das Adhikari, S.K. Dutta, A. Dutta, A.K. Guria, N. Pradhan, Angew. Chem. **56**, 8746–8750 (2017)
56. A. Ishii, T. Miyasaka, Adv. Sci. **7**, 1903142 (2020)
57. H. Shao, X. Bai, H. Cui, G. Pan, P. Jing, S. Qu, J. Zhu, Y. Zhai, B. Dong, H. Song, Nanoscale **10**, 1023–1029 (2018)
58. Z.J. Yong, S.Q. Guo, J.P. Ma, J.Y. Zhang, Z.Y. Li, Y.M. Chen, B.B. Zhang, Y. Zhou, J. Shu, J.L. Gu, L.R. Zheng, O.M. Bakr, H.T. Sun, J. Am. Chem. Soc. **140**, 9942–9951 (2018)
59. G.H. Ahmed, J.K. El-Demellawi, J. Yin, J. Pan, D.B. Velusamy, M.N. Hedhili, E. Alarousu, O.M. Bakr, H.N. Alshareef, O.F. Mohammed, ACS Energy Lett. **3**, 2301–2307 (2018)
60. J.-P. Ma, Y.-M. Chen, L.-M. Zhang, S.-Q. Guo, J.-D. Liu, H. Li, B.-J. Ye, Z.-Y. Li, Y. Zhou, B.-B. Zhang, O.M. Bakr, J.-Y. Zhang, H.-T. Sun, J. Mater. Chem. C **7**, 3037–3048 (2019)
61. T.K. Sham, D.T. Jiang, I. Coulthard, J.W. Lorimer, X.H. Feng, K.H. Tan, S.P. Frigo, R.A. Rosenberg, D.C. Houghton, B. Bryskiewicz, Nature, **363**, 331–334 (1993)
62. L. Armelao, F. Heigl, A. Jurgensen, R.I.R. Blyth, T. Regier, X.T. Zhou, T.K. Sham, J. Phys. Chem. C **111**, 10194–10200 (2007)
63. Y. Hu, A. Mcaclennan, T.K. Sham, J. Luminescence **166**, 143–147 (2015)
64. L. Liu, T.K. Sham, in *Titanium Dioxide: Material for a Sustainable Environment*, ed. by D. Yang, Luminescence from TiO2 Nanotubes And Related Nanostructures Investigated using Synchrotron X-Ray Absorption Near-Edge Structure and X-ray Excited Optical Luminescence (IntechOpen, 2017)
65. L. Liu, J. Li, T.-K. Sham, Can. J. Chem. **93**, 106–112 (2015)
66. A.K. Guria, S.K. Dutta, S.D. Adhikari, N. Pradhan, ACS Energy Lett. **2**, 1014–1021 (2017)
67. A.P. Hammersley, S.O. Svensson, M. Hanfland, A.N. Fitch, D. Hausermann, High Pressure Research **14**, 235–248 (1996)
68. L.B. McCusker, R.B. von Dreele, D.E. Cox, D. Louer, P. Scardi, J. Appl. Cryst. **32**, 36–50 (1999)
69. A.C. Larson, R.B. Von Dreele, Los Alamos National Laboratory Report, Los Alamos National Laboratory (2000)
70. B.H. Toby, J. Appl. Cryst. **34**, 210–213 (2001)
71. J. Yang, B.D. Siempelkamp, D. Liu, T.L. Kelly, ACS Nano **9**, 1955–1963 (2015)
72. J. Zhao, B. Cai, Z. Luo, Y. Dong, Y. Zhang, H. Xu, B. Hong, Y. Yang, L. Li, W. Zhang, C. Gao, Sci. Rep. **6**, 21976 (2016)
73. L. Zhang, Q. Zeng, K. Wang, J. Phys. Chem. Lett. **8**, 3752–3758 (2017)
74. P. Wang, J. Guan, D.T.K. Galeschuk, Y. Yao, C.F. He, S. Jiang, S. Zhang, Y. Liu, M. Jin, C. Jin, Y. Song, J. Phys. Chem. Lett. **8**, 2119–2125 (2017)
75. M. Szafranski, A. Katrusiak, J. Phys. Chem. Lett. **7**, 3458–3466 (2016)
76. P. Postorino, L. Malavasi, J. Phys. Chem. Lett. **8**, 2613–2622 (2017)

77. G. Liu, L. Kong, J. Gong, W. Yang, H.-K. Mao, Q. Hu, Z. Liu, R.D. Schaller, D. Zhang, T. Xu, Adv. Funct. Mater. **27**, 1604208 (2017)
78. A. Jaffe, Y. Lin, W.L. Mao, H.I. Karunadasa, J. Am. Chem. Soc. **139**, 4330–4333 (2017)
79. L. Wang, K. Wang, B. Zou, J. Phys. Chem. Lett. **7**, 2556–2562 (2016)
80. L. Wang, K. Wang, G. Xiao, Q. Zeng, B. Zou, J. Phys. Chem. Lett. **7**, 5273–5279 (2016)
81. M. Szafranski, A. Katrusiak, J. Phys. Chem. Lett. **8**, 2496–2506 (2017)
82. X. Lu, Y. Wang, C.C. Stoumpos, Q. Hu, X. Guo, H. Chen, L. Yang, J.S. Smith, W. Yang, Y. Zhao, H. Xu, M.G. Kanatzidis, Q. Jia, Adv. Mater. **28**, 8663–8668 (2016)
83. S. Jiang, Y. Fang, R. Li, H. Xiao, J. Crowley, C. Wang, T.J. White, W.A. Goddard 3rd., Z. Wang, T. Baikie, J. Fang, Angew. Chem. **55**, 6540–6544 (2016)
84. A. Jaffe, Y. Lin, C.M. Beavers, J. Voss, W.L. Mao, H.I. Karunadasa, ACS Central Sci. **2**, 201–209 (2016)
85. A. Jaffe, Y. Lin, W.L. Mao, H.I. Karunadasa, J. Am. Chem. Soc. **137**, 1673–1678 (2015)
86. Y. Lee, D. Mitzi, P. Barnes, T. Vogt. Phys. Rev. B, 68 (2003)
87. K. Gesi, Ferroelectrics **203**, 249–268 (1997)
88. K. Matsuishi, T. Ishihara, S. Onari, Y.H. Chang, C.H. Park, Phys. Stat. Sol. (b) **241**, 3328–3333 (2004)
89. I.P. Swainson, M.G. Tucker, D.J. Wilson, B. Winkler, V. Milman, Chem. Mater. **19**, 2401–2405 (2007)
90. A.D. Chijioke, W.J. Nellis, A. Soldatov, I.F. Silvera, J. Appl. Phys. **98**, 114905 (2005)
91. A.R. Milosavljevic, D.K. Bozanic, S. Sadhu, N. Vukmirovic, R. Dojcilovic, P. Sapkota, W. Huang, J. Bozek, C. Nicolas, L. Nahon, S. Ptasinska, J. Phys. Chem. Lett. **9**, 3604–3611 (2018)
92. L. Oesinghaus, J. Schlipf, N. Giesbrecht, L. Song, Y. Hu, T. Bein, P. Docampo, P. Müller-Buschbaum, Adv. Mater. Interfaces **3**, 1600403 (2016)
93. M. Kodur, R.E. Kumar, Y. Luo, D.N. Cakan, X. Li, M. Stuckelberger, D.P. Fenning. Adv. Energy Mater. 1903170 (2020)

# Chapter 7
# Perovskite Quantum Dot Photodetectors

**Xiangxing Xu and Linwei Yu**

**Abstract** In this chapter, we will review the developments of perovskite photodetectors in recent years. The application of perovskite nanomaterials—especially perovskite quantum dots (perovskite QDs)—in photodetectors. In Sect. 7.1, the background of perovskite QDs, and photodetectors are introduced. In Sect. 7.2, fundamentals of QD (beyond perovskite QD) photodetectors including their brief history are outlined. In Sect. 7.3, the perovskite based photodetectors, especially the perovskite QD photodetectors, are expounded in different categorizations on the detecting band type, working mechanism, mechanical performance, response feature, device structure, response wavelength and material design. The future of perovskite QD photodetectors is prospected in the last section.

## 7.1 Introduction to Perovskite QDs and Photodetectors

### *7.1.1 Quantum Dots*

Quantum dots (QDs) are small semiconductor nano particles showing quantum confinement effect: when the particle size is comparable or smaller than the corresponding exciton Bohr radius, the excitons in the particle are getting spatially confined in three dimensions, inducing increased bandgap versus its bulk state [1–3]. QDs are also sometimes called artificial atoms, emphasizing their size, singularity, surface bound and atomic-like discrete electronic states [4, 5]. Traditional QDs include IV, II–VI, IV–VI, III–V and I–III–VI group semiconductor materials, e.g.

X. Xu (✉)
School of Chemistry and Materials Science, Nanjing Normal University, Nanjing 210046, China
e-mail: xuxx@njnu.edu.cn

L. Yu
School of Electronic Science and Engineering, Nanjing University, Xianlin Campus, Nanjing 210046, China
e-mail: yulinwei@nju.edu.cn

 
Y. Zhou and Y. Wang (eds.), *Perovskite Quantum Dots*, Springer Series in Materials Science 303, https://doi.org/10.1007/978-981-15-6637-0_7

Si, CdSe, ZnO, PbS, InP, and $CuInS_2$ QDs, etc [6–8]. Physical, chemical, biological and combined methods have been developed to prepare QDs.

### *7.1.2 Perovskite QDs: Structure and Synthesis*

Perovskite structure is a type of crystal structure as calcium titanium oxide. Specifically, perovskite semiconductor materials have the chemical formula of $ABX_3$, where A can be a cation of organic methylammonium (MA), formamidinium (FA) or inorganic $Cs^+$ etc., sitting at cube corner positions of the unit cell; B is a second cation of $Pb^{2+}$ or $Sn^{2+}$, etc., sitting at body-center position; X is typically the halide ion of $Cl^-$, $Br^-$, $I^-$ or their mixtures, occupying face-centered positions. The $[BX_6]^{4-}$ octahedra shares the X with the neighboring $[BX6]^{4-}$ octahedral in the corner to corner manner. Structurally derived from the perovskite, there are also some perovskite-related structures. In the $A_4BX_6$ structure (or zero-dimensional (0D) structure), the X is no longer shared by the $[BX_6]^{4-}$ octahedras, which are completely decoupled in all dimensions [9]. The $AB_2X_5$ has a 2D featured structure, which is formed by alternating A and $[B_2X_5]^-$ layers [10]. The $A_2BX_4$ also has the 2D structure, formed by alternating of $[BX_6]^{4-}$ octahedras and large cations of A [11]. Other perovskite derivatives have the structures of $A_2BX_6$ (0D), [12] $A_2B^+B^{3+}X_6$ (3D) [13] and $A_3B_2X_9$ (2D) [14].

When the size of the perovskite perovskite semiconductor particles reduced to a few nanometers that comparable to corresponding Bohr radius, their optical and electronic properties could be tailored due to quantum mechanics. These quasi 0D particles are called perovskite QDs. In addition to the size, their electronic and optoelectronic properties also can be tuned by the component/doping, morphology, surface/ligands and hybrid structure with hetero materials. The organic-inorganic colloidal perovskite of $MAPbX_3(MA = CH_3NH_3, X = Cl/Br/I)$ QDs were first reported in 2014 by Schmidt et al. [15] The all inorganic colloidal perovskite $CsPbX_3(X = Cl/Br/I)$ QDs were first synthesized in 2015 by Protesescu et al., using a hot injection method [16]. To date, various methods have been developed to prepare perovskite QDs, mainly the top-down methods and the bottom-up methods. The former includes grinding, ball milling and exfoliation. The later is a big family containing mainly the solution phase synthesis of hot injection, heating up, ligand-assisted reprecipitation, microemulsion, microwave, microfluidic, ultrasonication and post-synthesis ion-exchange etc. [17].

In recent years, perovskite QDs have aroused great attention [15–18] and various applications of perovskite QDs have been demonstrated in fields of solar cells, [19] light emitting devices, [20] lasers, [21] sensors, [22] catalysis, [23] computing, [24] and photodetectors [25–30].

### 7.1.3 Types of Photodetectors

Photodetectors are also called photosensors. A photodetector absorbs visible or invisible electromagnetic radiation, transforms it into measurable signals. Based on the detection mechanisms, the photodetectors can be classified into different types.

1. **Photoemission**. When illuminated, the material emits free electrons into vacuum or gas. Typical devices include gaseous ionization detectors, phototubes and photomultiplier tubes.
2. **Thermal**. The electrons are excited to higher energy states and then transition back, during the decay process the energy transforms to detectable heat or phonon. Typical devices include bolometers, cryogenic detectors, pyroelectric detectors, thermopiles and Golay cells.
3. **Polarization**. When the polarization state of a material is responded to light, the change of polarization or refraction index can be detected or used. Typical applications include photorefractive materials for holographic data storage and linear/circular light detectors.
4. **Photochemical**. Chemical changes or reactions can be induced by light in proper materials. Photographic plates are such photodetectors. A typical biological example is that the photoreceptor cells in the retina detect light by photon-induced chemical cascade.
5. **Semiconductor**. Electrons are excited by light, generating conductive electrons and holes, and therefore the detectable light induced electrical current. Photodetectors based on semiconductors is a big family most widely used nowadays. Typical devices include photoconductors, photodiodes (specifically, solar cells are large area photodiodes generating electric power), phototransistors, photomultiplication-type photodetectors and photon drag detectors.

### 7.1.4 Photodetector Performance Metrics

Some important parameters are used to characterize the performance metrics of photodetectors. They are summarized as follows.

1. ***Dark current density*** ($J_D$, A cm$^{-2}$): The current flowing through a photodetector in the dark.
2. ***External quantum efficiency*** (*EQE*, %): The ratio of photo-generated charge carrier numbers that can be collected to the incident photon numbers:

$$EQE = \frac{I_{ph}/e}{P_{in}/h\nu} \tag{7.1}$$

where $I_{ph}$ is the photocurrent and $P_{in}$ is the incident light power on the active area.

3. ***Gain*** ($G$): The ratio of lifetime to transit time of charge carriers traversing through the device.
4. ***Internal quantum efficiency*** ($IQE$, %): The ratio of photo-generated charge carrier numbers that can be collected to the absorbed photon numbers by the active material.
5. ***Linear dynamic range*** ($LDR$, dB): The range of light powers for which the detector responds linearly:

$$LDR = 20\log\frac{I_{\text{upper}} - I_{\text{d}}}{I_{\text{lower}} - I_{\text{d}}} \tag{7.2}$$

   where $I_{\text{upper}}$ is the current value at which the response of the device deviates from linearity and $I_{\text{lower}}$ is the lower resolution limit, $I_{\text{d}}$ is the dark current.
6. ***Noise equivalent power*** ($NEP$, W $\text{Hz}^{-1/2}$): The lowest light power $P_1$ needed to generate a signal equivalent to the noise (the signal-to-noise ratio (SNR) is equal to 1) of the device as a function of frequency. It can be written as:

$$NEP = \frac{P_1}{\sqrt{\Delta f}} \tag{7.3}$$

   where $\Delta f$ is the bandwidth (Hz).
7. ***Noise spectrum*** ($I_{\text{n}}$, A $\text{Hz}^{-1/2}$): The intrinsic noise current as a function of frequency.
8. ***On/off ratio*** ($I_{\text{ph}}/I_{\text{d}}$): The ratio of photocurrent to dark current.
9. ***Photocurrent density*** ($J_{\text{L}}$, A $\text{cm}^{-2}$): The current flowing through a photodetector under light.
10. ***Rejection ratio*** ($R_1/R_2$): The ratio of responsivity at a detectable wavelength (in the detection spectrum window) to responsivity at a wavelength off the detection window.
11. ***Responsivity*** ($R$, A $\text{W}^{-1}$): The photogenerated current divided by total input light power on the photodetector. It is also called sensitivity.

$$R = \frac{I_{\text{ph}}}{P_{\text{in}}} = \frac{I_{\text{light}} - I_{\text{d}}}{P_{\text{in}}} \tag{7.4}$$

   where $I_{\text{light}}$ is the measured current.
12. ***Rise/fall (decay) time*** ($t_{\text{r}}/t_{\text{f}}$, s): The time taken by the detection signal change from 10 to 90% or from 90 to 10% of the maximum output step height.
13. ***Specific detectivity*** ($D^*$, Jones or cm $\text{Hz}^{1/2}$ $\text{W}^{-1}$): Briefly called detectivity. It is a key parameter and defined as:

$$D^* = \frac{\sqrt{A\Delta f}}{NEP} \tag{7.5}$$

   where $A$ is the active area of the device.

14. ***Spectral response***: The response of a photodetector as a function of photon frequency.
15. ***Stability***: The performance retention under operation condition.

## 7.2 QD Photodetectors

### 7.2.1 Fundamentals of QD Photodetectors

Application of QDs as photoactive materials in photodetectors can be traced back to 1990s [31]. In recent years, QD photodetectors attract growing attention, with the emerging of new QDs and novel device structures. A key driving force for the continuous development of this field is that some advantages of QDs are particularly suitable for photodetectors, e.g. the size and morphology tunable bandgap, tailorable surface chemistry, flexibility in trap/doping/composition engineering, and the feasibility to manufacture devices by inexpensive solution processing.

Sketches of simple lateral structured and vertical structured QD photodetectors are shown in Fig. 7.1. A thin layer of QDs is deposited between the electrodes. The electrodes can be Au, Ag, Pt, Al or transparent indium tin oxide (ITO), etc., and the substrate can be glass or flexible polymers, etc.

To illustrate how QDs determine the performance of photodetectors, an important parameter the photocurrent of photodetector need to be analyzed:

$$I_{ph} = IQE \cdot e \cdot N \cdot G \tag{7.6}$$

where $I_{ph}$ is the photocurrent of the device, $IQE$ is the internal quantum efficiency, $e$ is the elemental charge, $N$ is the number of photons absorbed by the active material per unit time, and $G$ is the gain, which defined as:

$$G = \tau / T \tag{7.7}$$

where $\tau$ is the carrier lifetime and $T$ is the transit time of carriers.

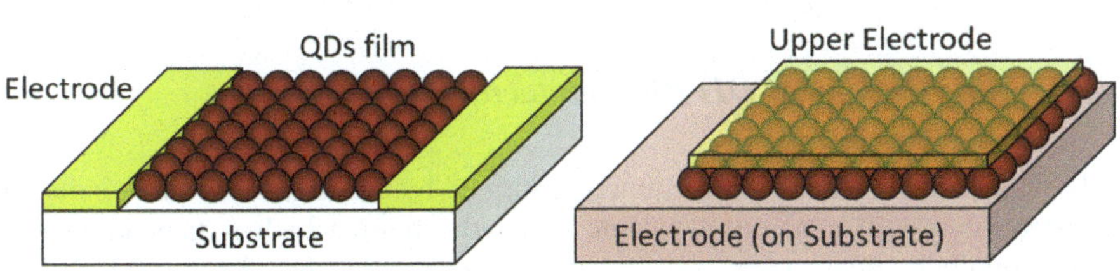


**Fig. 7.1** **a** A sketch of a typical lateral structured QD photodetector. **b** A sketch of a simple vertical structured QD photodetector. The QDs are not necessarily monolayer

*IQE* depends not only on the device structure, but also on the component, structure (including material structure and particle structures e.g. the core/shell structure), size, morphology (e.g. sphere, cubic, hollow etc.) and surface (including surface ligands or atoms, surface defects and surface reconstructions) of the QDs. It should be mentioned that the quantum yield (QY) or quantum efficiency (QE) of QD photoluminescence (PL) is different from the IQE or EQE of the photodetectors. Like *IQE*, *T* highly relies to the device structure and the QDs properties. For examples, the value of *T* could be much smaller in devices with the vertical structure than in lateral structured devices, because the distance between electrodes could be much smaller in the former than that in the later; *T* is also affected by the carrier transport mechanisms, e.g. hopping and tunneling mechanisms, which tightly related to the nature of the colloidal QDs assembled films. The carrier lifetime $\tau$ in photodetectors is controlled by the recombination routes, i.e. radiative and nonradiative, intrinsic and trap/defect states related recombinations. Specifically, $\tau$ is greatly controlled by the lifetime of trap states. Doping and surface induced strap states are two major contributions. The surface trap states of QDs can be caused by the unsaturated surface passivation of ligands because of the steric effect, leaving partial of dangling bonds. The surface trap states can capture photo-generated carriers or excitons. Among them, the long-living carrier traps (commonly the deep traps) are essential to $\tau$. Generally, the capture of one type of carriers (electrons or holes) by a certain kind of long-living traps would make this type of carrier to be the minority. With the presence of the long-living minority carrier traps, the majority carriers may transit through the detector for cycles before the recombination with the minority carriers. Therefore $\tau$ is greatly prolonged, resulting in a high *G* device. It means that a QD photodetector with high sensitivity or responsivity (*R*) would have long response or Rise/fall (decay) time ($t_r/t_f$). On the other hand, a short carrier lifetime (small $\tau$) leads to low *G* low *R* devise; however, a fast response can be achieved. Since the trap states of QDs can be tuned by defect engineering, [32–34] the responsivity and response time of QD photodetectors can be optimized according to different application requirements. For examples, the gain was reported > $10^5$ for $MoS_2$-ZnCdSe QD photodetectors with rise time of $\approx$ 0.3 s and fall time of $\approx$ 1.2 s; [35] While high-speed with rise time of 100 ns and fall time of 110 ns were achieved in PbSe QD photodetectors with gain <1 [36].

### 7.2.2 The History of QD Photodetectors

In 1839, Becquerel found that the junction formed by the contact of semiconductor and electrolyte will generate a voltage under the light, which is later known as the photovoltaic effect. Afterwards, many materials were studied and used in solar cells and photodetectors for their photoconductive property, e.g. Se, CdS, $Cu_2O$, Si, Ge, PbS etc. In these devices, photoactive semiconductor materials can be in bulk, film or powder state. In 1980s, physical theory of QDs were established and synthesis methods of colloidal QDs were appeared, [1–3] which prepared for the QDs application in the fields of photodetectors.

The pioneering study on photoconductivity of QDs is reported by Wang in 1992 [31]. It is a composite of polyvinylcarbazole (PVK) doped with CdS QDs with the diameter of $1.6 \pm 0.5$ nm. The photoconductivity is measured by the photo-induced discharge method [37, 38]. The film was deposited on an electrically grounded aluminum substrate. Then it was corona-charged with positive or negative charges in the dark. Under light illumination, the light generated carriers from CdS QDs migrated to the film surface, recombined with surface charges. The change of the surface charges was recorded in the form of surface potential measured by an electrostatic voltmeter. The gain spectra (wavelength dependent) of the sample for both positive and negative charging were measured. They are consistent with the absorption spectrum of CdS QDs in > 350 nm region (PVK absorbs light <350 nm) with the response fall time of ~1 s, demonstrating the effectiveness of CdS QDs. Other nanocrystals/nanoparticles such as InAs, $PbI_2$, HgS, $Ga_2S_3$, $In_2S_3$ and fullerenes were also doped into polymers of PVK, (phenylmethyl)polysilane (PMPS) or amine-doped polycarbonate to form photoconductive composites, and their charge generation and transport properties were investigated by similar methods [39–41].

With the development of synthesis techniques, colloidal QDs of many semiconductor materials with tunable size, good dispersability and tailored functional ligands came into a burst in 1990s. It brought about chances to fabricate photodetectors by using films of assembled colloidal QDs. The epitaxial QDs preparation methods of molecular beam epitaxy (MBE) and metal organic chemical vapor deposition (MOCVD) were appeared in the same decade of 1990s, [42, 43] and were also intensively studied for photodetectors [44]. Compared with the epitaxial QDs, the solution synthesized colloidal QDs have some significant advantages: low temperature and easy processing, low cost, large area, higher packing density and adaptable to many flexible substrates. The research on CdSe QDs based visible range photodetectors were carried out by Bawendi's group. In 2000, Bawendi and cooperators reported the photoconductivity of colloidal CdSe QD solids [45]. The relevance of QD size, excitation intensity, temperature, surface ligands and applied electric field to the photoconductivity were studied in detail. The results indicated that the photoexcited excitons were quantum confined in individual QDs. The ionization rate of the excitons (charge separation) in the lowest excited state (and therefore the photoconductivity) depended on the applied electric field, the size and surface passivation of the QDs. Separation of excitons confined to the core of the QD required significantly more energy than separation of carriers in trapped states, while both energies are much larger than the thermal energy at room temperature. The insulating ligands, inter QD spacing and high exciton binding energy (~200 meV for 2 nm CdSe QDs) resulted in the negligible contribution of separation of excitons by thermal. The charge separation proceeded via tunneling processes through the QD organic capping layer, which competed with radiative (photoluminescence) and nonradiative recombinations which decreased the internal charge generation efficiency. The photocurrent varied linearly over two orders of magnitude of excitation intensity. The dark current was below the noise level of 0.1 pA and about two orders of magnitude smaller than the photocurrent. However, even at high field of 250 kV/cm, the photoconductive gain (G) was only on the order of $10^{-4}$ charges/photon.

To improve the photoconductivity, Jarosz et al. used a variety of reagents of amines, tributylphosphine and sodium hydroxide to treat the CdSe QD films after deposition [46]. The photoconductivity of all the treated CdSe QD films increased. It confirmed that the improvement was largely a consequence of increased QD surface passivation due to the basic nature of amines and sodium hydroxide and decreased inter QD spacing, regardless of whether the new ligands were conjugated or able to cross-link the QDs. In 2005, Oertel et al. report vertical structured visible spectrum photodetectors with the active layer of 200 nm thick CdSe QDs between ITO/PEDOT:PSS and Ag electrodes [47]. The CdSe QDs were treated by n-butylamine, greatly increasing the exciton ionization and thus the photoconductivity. At 0 V bias, the on/off ratio was the highest up to $10^3$ with light wavelength of 514 nm. The 3 dB frequency was at 50 kHz. The detection limit was ~10 $\mu$W/cm$^2$. At −6 V bias, EQE ranged from 15 to 24% with the wavelength change from 350 to 575 nm. The photoconduction of CdSe QDs went further to the inorganic surface modification. Porter et al. studied the photoconductivity of close packed films of CdSe/ZnS core/shell QDs [48]. By replacing the organic ligands with an inorganic ZnS shell and the majority of the organic ligands of the core/shell QDs removed by thermal annealing, the photoconduction achieved near unity IQE at room temperature. It demonstrated that with the passivation of the CdSe core, the nonradiative recombination was reduced and the photoconductive mechanism changed from the carrier generation rate limit mode to the carrier mobility limit mode.

During the same period, the photodetection study by using colloidal QDs with the spectrum extending to infrared was first and mainly conducted by Sargent's group [49]. In 2004, they reported infrared (975–1300 nm) photoconductivity of a polymer(MEH-PPV)/PbS QDs (5 nm) composite film. It adopted a vertical structure of the composite film between ITO and Mg/Au electrodes. The photocurrent was attributed to PbS QDs with hole transfer by the polymer. The IQE was ~$5\times10^{-6}$ to $10^{-5}$ charges/photon at 5 V bias. The sensitivity of the composite was further tuned across the 800–2000 nm spectral region [50]. In 2006, a PbS QDs lateral photodetector was reported [51]. It showed large gain of $10^2$ to $10^4$ with responsivities greater than $10^3$ A/W. The device exhibited detectivity of $1.8 \times 10^{13}$ jones at 1.3 $\mu$m at room temperature. The high detectivity was attributed to the presence of long-lived surface traps states.

In the past 10 years, colloidal QDs based photodetectors with enhanced performance was achieved by many strategies. Among them, the QD surface engineering, device optimization, and new materials such as grapheme, $MoS_2$ and perovskite based materials are major contributions. The photodetection ranges beyond visible from infrared to solar bind, and even to X/gamma-ray.

## 7.3 Perovskite QD Photodetectors

### *7.3.1 Category Methods*

Perovskite photodetectors can be divided into various types according to different factors such as the detecting band type, working mechanism, device mechanical performance, response features, device structure, response wavelength and material system. Figure 7.2 summarizes some typical categories. In the following paragraphs, these types of perovskite photodetectors are illustrated in detail.

### *7.3.2 Band Types*

#### 7.3.2.1 Broadband Perovskite Photodetectors

The energy structure of the perovskite QDs can be seen directly from the absorption spectrum. Commonly, like most types of the QDs, in the axial of wavelength from the long wavelength side (low energy) to the short wavelength side (high energy), the absorption of perovskite QDs goes from zero to meet the absorption edge, and then fluctuates up and down with the energy structure (the density of state) of the QDs above zero. It means that perovskite QDs absorb light of the energy higher than

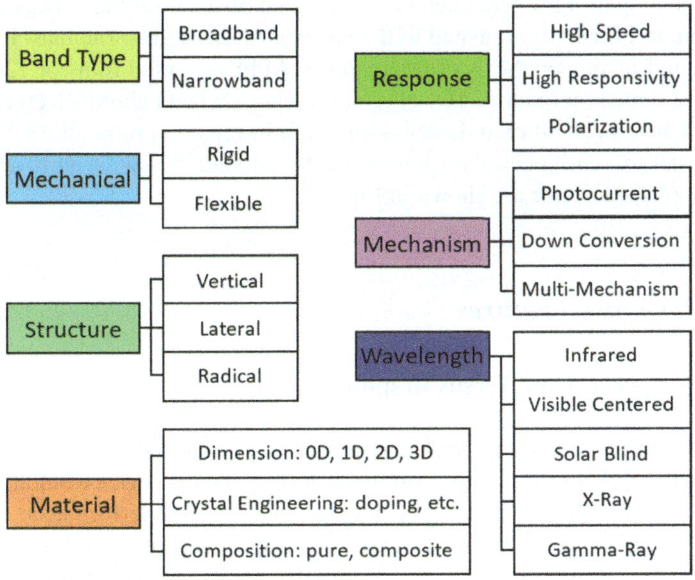


**Fig. 7.2** Categories of perovskite photodetectors

the bandgap and generate carriers. This feature indicates that the perovskite QDs are intrinsically applicable for broadband (panchromatic) photodetectors. The detection spectrum may range from ultraviolet (UV) to the band edge. The detection range depends on the bandgap of perovskite QDs. This kind of photodetectors are usually used for imaging that multicolor light detection is needed. High EQE covering the spectrum range is preferred. Most perovskite photodetectors are broadband.

#### 7.3.2.2 Narrowband Perovskite Photodetectors

Another type is called narrowband (wavelength selective) photodetectors. It is only sensitive for some specific wavelength with relatively narrow detectivity window. It often finds the applications that need detection window selectivity with suppressed background light, e.g. in imaging of multi-color labeling, coding systems for security or communication purpose, etc. In 2015, Armin et al. demonstrated a new concept of charge collection narrowing (CCN) in thick bulk heterojunctions to achieve narrowband photodetectors [52]. This concept was then adopted to perovskite materials. Fang et al. presented $MAPbX_3(X = Cl/Br/I/mixture)$ single crystal photodetectors with full-width at half-maximum (FWHM) of <20 nm as narrow for the response window [53, 54]. The response spectra can be continuously tuned in the visible range by the halide composition ratio. The device structure is Ga/perovskite/Au with the single crystal thickness of 0.1–1.5 mm. The narrowband photodetection can be explained by the mechanism that, the excess carriers close to the crystal surfaces generated by short wavelength light bear strong surface-charge recombination; while the below-bandgap light generated carriers locate away from the surfaces and can be much more efficiently transited to the electrodes assisted by the bias. Later, Rao et al. reported in situ synthesis of large-area $MAPbBr_3$ crystal films of 120 $cm^2$ on the fluorine-doped tin oxide (FTO) glass for the narrowband photodetectors; [55] Li et al. fabricated a self-filtered design with the light rejection ratio above 1000 [56]. Typical structures and spectral responses of a broadband [57] and a narrowband [53] perovskite photodetector are shown in Fig. 7.3.

### 7.3.3 *Response Features*

#### 7.3.3.1 Response Time Versus Responsivity

The response time ($t_r$ and $t_f$) and responsivity/sensitivity ($R$) are two interrelated properties. For QD photodetectors, a high $R$ suggests that the device wouldn't has very high speed; While a high speed device may lead to the low $R$ performance. However, since there are many factors that affect the performance, the usage of this seesaw rule is limited to devices of the same materials and structural parameters. In another word, the responsivity and response time can be improved simultaneously by: (i) perovskite design, e.g. the morphology engineering of QDs, nanowires, nanosheets etc. and

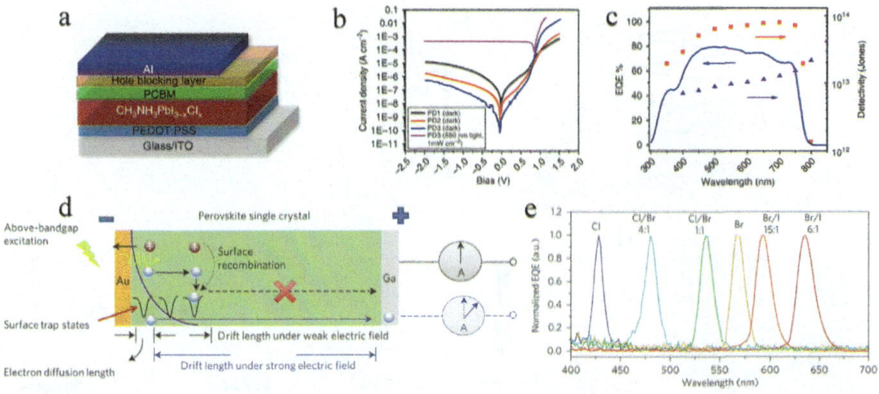


**Fig. 7.3** **a** Device structure of a broadband perovskite photodetector, and its **b** current density–voltage performance, **c** EQE and detectivity of the perovskite photodetector at different wavelength with purple triangles of a single-crystalline silicon diode. Reprinted with permission from ref [57]. Copyright 2014, Springer Nature. **d** Schematic illustration of the mechanism of the perovskite single crystal based narrowband photodetector, and its **e** ultra-narrow *EQE* peak and tunable spectral response under −1 V bias. Reprinted with permission from ref [53]. Copyright 2015, Springer Nature

surface/interface engineering; (ii) device structure optimization, e.g. shortening of the distance between electrodes in lateral structure or adopting vertical structures; (iii) introducing materials with high carrier transfer rate or to facilitate the carrier extraction.

#### 7.3.3.2 Linear and Circular Polarized Light Response

According to the physical interaction between electromagnetic wave and crystalline material, a single crystal can show strong polarization (linear or circular polarized light) dependent optical properties. When the material size comes down to micro/nano scale, the optoelectronic properties are also morphology and assembly pattern dependent [58–60]. Feng et al. fabricated polarization sensitive perovskite photodetectors by aligned single crystalline 1D $CsPbBr_3$ nanowire arrays (Fig. 7.4a, b) [61]. The photocurrent maximum appeared to linear polarization direction paralleled to the axial direction of the 1D nanowire arrays; while the minimum photocurrent emerged vertical to the axial direction in the plane. A polarization ratio of about 2.6:1 ($I_{max}$: $I_{min}$, the photocurrent maximum to minimum) was achieved. Sheng et al. developed a general method to synthesized size and thickness controllable colloidal $CsPbX_3$($X = Cl/Br/I$/mixture) perovskite 2D nanoplates [62]. These the 2D $CsPbX_3$ nanoplates exhibited polarized absorption and photoluminescence not only in liquid solvent and solid resin matrix, but also in self-assembled films. A general model was established to understand the absorption and emission polarization of nanocrystals. A photodetector of the $CsPbBr_3$ nanoplates assembled film was demonstrated sensitive

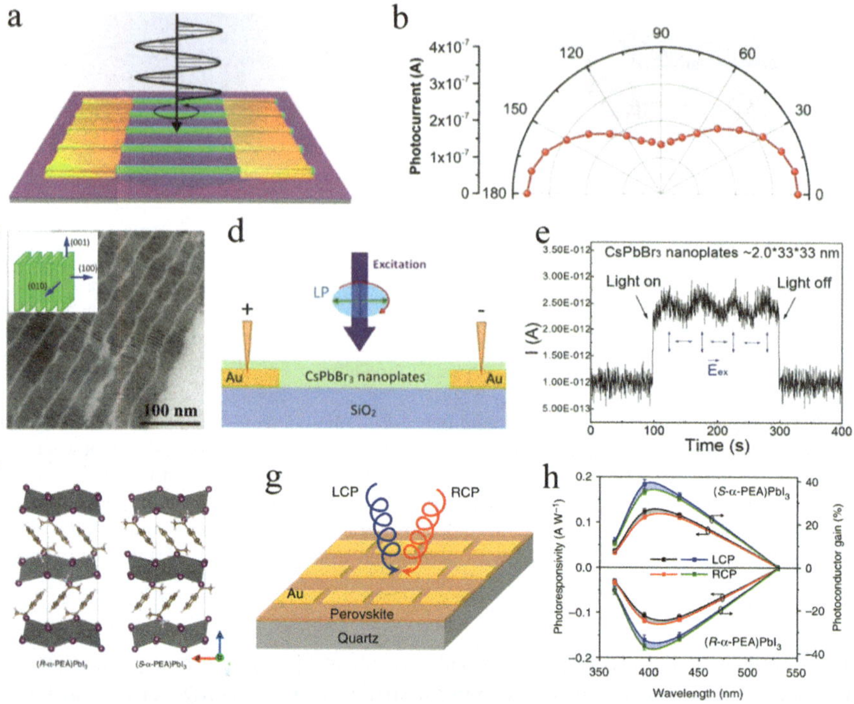


**Fig. 7.4** **a** Schematic illustration of polarization-sensitive photodetection on 1D perovskite arrays, and **b** the polarization dependence of photocurrent. Reprinted with permission from ref [61]. Copyright 2017, WILEY-VCH Verlag GmbH & Co. KGaA, Weinheim. **c** TEM image of the perovskite nanoplates and inset the assembly orientation. **d** Schematic of the photodetector. **e** Photocurrent response to linear polarized light. Reprinted with permission from ref [62]. Copyright 2018, WILEY-VCH Verlag GmbH & Co. KGaA, Weinheim. **f** The crystal structure of (R- and S-α-PEA)$PbI_3$. **g** Schematic diagram of the photodetector. **h** The responsivity and photoconductor gain of (R- and S-α–PEA)$PbI_3$ device under left-handed and right-handed circularly polarized light. Reprinted with permission from ref [63]. Copyright 2019, Springer Nature

for linear polarized light. Chen et al. demonstrated a flexible circular polarized light detector based on chiral organic-inorganic hybrid perovskites (R- and S–α–PEA)$PbI_3$ [63]. The device had responsivity of 797 mA/W, detectivity of $7.1 \times 10^{11}$ Jones and 3-dB frequency of 150 Hz. These work on the direct polarized light detection by perovskite shows potentials for future polarization imaging systems.

### *7.3.4 Mechanical Performance: Flexible or Not*

Common prototypes of perovskite photodetectors are manufactured on rigid substrate, such as $Si/SiO_2$, glass and ITO. In recent years, flexible electron technology aims to make electronic devices of organic or inorganic materials on flexible or ductile substrates for the growing demand for lightweight, portable electronics [64, 65]. Compared with the rigid ones, the flexible devices have more flexibility, adapt to different working environment and meet the deformation requirements.

In 2014, Hu et al. reported the deposition of $MAPbI_3$ on the ITO-coated polyethylene terephthalate (PET) sheet to fabricate the first flexible perovskite based broadband photodetectors. The wavelength covered 300-800 nm, showing responsivity of 3.49 A/W, 0.0367 A/W, EQE of $1.19 \times 103\%$, 5.84% at 365 nm and 780 nm with a bias of 3 V, respectively [66]. Zheng et al. reported a simple spin-coating combined co-electrospinning method of incorporating $MAPbI_3$ QDs with $TiO_2$ nanotubes (NTs) to fabricate flexible and transparent (85%) photodetectors on mica [67]. It revealed a relative fast and stable response in the 300–800 nm range with a responsivity of 0.2 A $W^{-1}$ at 700 nm. The performance retained well after 200 cycles of repeated bending at 90°. Bao et al. used patterned Au nanowires on polyethylene naphthalate to prepare flexible $MAPbI_3$ photodetectors [68]. The EQE of 60%, responsivity of 321 mA/W, linear dynamic range of ~84 dB and response time of ~4 $\mu$s were achieved. Li et al. reported photodetector based on the $CsPbBr_3$ nanosheet/carbon nanotube (CNT) composite films. The CNT greatly improved the conductivity. Consequently, a high *EQE* of 7488%, responsivity of 31.1 A/W and a rise time of 16 $\mu$s were achieved. Due to the flexibility of the ultrathin 2D perovskite and 1D CNTs composite, the device showed a remarkable bending—recovering stability up to 10,000 cycles (Fig. 7.5) [69].

## *7.3.5 Working Mechanisms*

### 7.3.5.1 Photocurrent Mode

Based on the photodetection mechanisms, the perovskite QD photodetectors may work under photoconductive or photoluminescent conditions. The former works in a direct way that the perovskite QDs absorb light, generate excitons and participate in the carrier separation and transmitting processes to form electrical signals. Most perovskite photodetectors take this mechanism, with the prototypes of photo conductors, photodiodes/solar cells and phototransistors.

### 7.3.5.2 Down Conversion Mode

Some perovskite QDs have excellent fluorescent property with the photoluminescence quantum yield (QY) >50% and even approaching 100% [70]. The photoluminescent life time is rather short, commonly 1–20 ns. Therefore, the perovskite QDs can be used as a light energy down conversion media. The absorbed photons with high energy are tuned to the lower emission with high QY. The photoluminescence emission is then entering a photodetector for sensing. The QD down conversion strategy has long been used for enlarging the effective spectrum range of solar cells.

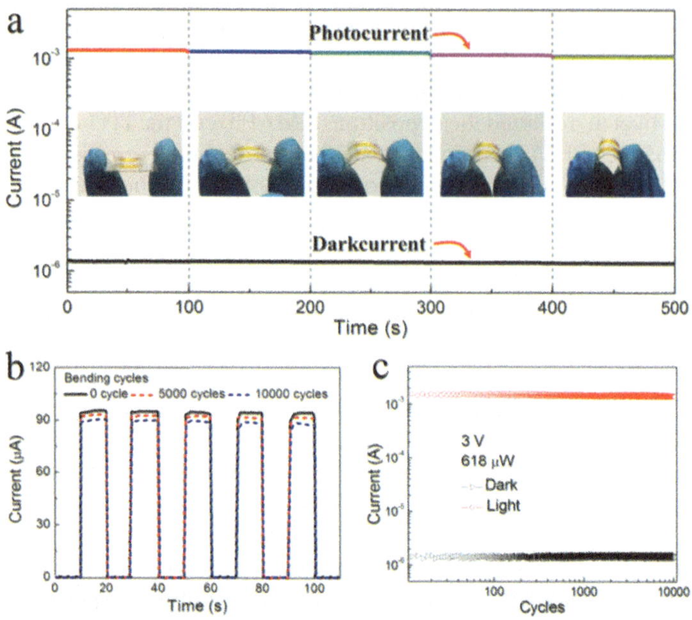


**Fig. 7.5** **a** Real time Current of the flexible device with bending at a bias of 3 V. **b** Performance of the device with bending—recovering cycles. **c** Photo and dark current during 10,000 cycles under 3 V bias with 442 nm light illumination (618 μW). Reprinted with permission from ref [69]. Copyright 2017, American Chemical Society

For examples, QDs of CdSe/CdS, $Cd_{0.5}Zn_{0.5}S$/ZnS, CdSe/ZnS and $CuInS_2$/ZnS have been applied as a down conversion function layer for solar cells [71–74]. In 2017, Lu et al. reported photodetectors by using $CsPbX_3$(X = Cl, Br, I or hybrid) perovskite QDs as effective down converters for the photovoltaic structures [75]. Zou et al. reported the combination of $CsPbX_3$ QDs with an organic–inorganic perovskite-based photodiode. The $CsPbX_3$ QDs not only acted as the UV down conversion layer, but also has the merit as a protection layer to prevent degradation of the organic–inorganic perovskite. An enhancement of one order magnitude for responsivity ($R$) and detectivity ($D^*$) was achieved [76].

#### 7.3.5.3 Multi-mechanism Modes

There were also reports of the conception that combine the photodetector with some other physical mechanisms, constructing new structured devices. Su et al. reported a $MAPbI_3$ perovskite based self-powered photodetector with dual sensing mechanisms of photoelectric and triboelectric effects. Both the photoconductivity and the surface triboelectric density of the device were significantly altered upon solar illumination [77]. Wang et al. fabricated a self-powered ZnO/$MAPbI_3$ heterostructured UV photodetector. The pyroelectric effect induced by ZnO nanowires on UV illumination was utilized for high-performance photon sensing. Compared with ZnO device without perovskite, the performance was improved with response rise time from 5.4 s to 53 μs and fall time from 8.9 s to 63 μs. The detectivity and the responsivity were both enhanced by 322% [78].

## 7.3.6 Device Structures

### 7.3.6.1 Vertical Structured Perovskite Photodetectors

In general, perovskite photodetectors can be divided into three types according to the structural features of the devices: the vertical, lateral and radial types. The vertical type has the sandwich structure: the plane electrode, function layers and counter electrode stack layer by layer. Sometimes there are interfacial layers on both sides of the perovskite layer, which is also referred as photodiode structured photodetectors. The distance between the electrodes commonly smaller than 1 μm. Therefore the carriers of electrons and holes can be extracted to corresponding electrodes easily, resulting fast response and low driving voltage. Additional merits including high responsivity, low dark current, high detectivity and broad linear dynamic range. In 2014, Dou et al. reported typical vertical type photodetectors of glass/ITO/PEDOT:PSS/$MAPbI_{3-x}Cl_x$/PCBM/Hole-blocking-layer/Al. The device exhibited the detectivity approaching $10^{14}$ Jones, a linear dynamic range over 100 dB and a fast photoresponse with 3−dB bandwidth up to 3 MHz [57]. Sutherland et al. fabricated perovskite photodetectors with the structure of glass/FTO/$TiO_2$/$Al_2O_3$/PCBM/$MAPbI_3$/Spriro/Au/Ag [79]. The response time was 1 μs as short. The photodetector was stable through a continuous pulsed-illumination over 7 billion light pulses. The responsivity approached 0.4 A $W^{-1}$ at 600 nm. The $Al_2O_3$/PCBM layer improved the operation stability in air, reduced the dark current and enhanced the responsivity to achieve a specific detectivity of $10^{12}$ Jones.

### 7.3.6.2 Lateral Structured Perovskite Photodetectors

The lateral type perovskite photodetectors may have the the photoconductor (Figs. 7.4 and 7.5) or field-effect phototransistor structures. In both structures, the electrode spacing is larger than that of the vertical type, commonly larger than 5 μm. Therefore, the driving voltage would be high and response time be prolonged. However, high gain ($G$) can be achieved through multiple carrier recirculation via long lived trap states, exhibiting high sensitivity for weak light.

***Photoconductor structures***. Dong et al. reported perovskite photodetector devices fabricated using the spin-coated films between two Au electrodes (on $SiO_x$ substrates) with the gap of 2 μm. $MAPbBr_3$, $MAPbBr_{3-x}Cl_x$ or $MAPbBr_{3-x}I_x$ colloidal nanocrystals with tunable bandgap were investigated. The on/off ratio of $10^5$ and response time <0.1 s were achieved [80]. Pang et al. fabricated a lateral photodetector with the channel width/length of 2000/5 μm. The effects of shape, size and surface passivation of $CsPbBr_3$ on the device performance were studied. It was found that the 6,6-phenyl-C61-butyric acid ethyl ester (PCBM) passivated $CsPbBr_3$ nanoribbons show better photo responsivity of 18.4 A/W, on/off ratio of $9.2 \times 10^3$ and rise/decay time of 8.7/3.5 ms, compared with those of unpassivated $CsPbBr_3$ QDs, nanocubes and nanoribbons [81]. The lateral photodetectors can also be made on bulk perovskite single crystals. The EQE for the $MAPbI_3$ single-crystal photodetector was found ~$10^3$ higher than that of the $MAPbI_3$ polycrystalline film photodetector [82].

***Phototransistor structures***. Due to the inherent amplification function of the phototransistors, it can work with an ultrahigh gain [83]. In 2015, Lee et al. reported a first perovskite($MAPbI_3$)-graphene phototransistor. This hybrid photodetector exhibited a photo responsivity of 180 A/W [84]. Then the responsivity was improved to 320 A/W by ambipolar phototransistors based on $MAPbI_3$ (Fig. 7.6a–c) [85]. By introducing high-mobility materials such as graphene, $MoS_2$, $WS_2$ etc. to the perovskite phototransistors, the responsivity was further enhanced to ~$10^6$ A/W [86–88]. In these photodetectors, the perovskite serves as the light absorbing media and the high-mobility materials provide charge transport channels. In 2017, Xie et al. reported an ultrasensitive phototransistor based on the multiheterojunction of $MAPbI_{3-x}Cl_x$ perovskite/poly(3-hexylthiophene) (P3HT)/graphene for the first time. The excitons were effectively separated by the P3HT. The perovskite trapped high-density electrons, resulting a strong photogating effect on the underlying graphene channel. An ultrahigh responsivity of ~$4.3 \times 10^9$ A/W and a gain of ~$10^{10}$ were achieved [89].

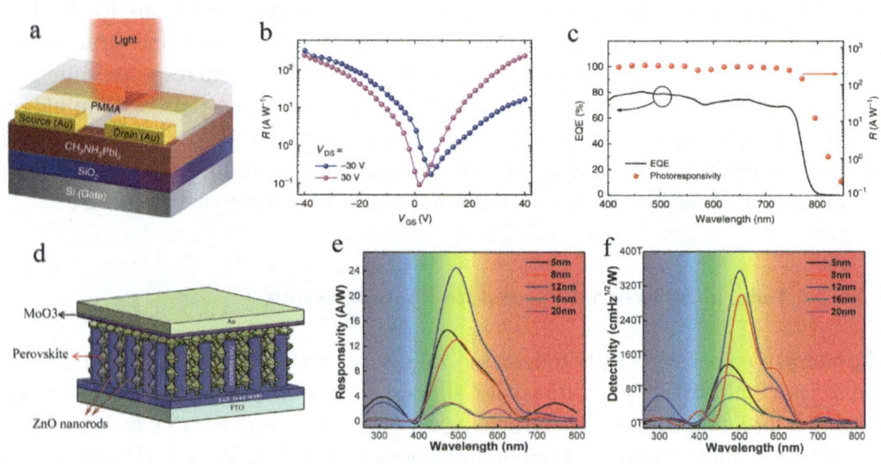


**Fig. 7.6** **a** Schematic of the perovskite phototransistor. **b** Photo responsivity of the phototransistor under white-light illumination (10 mW $cm^{-2}$). **c** *EQE* and responsivity at different wavelength with drain voltage ($V_{DS}$) of −30 V. Reprinted with permission from ref [85]. Copyright 2015, Springer Nature. **d** Schematic diagram, **e** responsivity and **f** detectivity of the radial structured perovskite photodetectors with the thicknesses of $MoO_3$ ranging from 5 to 20 nm. Reprinted with permission from ref [90]. Published by the Royal Society of Chemistry (2016)

#### 7.3.6.3 Radial Structures

The radial structures use vertical nanorod/nanowire arrays grown on substrate. The charge separation occurs radially and the carriers of either/both electron or hole may transport to the electrodes through the vertical nanorod/nanowire highway. Compared with flat lying nanorod/nanowire arrays, the vertical structure will trap light minimizing the reflection losses. In 2016, Yu et al. reported a self-powered photodetector based on the ZnO nanorod arrays/$MAPbI_3$ heterojunction. The photo responsivity of 24.3 A/W and detectivity of $3.56 \times 10^{14}$ Jones at 500 nm was achieved (Fig. 7.6d–f). A $MoO_3$ layer was deposited as a hole-transport layer. A 3-month illumination test indicated a good stability with only 9.3% response current decay [90]. The same group also reported a similar photodetector based on ZnO nanorod arrays/perovskite heterojunction structure by using spiro-OMeTAD instead of $MoO_3$.[91]. They showed the tunable performance of the photodetector by thermal treatment of ZnO nanorods. Optimized responsivity of 7.8 A/W and detectivity of $10^{14}$ Jones were obtained. Ga-doped ZnO nanorods and $TiO_2$ nanorods arrays for radial structured photodetectors were also investigated [92, 93]. The radial junction structured photodetector combining silicon nanowire (SiNW) arrays with perovskite QDs was revealed by Lu et al. [75] The SiNWs served as the self-driven detector by photovoltaic mechanism and $CsPbX_3$ QDs functioned as down converters.

### 7.3.7 *Detection Ranges of Wavelength*

#### 7.3.7.1 Visible Centered Perovskite Photodetectors (300–800 Nm)

There are several general factors that determine the detection ranges of wavelength. The first one is the material. The bandgap of perovskite QDs can be tuned by its size due to the quantum confinement effect. It also can be tuned by the perovskite component, such as the halogen anion ratio of Br:Cl or Br:I in $APbX_3$($A$ = MA/FA/Cs; $X$ = Br/Cl/I) QDs [94, 95]. The device structure and operating condition are additional dominating parameters for the response range, especially for the narrow band perovskite photodetectors.

#### 7.3.7.2 Perovskite Infrared Photodetectors (>800 Nm)

There are several strategies to extend the photodetection spectrum range to the Infrared wavelength beyond 800 nm.

***To use conventional pure lead based perovskite.*** The pure perovskite $MAPbI_3$ and $FAPbI_3$ have the bandgap of 1.4−1.5 eV. Therefore, the the near infrared light of the 800–900 range can be measured directly by corresponding photodetectors. Due to the sharp drop of the absorption spectrum at the bandgap edge, the deleterious decline of the device performance at this end is the main challenge toward practical utilization.

This obstacle was overcome by a phototransistor based on the multiheterojunction of $CH_3NH_3PbI_{3-x}Cl_x$ perovskite/P3HT/graphene structure (Fig. 7.7a–c) [89]. Though the $CH_3NH_3PbI_{3-x}Cl_x$ perovskite has the bandgap of ~1.55 eV (800 nm), under 895 nm light illumination, the responsivity was as high as $1.1 \times 10^9$ A/W with a gain

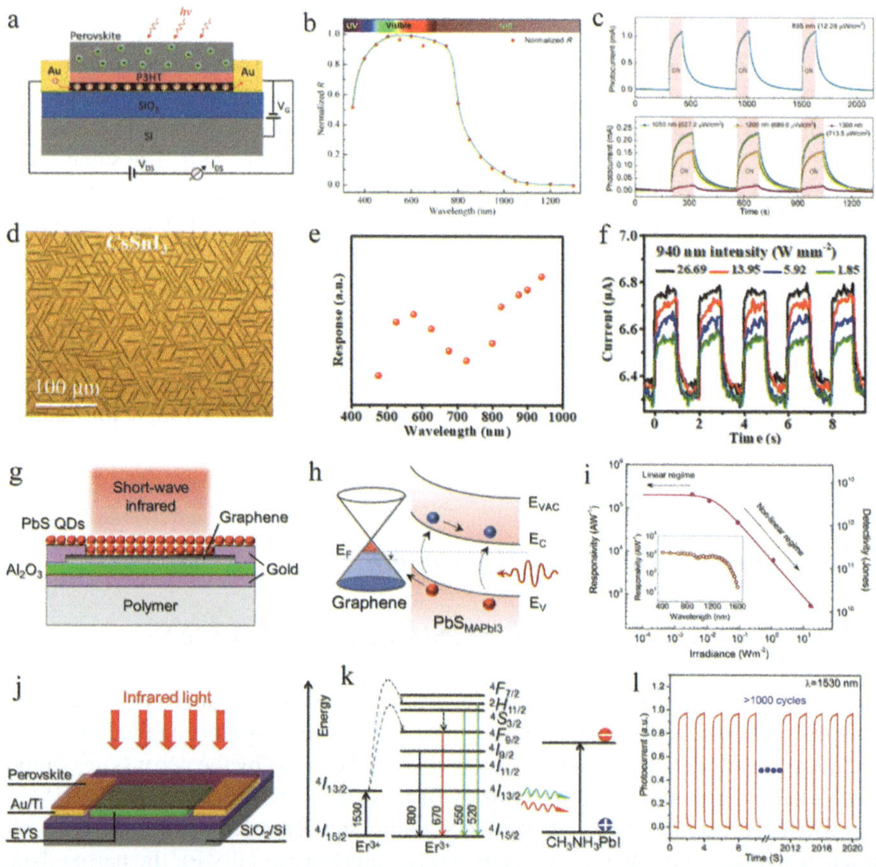


**Fig. 7.7** **a** A schematic side view of the perovskite phototransistor, **b** its normalized spectral responsivity, and **c** photoresponses under infrared illumination on/off cycles with different wavelengths. Reprinted with permission from ref [89]. Copyright 2017, American Chemical Society. **d** The optical images of the as-grown $CsSnI_3$ perovskite nanowire array. **e** Spectral response and **f** time–response curves of the $CsSnI_3$ photodetector irradiated with 940 nm light. Reprinted with permission from ref [99]. Copyright 2019, American Chemical Society. **g** Schematic cross-section view of photodetector structure. **h** Schematic energy band diagram and the charge transfer directions under illumination. **i** Responsivity and detectivity as a function of irradiance ($\lambda = 520$ nm). The inset demonstrates uniform spectral responsivity from 400 to 1500 nm. Reprinted with permission from ref [101]. Copyright 2017, American Chemical Society. **j** Schematic of the device, and corresponding **k** the radiative energy transfer and **l** performance of alternating dark and light illumination at 1530 nm. Reprinted with permission from ref [109]. Copyright 2017 WILEY-VCH Verlag GmbH & Co. KGaA, Weinheim

of $1.5 \times 10^9$. Even under 1300 nm light, it still showed a photocurrent of ~25 μA and responsivity of ~876 A/W. This unexpected strong photoresponse in the near infrared region beyond the absorption edge was attributed to the exciton formed by the valence band to traps states excitation within the perovskite bandgap. Different from the band/defect engineering, two-photon absorption was also proved to be an alternative way to detect light >800 nm, which was successfully reveal in $MAPbBr_3$ and $CsPbBr_3$ bulk single crystals photodetectors [96, 97].

***To dope or totally substitute lead with hetero ions in perovskite.*** The partial or total substitution of the Pb by Sn has been proved to be an effective way. Xu et al. reported a near infrared photodetector by using $MA_{0.5}FA_{0.5}Pb_{0.5}Sn_{0.5}I_3$ perovskite [98]. It showed high responsivity with a detectivity of over $10^{12}$ Jones ranging from 800 to 970 nm. Han et al. demonstrated the near infrared photodetector based on $CsSnI_3$ perovskite (bandgap of 1.34 eV) nanowire arrays with the responsivity of 54 mA/W at 940 nm light illumination, prepared by a solid-source chemical vapor deposition method (Fig. 7.7d–f) [99]. Density functional theory calculations suggested that, Ge/Sn-doped $MAPbI_3$ perovskite may have the optical absorption coefficient significantly enhanced, and the tunable absorption spectra would shift toward the near infrared and even the middle infrared [100].

***To introduce infrared sensitive materials.*** This strategy combines perovskite with infrared sensitive materials. The former provides good photodetective performance in the visible range, and the later extends the infrared end. The infrared sensitive materials can be inorganic semiconductors and organic materials. Bessonov et al. demonstrated that by surface passivation of PbS QDs with $MAPbI_3$ perovskite ($PbS_{MAPbI3}$), the photodetector exhibited detectivity of $5 \times 10^{12}$ Jones and uniform responsivity from 400 to 1500 nm (Fig. 7.7g–i) [101]. $PbS_{MAPbI3}$ photodetector is attractive for video-rate applications, whereas PbS QDs embedded into $MAPbI_3$ matrix ($MAPbI_3 \cdot PbS$) leading to ultrahigh gain. Yu et al. reported a hybrid phototransistors based on $CsPbBr_3$ QDs and PbS QDs. Responsivity of $4.5 \times 10^5$ A/W and detectivity of $7 \times 10^{13}$ Jones were obtained, showing photo detectivity between 400 and 1500 nm [102]. Lin et al. reported the coupling dyes of small bandgap ($CyPF_6$ and $Cy1BF_4$) to the perovskite photodetector, extends the spectrum range to 1600 nm. High responsivity, low dark current and fast response at room temperature were achieved [103]. Low bandgap polymers, e.g. PDPP3T, PDPPTDTPT:PCBM, PDPP3T:PC71BM, etc. were reported effective in this type of composites for infrared photodetection devices [104–106].

***To integrate up conversion materials.*** Up conversion materials can absorb long wavelength light and convert the energy to short wavelength emission, which can be detected by the perovskite. The composite of up conversion material $NaYF_4$:Yb/Er nanocrystals and $MAPbI_3$ perovskite exhibited near infrared photoresponse with high responsivity of 0.27 A/W and detectivity of $0.76 \times 10^{12}$ Jones at 980 nm [107]. Zhang et al. fabricated photodetectors based on the composite of $CsPbI_3$ QDs and $NaYF_4$:Yb,Er nanocrystals expanding optical response to 1100 nm [108]. The photodetector that integrate $MAPbI_3$ perovskite with erbium ytterbium silicate (EYS) nanosheets, an up conversion material, showed a fast photoresponse speed of ≈900 μs under 1530 nm light illumination (Fig. 7.7j–l) [109].

#### 7.3.7.3 Perovskite Solar Blind Photodetectors (200–300 Nm)

***Background***. Due to the strong absorption and reflection of ozone, fine particles, water vapor and other substances in the atmosphere to the deep ultraviolet, there is almost no solar radiation with wavelength less than 280 nm on the earth's surface, which is like a natural dark room. Therefore, the light of 200–300 nm (or 200–280 in some literatures) is called solar blind light. Because of the low background noise in the atmosphere, the solar blind UV detector usually has the advantages of high signal to noise ratio, low false alarm rate, high accuracy and high sensitivity. It can be used to detect non solar radiation sources, such as missile alarms, flame detection, explosion, ozone detection, free-space communication and leakage of high-voltage power [110]. Because of the importance in military and civil aspects, it has become the cutting-edge technology in the photodetection field. Wide band-gap semiconductor materials, such as AlGaN, [111] MgZnO, [112] $Ga_2O_3$, [113] diamond, [114] BN, [115] ZnS [116] and polymers [117] have been employed as sensitive materials for solar-blind photodetectors. However, the fabrication of functional films of these inorganic materials with high quality requires expensive equipment and complicated control over the process, such as molecular beam epitaxy (MBE), metal organic vapor deposition (MOCVD) and magnetron sputtering; while the organic materials may suffer from slow response time (≥10 V), relative high operation voltage (≥10 V) and degradation under UV light [118–124]. Therefore, easy processing, low cost and stable materials for solar blind photodetectors are major challenges. As the highlighted materials in the last decade, perovskite semiconductors gained great attention for their potentials in solving the above challenges for solar blind photodetectors. They could be solution processed and readily for fabricate large area, flexible and high performance devices. By the device structure and operating mechanism, the solar blind photodetectors can be classified into 3 types as follows.

***Lateral structure***. Though perovskite photodetectors were fabricated and studied in some early reports, they gave no measurement on the solar blind range by knowing the perovskite absorption keeping high and extending to 200 nm [66, 84]. In 2015, Guo et al. reported lateral photoconductive structured $MAPbI_{3-x}Cl_x$ photodetectors protected by a water-resistant fluorous polymer CYTOP (Fig. 7.8a–c) [125]. Both the perovskite and CYTOP films were solution processed (spin-coating). It was operated at a low voltage (≤2 V) with responsivity of 7.85 A/W and EQE = 3832% to solar blind UV light ($\lambda$ = 254 nm; 0.85 $\mu$W/cm$^2$). The device maintained 75% of initial performance after 100 days in air. The effectiveness of the simple lateral photoconductive structure was also proved for the all inorganic perovskite $CsPbB_{r3}$ film based broadband photodetector. The on/off ration of $10^3$, rise time of 260 ms and responsivity of 0.24 mA/W was achieved for solar blind 254 nm light [126]. The PET substrate provided the mechanical flexibility with a long-term environmental stability for two months without encapsulation, which was greatly enhanced than the organic-inorganic perovskite materials. However, the broadband response extending to the visible range may raise the noise for detecting solar blind objects. Zhang et al. successfully solved this problem by using a wide bandgap (3.8 eV) perovskite $Cs_3Cu_2I_5$ crystalline film prepared by a solution process [127]. The photodetector

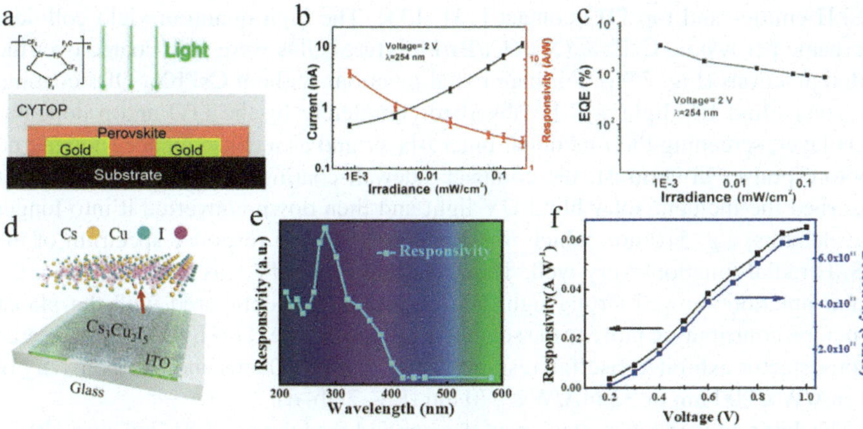

**Fig. 7.8** **a** Device configuration of the CYTOP-protected photodetector based on $MAPbI_{3-x}Cl_x$. **b** The responsivity and **c** *EQE* of the device under 254 nm irradiance. Reprinted with permission from ref [125]. Copyright 2015, American Chemical Society. **d** Schematic of the $Cs_3Cu_2I_5$ photodetector. **e** The spectral response. **f** Responsivity and detectivity under 265 nm irradiation. Reprinted with permission from ref [127]. Copyright 2019, American Chemical Society

was nearly blind to visible light of wavelength >405 nm, but showing good responsivity within 200–400 nm (Fig. 7.8d–f). The rejection ratio of the photocurrent under 265 nm light over that under 405 nm light was 17.8. The on/off ratio increased linearly with the deep UV intensity. It reached a maximum value of 142 with the incident 265 nm light of 2.74 $mW/cm^{-2}$, and the responsivity of 64.9 mA/W, detectivity of $6.9 \times 10^{11}$ Jones, rise/fall time of 26.2/49.9 ms and air stability were demonstrated. Xu et al. reported a phototransistor based on the organic material (C8-BTBT) single-crystal array coated with $MAPbI_3$ nanocrystals synthesized by a solution method. It exhibited a high responsivity of $>1.72 \times 10^4$ A/W in the 252–780 nm broadband region with a robust stability of >50 days in air [128].

***Vertical structure***. Zheng et al. reported a solar blind photodetector based on $MAPbCl_3$ perovskite film with the vertical structure of ITO/$MAPbCl_3$/poly(triaryl amine) (PTAA)/Al. At bias voltage of −1 V, the responsivity was measured 4–10 mA/W for the 200–280 nm solar blind UV light [129]. Another vertical structure combined a wide bandgap (4.9 eV) material the monoclinic β–$Ga_2O_3$ with $MAPbI_3$ perovskite. Under 254 nm UV light irradiation, the device exhibited the photo-to-dark current ratio of 1460, responsivity of 85 mA/W, detectivity of $10^{11}$ Jones, rise/decay time of 0.11/0.45 s, rejection ratio $I_{254nm}/I_{325nm}$ of 221 were achieved [130].

***Down conversion type (Radial and Planar structures)***. A novel radial solar blind photodetector was constructed over SiNWs on the substrate by Lu et al. [75] The SiNW were fabricated via a tin nanoparticle catalyzed vapor–liquid–solid (VLS) growth on AZO-coated (AZO:aluminum-doped zinic oxide) glass by plasma enhanced chemical vapor deposition (PECVD). The radial junctions composed of p-doped SiNWs in the core, radially coated layers of intrinsic a-Si:H absorber, n-doped a-Si:H emitter and top ITO contact [131–133]. The high quantum yield colloidal inorganic perovskite $CsPbX_3$(X = Cl/Br/I/mixture) QDs were spin-coated over the radial junctions (Fig. 7.9a). For bare radial junctions without $CsPbX_3$ QDs coating, the solar blind UV light will be absorbed completely by the ITO and n-doped a-Si:H layer, screening the functional inner i-layer and as a consequence showing no photoresponse. In contrast, the compact sidewall-coating of perovskite QD layer absorbed the incident solar blind UV light and then down-converted it into longer wavelengths, e.g. 510 nm, which matches with the EQE response spectrum of the a-Si:H radial junctions very well. Thus the solar blind UV was detected. Also, the radial junction showed strong light confinement effect compared with the planar structure, contributing more to the solar blind UV response. This hybrid self-powered photodetector exhibited rise/fall response time of 0.48/1.03 ms and a responsivity of 54 mA/W@200 nm or 32 mA/W@270 nm (Fig. 7.9b–d).

The down conversion strategy was also applied to planner solar blind photodetectors. Zhang et al. reported the polyvinylidene fluoride composite films embedded with perovskite $MAPbBr_3$ QDs as downshifting materials to successfully enhance the UV response (200–400 nm) of silicon photodetectors. The EQE of the Si photodiodes were greatly improves from near 0% to at most of 50.6% ± 0.5%@290 nm. A solar-blind UV imaging was demonstrated by integrating this down conversion composite to the electron multiplying charge coupled device EMCCD [134]. In addition to the polymer composites, it was found that the $Cs_4PbBr_6$ in an inorganic dual-phase $CsPbBr_3$–$Cs_4PbBr_6$ perovskite would also enhanced the capture and down conversion of the incident light of 200–350 nm. A fast response rise/fall time of 7.8/33.6 μs

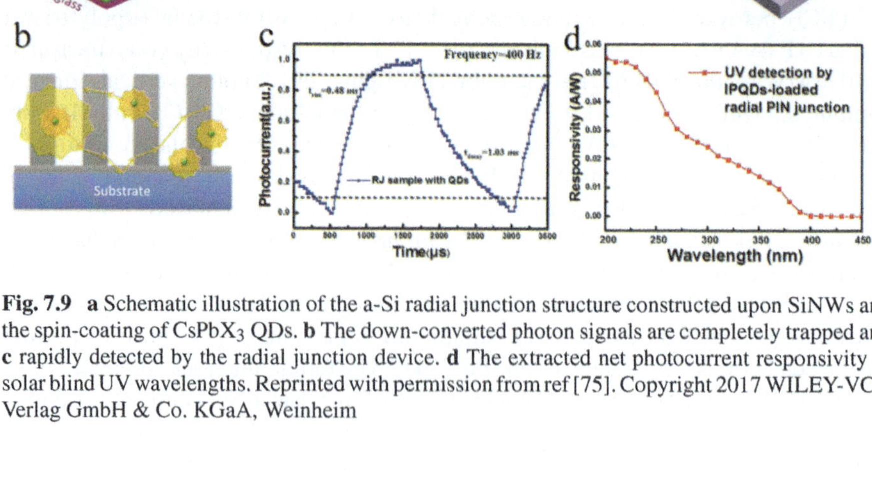


**Fig. 7.9** **a** Schematic illustration of the a-Si radial junction structure constructed upon SiNWs and the spin-coating of $CsPbX_3$ QDs. **b** The down-converted photon signals are completely trapped and **c** rapidly detected by the radial junction device. **d** The extracted net photocurrent responsivity to solar blind UV wavelengths. Reprinted with permission from ref [75]. Copyright 2017 WILEY-VCH Verlag GmbH & Co. KGaA, Weinheim

and a high responsivity of 49.4 mA/W at 254 nm were achieved for the self-powered perovskite photodetector [135]. It was reported that by using $CsPbBr_3$ QDs as down converter, an EQE increase of ~25% in the solar-blind region were observed for a commercial silicon photodetector with a high data rate of 34 Mbps in solar blind (278 nm) communication was achieved; [136] and the performance (200–280 nm) of a perovskite film based photodiode structured photodetector enhanced by 1 order in magnitude with responsibility of 1.4 mA/W and detectivity of $2.4 \times 10^{11}$ Jones [76].

#### 7.3.7.4 Perovskite X-Ray and Gamma-Ray Detectors

***Perovskite X-ray detectors***. Since X-rays were discovered by Röntgen at 1895, [137] their wide applications have been invented in many fields, e.g. medical radiography, crystal structure determinations, astronomy and industrial non-destructive testing. In 2013, Stoumpos et al. demonstrated that the inorganic $CsPbBr_3$ perovskite single crystal has significant potential for high-energy detectors, for it is a direct bandgap semiconductor with high attenuation, high resistivity and significant photoconductivity response that meets most of the requirements for X-ray and gamma-ray detection [138]. Two years later, Yakunin et al. reported that the organic-inorganic perovskite ($MAPbI_3$) with the architecture of ITO/PEDOT:PSS/$MAPbI_3$/PCBM/ZnO/Ag many not only be used as a solar cell, but also applicable for X-ray detection [139]. Owing to the combination of fast photoresponse and a high absorption cross-section for X-rays of the heavy Pb and I atoms, this detector with a $MAPbI_3$ thickness of 600 nm exhibited a sensitivity of 25 $\mu C\ Gy_{air}^{-1}\ cm^{-2}$ and responsivity $1.9 \times 10^4$ carriers/photon, commensurate with those obtained by the current solid-state technology. Wei et al. fabricated the first X-ray detector made of $MAPbBr_3$ single crystal [140]. Due to the reduced bulk defects and passivated surface traps, the single crystal X-ray detector with the thickness of 2–3 mm showed 16.4% detection efficiency at near zero bias under irradiation with continuum X-ray energy up to 50 keV. A high sensitivity of 80 $\mu C\ Gy_{air}^{-1}\ cm^{-2}$ was achieved at a low X-ray dose rate of 0.5 $\mu Gy_{air}^{-1}\ s^{-1}$, which allows applications to medical and security check. Later, $CsPbBr_3$ QDs were applied to make a soft-X-ray detector [141]. It exhibited a high sensitivity of 1450 $\mu C\ Gy_{air}^{-1}\ cm^{-2}$ under an X-ray dose rate of 0.0172 $\mu Gy_{air}^{-1}\ s^{-1}$ with 0.1 V bias voltage. Due to the X-ray-induced emissions can be easily tunable across the visible spectrum by tailoring the anionic component of colloidal $CsPbX_3$(X = Cl/Br/I/mixture) QDs, they were used to fabricate flexible scintillator-type X-ray detectors [142]. Figure. 7.10a shows a proposed mechanism of X-ray scintillation in a lead halide perovskite QD. Upon X-ray irradiation, an electron is ejected from a lattice atom through photoelectric ionization to form a high-energy electron, which will produce secondary high-energy electrons. These generated hot charge carriers then undergo thermalization and produce low-energy excitons. Next, recombination of the excitons either from a singlet (S) or triplet (T) state produces luminescence at the electronic band edge. A detection limit of 13 $\mu Gy_{air}\ s^{-1}$, which is about 400 times lower than typical medical imaging doses,

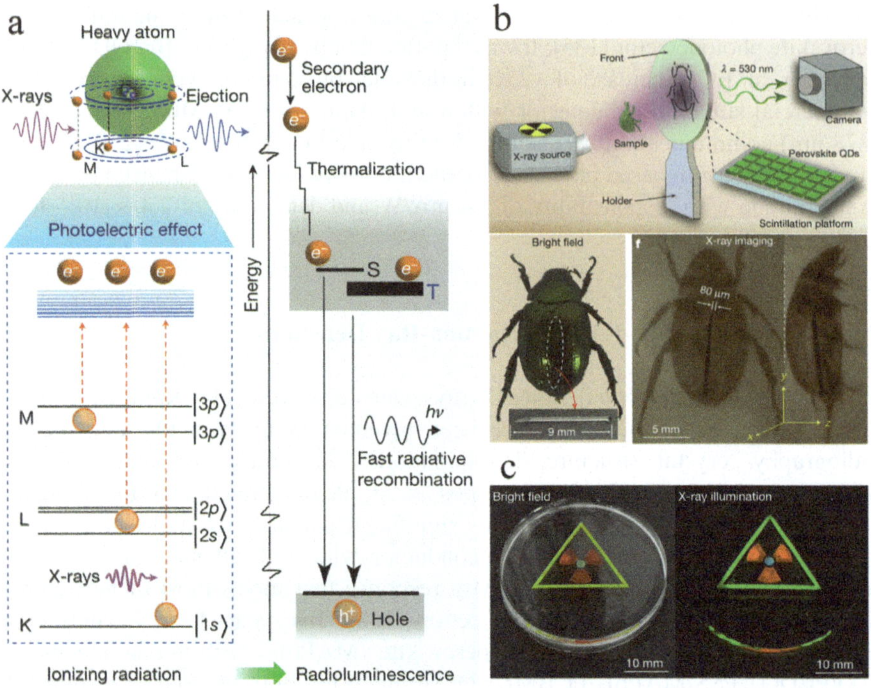


**Fig. 7.10** **a** Proposed mechanism of X-ray scintillation in a lead halide perovskite QD. **b** Schematic of the real-time X-ray diagnostic imaging of biological samples. Bright-field and the X-ray images of a beetle, recorded with a digital camera. **c** Multicolour X-ray scintillation from perovskite QDs of $CsPbBr_2I$ (orange), $CsPbBr_3$ (green), $CsPbClBr_2$ (blue). (X-ray illumination at a voltage of 50 kV). Reprinted with permission from ref [142]. Copyright 2018, Springer Nature

was reported. It provided a convenient visualization tool for X-ray radiography, for the images can be recorded by standard digital cameras (Fig. 7.10b, c). It also can be integrated with commercial flat-panel imagers to serve in examining electronic circuit boards. Lead-free solution-processed $Cs_2AgBiBr_6$ single crystals with double perovskite structure were also reported for the X-ray detection [143].

***Perovskite gamma-ray detectors***. Emission of gamma photons with energies of ~50 keV to 10 MeV is commonly companied with the decay of radioactive isotopes. Utilization of the gamma-ray is of great importance in medical and security applications. Developing low-cost, highly sensitive and easy operating are research trends nowadays. In 2015, Dong et al. reported the application of $MAPbI_3$ single crystals to gamma-ray detection [144]. The long carrier diffusion lengths enabled an efficiency of 3.9% through the gammavoltaic effect by using an intense cesium-137 source. Later, Yakunin et al. demonstrate that 0.3–1 cm single crystalline perovskite ($MAPbI_3$, $FAPbI_3$ and I-treated $MAPbBr_3$) can serve as solid-state gamma-detecting materials [145]. It was successfully used to test he radiopurity of medical radiotracer compounds such as $^{18}$F-fallypride. For the performance of gamma-ray detector depends largely on the perovskite carrier mobility and lifetime, the research focuses mainly on bulk perovskite single crystals.

### 7.3.8 *Material-System Design*

As highly tailorable materials for their photo-electronic properties, the perovskite materials can be well designed and prepared in manners of the dimension control, crystal engineering and composite design.

#### 7.3.8.1 Dimension Control: 0D, 1D, 2D and 3D Perovskite

Materials can be classified as zero, one, two or three dimensional (0D, 1D, 2D and 3D) materials based on the comparative size feature of the outer shapes in all three dimensions. Nanoparticles or nanocrystals can be regarded as a (quasi-) 0D materials in case they are very small, typically the QDs of several nanometers, and isotropy of the shape/property or the anisotropy can be ignored in some conditions. Sometime the prefix quasi is used suggesting that the particles are not ideal 0D physically. Similar definitions also apply to (quasi-) 1D, 2D, and 3D materials for nanowire/nanorod, nanoplate/nanosheet/nanoflake, and bulk materials, respectively. It should be pay attention to that under some circumstances, the phrase 0D, 2D and 3D perovskite were used to emphasize the stoichiometry related crystal structures as introduced in Sect. 1.1, different from the discussion here.

The photodetector based on entirely the 0D inorganic $CsPbI_3$ nanocrystals of ~10 nm was first reported by Ramasamy et al. [95] The lateral photoconductive photodetector was prepared by drop-coating the colloidal perovskite nanocrystals on Au electrodes with the gap of 3 $\mu$m. It showed a good on/off photocurrent ratio of $10^5$ and response rise/fall time of 24/29 ms. 1D perovskite of e.g. nanorods and nanowires have larger surface-to-volume ratios compared with 0D perovskite of the same diameter, which may lead to lower trap state density, longer charge-carrier lifetime and reduced carrier recombination rate. For the same reason, it is also true and more superior for 2D and 3D perovskite [146]. The major advantage of 3D perovskite lies in the greatly eliminated trap states. The merits of aligned 1D perovskite include high responsivity and detectivity, with polar light detection ability and superior flexibility. The solution synthesis and vapor growth methods were widely used for fabrication of all the 1D, 2D, 3D perovskite photodetectors [69, 147]. The solution-coating-evaporation [148, 149] and template [150] strategies were successfully applied to prepare 1D perovskite array photodetectors.

#### 7.3.8.2 Crystal Engineering: Crystal Structure, Doping and Lead Free

The principle of crystal design depends on the photodetector service requirements. One of the major purposes is to control the bandgap with the perovskite morphology the same or remained, e.g. the halogen anion ratio tuning for $MAPbBr_{3-x}Cl_x$ and $MAPbBr_{3-x}I_x$ QDs or single crystalline perovskite arrays, [80, 151] and the cation ratio tuning for $Cs_x(MA)_{1-x}PbI_3$ perovskite [152]. The crystal structure or stoichiometric component tuning is a powerful tool to tailor photodetection performances, e.g. the $CsPbBr_3$ vs $Cs_4PbBr_6$ perovskite single crystals, [153] layer structured perovskite materials of the general formula of $(RNH_3)_2A_{n-1}M_nX_{3n+1}$, where n is the number of $MX_6$ octahedra in the perovskite layers between adjacent insulating organic cation layers, $RNH_3^+$ is a large organic cation such as phenylethylamine (PEA), n-butylamine, or oleylamine [154, 155]. A layer structured perovskite in photodetectors is not necessarily taking the nanoplate/nanosheet morphology, it can be appeared even as nanowires [156].

Another important topic in perovskite crystal engineering is to substitute Pb with more environment friendly metal elements such as Sn, Ge, Cu, In, Bi, and Sb etc., e.g., lead free perovskite of $MASnI_3$, [157] $(PEA)_2SnI_{4-x}Br_x (x = 0,1,2)$, [158] $Cs_3Bi_2I_9$, [159, 160] $Cs_2AgBiBr_6$, [161] $MA_2FeCuI_4Cl_2$ and $MA_2InCuI_6$ [162] were reported for their application in photodetectors.

#### 7.3.8.3 Composition Design: Pure Versus Composite

Perovskite materials as photon-gaining medium used in photodetectors follow two composition status, the pure phase or in composite with hetero materials. Commonly, photodetectors can be based on pure perovskite single crystals bigger than micrometers or composites containing nano perovskite crystals. The pure crystalline phase aims to reduce the surface states, thus enhance the performance. While for perovskite nanomaterials prepared by solution synthesis, there commonly exist abundant surface states and long-chain surface ligands. When thin films are fabricated by these perovskite nanocrystals, the charge separation and transfer are limited. Therefore, thermal annealing to crystallize the films, ligands exchange with small molecules, and combine with other conductor/semiconductor materials to form composites can be used to break the limit. Polymers, [104, 106] organic semiconductors, [163] Au nanocrystals, [164] Ag nanocrystals, [165] graphene, [84, 94] CNTs, [166] 2D $MoS_2$, [167] monolayer $WS_2$, [88] MXene, [168] etc. have been used to prepare composites with perovskite to enhance the performance of photodetectors. Chen et al. reported the use of DNTT, a conjugate organic molecule with high charge mobility and stability, to hybrid with $CsPbBr_3$ QDs, which significantly improve the transportation of the photoexcited charges. The devices exhibited outstanding properties in terms of responsivity of $1.7 \times 10^4$ A/W, detectivity of $2.0 \times 10^{14}$ Jones, EQE of 67,000%, $I_{photo}/I_{dark}$ ratio of $8.1 \times 10^4$ and stability of 100 days in air [163]. The incorporation of CNTs and graphene with perovskite can offer the pathways for high charge-carrier mobility, which tremendously boosts the charge transfer and the photocurrent. Kwak et al. prepared the graphene-$CsPbBr_{3-x}I_x$ perovskite nanocrystal based photodetector with an ultrahigh responsivity of $8.2 \times 10^8$ A/W and detectivity of ~$10^{16}$ Jones [94]. When Au or Ag nanocrystals are mixed with the perovskite rather as an electrode, the surface plasmonic effect of the metal introduces near-field enhancement and consequently facilitate the light-harvesting of nearby perovskite. Wu et al. reported an all-inorganic perovskite QD-monolayer $MoS_2$ mixed-dimensional van der Waals heterostructure for ultrasensitive photodetector [169]. A favorable energy band alignment facilitated the interfacial photocarrier separation and efficient carrier injection into the $MoS_2$. Owing to the synergistic effect of the photogating mechanism and the modulation of Schottky barriers, the corresponding phototransistor exhibited high photo responsivity of $7.7 \times 10^4$ A/W, detectivity of $5.6 \times 10^{11}$ Jones, and an EQE exceeding 107%.

## 7.4 Summary and Perspective

Due their fascinating properties, perovskite materials have presented competitiveness for new-generation of optoelectronic devices. We have focused the sight over the development progress of perovskite, particularity the perovskite QDs, for their application in photodetectors. The near future developments of QD photodetectors are prospected in the following aspects.

1. ***New perovskite materials***. A major theme of new perovskite materials is lead free perovskite semiconductor materials, their crystal structure, synthesis, stability, optical and electronic properties [170]. New perovskite materials emerging from crystal engineering and surface engineering will bring about tailorable functions, e.g. circular polarized light coding/decoding, wide or narrow bandgap for solar blind or infrared light detection, respectively.
2. **Self-assembly of perovskite nanomaterials**. Theory and techniques of self-assembly of nanomaterials as building blocks to form 2D or 3D supper lattices have kept developing for unary, binary and ternary nanocrystals/nanoclusters in recent years [171–175]. Well assembled perovskite-containing supper lattices of unary, binary and ternary nanoparticle systems are thus highly expected for novel collecting properties.
3. **New device structures**. New devices structures of perovskite photodetectors are boosting in recent years, e.g. radial structures, multi-layer structures, multi-functional structures. These device structures contribute to the knowledge base and hybridization in them will certainly bring devices with unexpected performances.
4. **New mechanism combinations**. Like perovskite photodetectors having photoconductivity integrated with triboelectric and pyroelectric effects, similar combinations could be found in many aspects, e.g. combined with photoacoustic or biomimetic conceptions [176].

5. **Stability**. It remains a challenge for perovskite photodetectors. The device stability can be improved by optimizations of the material design (component, structure, surface, shape, size, composition) and device structure. It is the last step toward practical use, and continuous improving is expected.
6. **Photodetection range**. Most researches have been spent on the visible range performance. For the importance of photodetection for solar blind and infrared light. More attentions have now been attracted in both ends of the spectrum.
7. **New composites and perovskite nano heterojunctions**. Perovskite based new composites can combine the superiorities properties of perovskite QDs with the hetero compositions, such as facilitate the charge transit, expand the detection range, etc. In recent years, perovskite based nano heterojunctions have raised more and more interest and their novel properties discovered, e.g. perovskite-PbS QDs showed remarkable dual emission within the visible and near infrared range, and exciton energy transfer was demonstrated [177].
8. ***From bulk to nano***. The concepts of some perovskite photodetectors are only realized on bulk single crystal or bulk film perovskite, to migrate them to the corresponding perovskite QDs will give new opportunities.
9. ***Commercialization of perovskite photodetector products***. Some perovskite photodetectors prototypes have shown great advantages in competition with commercial photodetectors. It can be predicted that the commercialization process of the perovskite photodetector products will come into accelerating development.

In summary, the times of perovskite semiconductors is unstoppable in ways of their tremendous application potentials in many territories. Among them, perovskite QDs have exhibited great potentials in photodetectors for flexible, transparent, wearable, self-powered, high sensitive, fast response, low cost and multifunctional device applications, and it is believed that some of them will come into our life in the near future.

## References


1. L.E. Brus, A simple model for the ionization potential, electron affinity, and aqueous redox potentials of small semiconductor crystallites. J. Chem. Phys. **79**(11), 5566–5571 (1983)
2. A.L. Efros, A.L. Efros, Interband light absorption in a semiconductor sphere. Sov. Phys. Semicond. **16**(7), 772–775 (1982)
3. A.I. Ekimov, A.A. Onushchenko, Quantum size effect in the optical-spectra of semiconductor micro-crystals. Sov. Phys. Semicond. **16**(7), 775–778 (1982)
4. R.C. Ashoori, Electrons in artificial atoms. Nature **379**(6564), 413–419 (1996)
5. U. Banin, Y.W. Cao, K. David, O. Millo, Identification of atomic-like electronic states in indium arsenide nanocrystal quantum dots. Nature **400**(6744), 542–544 (1999)
6. A.P. Alivisatos, Semiconductor clusters, nanocrystals, and quantum dots. Science **271**(5251), 933–937 (1996)
7. D.V. Talapin, J.S. Lee, M.V. Kovalenko, E.V. Shevchenko, Prospects of colloidal nanocrystals for electronic and optoelectronic applications. Chem. Rev. **110**(1), 389–458 (2010)

8. Y. Pu, F.H. Cai, D. Wang, J.X. Wang, J.F. Chen, Colloidal synthesis of semiconductor quantum dots toward large scale production: a review. Ind. Eng. Chem. Res. **57**(6), 1790–1802 (2018)
9. Q.A. Akkerman, A.L. Abdelhady, L. Manna, Zero-dimensional cesium lead halides: history, properties, and challenges. J. Phys. Chem. Lett. **9**(9), 2326–2337 (2018)
10. E.F.S. Iyikanat, H. Sahin, Thinning $CsPb_2Br_5$ perovskite down to monolayers: cs-dependent stability. Phys. Rev. B **96**(15), 155442 (2017)
11. L.T. Dou, A.B. Wong, Y. Yu, M.L. Lai, N. Kornienko, S.W. Eaton, A. Fu, C.G. Bischak, J. Ma, T.N. Ding, N.S. Ginsberg, L.W. Wang, A.P. Alivisatos, P.D. Yang, Atomically thin two-dimensional organic-inorganic hybrid perovskites. Science **349**(6255), 1518–1521 (2015)
12. Y. Cai, W. Xie, H. Ding, Y. Chen, K. Thirumal, L.H. Wong, N. Mathews, S.G. Mhaisalkar, M. Sherburne, M. Asta, Computational study of halide perovskite-derived $A_2BX_6$ inorganic compounds: chemical trends in electronic structure and structural stability. Chem. Mater. **29**(18), 7740–7749 (2017)
13. X.G. Zhao, J.H. Yang, Y.H. Fu, D.W. Yang, Q.L. Xu, L.P. Yu, S.H. Wei, L.J. Zhang, Design of lead-free inorganic halide perovskites for solar cells via cation-transmutation. J. Am. Chem. Soc. **139**(7), 2630–2638 (2017)
14. J. Zhang, Y. Yang, H. Deng, U. Farooq, X.K. Yang, J. Khan, J. Tang, H.S. Song, High quantum yield blue emission from lead free inorganic antimony halide perovskite colloidal quantum dots. ACS Nano **11**(9), 9294–9302 (2017)
15. L.C. Schmidt, A. Pertegas, S. Gonzalez-Carrero, O. Malinkiewicz, S. Agouram, G. Minguez Espallargas, H.J. Bolink, R.E. Galian, J. Perez-Prieto, Nontemplate synthesis of $CH_3NH_3PbBr_3$ perovskite nanoparticles. J. Am. Chem. Soc. **136**(3), 850–853 (2014)
16. L. Protesescu, S. Yakunin, M.I. Bodnarchuk, F. Krieg, R. Caputo, C.H. Hendon, R.X. Yang, A. Walsh, M.V. Kovalenko, Nanocrystals of cesium lead halide perovskites ($CsPbX_3$, X = Cl, Br, and I): novel optoelectronic materials showing bright emission with wide color gamut. Nano Lett. **15**(6), 3692–3696 (2015)
17. J. Shamsi, A.S. Urban, M. Imran, L. De Trizio, L. Manna, Metal halide perovskite nanocrystals: synthesis, post-synthesis modifications, and their optical properties. Chem. Rev. **119**(5), 3296–3348 (2019)
18. M.V. Kovalenko, L. Protesescu, M.I. Bodnarchuk, Properties and potential optoelectronic applications of lead halide perovskite nanocrystals. Science **358**(6364), 745–750 (2017)
19. E.M. Sanehira, A.R. Marshall, J.A. Christians, S.P. Harvey, P.N. Ciesielski, L.M. Wheeler, P. Schulz, L.Y. Lin, M.C. Beard, J.M. Luther, Enhanced mobility $CsPbI_3$ quantum dot arrays for record-efficiency, high-voltage photovoltaic cells. Sci. Adv. **3**(10), eaao4204 (2017)
20. T. Chiba, Y. Hayashi, H. Ebe, K. Hoshi, J. Sato, S. Sato, Y.J. Pu, S. Ohisa, J. Kido, Anion-exchange red perovskite quantum dots with ammonium iodine salts for highly efficient light-emitting devices. Nat. Photonics **12**(11), 681–687 (2018)
21. Y. Wang, X. Li, J. Song, L. Xiao, H. Zeng, H. Sun, All-inorganic colloidal perovskite quantum dots: a new class of lasing materials with favorable characteristics. Adv. Mater. **27**(44), 7101–7108 (2015)
22. X. Sheng, Y. Liu, Y. Wang, Y. Li, X. Wang, X. Wang, Z. Dai, J. Bao, X. Xu, Cesium lead halide perovskite quantum dots as a photoluminescence probe for metal ions. Adv. Mater. **29**(37), 1700150 (2017)
23. Y.F. Xu, M.Z. Yang, B.X. Chen, X.D. Wang, H.Y. Chen, D.B. Kuang, C.Y. Su, A CsPbBr 3 perovskite quantum dot/graphene oxide composite for photocatalytic $CO_2$ reduction. J. Am. Chem. Soc. **139**(16), 5660–5663 (2017)
24. Y. Wang, Z. Lv, J. Chen, Z. Wang, Y. Zhou, L. Zhou, X. Chen, S.T. Han, Photonic synapses based on inorganic perovskite quantum dots for neuromorphic computing. Adv. Mater. **30**(38), 1802883 (2018)
25. Y. Wang, L. Song, Y. Chen, W. Huang, Emerging new-generation photodetectors based on low-dimensional halide perovskites. ACS Photonics **7**(1), 10–28 (2020)
26. J.L. Miao, F.J. Zhang, Recent progress on highly sensitive perovskite photodetectors. J. Mater. Chem. C **7**(7), 1741–1791 (2019)

27. T. Yang, F. Li, R. Zheng, Recent progress on cesium lead halide perovskites for photodetection applications. ACS Appl. Electron. Mater. **1**(8), 1348–1366 (2019)
28. P. Wangyang, C. Gong, G. Rao, K. Hu, X. Wang, C. Yan, L. Dai, C. Wu, J. Xiong, Recent advances in halide perovskite photodetectors based on different dimensional materials. Adv. Opt. Mater. **6**(11), 1701302 (2018)
29. Y. Dong, Y. Zou, J. Song, X. Song, H. Zeng, Recent progress of metal halide perovskite photodetectors. J. Mater. Chem. C **5**(44), 11369–11394 (2017)
30. H. Wang, D.H. Kim, Perovskite-based photodetectors: materials and devices. Chem. Soc. Rev. **46**(17), 5204–5236 (2017)
31. Y. Wang, N. Herron, Photoconductivity of CdS nanocluster-doped polymers. Chem. Phys. Lett. **200**(1), 71–75 (1992)
32. H. Wei, Y. Fang, Y. Yuan, L. Shen, J. Huang, Trap engineering of CdTe nanoparticle for high gain, fast response, and low noise P3HT:CdTe nanocomposite photodetectors. Adv. Mater. **27**(34), 4975–4981 (2015)
33. K. Deng, L. Li, CdS nanoscale photodetectors. Adv. Mater. **26**(17), 2619–2635 (2014)
34. C. Giansante, I. Infante, Surface traps in colloidal quantum dots: a combined experimental and theoretical perspective. Phys. Chem. Lett. **8**(20), 5209–5215 (2017)
35. S. Zhang, X. Wang, Y. Chen, G. Wu, Y. Tang, L. Zhu, H. Wang, W. Jiang, L. Sun, T. Lin, H. Shen, W. Hu, J. Ge, J. Wang, X. Meng, J. Chu, Ultrasensitive hybrid $MoS_2$-ZnCdSe quantum dot photodetectors with high gain. ACS Appl. Mater. Interfaces **11**(26), 23667–23672 (2019)
36. M. Dolatyari, A. Rostami, S. Mathur, A. Klein, Trap engineering in solution processed PbSe quantum dots for high-speed mid-infrared photodetectors. J. Mater. Chem. C **7**(19), 5658–5669 (2019)
37. P.J. Regensburger, Optical sensitization of charge carrier transport in poly(N-Vinyl carbazole)*. Photochem. Photobiol. **8**(5), 429–440 (1968)
38. I. Chen, J. Mort, Xerographic discharge characteristics of photoreceptors. J. Appl. Phys. **43**(3), 1164–1170 (1972)
39. Y. Wang, N. Herron, Semiconductor nanocrystals in carrier-transporting polymers: charge generation and charge transport. J. Lumines. **70**(1), 48–59 (1996)
40. Y. Wang, Photoconductivity of fullerene-doped polymers. Nature **356**, 585–587 (1992)
41. Y. Wang, R. West, C.H. Yuan, Fullerene-doped polysilane photoconductor. J. Am. Chem. Soc. **115**(9), 3844–3845 (1993)
42. D. Leonard, M. Krishnamurthy, C.M. Reaves, S.P. Denbaars, P.M. Petroff, Direct formation of quantum-sized dots from uniform coherent Islands of InGaAs on GaAs surfaces. Appl. Phys. Lett. **63**(23), 3203–3205 (1993)
43. P.M. Petroff, S.P. DenBaars, MBE and MOCVD growth and properties of self-assembling quantum dot arrays in III-V semiconductor structures. Superlattices Microstr. **15**(1), 15 (1994)
44. A. Kazemi, M. Zamiri, J.O. Kim, T. Schuler-Sandy, S. Krishna, Colloidal and epitaxial quantum dot infrared photodetectors: growth, performance, and comparison. Wiley Encycl. Electr Electron Eng 1–26 (1999)
45. C.A. Leatherdale, C.R. Kagan, N.Y. Morgan, S.A. Empedocles, M.A. Kastner, M.G. Bawendi, Photoconductivity in CdSe quantum dot solids. Phys. Rev. B **62**(4), 2669–2680 (2000)
46. M.V. Jarosz, V.J. Porter, B.R. Fisher, M.A. Kastner, M.G. Bawendi, Photoconductivity studies of treated CdSe quantum dot films exhibiting increased exciton ionization efficiency. Phys. Rev. B **70**(19), 195327 (2004)
47. D.C. Oertel, M.G. Bawendi, A.C. Arango, V. Bulovic, Photodetectors based on treated CdSe quantum-dot films. Appl. Phys. Lett. **87**(21), 213505.1–213505.3 (2005)
48. V.J. Porter, S. Geyer, J.E. Halpert, M.A. Kastner, M.G. Bawendi, Photoconduction in annealed and chemically treated CdSe/ZnS inorganic nanocrystal films. J. Phys. Chem. C **112**(7), 2308–2316 (2008)
49. S. McDonald, P. Cyr, L. Levina, E. Sargent, Photoconductivity from PbS-Nanocrystal/semiconducting polymer composites for solution-processible, quantum-size tunableinfrared photodetectors. Appl. Phys. Lett. **85**(11), 2089–2091 (2004)

50. S.A. McDonald, G. Konstantatos, S. Zhang, P.W. Cyr, E.J. Klem, L. Levina, E.H. Sargent, Solution-processed PbS quantum dot infrared photodetectors and photovoltaics. Nat. Mater. **4**(2), 138–142 (2005)
51. G. Konstantatos, I. Howard, A. Fischer, S. Hoogland, J. Clifford, E. Klem, L. Levina, E.H. Sargent, Ultrasensitive solution-cast quantum dot photodetectors. Nature **442**(7099), 180–183 (2006)
52. A. Armin, R.D. Jansen-van Vuuren, N. Kopidakis, P.L. Burn, P. Meredith, Narrowband light detection via internal quantum efficiency manipulation of organic photodiodes. Nat. commun. **6**(1), 1–8 (2015)
53. Y. Fang, Q. Dong, Y. Shao, Y. Yuan, J. Huang, Highly narrowband perovskite single-crystal photodetectors enabled by surface-charge recombination. Nat. Photonics **9**(10), 679–686 (2015)
54. M.B. Johnston, Optoelectronics: colour-selective photodiodes. Nat. Photonics **9**(10), 634 (2015)
55. H.S. Rao, W.G. Li, B.X. Chen, D.B. Kuang, C.Y. Su, In situ growth of 120 $cm^2$ $CH_3NH_3PbBr$ 3 perovskite crystal film on FTO glass for narrowband-photodetectors. Adv. Mater. **29**(16), 1602639 (2017)
56. L.L. Li, Y.H. Deng, C.X. Bao, Y.J. Fang, H.T. Wei, S. Tang, F.J. Zhang, J.S. Huang, Self-filtered narrowband perovskite photodetectors with ultrafast and tuned spectral response. Adv. Opt. Mater. **5**(22), 1700672 (2017)
57. L.T. Dou, Y. Yang, J.B. You, Z.R. Hong, W.-H. Chang, G. Li, Y. Yang, Solution-processed hybrid perovskite photodetectors with high detectivity. Nat. Commun. **5**(1), 5404–5409 (2014)
58. X. Peng, L. Manna, W. Yang, J. Wickham, E. Scher, A. Kadavanich, A.P. Alivisatos, Shape control of CdSe nanocrystals. Nature **404**(6773), 59–61 (2000)
59. T. Wang, J. Zhuang, J. Lynch, O. Chen, Z. Wang, X. Wang, D. LaMontagne, H. Wu, Z. Wang, Y.C. Cao, Self-assembled colloidal superparticles from nanorods. Science **338**(6105), 358–363 (2012)
60. Y. Tong, B.J. Bohn, E. Bladt, K. Wang, P. Müller-Buschbaum, S. Bals, A.S. Urban, L. Polavarapu, J. Feldmann, From precursor powders to $CsPbX_3$ perovskite nanowires: one-pot synthesis, growth mechanism, and oriented self-assembly. Angew. Chem. Int. Ed. **56**(44), 13887–13892 (2017)
61. J. Feng, X. Yan, Y. Liu, H. Gao, Y. Wu, B. Su, L. Jiang, Crystallographically aligned perovskite structures for high-performance polarization-sensitive photodetectors. Adv. Mater. **29**(16), 1605993 (2017)
62. X. Sheng, G. Chen, C. Wang, W. Wang, J. Hui, Q. Zhang, K. Yu, W. Wei, M. Yi, M. Zhang, Polarized optoelectronics of $CsPbX_3$ ($X = Cl, Br, I$) perovskite nanoplates with tunable size and thickness. Adv. Funct. Mater. **28**(19), 1800283 (2018)
63. C. Chen, L. Gao, W. Gao, C. Ge, X. Du, Z. Li, Y. Yang, G. Niu, J. Tang, Circularly polarized light detection using chiral hybrid perovskite. Nat. Commun. **10**(1), 1–7 (2019)
64. Z. Liu, J. Xu, D. Chen, G. Shen, Flexible electronics based on inorganic nanowires. Chem. Soc. Rev. **44**(1), 161–192 (2015)
65. Y.J. Liu, Y.D. Liu, S.C. Qin, Y.B. Xu, R. Zhang, F.Q. Wang, Graphene-carbon nanotube hybrid films for high-performance flexible photodetectors. Nano Res. **10**(6), 1880–1887 (2017)
66. X. Hu, X. Zhang, L. Liang, J. Bao, S. Li, W. Yang, Y. Xie, High-performance flexible broadband photodetector based on organolead halide perovskite. Adv. Funct. Mater. **24**(46), 7373–7380 (2014)
67. Z. Zheng, F. Zhuge, Y. Wang, J. Zhang, L. Gan, X. Zhou, H. Li, T. Zhai, Decorating perovskite quantum dots in $TiO_2$ nanotubes array for broadband response photodetector. Adv. Funct. Mater. **27**(43), 1703115 (2017)
68. C. Bao, W. Zhu, J. Yang, F. Li, S. Gu, Y. Wang, T. Yu, J. Zhu, Y. Zhou, Z. Zou, Highly flexible self-powered organolead trihalide perovskite photodetectors with gold nanowire networks as transparent electrodes. ACS Appl. Mater. Inter. **8**(36), 23868–23875 (2016)
69. X.M. Li, D.J. Yu, J. Chen, Y. Wang, F. Cao, Y. Wei, Y. Wu, L. Wang, Y. Zhu, Z.G. Sun, J.P. Ji, Y.L. Shen, H.D. Sun, H.B. Zeng, Constructing fast carrier tracks into flexible perovskite photodetectors to greatly improve responsivity. ACS Nano **11**(2), 2015–2023 (2017)

70. F. Liu, Y. Zhang, C. Ding, S. Kobayashi, T. Izuishi, N. Nakazawa, T. Toyoda, T. Ohta, S. Hayase, T. Minemoto, Highly luminescent phase-stable $CsPbI_3$ perovskite quantum dots achieving near 100% absolute photoluminescence quantum yield. ACS Nano **11**(10), 10373–10383 (2017)
71. R. Lopez-Delgado, Y. Zhou, A. Zazueta-Raynaud, H. Zhao, J. Pelayo, A. Vomiero, M. Álvarez-Ramos, F. Rosei, A. Ayon, Enhanced conversion efficiency in si solar cells employing photoluminescent down-shifting CdSe/CdS Core/Shell quantum dots. Sci. Rep.-UK **7**(1), 1–8 (2017)
72. S.W. Baek, J.H. Shim, J.G. Park, The energy-down-shift effect of $Cd_{0.5}Zn_{0.5}S$–ZnS core–shell quantum dots on power-conversion-efficiency enhancement in silicon solar cells. Phys. Chem. Chem. Phys. 16(34), 18205–18210 (2014)
73. S. Gardelis, A.G. Nassiopoulou, Evidence of significant down-conversion in A Si-based solar cell using $CuInS_2$/ZnS core shell quantum dots. Appl. Phys. Lett. **104**(18), 183902 (2014)
74. S.W. Baek, J.-H. Shim, H.M. Seung, G.S. Lee, J.P. Hong, K.S. Lee, J.G. Park, Effect of core quantum-dot size on power-conversion-efficiency for silicon solar-cells implementing energy-down-shift using CdSe/ZnS Core/Shell quantum dots. Nanoscale **6**(21), 12524–12531 (2014)
75. J.W. Lu, X.X. Sheng, G.Q. Tong, Z.W. Yu, X.L. Sun, L.W. Yu, X.X. Xu, J.Z. Wang, J. Xu, Y. Shi, K.J. Chen, Ultrafast solar-blind ultraviolet detection by inorganic perovskite $CsPbX_3$ quantum dots radial junction architecture. Adv. Mater. **29**(23), 1700400 (2017)
76. T.Y. Zou, X.Y. Liu, R.Z. Qiu, Y. Wang, S.Y. Huang, C. Liu, Q. Dai, H. Zhou, Enhanced UV-C detection of perovskite photodetector arrays via inorganic CsPbBr 3 quantum dot down-conversion layer. Adv. Opt. Mater. **7**(11), 1801812 (2019)
77. L. Su, Z.X. Zhao, H.Y. Li, J. Yuan, Z.L. Wang, G.Z. Cao, G. Zhu, High-performance organolead halide perovskite-based self-powered triboelectric photodetector. ACS Nano **9**(11), 11310–11316 (2015)
78. Z.N. Wang, R.M. Yu, C.F. Pan, Z.L. Li, J. Yang, F. Yi, Z.L. Wang, Light-induced pyroelectric effect as an effective approach for ultrafast ultraviolet nanosensing. Nat. Commun. **6**(1), 8401–8407 (2015)
79. B.R. Sutherland, A.K. Johnston, A.H. Ip, J.X. Xu, V. Adinolfi, P. Kanjanaboos, E.H. Sargent, Sensitive, fast, and stable perovskite photodetectors exploiting interface engineering. ACS Photonics **2**(8), 1117–1123 (2015)
80. D.M. Jang, K. Park, D.H. Kim, J. Park, F. Shojaei, H.S. Kang, J.-P. Ahn, J.W. Lee, J.K. Song, Reversible halide exchange reaction of organometal trihalide perovskite colloidal nanocrystals for full-range band gap tuning. Nano Lett. **15**(8), 5191–5199 (2015)
81. L.Q. Pang, Y.Yao, Q. Wang, X.S. Zhang, Z. W. Jin, S.Z. (Frank) Liu, Shape- and trap-controlled nanocrystals for giant-performance improvement of all-inorganic perovskite photodetectors. Part. Part. Syst. Charact. 35(3), 1700363 (2018)
82. Z.P. Lian, Q.F. Yan, Q.R. Lv, Y. Wang, L.L. Liu, L.J. Zhang, S.L. Pan, Q. Li, L.D. Wang, J.-L. Sun, High-performance planar-type photodetector on (100) facet of $MAPbI_3$ single crystal. Sci. Rep.-UK 5(1), 16563–16572 (2015)
83. G. Konstantatos, M. Badioli, L. Gaudreau, J. Osmond, M. Bernechea, F.P. Garcia de Arquer, F. Gatti, F.H.L. Koppens, Hybrid graphene-quantum dot phototransistors with ultrahigh gain. Nat. Nanotechnol. **7**(6), 363–368 (2012)
84. Y. Lee, J. Kwon, E. Hwang, C.-H. Ra, W.J. Yoo, J.-H. Ahn, J.H. Park, J.H. Cho, High-performance perovskite-graphene hybrid photodetector. Adv. Mater. **27**(1), 41–46 (2015)
85. F. Li, C. Ma, H. Wang, W.J. Hu, W.L. Yu, A.D. Sheikh, T. Wu, Ambipolar solution-processed hybrid perovskite phototransistors. Nat. Commun. 6(1), 8238–8245 (2015)
86. M. Spina, M. Lehmann, B. Náfrádi, L. Bernard, E. Bonvin, R. Gaál, A. Magrez, L. Forró, E. Horváth, Microengineered $CH_3NH_3PbI_3$ nanowire/graphene phototransistor for low-intensity light detection at room temperature. Small **11**(37), 4824–4828 (2015)
87. D.-H. Kang, S.R. Pae, J. Shim, G. Yoo, J. Jeon, J.W. Leem, J.S. Yu, S. Lee, B. Shin, J.-H. Park, An ultrahigh-performance photodetector based on a perovskite-transition-metal-dichalcogenide hybrid structure. Adv. Mater. **28**(35), 7799–7806 (2016)

88. C. Ma, Y.M. Shi, W.J. Hu, M.-H. Chiu, Z.X. Liu, A. Bera, F. Li, H. Wang, L.-J. Li, T. Wu, Heterostructured $WS_2/CH_3NH_3PbI_3$ photoconductors with suppressed dark current and enhanced photodetectivity. Adv. Mater. **28**(19), 3683–3689 (2016)
89. C. Xie, F. Yan, Perovskite/Poly(3-hexylthiophene)/Graphene multiheterojunction phototransistors with ultrahigh gain in broadband wavelength region. ACS Appl. Mater. Inter. **9**(2), 1569–1576 (2017)
90. J.C. Yu, X. Chen, Y. Wang, H. Zhou, M.N. Xue, Y. Xu, Z.S. Li, C. Ye, J. Zhang, P.A. van Aken, P.D. Lund, H. Wang, A high-performance self-powered broadband photodetector based on a $CH_3NH_3PbI_3$ Perovskite/ZnO nanorod array heterostructure. J. Mater. Chem. C **4**(30), 7302–7308 (2016)
91. H. Si, Z. Kang, Q. Liao, Z. Zhang, X. Zhang, L. Wang, Y. Zhang, Design and tailoring of patterned ZnO nanostructures for energy conversion applications. Sci. China-Mater. **60**(5), 407–414 (2017)
92. H. Zhou, L. Yang, P.B. Gui, C.R. Grice, Z.H. Song, H. Wang, G.J. Fang, Ga-doped ZnO nanorod scaffold for high-performance, hole-transport-layer-free, self-powered $CH_3NH_3PbI_3$ perovskite photodetectors. Sol. Energ. Mat. Sol. C. **193**, 246–252 (2019)
93. H. Zhou, Z.H. Songa, P. Taoa, H.W. Leib, P.B. Guia, J. Meia, H. Wang, G.J. Fang, Self-powered, ultraviolet-visible perovskite photodetector based on $TiO_2$ nanorods. RSC Adv. **6**(8), 6205–6208 (2016)
94. D.-H. Kwak, D.-H. Lim, H.-S. Ra, P. Ramasamy, J.-S. Lee, High performance hybrid Graphene-CsPbBr 3 − xIx perovskite nanocrystal photodetector. RSC Adv. **6**(69), 65252–65256 (2016)
95. P. Ramasamy, D.-H. Lim, B. Kim, S.-H. Lee, M.-S. Lee, J.-S. Lee, All-inorganic cesium lead halide perovskite nanocrystals for photodetector applications. Chem. Commun. **52**(10), 2067–2070 (2016)
96. G. Walters, B.R. Sutherland, S. Hoogland, D. Shi, R. Comin, D.P. Sellan, O.M. Bakr, E.H. Sargent, Two-photon absorption in organometallic bromide perovskites. ACS Nano **9**(9), 9340–9346 (2015)
97. J.Z. Song, Q.Z. Cui, J.H. Li, J.Y. Xu, Y. Wang, L.M. Xu, J. Xue, Y.H. Dong, T. Tian, H.D. Sun, H.B. Zeng, Ultralarge all-inorganic perovskite bulk single crystal for high-performance visible-infrared dual-modal photodetectors. Adv. Opt. Mater. **5**(12), 1700157 (2017)
98. X.B. Xu, C.-C. Chueh, P.F. Jing, Z.B. Yang, X.L. Shi, T. Zhao, L.Y. Lin, A.K.-Y. Jen, High-performance near-IR photodetector using low-bandgap $MA_{0.5}FA_{0.5}Pb_{0.5}Sn_{0.5}I_3$ perovskite. Adv. Funct. Mater. 27(28), 1701053 (2017)
99. M.M. Han, J.M. Sun, M. Peng, N. Han, Z.H. Chen, D. Liu, Y.N. Guo, S. Zhao, C.X. Shan, T. Xu, X.T. Hao, W.D. Hu, Z.X. Yang, Controllable growth of lead-free all-inorganic perovskite nanowire array with fast and stable near-infrared photodetection. J. Phys. Chem. C **123**(28), 17566–17573 (2019)
100. J.L. Chang, G.Z. Wang, Y.H. Huang, X.K. Luo, H. Chen, New insights into the electronic structures and optical properties in the orthorhombic perovskite $MAPbI_3$: a mixture of Pb and Ge/Sn. New J. Chem. **41**(19), 11413–11421 (2017)
101. A.A. Bessonov, M. Allen, Y.L. Liu, S. Malik, J. Bottomley, A. Rushton, I. Medina-Salazar, M. Voutilainen, S. Kallioinen, A. Colli, C. Bower, P. Andrew, T. Ryhanen, Compound quantum dot-perovskite optical absorbers on graphene enhancing short-wave infrared photodetection. ACS Nano **11**(6), 5547–5557 (2017)
102. Y. Yu, Y.T. Zhang, X.X. Song, H.T. Zhang, M.X. Cao, Y.L. Che, H.T. Dai, J.B. Yang, H. Zhang, J.Q. Yao, High performances for solution-pocessed 0D-0D heterojunction phototransistors. Adv. Opt. Mater. **5**(24), 1700565 (2017)
103. W.H. Li, K. Ding, H.R. Tian, M.S. Yao, B. Nath, W.H. Deng, Y.B. Wang, G. Xu, Conductive metal-organic framework nanowire array electrodes for high-performance solid-state supercapacitors. Adv. Funct. Mater. **27**(27), 1702485 (2017)
104. S. Chen, C. Teng, M. Zhang, Y. Li, D. Xie, G. Shi, A flexible UV-Vis-NIR photodetector based on a perovskite/conjugated-polymer composite. Adv. Mater. **28**(28), 5969–5974 (2016)

105. L. Shen, Y.Z. Lin, C.X. Bao, Y. Bai, Y.H. Deng, M.M. Wang, T. Li, Y.F. Lu, A. Gruverman, W.W. Li, J.S. Huang, Integration of perovskite and polymer photoactive layers to produce ultrafast response, ultraviolet-to-near-infrared. Sensitive photodetectors. Mater. Horiz. 4(2), 242–248 (2017)
106. Y.K. Wang, D.Z. Yang, X.K. Zhou, D.G. Ma, A. Vadim, T. Ahamad, S.M. Alshehri, Perovskite/polymer hybrid thin films for high external quantum efficiency photodetectors with wide spectral response from visible to near-infrared wavelengths. Adv. Opt. Mater. **5**(12), 201700213 (2017)
107. J.B. Li, Y.L. Shen, Y.C. Liu, F. Shi, X.D. Ren, T.Q. Niu, K. Zhao, S.Z. Liu, Stable high-performance flexible photodetector based on upconversion nanoparticles/perovskite microarrays composite. ACS Appl. Mater. Inter. **9**(22), 19176–19183 (2017)
108. X.S. Zhang, Q. Wang, Z.W. Jin, S.Z. Liu, Stable ultra-fast broad-bandwidth photodetectors based on $\alpha$-$CsPbI_3$ perovskite and $NaYF_4$:Yb, Er Quantum Dots. Nanoscale **9**(19), 6278–6285 (2017)
109. X.H. Zhang, S.Z. Yang, H. Zhou, J.W. Liang, H.W. Liu, H. Xia, X.L. Zhu, Y. Jiang, Q.L. Zhang, W. Hu, X.J. Zhuang, H.J. Liu, W.D. Hu, X. Wang, A.L. Pan, Perovskite-erbium silicate nanosheet hybrid waveguide photodetectors at the near-infrared telecommunication band. Adv. Mater. **29**(21), 1604431 (2017)
110. L.W. Sang, M.Y. Liao, M. Sumiya, A comprehensive review of semiconductor ultraviolet photodetectors: from thin film to one-dimensional nanostructures. Sensors **13**(8), 10482–10518 (2013)
111. G.H. Tang, F. Yang, X.L. Chen, W.K. Luo, High-quantum-efficiency ultraviolet solar-blind AlGaN photocathode detector with a sharp spectral sensitivity threshold at 300 nm. Appl. Opt. **57**(27), 8060–8064 (2018)
112. H.Y. Chen, P.P. Yu, Z.Z. Zhang, F. Teng, L.X. Zheng, K. Hu, X.S. Fang, Ultrasensitive self-powered solar-blind deep-ultraviolet photodetector based on all-solid-state Polyaniline/MgZnO bilayer. Small **12**(42), 5809–5816 (2016)
113. X.H. Chen, F.F. Ren, S.L. Gu, J.D. Ye, Review of gallium-oxide-based solar-blind ultraviolet photodetectors. Photonics Res. **7**(4), 11–45 (2019)
114. C.N. Lin, Y.J. Lu, X. Yang, Y.Z. Tian, C.J. Gao, J.L. Sun, L. Dong, F. Zhong, W.D. Hu, C.X. Shan, Diamond-based all-carbon photodetectors for solar-blind imaging. Adv. Opt. Mater. **6**(15), 1800068 (2018)
115. C.H. Lin, H.C. Fu, B. Cheng, M.L. Tsal, W. Luo, L.H. Zhou, S.H. Jang, L.B. Hu, J.H. He, A flexible solar-blind 2D boron nitride nanopaper-based photodetector with high thermal resistance. npj 2D Mater. Appl. **2**, 23 (2018)
116. Y. Xia, G.M. Zhai, Z. Zheng, L.Y. Lian, H. Liu, D.L. Zhang, J.B. Gao, T.Y. Zhai, J.B. Zhang, Solution-processed solar-blind deep ultraviolet photodetectors based on strongly quantum confined ZnS quantum dots. J. Mater. Chem. C **6**(42), 11266–11271 (2018)
117. L.Z. Qiu, S.Y. Wei, H.S. Xu, Z.X. Zhang, Z.Y. Guo, X.G. Chen, S.Y. Liu, D. Wu, L.B. Luo, Ultrathin polymer nanofibrils for solar-blind deep ultraviolet light photodetectors application. Nano Lett. **20**(1), 644–651 (2020)
118. K.J. Baeg, M. Binda, D. Natali, M. Caironi, Y.Y. Noh, Organic light detectors: photodiodes and phototransistors. Adv. Mater. **25**(31), 4267–4295 (2013)
119. P. Peumans, A. Yakimov, S.R. Forrest, Small molecular weight organic thin-film photodetectors and solar cells. J. Appl. Phys. **93**(7), 3693–3723 (2003)
120. Q.X. Tang, L.Q. Li, Y.B. Song, Y.L. Liu, H.X. Li, W. Xu, Y.Q. Liu, W.P. Hu, D.B. Zhu, Photoswitches and phototransistors from organic single-crystalline sub-micro/nanometer ribbons. Adv. Mater. **19**(18), 2624–2628 (2007)
121. K. Xiao, J. Tao, Z.W. Pan, A.A. Puretzky, I.N. Ivanov, S.J. Pennycook, D.B. Geohegan, Single-crystal organic nanowires of copper-tetracyanoquinodimethane: synthesis, patterning, characterization, and device applications. Angew. Chem. Int. Ed. **46**(15), 2650–2654 (2007)
122. Y. Zhou, L. Wang, J. Wang, J. Pei, Y. Cao, Highly sensitive, air-stable photodetectors based on single organic sub-micrometer ribbons self-assembled through solution processing. Adv. Mater. **20**(19), 3745–3749 (2008)

123. X. Gong, M.H. Tong, Y.J. Xia, W.Z. Cai, J.S. Moon, Y. Cao, G. Yu, C.L. Shieh, B. Nilsson, A.J. Heeger, High-detectivity polymer photodetectors with spectral response from 300 to 1450 nm. Science **325**(5948), 1665–1667 (2009)
124. Y.L. Guo, G. Yu, Y.Q. Liu, Functional organic field-effect transistors. Adv. Mater. **22**(40), 4427–4447 (2010)
125. Y.L. Guo, C. Liu, H. Tanaka, E. Nakamura, Air-stable and solution-processable perovskite photodetectors for solar-blind UV and visible light. J. Phys. Chem. Lett. **6**(3), 535–539 (2015)
126. T. Zhang, F. Wang, P. Zhang, Y.F. Wang, H. Chen, J. Li, J. Wu, L. Chen, Z.D. Chen, S.B. Li, Low-temperature processed inorganic perovskites for flexible detectors with a broadband photoresponse. Nanoscale **11**(6), 2871–2877 (2019)
127. Z.X. Zhang, C. Li, Y. Lu, X.W. Tong, F.X. Liang, X.Y. Zhao, D. Wu, C. Xie, L.B. Luo, Sensitive deep ultraviolet photodetector and image sensor composed of inorganic lead-free $Cs_3Cu_2I_5$ perovskite with wide bandgap. J. Phys. Chem. Lett. **10**(18), 5343–5350 (2019)
128. X.Z. Xu, W. Deng, X.J. Zhang, L.M. Huang, W. Wang, R.F. Jia, D. Wu, X.H. Zhang, J.S. Jie, S.T. Lee, Dual-band, high-performance phototransistors from hybrid perovskite and organic crystal array for secure communication applications. ACS Nano **13**(5), 5910–5919 (2019)
129. E.R. Zheng, B. Yuh, G.A. Tosado, Q.M. Yu, Solution-processed visible-blind UV-A photodetectors based on $CH_3NH_3PbCl_3$ perovskite thin films. J. Mater. Chem. C. **5**(15), 3796–3806 (2017)
130. L.P. Dong, T.Q. Pang, J.G. Yu, Y.C. Wang, W.G. Zhu, H.D. Zheng, J.H. Yu, R.X. Jia, Z. Chen, Performance-enhanced solar-blind photodetector based on a $CH_3NH_3PbI_3/\beta\text{-}Ga_2O_3$ hybrid structure. J. Mater. Chem. C. **7**(45), 14205–14211 (2019)
131. L.W. Yu, S. Misra, J.Z. Wang, S.Y. Qian, M. Foldyna, J. Xu, Y. Shi, E. Johnson, P.R. Cabarrocas, Understanding light harvesting in radial junction amorphous silicon thin film solar cells. Sci. Rep.-UK 4, 4357 (2014)
132. J.W. Lu, S.Y. Qian, Z.W. Yu, S. Misra, L.W. Yu, J. Xu, Y. Shi, P.R.i. Cabarrocas, K.J. Chen, How tilting and cavity-mode-resonant absorption contribute to light harvesting in 3D radial junction solar cells. Opt. Express **23**, A1288–A1296, (2015)
133. L.W. Yu, F. Fortuna, B. O'Donnell, T. Jeon, M. Foldyna, G. Picardi, P.R.I. Cabarrocas, Bismuth-catalyzed and doped silicon nanowires for one-pump-down fabrication of radial junction solar cells. Nano Lett. **12**(8), 4153–4158 (2012)
134. M.J. Zhang, L.X. Wang, L.H. Meng, X.G. Wu, Q.W. Tan, Y.J. Chen, W.Y. Liang, F. Jiang, Y. Cai, H.Z. Zhong, Perovskite quantum dots embedded composite films enhancing UV response of silicon photodetectors for broadband and solar-blind light detection. Adv. Opt. Mater. **6**(16), 1800077 (2018)
135. G.Q. Tong, H. Li, Z.F. Zhu, Y. Zhang, L.W. Yu, J. Xu, Y. Jiang, Enhancing hybrid perovskite detectability in the deep ultraviolet region with down-conversion dual-phase (CsPbBr 3-Cs4PbBr6) films. J. Phys. Chem. Lett. **9**(7), 1592–1599 (2018)
136. C.H. Kang, I. Dursun, G. Liu, L. Sinatra, X.B. Sun, M.W. Kong, J. Pan, P. Maity, E.N. Ooi, T.K. Ng, O.F. Mohammed, O.M. Bakr, B.S. Ooi, High-speed colour-converting photodetector with all-inorganic CsPbBr 3 perovskite nanocrystals for ultraviolet light communication. Light Sci. Appl. **8**, 94 (2019)
137. P.K. Spiegel, The first clinical x-ray made in America–100 years. Am. J. Roent. **164**(1), 241–243 (1995)
138. C.C. Stoumpos, C.D. Malliakas, J.A. Peters, Z.F. Liu, M. Sebastian, J. Im, T.C. Chasapis, A.C. Wibowo, D.Y. Chung, A.J. Freeman, B.W. Wessels, M.G. Kanatzidis, Crystal growth of the perovskite semiconductor CsPbBr 3: a new material for high-energy radiation detection. Cryst. Growth Des. **13**(7), 2722–2727 (2013)
139. S. Yakunin, S.M. Sytnyk, D. Kriegner, S. Shrestha, M. Richter, G.J. Matt, H. Azimi, C.J. Brabec, J. Stangl, M.V. Kovalenko, W. Heiss, Detection of X-Ray photons by solution-processed organic-inorganic perovskites. Nat. Photonics 9(7), 444–449 (2015)
140. H.T. Wei, Y.J. Fang, P. Mulligan, W. Chuirazzi, H.H. Fang, C.C. Wang, B.R. Ecker, Y.L. Gao, M.A. Loi, L. Cao, J.S. Huang, Sensitive X-ray detectors made of methylammonium lead tribromide perovskite single crystals. Nat. Photonics **10**(5), 333–339 (2016)

141. J.Y. Liu, B. Shabbir, C.J. Wang, T. Wan, Q.D.Y.M. Dong, J. Jasieniak, Y.P. Zhang, Q.L. Bao Flexible, Printable Soft-X-Ray detectors based on all-inorganic perovskite quantum dots. Adv. Mater. **31**(30), 1901644 (2019)
142. Q.S. Chen, Q.J. Wu, X.Y. Ou, B.L. Huang, J. Almutlaq, A.A. Zhumekenov, X.W. Guan, S.Y. Han, L.L. Liang, Z.G. Yi, J. Li, X.J. Xie, Y. Wang, Y. Li, D.Y. Fan, D.B.L. The, A.H. All, O.F. Mohammed, O.M. Bakr, T. Wu, M. Bettinelli, H.H. Yang, W. Huang, X.G. Liu, All-inorganic perovskite nanocrystal scintillators. Nature **561**(7721), 88–93 (2018)
143. W. Pan, H. Wu, J. Luo, Z. Deng, C. Ge, C. Chen, X. Jiang, W.J. Yin, G. Niu, L. Zhu, L. Yin, Y. Zhou, Q. Xie, X. Ke, M. Sui, J. Tang, Cs2AgBiBr 6 single-crystal x-ray detectors with a low detection limit. Nat. Photonics **11**(11), 726–732 (2017)
144. Q.F. Dong, Y.J. Fang, Y.C. Shao, P. Mulligan, J. Qiu, L. Cao, J.S. Huang, Electron-hole diffusion lengths >175 μm in solution-grown $CH_3NH_3PbI_3$ single crystals. Science **347**, 967–970 (2015)
145. S. Yakunin, D.N. Dirin, Y. Shynkarenko, V. Morad, I. Cherniukh, O. Nazarenko, D. Kreil, T. Nauser, M.V. Kovalenko, Detection of gamma photons using solution-grown single crystals of hybrid lead halide perovskites. Nat. Photonics **10**(9), 585–589 (2016)
146. J. Song, L. Xu, J. Li, J. Xue, Y. Dong, X. Li, H. Zeng, Monolayer and few-layer all-inorganic perovskites as a new family of two-dimensional semiconductors for printable optoelectronic devices. Adv. Mater. **28**(24), 4861–4869 (2016)
147. Y. Li, Z. Shi, L. Lei, Z. Ma, F. Zhang, S. Li, D. Wu, T. Xu, X. Li, C. Shan, G. Du, Controllable vapor-phase growth of inorganic perovskite microwire networks for high-efficiency and temperature-stable photodetectors. ACS Photonics **5**(6), 2524–2532 (2018)
148. P. Zhu, S. Gu, X. Shen, N. Xu, Y. Tan, S. Zhuang, Y. Deng, Z. Lu, Z. Wang, J. Zhu, Direct conversion of perovskite thin films into nanowires with kinetic control for flexible optoelectronic devices. Nano Lett. **16**(2), 871–876 (2016)
149. W. Deng, X. Zhang, L. Huang, X. Xu, L. Wang, J. Wang, Q. Shang, S.T. Lee, J. Jie, Aligned single-crystalline perovskite microwire arrays for high-performance flexible image sensors with long-term stability. Adv. Mater. **28**(11), 2201–2208 (2016)
150. W. Deng, L. Huang, X. Xu, X. Zhang, X. Jin, S.T. Lee, J. Jie, Ultrahigh-responsivity photodetectors from perovskite nanowire arrays for sequentially tunable spectral measurement. Nano Lett. **17**(4), 2482–2489 (2017)
151. H. Gao, J. Feng, Y. Pi, Z. Zhou, B. Zhang, Y. Wu, X. Wang, X. Jiang, L. Jiang, Bandgap engineering of single-crystalline perovskite arrays for high-performance photodetectors. Adv. Funct. Mater. **28**(46), 1804349 (2018)
152. D. Dong, H. Deng, C. Hu, H. Song, K. Qiao, X. Yang, J. Zhang, F. Cai, J. Tang, H. Song, Bandgap tunable $Csx(CH_3NH_3)1\text{-}xPbI_3$ perovskite nanowires by aqueous solution synthesis for optoelectronic devices. Nanoscale **9**(4), 1567–1574 (2017)
153. J.H. Cha, J.H. Han, W. Yin, C. Park, Y. Park, T.K. Ahn, J.H. Cho, D.Y. Jung, Photoresponse of CsPbBr 3 and Cs4PbBr6 perovskite single crystals. J. Phys. Chem. Lett. **8**(3), 565–570 (2017)
154. D.J. Yu, F. Cao, Y.L. Shen, X.H. Liu, Y. Zhu, H.B. Zeng, Dimensionality and Interface engineering of 2D homologous perovskites for boosted charge-carrier transport and photodetection performances. J. Phys. Chem. Lett. **8**(12), 2565–2572 (2017)
155. B.S. Zhu, Z. He, J.S. Yao, C. Chen, K.H. Wang, H.B. Yao, J.W. Liu, S.H. Yu, Potassium Ion assisted synthesis of organic-inorganic hybrid perovskite nanobelts for stable and flexible photodetectors. Adv. Opt. Mater. **6**(3), 1701029 (2018)
156. J.G. Feng, C. Gong, H.F. Gao, W. Wen, Y.J. Gong, X.Y. Jiang, B. Zhang, Y.C. Wu, Y.S. Wu, H.B. Fu, L. Jiang, X. Zhang, Single-crystalline layered metal-halide perovskite nanowires for ultrasensitive photodetectors. Nat. Electron. **1**(7), 404–410 (2018)
157. A. Waleed, M.M. Tavakoli, L.L. Gu, Z.Y. Wang, D.Q. Zhang, A. Manikandan, Q.P. Zhang, R.J. Zhang, Y.L. Chueh, Z.Y. Fan, Lead-free perovskite nanowire array photodetectors with drastically improved stability in nanoengineering templates. Nano Lett. **17**(1), 523–530 (2017)
158. L. Qian, Y.L. Sun, M.X. Sun, Z.M. Fang, L. Li, D. Xie, C. Li, L.M. Ding, 2D perovskite microsheets for high-performance photodetectors. J. Mater. Chem. C. **7**(18), 5353–5358 (2019)

159. Z.Y. Qi, X.W. Fu, T.F. Yang, D. Li, P. Fan, H.L. Li, F. Jiang, L.H. Li, Z.Y. Luo, X.J. Zhuang, A. Pan, Highly stable lead-free $Cs_3Bi_2I_9$ perovskite nanoplates for photodetection applications. Nano Res. **12**(8), 1894–1899 (2019)
160. W.G. Li, X.D. Wang, J.F. Liao, Y. Jiang, D. B. Kuang, Enhanced on-off ratio photodetectors based on lead-free $Cs_3Bi_2I_9$ Single crystal thin films. Adv. Funct. Mater. 1909701 (2020)
161. Y.Y. Dang, G.Q. Tong, W.T. Song, Z.H. Liu, L.B. Qiu, L.K. Ono, Y.B. Qi, Interface engineering strategies towards Cs2AgBiBr 6 single-crystalline photodetectors with good ohmic contact behaviours. J. Mater. Chem. C. **8**(1), 276–284 (2020)
162. S.W. Tao, Y. Chen, J.W. Cui, H.W. Zhou, N. Yu, X.R. Gao, S.T. Cui, C. Yuan, M.Q. Liu, M.M. Wang, X.J. Wang, H.Y. Gong, Y.M. Li, T. Liu, X. Sun, J. Yin, X.X. Zhang, M.X. Wu, Organic-inorganic hybrid $(CH_3NH_3)_2FeCuI_4Cl_2$ and $(CH_3NH_3)_2InCuI_6$ for ultraviolet light photodetectors. Chem. Commun. **56**(12), 1875–1878 (2020)
163. Y.T. Chen, Y.L. Chu, X.H. Wu, W. Ou-Yang, J. Huang, High-performance inorganic perovskite quantum dot-organic semiconductor hybrid phototransistors. Adv. Mater. **29**(44), 1704062 (2017)
164. Y.H. Dong, Y. Gu, Y.S. Zou, J.Z. Song, L.M. Xu, J.H. Li, J. Xue, X.M. Li, H.B. Zeng, Improving all-inorganic perovskite photodetectors by preferred orientation and plasmonic effect. Small **12**(40), 5622–5632 (2016)
165. D.Y. Li, D.L. Zhou, W. Xu, X. Chen, G.C. Pan, X.Y. Zhou, N. Ding, H.W. Song, Plasmonic photonic crystals induced two-order fluorescence enhancement of blue perovskite nanocrystals and its application for high-performance flexible ultraviolet photodetectors. Adv. Funct. Mater. **28**(41), 1804429 (2018)
166. W.Z. Xu, Y.K. Guo, X.T. Zhang, L.Y. Zheng, T. Zhu, D.H. Zhao, W.P. Hu, X. Gong, Room-temperature-operated ultrasensitive broadband photodetectors by perovskite incorporated with conjugated polymer and single-wall carbon nanotubes. Adv. Funct. Mater. **28**(7), 1705541 (2018)
167. Y. Wang, R. Fullon, M. Acerce, C.E. Petoukhoff, J. Yang, C.G. Chen, S.N. Du, S.K. Lai, S.P. Lau, D. Voiry, D. O'Carroll, G. Gupta, A.D. Mohite, S.D. Zhang, H. Zhou, M. Chhowalla, Solution-processed $MoS_2$/organolead trihalide perovskite photodetectors. Adv. Mater. **29**(4), 1603995 (2017)
168. W. Deng, H.C. Huang, H.M. Jin, W. Li, X. Chu, D. Xiong, W. Yan, F.J. Chun, M.L. Xie, C. Luo, L. Jin, C.Q. Liu, H.T. Zhang, W.L. Deng, W.Q. Yang, All-sprayed-processable, large-area, and flexible perovskite/mxene-based photodetector arrays for photocommunication. Adv. Opt. Mater. **7**(6), 1801521 (2019)
169. H.L. Wu, H.N. Si, Z.H. Zhang, Z. Kang, P.W. Wu, L.X. Zhou, S.C. Zhang, Z. Zhang, Q.L. Liao, Y. Zhang, All-inorganic perovskite quantum dot-monolayer $MoS_2$ mixed-dimensional van der waals heterostructure for ultrasensitive photodetector. Adv. Sci. **5**(12), 1801219 (2018)
170. Q.Q. Fan, G.V. Biesold-McGee, J.Z. Ma, Q.N. Xu, S. Pan, J. Peng, Z.Q. Lin, Lead-free halide perovskite nanocrystals: crystal structures, synthesis, stabilities, and optical properties. Angew. Chem. Int. Ed. **59**(3), 1030–1046 (2020)
171. X.X. Xu, X. Wang, A. Nisar, X. Liang, J. Zhuang, S. Hu, Y. Zhuang, Combinatorial hierarchically ordered 2D architectures self-assembled from nanocrystal building blocks. Adv. Mater. **20**(19), 3702–3708 (2008)
172. S.M. Zhang, W.X. Shi, T.D. Siegler, X.Q. Gao, F. Ge, B.A. Korgel, Y. He, S.Z. Li, X. Wang, An all-inorganic colloidal nanocrystal flexible polarizer. Angew. Chem. Int. Ed. **58**(26), 8730–8735 (2019)
173. J.L. Liu, W.X. Shi, B. Ni, Y. Yang, S.Z. Li, J. Zhuang, X. Wang, Incorporation of clusters within inorganic materials through their addition during nucleation steps. Nat. Chem. **11**(9), 839–845 (2019)
174. E.V. Shevchenko, D.V. Talapin, N.A. Kotov, S. O'Brien, C.B. Murray, Structural diversity in binary nanoparticle superlattices. Nature **439**(7072), 55–59 (2006)
175. A.M. Kalsin, M. Fialkowski, M. Paszewski, S.K. Smoukov, K.J. Bishop, B.A. Grzybowski, Electrostatic self-assembly of binary nanoparticle crystals with a diamond-like lattice. Science **312**(5772), 420–424 (2006)

176. F. Yang, J.Z. Wang, J.W. Lu, Z. W. Yu, L. W. Yu, J. Xu, Y. Shi, K. J. Chen, P. Roca i cabarrocas, biomimetic radial tandem junction photodetector with natural RGB color discrimination capability. Adv. Opt. Mater. **5**(19), 1700390 (2017)
177. X.J. Zhang, X.X. Wu, X.Y. Liu, G.Y. Chen, Y.K. Wang, J.C. Bao, X.X. Xu, X.F. Liu, Q. Zhang, K.H. Yu, W. Wei, J.J. Liu, J. Xu, H. Jiang, P. Wang, X. Wang, Heterostructural $CsPbX_3$-PbS (X = Cl, Br, I) quantum dots with tunable Vis-NIR dual emission. J. Am. Chem. Soc. **142**(9), 4464–4471 (2020)

# Chapter 8
# Perovskite Quantum Dots Based Luminescent Solar Concentrators

**Haiguang Zhao**

**Abstract** Luminescent solar concentrators (LSCs) are emerging technology of current interest as a platform for solar energy harvesting. LSCs can serve as large-area sunlight collectors for photovoltaic cells to reduce the cost of electricity by decreasing the use of expensive photovoltaic materials. Among various types of fluorophores, perovskite quantum dots have been used as light-converters due to their size/shape/composition-tunable wide absorption spectrum, narrow emission spectrum, high quantum yield and structure-engineered large Stokes shift. In this Chapter, we summarized the use of various types of perovskites and different configuration to fabricate high efficiency and large-area LSCs. The optical properties of perovskite are strongly depending on their size, shape and composition. Therefore, one can optimize their structure to obtain the suitable emitters for LSCs. Among various types of perovskites, doped quantum dots offer a great opportunity to synthesize high quality perovskites with both high quantum yield and large Stokes shift, indicating the doped perovskites may be a great candidate as emitters for high efficiency LSCs. We conclude with a detailed account of the latest research progress in synthesis, structure, materials, and performance of LSCs based on perovskite quantum dots and a further perspective on the remaining key issues and open opportunities in the field.

## 8.1 Introduction

Solar technologies represent a great opportunity towards addressing the increasing demand for clean and renewable energy [1–7]. Radiation impinging from the sun in an hour could power the planet for one year. However, the cost of the electricity generated by commercial silicon photovoltaic (PV) devices is still high compared to traditional power plants [8–10]. Luminescent solar concentrators (LSCs) are emerging

H. Zhao (✉)
College of Physics & State Key Laboratory of Bio-Fibers and Eco-Textiles, Qingdao University, No. 308 Ningxia Road, Qingdao 266071, People's Republic of China
e-mail: hgzhao@qdu.edu.cn



Y. Zhou and Y. Wang (eds.), *Perovskite Quantum Dots*, Springer Series in Materials Science 303, https://doi.org/10.1007/978-981-15-6637-0_8

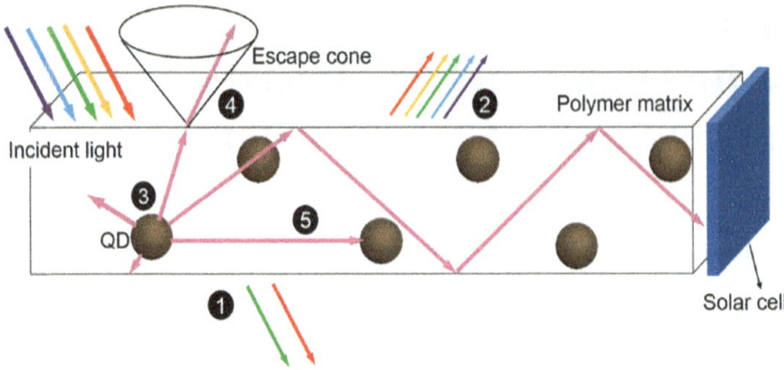


**Fig. 8.1** Schematic representation of a QD-based LSC. The QDs are embedded inside the polymer matrix. Reproduced with permission from Ref. [47]. Copyright 2016, Wiley–VCH

technology of current interest as a platform for solar energy harvesting [3, 9–34, 36–45]. LSCs can serve as large-area sunlight collectors for PVs to reduce the cost of electricity by decreasing the use of expensive PV materials. A typical LSC consists of an optical waveguide embedded or spin-coated with fluorophores [e.g. dyes, polymers or quantum dots (QDs)]. The waveguide could be transparent/semitransparent polymers or glass with high refractive index. Fluorophores partially absorb sunlight and re-emit concentrated emissions at a shorter (by upconversion fluorophores [46]) or longer wavelength (by down-shifting fluorophores) [47]. The concentrated emissions on the edges of the LSC are collected by PV cells placed at the edges of the waveguide, where it is converted into electric power (Fig. 8.1) [47]. The LSCs could be widely used for building integrated PVs (BIPVs) due to its adjustable transparent nature by tuning the type and concentration of fluorophores, light weight, and low-light intensity PV response [3, 8–10, 20, 30, 35, 38, 47, 48]. So far, the research direction on LSCs focuses on the improvement of their external optical efficiency (defined as the ratio between the optical power of re-emitted photons reaching the edges of an LSC and the optical power of incident photons) by improving the optical properties of the fluorophones and configuration of the LSCs. We take the example of colloidal QDs as fluorophones to explain the energy loss process in LSCs. As shown in Fig. 8.1, sunlight is lost in the LSC based on colloidal QDs in the following ways: (1) only a fraction of incident sunlight could be absorbed because of the limited overlap between the QD's absorption and solar spectrum; (2) part of the sunlight is reflected at the top surface of the LSC; (3) sunlight is lost during re-emission process, as the fluorescent quantum yield (QY, defined as the ratio between the number of emitted photons to the number of the absorbed photons) is below 100%; (4) the fraction of radiation, which is re-emitted with an emission angle falling into the escape cone cannot undergo total internal reflection (TIR) inside the waveguide and is lost. The escape cone is determined by the refractive index of an LSC and air; (5) part of the sunlight is lost due to re-absorption, which is caused by the overlap between the absorption and emission spectra of the QDs. Usually, energy losses from re-absorption of the emitters occur repetitively, because photons traveling within the waveguide can be absorbed by other emitters and re-emitted with similar energy band edges multiple times before reaching an edge. For example, in practical LSC implementations (1 $m^2$), even very small overlap between absorption and emission spectra may result in strong energy losses, leading to a low external optical efficiency [3, 8]. It is still a challenge to obtain a fluorophore with overlap between absorption and emission spectra. Other minor light losses could be due to the absorption of emitted light by the polymer matrix or the emissions are quenched by the defects in the polymer matrix or glass. Preparation of the LSC with high external optical efficiency is still very challenging so far.

The theoretical optical efficiency (defined as the ratio between the optical power of re-emitted photons of an LSC and the optical power of absorbed photons) of an LSC can be calculated as using Eq. (8.1) [8].

$$\eta_{\text{opt}} = (1 - R)P_{\text{TIR}} \cdot \eta_{\text{Abs}} \cdot \eta_{\text{QY}} \cdot \eta_{\text{Stokes}} \cdot \eta_{\text{Host}} \cdot \eta_{\text{TIR}} \cdot \eta_{\text{Self}} \quad (8.1)$$

where $R$ is the reflection of light from the LSC surface, $P_{\text{TIR}}$ is the TIR efficiency, $\eta_{\text{Abs}}$ is the fraction of absorbed sunlight, $\eta_{\text{QY}}$ is the QY of emitters, $\eta_{\text{Stokes}}$ is the energy lost due to heat generation, $\eta_{\text{Host}}$ is the transport efficiency of the wave-guided photons through the waveguide, $\eta_{\text{TIR}}$ is the reflection efficiency of the waveguide determined by the smoothness of the waveguide surface, and $\eta_{\text{Self}}$ is related to self-absorption due to the spectral overlap of absorption and emission. The optical efficiency of the LSCs is different to the external optical efficiency, which is same as the power conversion efficiency (PCE) [10].

A common method to estimate $\eta_{\text{opt}}$ is to couple a PV cell at the end of the LSC. $\eta_{\text{opt}}$ can be calculated as [10]:

$$\eta_{\text{opt}} = \frac{\text{I}_{\text{LSC}} \cdot \text{A}_{\text{PV}}}{\text{I}_{\text{SC}} \cdot \text{A}_{\text{LSC}}} = \frac{\text{I}_{\text{LSC}}}{\text{I}_{\text{SC}} \cdot \text{G}}$$

where $I_{\text{LSC}}$ is the short circuit current generated by the cell coupled to the LSC, $I_{\text{SC}}$ is the short circuit current of the same PV cell under direct illumination, $A_{\text{PV}}$ is the area of the edge of the LSC (or the active area on the PV cell, if the area is smaller) and $A_{\text{LSC}}$ is the area of the top of the LSC.

During the last thirty years, various types of fluorescent materials have been used as light converters in LSCs, including dyes, polymers, QDs (e.g. CdSe/CdS, PbS/CdS, and perovskite), and upconversion nanocrystals (NCs) etc. [3, 20, 27, 28, 49–51]. Among them, inorganic perovskite [e.g. $CsPbX_3$ (X = Br/Cl, Br/I)] QDs have been used for LSCs in view of their size/compositional dependent absorption and emission spectra, wide absorption spectrum, high brightness with a QY up to 100%, and easy solution phase synthesis [31, 52–57]. Inorganic perovskite QDs may represent a new promising candidate for large-area and high-efficiency LSCs. In this Chapter, we will focus on the introduction for the various configurations of the LSCs based on the perovskites, including mixed-halide perovskite QDs; doped inorganic perovskites; zero-dimensional (0D) perovskite NCs and perovksite nanoplatelets (NPLs).

## 8.2 Single-Layer LSC

### *8.2.1 LSC Based on Mixed-Halide Perovskite QDs*

Recently, organic–inorganic hybrid perovskite thin films ($CH_3NH_3PbI_3$ or $CH_3NH_3PbI_{3-x}Cl_x$) coated on the glass substrate have been used as emitters for LSCs [19, 58]. In the organic–inorganic hybrid perovskite, the strong overlap between the emission and absorption spectra lead to strong reabsorption energy loss, indicating that this type of organic–inorganic hybrid perovskites is not suitable for large-scale high-efficiency LSCs. In addition, without the surface protection (e.g. polymer), the perovskite thin film is very sensitive to the moisture, limiting their long-term stability. Zhao et al. first demonstrated the application of mixed-halide $CsPb(Br_xI_{1-x})_3$ inorganic perovskite QDs as highly emissive fluorophores in high performance LSCs [26]. By only tuning the molar ratio of Br/I or Cl/Br, $CsPb(Br_xI_{1-x})_3$ or $CsPb(Br_xCl_{1-x})_3$ perovskite QDs were synthesized via a hot-injection approach [26, 34]. The as-produced mixed-halide perovskite QDs exhibit a high QY of 40–80% with a broad absorption ranging from 300 to 650 nm [26]. A single-layer LSC was further prepared by using polymer as waveguide [26]. In details, the as-synthesized mixed-halide perovskite QDs were dispersed into monomers of lauryl methacrylate and crosslinking agent of ethylene glycol dimethacrylate (wt% of 5:1) and a UV initiator diphenyl(2,4,6-trimethylbenzoyl)phosphine oxide [26]. Crosslinked poly(lauryl methacrylate) (PLMA) has long side chains that can interact with surface long-chain ligands (oleic acid) on the surface of QDs to prevent agglomeration of the mixed-halide perovskite QDs during polymerization. As shown in Fig. 8.2a, the LSC ($6 \times 6 \times 0.2$ cm$^3$) appears semi-transparent in ambient environment, and a clear concentrated red light can be seen from their edges (Fig. 8.2b), when the top surface of the LSC is partially placed upon simulated one sun illumination (100 mW/cm$^2$). The as-prepared LSC has a geometrical gain factor of 7.5 (*G*, defined as the ratio between the area of the top of the LSC and the area of the edges of the LSC [8]). Figure 8.2c presents a photograph of two overlapping LSCs using $CsPbBr_3$ QDs (green) and $CsPb(Br_{0.2}I_{0.8})_3$ (red) QDs under UV illumination. As shown in Fig. 8.2d, LSCs based on mixed-halide perovskite QDs exhibit an absorption spectrum ranging from 300 to 650 nm. As the overlap between absorption and emission spectra is significantly larger in $CsPbBr_3$ compared to that of the $CsPb(Br_{0.2}I_{1-0.8})_3$ (Fig. 8.2d), by plotting the integrated PL area as a function of the optical paths, Zhao et al. found that the LSC based on $CsPb(Br_{0.2}I_{1-0.8})_3$ perovskite QDs have a less energy loss compared to $CsPbBr_3$ QDs. For example, the integrated PL area of $CsPb(Br_{0.2}I_{1-0.8})_3$ QDs remains 0.5 in 6-cm LSC, which is higher than 0.3 in 6-cm LSC based on $CsPbBr_3$ QDs due to less photon escape and reabsorption

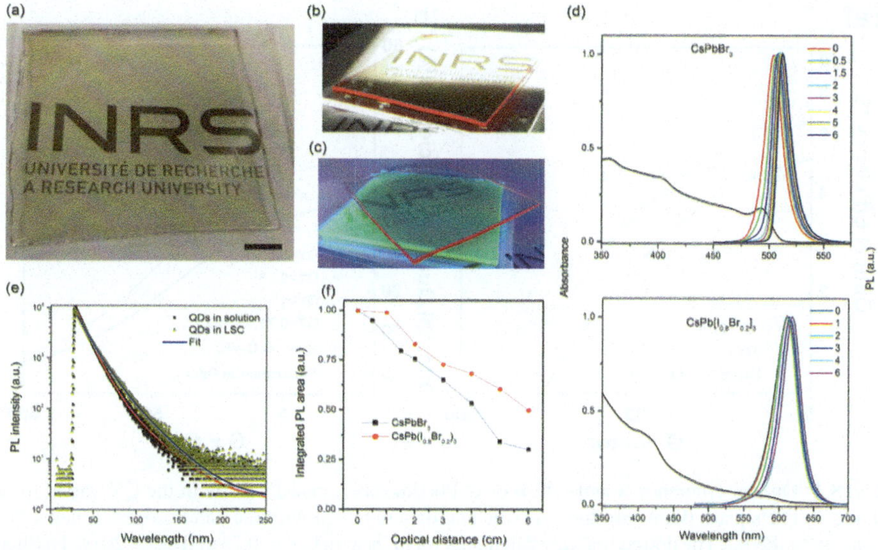


**Fig. 8.2** Photographs of the LSC comprising $CsPb(Br_{0.2}I_{0.8})_3$ QDs under ambient **a**, one sun (100 $mW/cm^2$) **b**, and UV illumination **c**. LSC dimensions, $6 \times 6 \times 0.2$ $cm^3$. **d** Absorption and PL spectra measured at different optical paths for the $CsPb(Br_{0.2}I_{0.8})_3$ QD based LSCs. **e** PL decay curves for $CsPb(Br_{0.2}I_{0.8})_3$ QDs in toluene and polymer matrix (excitation at 440 nm). **f** Integrated PL area at different optical paths. Reproduced with permission from Ref. [26]. Copyright 2017, Elsevier Ltd

energy loss (Fig. 8.2f). By increasing the optical path, the PL spectra of both types of QDs exhibit a red-shift up to 8 nm with a narrowing of the photoluminescence (PL) peak width, which confirms that the energy loss is due to light reabsorption (Fig. 8.2d). The preparation process for the LSCs based on perovskite QDs is very sensitive to the moisture. With high humidity (>40%), there is almost no concentrated emission on the edge of the LSC, indicating the degradation of the perovskite materials, due to the fast reaction of the perovskite QDs with the moisture [26]. Therefore, further a stable polymer protection layer coated on the surface of perovskite QDs may improve the stability of the LSCs.

As shown in Fig. 8.3 a, $\eta_{opt}$ of the LSC using $CsPb(Br_{0.2}I_{0.8})_3$ QDs drops exponentially with the increase of the G factor of the LSCs, which is consistent with simulation. A maximum optical efficiency of 2% is achieved with G of 45 (Fig. 8.3a). Increasing the G up to 55 (corresponding to a LSC length of 11 cm), $\eta_{opt}$ decreased to ~1.45% due to reabsorption and scattering loss. The $\eta_{opt}$ of 2% obtained with the LSCs based on mixed-halide perovskite QDs is comparable to other recent inorganic QDs-based LSCs (e.g. PbS/CdS or CdSe/CdS). With the ideal $CsPb(Br_{0.2}I_{0.8})_3$ QDs with a QY of 100%, the optical efficiency could be up to 9% and with an internal quantum efficiency (IQE) of 70% (Fig. 8.3b). An ideal $CsPb(Br_{0.2}I_{0.8})_3$ QDs, with QY of 1, would have a maximum optical efficiency of more than 6% at $G = 45$ [26].

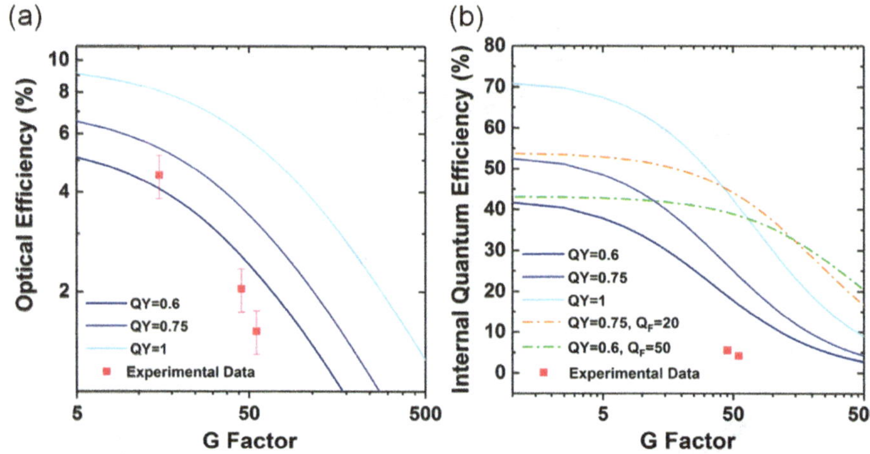


**Fig. 8.3** Optical efficiency **a** and IQE **b** of $CsPb(Br_{0.2}I_{0.8})_3$ based LSC with the QY varied from 0.6 to 1. The square points are the experimental data obtained with the electro-optical method of comparing the $J_{sc}$. The dotted lines are calculated for QDs with QY = 0.75 and $Q_F = 20$ (red dashed line), QY = 0.6 and $Q_F = 50$ (green dashed line). $Q_F$ is quality factor defined as the ratio between the absorption coefficient ($\alpha_1$) at the wavelength $\lambda_1$ of collected light and the absorption coefficient ($\alpha_2$) at the emission peak). Reproduced with permission from Ref. [26]. Copyright 2017, Elsevier Ltd

Even the single-layer LSC based on the perovskite $CsPb(Br_xI_{1-x})_3$ QDs exhibited an optical efficiency of 2%, this optical efficiency is still far from the value (6%) for potential commercialization [10]. In addition, the smaller spectral overlap in mixed-halide perovskite QDs compared to that of $CsPbI_3$ or $CsPbBr_3$ is due to the energy transfer between the QDs, which leads to the decrease of QYs [26]. Further optimization of the size distribution of QDs and their QY is required to further improve the optical efficiency of LSCs based on mixed-halide perovskite QDs. Another issue in mixed-halide lead based perovskite QDs is their toxicity. Further research direction should focus on the replacement of Pb by Sn or other low-toxic/nontoxic elements (e.g. Ag, In, Sb).

### *8.2.2 LSC Based on Doped Perovskite QDs*

Several strategies have been used to decrease the spectral overlap between the emission and absorption, including structure engineering and doping the QDs with ions [3, 18, 38]. Especially in doped QDs, there could be not any spectral overlap between the doped ion emission and absorption spectrum, by creating the energy level between the conduction and valence bands [20, 21, 29, 40, 46, 59–70]. Variety of transition metal atoms, such as manganese (Mn) and ytterbium (Yb) have been demonstrated for perovskite NCs [21, 37]. Meinardi et al. demonstrated the use of Mn-doped $CsPbCl_3$

QDs ($Mn^{2+}$:$CsPbCl_3$) for large-area LSCs [21]. The LSCs were fabricated via mass copolymerization of acrylate monomers. The as-synthesized doped perovskite QDs have a very narrow size distribution ($8.6 \pm 0.6$ nm) as shown in Fig. 8.4 a. The typical Mn concentration in the QDs is around 3.9%. High resolution TEM (HRTEM) image of the $Mn^{2+}$:$CsPbCl_3$ with 5% Mn shows a $Mn^{2+}$ ion in a substitutional Pb site with essentially identical octahedral coordination with six halide atoms (Fig. 8.4a). The X-ray diffraction (XRD) pattern indicates a match with cubic $CsPbCl_3$ (Fig. 8.4a). Typical dual emissions were found in the $Mn^{2+}$:$CsPbCl_3$ QDs (Fig. 8.4 a). The

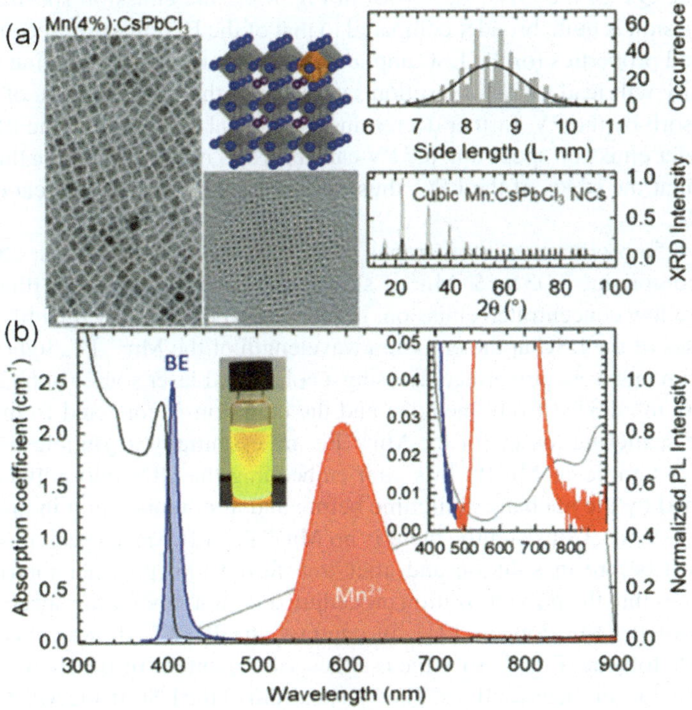


**Fig. 8.4** **a** Structure of $Mn^{2+}$:$CsPbCl_3$ QDs. Structural data of 4% Mn-doped $CsPbCl_3$ QDs with TEM and HRTEM (the scale bar corresponds to 50 and 2 nm in the inset) and related size histogram extracted from analysis of 400 particles. XRD pattern of $Mn^{2+}$:$CsPbCl_3$ perovskite QDs at room temperature (red line) compared to the XRD patterns of bulk cubic $CsPbCl_3$ perovskite (reference pattern 98-020-1251). **b** Optical absorption (black line) and PL spectra of $Mn^{2+}$:$CsPbCl_3$ QDs (4% Mn) under 390 nm excitation in toluene solution. The BE and Mn-related PL bands are highlighted by blue and red shading, respectively. A photograph of the solution under 365 nm excitation highlights the dominant yellow color due to $Mn^{2+}$ emission with respect to BE PL. Inset: enlargement of the main panel between 400 and 800 nm emphasizing the small absorption contribution due to the C–H vibrational modes of the ligand molecules (the measurement is corrected for solvent effects) between 650 and 900 nm and by direct absorption between intragap Mn levels. The inset further highlights the near-complete absence of resonance between the QD absorption tail and the high-energy portion of the Mn PL. Reproduced with permission from Ref. [21]. Copyright 2017, American Chemical Society

$Mn^{2+}$:$CsPbCl_3$ QDs have sharp absorption edge at ~395 nm and the respective narrowline band energy (BE) PL peaked at ~405 nm. The BE PL emission accounts for ~20% of the total emission and the remaining 80% of emitted photons are due to the $4T_1 \rightarrow 6A_1$ optical transition of the $Mn^{2+}$ dopants, showing a characteristic broad peak at ~590 nm, resulting in a well separation of the absorption edge and the $Mn^{2+}$ emission. While in the doped QDs, the QY is still quite low, $5 \pm 1\%$ for the BE emission and $10 \pm 2\%$ for the $Mn^{2+}$ emission. The low QY is a strong limited factor for high efficiency large area LSCs. A possible synthetic approach may use core/shell structure to improve the surface passivation of the doped QDs, in order to improve the QY of the QDs. As shown in Fig. 8.4b, the emission spectrum of the doped emission is quite broader compared to that of the BE emission spectrum. This is very good properties for the LSC application. The narrow emission line with high light density will lead to the absorption saturation of the PV, and some of emission cannot absorb by the PV, further decreasing the optical efficiency of the LSCs [10]. With broader emission spectrum, the PV can absorb as much as possible the concentrated light at the edges of the LSC, thus leading to the improved optical efficiency of the LSC.

Figure 8.5a reports two photographs of the LSC (LSC dimensions: $25 \times 20 \times 0.5$ cm$^3$) comprising 0.03 wt % $Mn^{2+}$:$CsPbCl_3$ QDs under 365 nm UV-illumination, showing yellow concentrated emission. In order to experimentally quantify the scattering losses of the LSC at the emission wavelength of the $Mn^{2+}$ PL, light propagation experiments were performed by using a collimated laser source at 633 nm. The laser enters the device from one edge and the emission is collected from the LSC faces. The scattering losses for the $Mn^{2+}$ PL are essentially negligible (Fig. 8.5b). There is no significant Mn-PL loss after embedding the QDs into polymer matrix as confirmed by the unchanged lifetime before and after embedding in the polymer matrix (Fig. 8.5d). Consistently, there is no $Mn^{2+}$ PL difference (peak position and peak width) before in solution and after transferred into polymer matrix, which demonstrates that the polymerization procedure does not modify the spectral optical properties of the Mn dopants [21]. In contrast, the BE PL shows a faster decay compared to toluene (Fig. 8.5e), which suggests that excitons of BE are more subject than the $Mn^{2+}$ to surface reactions. The PL spectrum of the LSC using $Mn^{2+}$:$CsPbCl_3$ was collected from one of the slab edges ($20 \times 0.5$ cm$^2$) using 395 nm laser excitation positioned at different distances, $d$, from the waveguide edge. There is slight PL spectrum shift in the Mn-PL as there is no spectral overlap between the absorption spectrum and the Mn-PL spectrum. While a significant shift was found in BE emission because of strongly overlap between the absorption spectrum and BE-PL emission (Fig. 8.5f). In Fig. 8.5g, the relative output power extracted from one of the LSC edges ($20 \times 0.5$ cm$^2$) was measured by coupling silicon solar cells to one edge and by progressively exposing increasingly larger portions of the LSC area to the solar simulator. The calculated curve by Monte Carlo ray tracing simulation matches well the experimental data as large as ~80%, indicating that suppression of reabsorption in $Mn^{2+}$:$CsPbCl_3$ perovskite QDs is effective in rendering the device behavio.

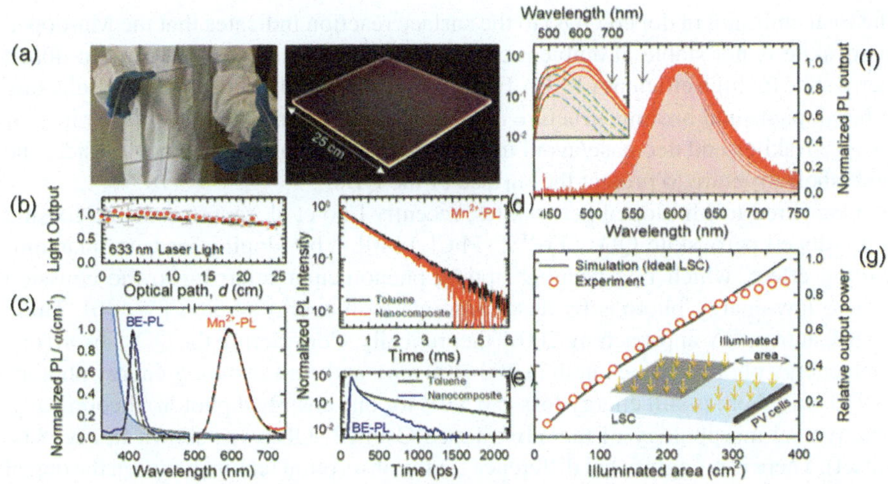


**Fig. 8.5** **a** Photograph of the LSC comprising Mn-doped $CsPbCl_3$ QDs (LSC dimensions: 25 × 20 × 0.5 $cm^3$; QD concentration 0.03 wt %) under ambient (left) and UV illumination (365 nm, right). **b** Intensity of 633 nm laser light linearly propagating inside of the LSC in (**a**) as a function of propagation path *d*. The light intensity versus *d* calculated using the Lambert – Beer equation for linear propagation (absorption coefficient α (633 nm) = 1 × $10^{-2}$ $cm^{-1}$) is reported as a black dashed line. **c** PL spectra of the BE and $Mn^{2+}$ PL of Mn-doped $CsPbCl_3$ QDs excited at 395 nm in P(MMA/LMA) (blue and red lines, respectively) and in toluene solution (dashed black lines). The PL decay curves collected at **d** 600 nm for the $Mn^{2+}$ PL for QDs and **e** 405 nm for the BE emission in toluene solution (black lines) and embedded in the P(MMA/LMA) matrix (red and blue lines, respectively). **f** Normalized PL spectra (excitation at 395 nm) collected at the edge of the LSC when the excitation spot is located at increasing distances d from 0 to 25 cm from the edge (as indicated by the black arrow) after subtraction of the corresponding emission contribution by the polymer matrix. The raw PL data under site-selective excitation are shown in the inset; the red lines are excited at 395 nm and show contribution of both the polymer and the QDs. The green dashed lines are excited in the low-energy tail of the polymer defects at 410 nm and are due exclusively to the polymer matrix. The black arrow indicates increasing *d*. **g** Relative optical output power measured from c-Si PVs coupled to one perimeter edge of the QD-LSC as a function of the device area illuminated by a calibrated solar simulator (100 mW/$cm^2$, circles). The theoretical trend obtained through Monte Carlo ray tracing simulation of an ideal LSC with no scattering or reabsorption losses is reported as a black curve. The inset shows the schematic depiction of the experimental configuration. Reproduced with permission from Ref. [21]. Copyright 2017, American Chemical Society

Compared to $CsPbCl_3$ or mixed-halide perovskite QDs, Mn-doped perovskite open a new way to avoid the energy loss due to the reabsorption in the large-area LSC based on perovskites, showing potential for using perovskite in the community of LSCs [21]. While in the Mn-doped $CsPbCl_3$ perovskite, the absorption spectrum is only ranging from 300 to 400 nm (5% of solar irradiation) and the QY is relatively low (10%), limiting their further application as building blocks for high-efficiency large-area LSCs. In addition, even there is no energy loss in Mn-PL, the strong overlap between the absorption and the BE emission spectra leads to significant energy loss due to the reabsorption. The sensitivity of the BE emission (20% over the total emission in doped QDs) to the surface reaction indicates that the Mn-doped perovskite is not stable at ambient condition. All these issues limit the Mn-doped perovskite for high efficiency LSCs. Further research direction may focus on selecting other dopants and hosting structure of perovskite to extend the absorption range of the perovskites and decrease/avoid the BE PL. Improving the QY may consider the core/shell structure to protect the surface of the QDs.

Compared to Mn-doped perovskites, recently Luo et al. reported the LSC using $Yb^{3+}$-doped perovskite QDs ($Yb^{3+}$:$CsPbCl_3$) [40]. $Yb^{3+}$ doping leads to quantum-cutting effect, which is a peculiar optical phenomenon referring to the emission of two low-energy photons by absorbing one high-energy photon [37, 40]. These QDs feature a QY approaching 200% theoretically. Considering the QDs absorb one high energy photon (wavelength in 300–400 nm), and emit two long-energy photons (~990 nm), there is still energy loss (>0.5 eV for one absorbed photon) theoretically. The typical morphology of the $Yb^{3+}$-doped $CsPbCl_3$ QDs was shown in Fig. 8.6a (inset). There is no significant difference for the absorption feature between the doped and undoped $CsPbCl_3$ QDs (Fig. 8.6a), while their PL spectra are different. Due to effective excitation energy transfer from $CsPbCl_3$ host to $Yb^{3+}$-dopants, the BE PL of QDs is strongly quenched in the doped sample and is replaced by a sharp PL band (Yb emission) at ~990 nm arising from the $2F_{5/2} \rightarrow 2F_{7/2}$ ff emission of $Yb^{3+}$-dopants. This behavior is very different to Mn doped perovskites, in which a clear dual emission was observed [66]. The QY of the dopant emission is ~164 ± 7%, an indication of quantum-cutting effect. $Yb^{3+}$-doped $CsPbCl_3$ QDs were incorporated

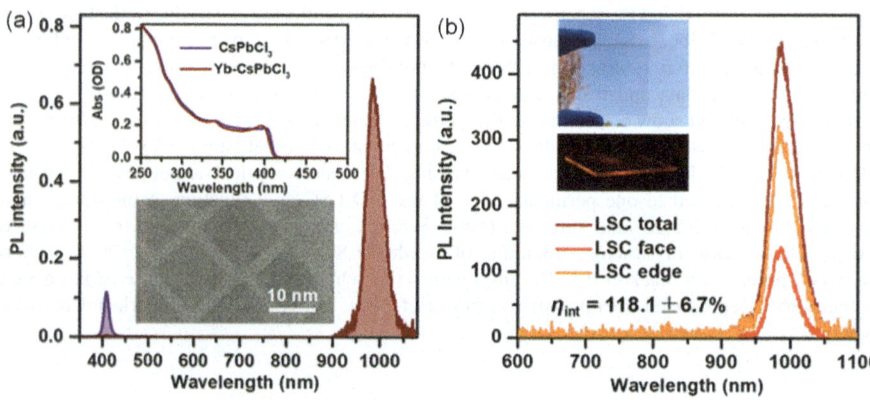


**Fig. 8.6** **a** PL spectra (solid lines with shading) of undoped (purple) and $Yb^{3+}$-doped (dark red) $CsPbCl_3$ QDs excited with a 365 nm light source. The top inset shows their absorption spectra. The bottom inset is a TEM image of doped QDs. **b** Total (dark red), face (light red), and edge (orange) emissions measured for a 5 × 5 $cm^2$ LSC using $Yb^{3+}$-doped $CsPbCl_3$ NCs. The IQE of this LSC is measured to be 118.1 ± 6.7%. The top inset is the picture of the LSC under sunlight; the bottom inset shows the edge emission from the LSC under UV illumination and with a 570 nm long-pass filter taken using a cell phone (MI 8 SE) camera. Reproduced with permission from Ref. [40]. Copyright 2019, American Chemical Society

into LSCs by coating QD-poly(methyl methacrylate) (PMMA) mixtures as thin-films onto borosilicate glass substrates using doctor-blade or spin-coating methods. The dimensions of the LSC are $5 \times 5 \times 0.2$ cm$^3$ (G of 6.25). LSC under sunlight shows a transparent appearance (Fig. 8.6b). By the application of UV illumination in combination with a 570 nm long-pass filter, strong near-infrared (NIR) emissions from the edges can be seen using a cell phone camera (inset, Fig. 8.6b). The IQE of the LSC is calculated to be $118.1 \pm 6.7\%$. Luo et al. further predict that if using $CsPbCl_xBr_{3-x}$ QDs capable of absorbing ~7.6% of solar photons, the projected optical efficiency of quantum cutting-LSCs can exceed 10% for >100 cm$^2$ devices [40]. The advantage of quantum-cutting LSCs over conventional QD-LSCs could be especially obvious with increasing LSC sizes due to the lack of energy loss because of emission and absorption spectral overlap.

### *8.2.3 LSC Based on Zero-Dimensional Perovskite NCs*

Compared to above mentioned three dimensional (3-D) undoped perovskites (e.g. $CsPbX_3$) having a large emission/absorption spectral overlap, zero-dimensional (0-D) inorganic perovskites [e.g. $Cs_4PbX_6$, (X = Cl, Br/Cl, Br), $Cs_4SnBr_6$, $Cs_{4-x}A_xSn(Br_{1-y}I_y)_6$ (A = Rb, K) and $Cs_3Cu_2I_5$] allow a minimal absorption and emission spectral overlap (large Stokes shift, defined as the energy difference between positions of the band maxima of the first-excitonic absorption and emission spectra) [52, 54, 55, 57, 71–76]. There is no obvious absorption and emission spectral overlap in the case of $Cs_4SnBr_6$ perovskite NCs [72]. There is no quantum confinement effect in 0-D perovskite, thus this type of perovskites does not belong to the family of QDs [52]. In another word, one cannot tune the optical properties of the 0-D perovskite by tuning the size of the perovskite NCs. Zhao et al. reported the fabrication of the LSC based on 0-D perovskite by spin-coating or drop-casting perovskite/polymer solution on a glass substrate (Fig. 8.7a, top) [38]. The PL signals from the LSC were recorded at a certain distance, *d*, between the edge and vertically illuminated spot at excitation wavelength of 400 nm. There is very slightly spectral variation as a function of the optical paths for the LSC based on $Cs_4PbBr_6$ NCs (Fig. 8.7b, d). In contrast, $CsPbBr_3$ NCs exhibit a significant PL red-shift. The unchanged PL peak in 0-D perovskite NCs may be due to the narrow emission spectrum and broad absorption spectrum of the 0-D NCs. In a contrast, for 3-D $CsPbBr_3$ NCs, the PL peak shows a strong red-shift as both the emission and absorption spectra in the overlapping range are very sharp (Fig. 8.7d). This result is the straightforward evidence that 0-D perovskite NCs may be better candidate as emitter for LSCs applications compared to 3-D perovskite NCs. While compared to doped perovskite QDs, $Cs_4SnBr_6$ perovskite NCs still have a sight overlap between the absorption and emission spectra.

The LSC (dimensions: $10 \times 10 \times 0.4$ cm$^3$, G of 6.25) was illuminated perpendicular to its surface by a 1.5 AM global solar simulator (100 mW/cm$^2$) or natural sunlight (30 mW/cm$^2$). With the increase of the $Cs_4PbBr_6$ concentration, the external optical efficiency of the LSCs increases from $0.8 \pm 0.15$ (1.5 wt %) to $1.4 \pm 0.15$ (3 wt %) and $2.2 \pm 0.2$ (4.5 wt %), due to the increase of the absorption of sunlight.

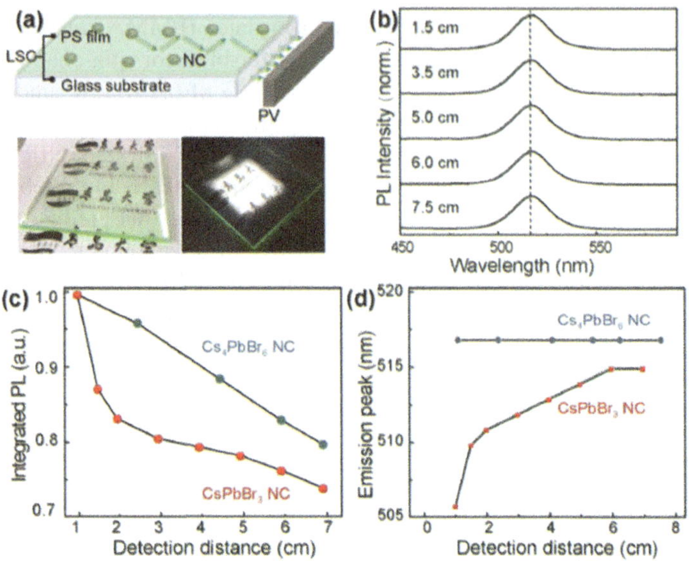


**Fig. 8.7** **a** Scheme of a LSC and photographs of the LSC comprising 0-D perovskite NCs under ambient (left) and one sun (100 mW/cm$^2$) illumination (right). LSC dimensions, $10 \times 10 \times 0.4$ cm$^3$. **b** PL spectra measured at different optical paths for the $Cs_4PbBr_6$ NCs based LSCs. Integrated PL intensity **c** and Emission peak positions **d** as a function of detection distance. The excitation wavelength is 400 nm. Reproduced with permission from Ref. [38]. Copyright 2019, Wiley–VCH

The absorption of sunlight of perovskites with tunable concentrations was calculated to be 4.3% (1.5 wt %), 7.6% (3 wt %) and 11.6% (4.5 wt %), respectively. As the absorption of the $Cs_4PbBr_6$ perovskite NCs is very strong in the range of 300–400 nm compared to 400–500 nm, the increase of concentration can only lead to the increase of the absorption in the long-wavelength range, which explains that the increase of absorption is not proportional to the concentration. The current obtained highest external optical efficiency (2.4% under one sun illumination) is comparable with those reported for the LSCs based on perovskite NCs (Table 8.1). The higher optical efficiency in 0-D perovskites compared to 3-D perovskites is due to their less absorption and emission overlap. At low-light intensity (30 mW/cm$^2$), the LSC based on $Cs_4PbBr_6$ NCs (3C$_0$, $10 \times 10 \times 0.4$ cm$^3$) exhibit a PCE of 1.8% (Fig. 8.8a), which indicates that the $Cs_4PbBr_6$ NCs is good absorbers for the LSC, as this system could operate at lower light intensity condition [38]. The predicted optical efficiency of the LSCs based on 0-D perovskite is strongly dependent on the concentration and QY of the perovskite, and the highest optical efficiency could be up to 9% (Fig. 8.8b, c). In addition, the LSCs based on 0-D perovskite have a good photostability. In view of the simple synthesis procedure and excellent optical properties, 0-D perovskites, [e.g. $Cs_4PbBr_6$, $Cs_4Sn(Br_xI_{1-x})_6$, $Cs_{4-x}RbSn(Br_{1-y}I_y)_6$ and $Cs_3Cu_2I_5$] with broad absorption represent a practical emitter for large-area, high efficiency LSCs.

**Table 8.1** Performance in LSCs based on various types of perovskites

| Perovskites | Stokes shift (meV) | λ range (nm) | QY (%) | Area ($cm^2$) | $\eta_{opt}$ (%) | References |
|---|---|---|---|---|---|---|
| $CH_3NH_3PbI_3$ thin film | 60 | 300–760 | 80 | 1.5 × 1.5 | 29 | [19] |
| $CsPb(Br_xI_{1-x})_3$ QDs | 25 | 300–650 | 60 | 1.5 × 9 | 2 | [26] |
| $Yb^{3+}$:$CsPb(Br_xI_{1-x})_3$ QDs | 1850 | 300–400 | 118 | 5 × 5 | 3.7 | [41] |
| $Mn^{2+}$:$CsPb(Br_xI_{1-x})_3$ QDs | 1000 | 300–400 | 20 | 20 × 0.5 | 0.5 | [21] |
| $Cs_4PbBr_6$ NCs | 1500 | 300–500 | 58 | 10 × 10 | 2.4 | [39] |
| $HA_2MA_{n-1}Pb_nX_{3n+1}$ NPLs | 275 | 300–500 | 80 | 10 × 10 | 26 | [78] |

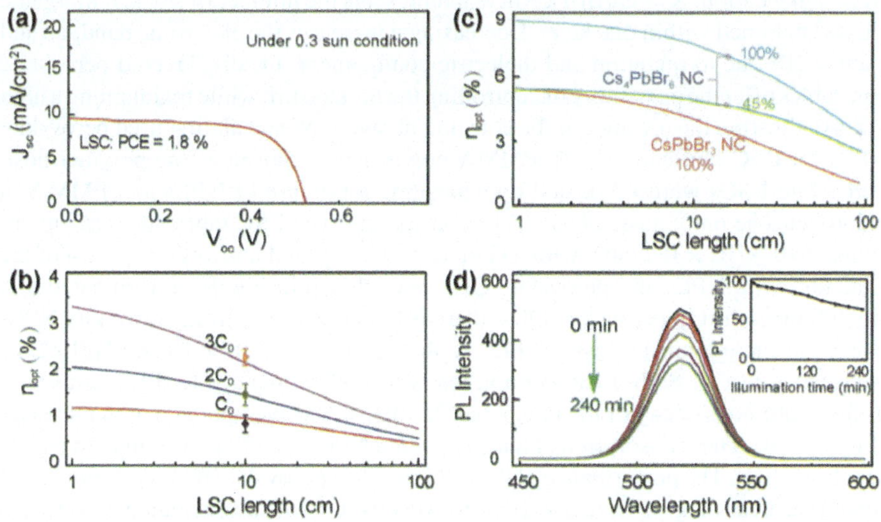


**Fig. 8.8** **a** J-V response of silicon PVs attached on the edge of the LSC under the natural sunlight (30 mW/$cm^2$). **b** The calculated and measured external optical efficiency of $Cs_4PbBr_6$ based LSC with the QY of 45% and different $Cs_4PbBr_6$ concentrations ($C_0$–$3C_0$). The round points are the experimental data obtained with the power meter. **c** The calculated external optical efficiency of $Cs_4PbBr_6$ and $CsPbBr_3$ NCs based LSCs with the QY varied from 0.45 to 1. The NC concentration is $3C_0$ (4.5 wt %) for both types of perovskites. **d** PL spectra of $Cs_4PbBr_6$ NCs based LSCs upon 365 nm light exposure (400 mW/$cm^2$) for different hours. Insetis the integrated PL area of the LSC as a function of illumination time Reproduced with permission from Ref. [38]. Copyright 2019, Wiley–VCH

### 8.2.4 LSC Based on Perovskite Nanoplatelets

To further enhance the external optical efficiency of the LSC based on perovskites, Wei et al. engineered energy transfer within metal halide perovskite nanoplatelets in order to prepare high-efficiency LSCs [77]. Compared to spherical-shape perovskite QDs (or NCs), layered, or quasi-two-dimensional (2-D) perovskites, consist of sheets of 3-D perovskites sandwiched by bulky cations that divide the corner-sharing inorganic octahedral 3-D framework [77]. The thickness of 2-D nanoplatelets (NPLs) can be well controlled through the reaction conditions. For example, $PbBr_2$, MABr and HEABr were dissolved in N,N-dimethylformamide with the identical concentration of $PbBr_2$ at 0.06 M. Molar ratios of $PbBr_2$, MABr and HABr were tuned to control domain distributions [77]. Then the stock precursor solution was dipped into chlorobenzene to form perovskite NPLs. The typical size of the perovskite NPLs is several hundred nanometers. The perovskites have a formula $R_2(MA)_{n-1}Pb_nX_{3n+1}$ (R = bulky cation, X = Cl/Br/I), where n represents the number of stacked inorganic sheets contained within one layer. Decreasing n increases the electronic bandgap and exciton BE due to quantum and dielectric confinement. Ideally, layered perovskite ensembles offer the possibility of controlling the Stokes shift while maintaining a high QY by adjusting the n values of the constituent layers. Wei et al. first used perovskite NPLs for LSC fabrication [77]. PMMA polymer was chosen as the polymer host. Perovskite LSCs were fabricated by depositing a mixture of NPLs and PMMA in chlorobenzene onto glass substrates by rod coating. The LSC fabricated has dimensions of $10 \times 10 \times 0.2$ cm$^3$, with a G of 12.5. The PL and absorption spectra of the LSC film with NPLs and perovskite QDs were shown in Fig. 8.9a. Compared to a small Stokes shift in perovskite QDs, there is larger Stokes shift in perovskite NPLs due to the energy transfer inside NPLs. A photograph of the $10 \times 10$ cm$^2$ NPLs LSC is provided in Fig. 8.9b. Reabsorption loss was both evidenced by PL loss as well as the continuous decay of the integrated PL intensity as the optical length increases (Fig. 8.9c, d). The reabsorption of the engineered NPLs is sufficiently small to enable low-loss LSCs. The performance of the LSC based on perovskite NPLs was measured for device areas ranging from $2 \times 2$ cm$^2$ to $10 \times 10$ cm$^2$ with an excitation wavelength of 400 nm as shown in Fig. 8.9e. The average QY and optical efficiency decreases as the LSC length increases from 2 to 5 cm. To evaluate the potential of perovskite LSCs for real-life applications, electro-optical measurements were conducted for the LSC film ($10 \times 10$ cm$^2$) under outdoor conditions. An optical efficiency of 25% was obtained for the LSC based on NPLs. The high optical efficiency is a combination effect of high QY and relatively low reabsorption losses [77]. The perovskite NPL LSCs showed a 20% decrease in PL intensity after 8-h illumination, comparable to the QD LSCs (Fig. 8.9f).

Although the overlap between the absorption and emission spectra of the NPLs is smaller compared to that of perovskite QDs, the presence of spectral overlap still leads to energy loss, especially in large-area LSC (~m$^2$). In view of the use of organic cations in perovskite NPLs, the stability of the LSC is still not excellent for long-term

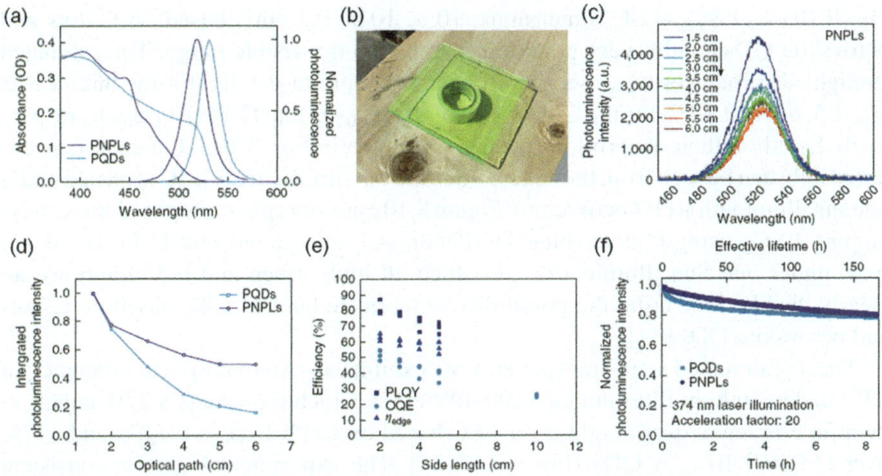

Fig. 8.9 **a** Absorption and emission spectra of LSCs of perovskite NPLs and perovskite QDs. Spectral overlap is greatly reduced for perovskite NPLs compared with QDs. **b** Photograph of a 10 × 10 $cm^2$ NPL-based LSC. **c** PL spectra of perovskite NPLs emerging from the LSC edge as a function of the distance between the excitation spot and the collection edge. Reabsorption in the blue side of the spectra is observed. **d** Spectrally integrated intensity of PL collected at the LSC edge as a function of propagation length. Stronger PL losses are observed for perovskite QDs. **e**, The QY, edge coupling efficiency and the OQE (external optical efficiency) of the square NPL LSC as a function of LSC side length. From 2 × 2 $cm^2$ to 5 × 5 $cm^2$ device area, optical measurements are conducted. For each area, three independent measurements were carried out and their results are displayed in the panel as a function of side length. For the 10 × 10 $cm^2$ device area, an electro-optical method was utilized to obtain two independent measurements. **f** An accelerated light-soaking stability test of perovskite LSCs. The excitation source is a 374 nm laser diode, which corresponds to an acceleration factor of 20 Adapted with permission from Ref. [77]. Copyright 2019, Nature Publishing Group

application. In addition, the toxic lead may be replaced by other metals, such as Sn or Ge.

## 8.3 Tandem LSC Based on Perovskite QDs

Single-layer LSCs fabricated using colloidal perovskite QDs (or NCs) still face major challenges such as low optical efficiency and limited long-term stability. A tandem structured LSC can improve the light absorption efficiency and decreasing the energy loss due to the light emitted because of the escape cone. Tandem LSCs can absorb different portions of the solar spectrum using different layers of emitters allows to realize high efficiency multi-layered LSCs compared with standard single-layer LSCs [28, 33, 34, 78]. Zhao et al. reported a large-area (~100 $cm^2$) tandem LSC based on carbon dots (C-dots) and inorganic mixed-halide perovskite QDs [34]. Generally, there would be two layers or three layers in tandem structured LSCs. As shown in

Fig. 8.10a–c, LSCs (LSC dimensions: $10 \times 10 \times 0.2\ cm^3$) based on C-dots and perovskite QDs exhibit very good transparency in the visible range. The estimated sunlight absorptances ($\eta_{abs}$, see experimental section for details) for the three LSCs are: 1.5, 6.1 and 7.1% for the C-dots, $CsPb(Br_{0.8}Cl_{0.2})3$ and $CsPb(Br_{0.2}I_{0.8})_3$, respectively based on their absorption spectra. As shown in Fig. 8.10d–f, a clear concentrated light can be seen from the edges, when the top surface of the LSC is placed under one sun illumination ($100\ mW/cm^2$). Figure 8.10g presents photographs of three overlapping LSCs using C-dots (blue), $CsPb(Br_{0.8}Cl_{0.2})_3$ (green) and $CsPb(Br_{0.2}I_{0.8})_3$ (red) under one Sun illumination, in which all blue, green and red emissions are clearly visible, indicating the possibility to fabricate tandem LSC based on C-dots and perovskite QDs [34].

The as-fabricated semi-transparent LSC exhibits an external optical efficiency of ~3% under sunlight illumination ($100\ mW/cm^2$), which represents a 27% enhancement in efficiency over single layer LSC based on $CsPb(Br_xI_{1-x})_3$ QDs and 117% over $CsPb(Cl_xBr_{1-x})_3$ QDs (Fig. 8.11) [34]. The experimental data is consistent with the simulated optical efficiency, indicating that the tandem structure can significantly enhance the efficiency of the LSCs. In fact, as the structure of mixed-halide perovskite QDs can be tuned simply by controlling the molar ratio of Cl/Br or Br/I, one can optimize the absorption range of the QDs and concentration of the QDs to further enhance the efficiency of the tandem LSCs. The expected optical efficiency in tandem LSCs could be up to 10% (Fig. 8.10d).

The drawback of perovskite QDs is their strong sensitivity to the moisture and ultraviolet light. Zhao et al. used the top layer of the LSC based on C-dots on the LSCs based on perovskite QDs to dramatically enhance the long-term durability of LSC devices because C-dots have a very good photostability under UV light [34]. The first C-dots layer filters the UV light, and then the perovskite QDs avoid the long-term operation under UV light.

In summary, tandem structure LSCs have the following advantages compared to the single layer LSCs: (1) the optical efficiency can be enhanced significantly; (2)

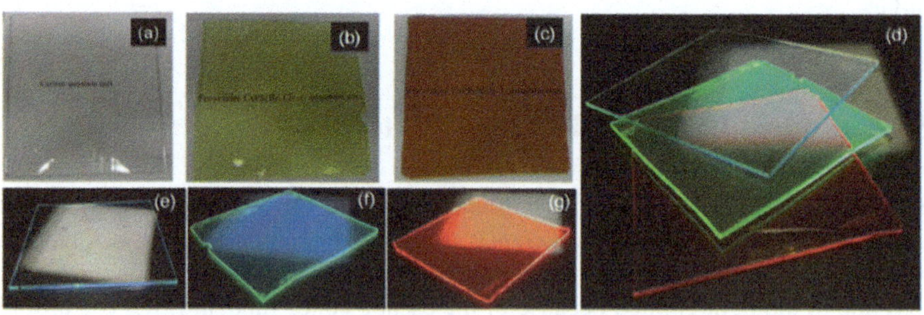


**Fig. 8.10** Photographs of the LSC comprising C-dots under ambient **a** and under one Sun illumination ($100\ mW/cm^2$) **e**, $CsPb(Cl_{0.2}Br_{0.8})_3$ QDs under ambient **b** and one Sun illumination **f**, and $CsPb(Br_{0.2}I_{0.8})_3$ QDs under ambient **c** and one Sun illumination **g**. **d** LSCs based on QDs under one Sun illumination. LSC dimensions, $10 \times 10 \times 0.2\ cm^3$. The weight concentration of perovskite QDs in PLMA is 1.3%. Reproduced with permission from Ref. [34]. Copyright 2018, Elsevier Ltd

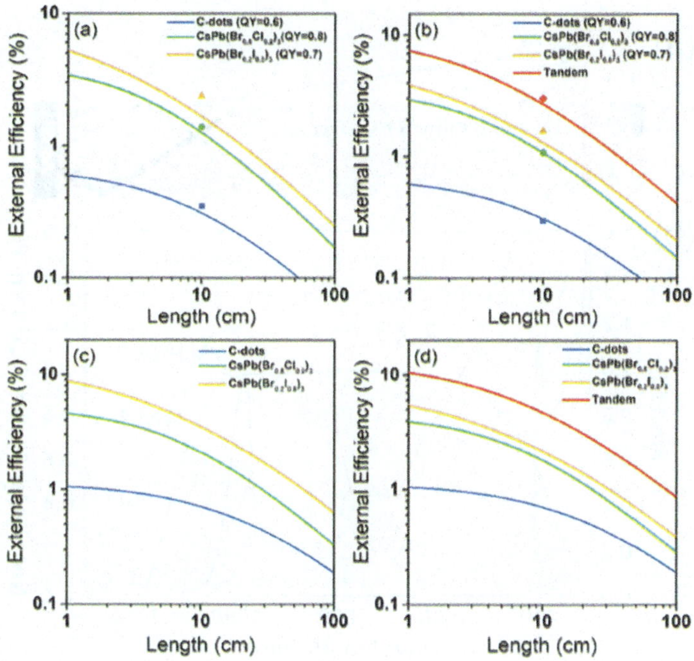


**Fig. 8.11** Analytical model of the performance of the different LSCs. **a** External efficiency of the single LSC under one Sun illumination, with the experimental QY. The square points are the experimental data obtained with the power meter. **b** External efficiency of the tandem LSC solution (red line) and of the single LSC under spectrum filtered by the different layers. External optical efficiencies for single LSCs (c) and tandem LSC (d) with ideal QY = 1. Reproduced with permission from Ref. [34]. Copyright 2018, Elsevier Ltd

the photostability is much better than that of single layer LSCs; (3) other than only optimization of one type fluorophores, in the tandem structure LSCs, one can select different types of fluorophores to design and prepare tandem LSCs with high optical efficiency. However, in the tandem configuration, the use of PV on the edges of the LSCs is several-time larger than that in the single layer LSC. In addition, the cost for the installation of the tandem LSCs is higher than that of single layer LSC. Last but not least, without well sealing, the LSCs based on perovskite are still sensitive to moisture, limiting their further long-term use in high efficiency LSCs.

## 8.4 Bilayer LSCs Based on Perovskite QDs

Another strategy to improve the optical efficiency of the LSCs based on perovskites is to use the biolayer configuration (Fig. 8.12a) [37]. The key for this type of LSC is that

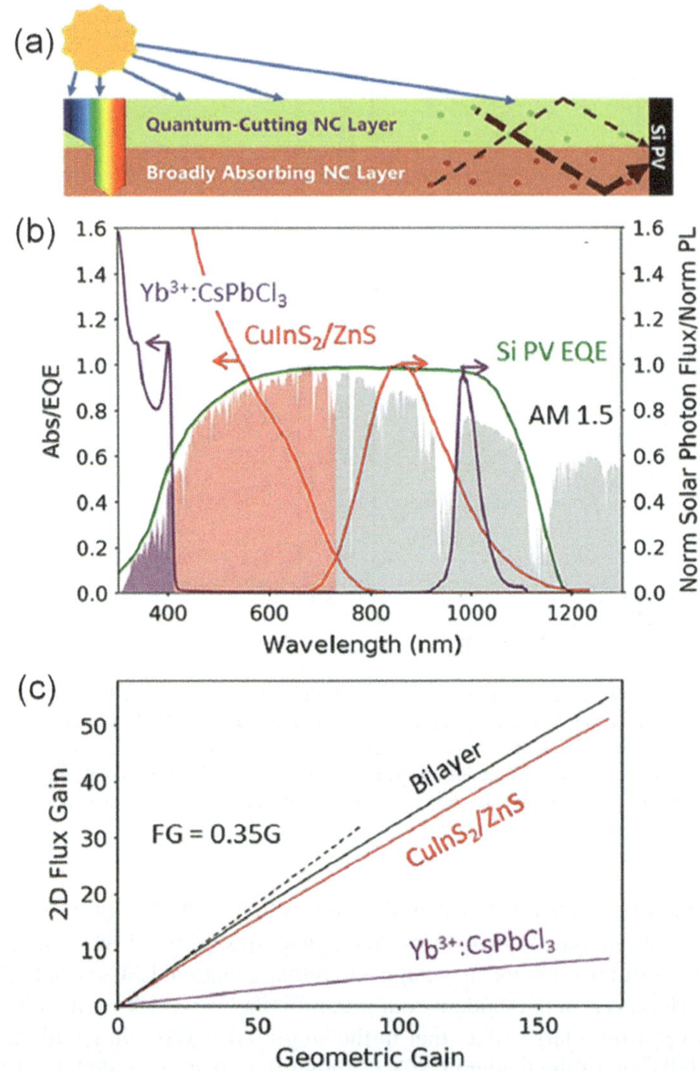


**Fig. 8.12** **a** Schematic of the proposed monolithic bilayer LSC. **b** Absorption and normalized PL spectra of $Yb^{3+}$:$CsPbCl_3$ NCs (purple) and $CuInS_2$/ZnS NCs (red) overlaid with the AM 1.5 solar spectrum (shaded area) and the external quantum efficiency (EQE) of a NIR enhanced Si HIT PV (green). **c** Projected 2D flux gain of a $Yb_{3+}$:$CsPbCl_3$ NC LSC (purple), a $CuInS_2$/ZnS NC LSC (red), and the monolithic, bilayer device shown in (**a**) (black). Adapted with permission from Ref. [37]. Copyright 2019, Royal Society of Chemistry

the emission spectrum of the first layer does not overlap with the absorption spectrum of the second layer. Cohen et al. used the Yb doped $CsPbX_3$ NCs as the emitters in the first layer and $CuInS_2$ QDs in the second layer (Fig. 8.12a) [37]. In this configuration, high-energy sunlight is first absorbed by the quantum-cutting $Yb^{3+}$:$CsPbCl_3$ NCs. Lower-energy light is transmitted by the top layer and absorbed in the second layer. $CuInS_2$ QDs with a broad NIR absorption was selected as the emissive materials for the second layer [37]. There is no any overlap between the absorption spectrum of the $CuInS_2$ and emission spectrum of the $Yb^{3+}$:$CsPbX_3$ (Fig. 8.12b) [37]. In addition, there is also no overlap between the absorption spectrum of the $Yb^{3+}$:$CsPbX_3$ and the PL spectrum of the $CuInS_2$ QDs. In this way, during the light travelling, there is no PL loss due to the reabsorption, because there is no overlap of the absorption and emission spectra. This bilayer structure improves the optical efficiency of the bilayer LSC by eliminating thermalization losses associated with the blue photons. Unlike tandem LSCs, this bilayer structure can avoid the use of two separate LSCs. Modeling suggests that a bilayer LSC using $Yb^{3+}$ doped $CsPb(Cl_{0.25}Br_{0.75})_3$ NCs for the top layer could improve the performance of an idealized $CuInS_2$/ZnS NC LSC bottom layer by at least 19% (Fig. 8.12c). Overall, Cohen et al. suggested a new device structure for the development of next-generation NC LSCs [37]. For large-area bilayer LSCs, it is still a challenge for LSC fabrication as there should be no any air gap between the two layers. Further research direction may focus on designing the biolayer LSCs by selecting various types of suitable emitters for high efficiency LSCs.

## 8.5 Conclusion and Future Directions

In this Chapter, we summarized the use of various types of perovskites and different configuration to fabricate high efficiency and large-area LSCs. The optical properties of perovskite are strongly depending on their size, shape and composition. Therefore, one can optimize their structure to obtain the suitable emitters for LSCs. Among various types of perovskites, doped perovskite QDs offer a great opportunity to synthesize high-quality perovskites with both high QY and large Stokes shift, indicating the doped perovskites may be a great candidate as emitters for high efficiency LSCs.

Although the current techniques have lead to high quality LSCs, there are still some issues needed to be solved: (1) there is still no report for producing doped perovskites with wide absorption. As both Mn and Yb doped $CsPbCl_3$ perovskite have a limited absorption (less than 400 nm), the maximum optical efficiency of the LSCs is far from the commercial value (6%). It is still very challenge to obtain high efficiency LSCs based on perovskites; (2) the above mentioned perovskites used for LSCs contains heavy-metal Pb, which limits these perovskite emitters used for further potential commercialization, unless these materials can be sealed very well during the LSC fabrication, use and further disposal; (3) the long-term stability of the perovskites is not good compared to other reported high-quality emitters (e.g. C-dots, CdSe/CdS

QDs), as the perovskites are very sensitive to the moisture and UV light. Technically, it is still very difficult to prepare core/shell structure, in which the shells typically give a better surface protection for the core; (4) the theoretical simulation predicts the high optical efficiency in large area LSCs (10%), while experimentally there is still lack of reliable technique to prepare a large area (~1 $m^2$), high performance LSCs.

Further research direction for fabricating high-efficiency LSCs based on perovskites may focus on (1) developing a technique (e.g. core/shell structure) to synthesize high quality emitters for LSCs, including wide absorption, high QY (up to 100% [74, 79]), large Stokes shift and high stability; (2) replacing the Pb with Sn or other low-/non-toxic metals; [56] (3) other than tandem or biolayer configuration, a laminated structure should be developed to protect the perovskite against the moisture or other chemicals. In addition, the laminated structure can engage with the industrial technique to produce large sized LSCs with low-cost.

## References

1. H. Zhao, F. Rosei, Colloidal quantum dots for solar technologies. Chem **3**, 229–258 (2017). https://doi.org/10.1016/j.chempr.2017.07.007
2. X. Tong et al., Heavy metal-free, near-infrared colloidal quantum dots for efficient photoelectrochemical hydrogen generation. Nano Energy **31**, 441–449 (2017). https://doi.org/10.1016/j.nanoen.2016.11.053
3. Y. Zhou et al., Near infrared, highly efficient luminescent solar concentrators. Adv. Energy Mater. **6** (2016). Doi: https://doi.org/10.1002/aenm.201501913
4. L. Jin et al., Near-infrared colloidal quantum dots for efficient and durable photoelectrochemical solar-driven hydrogen production. Adv. Sci. **3** (2016). Doi: https://doi.org/10.1002/advs.201500345
5. F. Navarro-Pardo, H. Zhao, Z.M. Wang, F. Rosei, Structure/property relations in "giant" semiconductor nanocrystals: opportunities in photonics and electronics. Acc. Chem. Res. (2017). https://doi.org/10.1021/acs.accounts.7b00467
6. G.S. Selopal et al., Highly stable colloidal "giant" quantum dots sensitized solar cells. Adv. Func. Mater. **27**, 1701468 (2017). https://doi.org/10.1002/adfm.201701468
7. X. Tong et al., Near-infrared, heavy metal-free colloidal "giant" core/shell quantum dots. Adv. Energy Mater.**8** (2018). Doi: https://doi.org/10.1002/aenm.201701432
8. M.G. Debije, P.P.C. Verbunt, Thirty years of luminescent solar concentrator research: solar energy for the built environment. Adv. Energy Mater. **2**, 12–35 (2012). https://doi.org/10.1002/aenm.201100554
9. R. Mazzaro, A. Vomiero, The renaissance of luminescent solar concentrators: the role of inorganic nanomaterials. Adv. Energy Mater. **8** (2018). Doi: https://doi.org/10.1002/aenm.201801903
10. Y.F. Zhou, H.G. Zhao, D.L. Ma, F. Rosei, Harnessing the properties of colloidal quantum dots in luminescent solar concentrators. Chem. Soc. Rev. **47**, 5866–5890 (2018). https://doi.org/10.1039/c7cs00701a
11. P.P.C. Verbunt et al., Controlling light emission in luminescent solar concentrators through use of dye molecules aligned in a planar manner by liquid crystals. Adv. Func. Mater. **19**, 2714–2719 (2009). https://doi.org/10.1002/adfm.200900542
12. I. Coropceanu, M.G. Bawendi, Core/shell quantum dot based luminescent solar concentrators with reduced reabsorption and enhanced efficiency. Nano Lett. **14**, 4097–4101 (2014). https://doi.org/10.1021/nl501627e

13. C.S. Erickson et al., Zero-reabsorption doped-nanocrystal luminescent solar concentrators. ACS Nano **8**, 3461–3467 (2014). https://doi.org/10.1021/nn406360w
14. F. Meinardi et al., Large-area luminescent solar concentrators based on 'stokes-shift-engineered' nanocrystals in a mass-polymerized PMMA matrix. Nat. Photonics **8**, 392–399 (2014). https://doi.org/10.1038/nphoton.2014.54
15. L.R. Bradshaw, K.E. Knowles, S. McDowall, D.R. Gamelin, Nanocrystals for luminescent solar concentrators. Nano Lett. **15**, 1315–1323 (2015). https://doi.org/10.1021/nl504510t
16. F. Meinardi et al., Highly efficient large-area colourless luminescent solar concentrators using heavy-metal-free colloidal quantum dots. Nat. Nanotechnol. **10**, 878–885 (2015). https://doi.org/10.1038/nnano.2015.178
17. V.I. Klimov, T.A. Baker, J. Lim, K.A. Velizhanin, H. McDaniel, Quality factor of luminescent solar concentrators and practical concentration limits attainable with semiconductor quantum dots. Acs Photonics **3**, 1138–1148 (2016). https://doi.org/10.1021/acsphotonics.6b00307
18. H.B. Li, K.F. Wu, J. Lim, H.J. Song, V.I. Klimov, Doctor-blade deposition of quantum dots onto standard window glass for low-loss large-area luminescent solar concentrators. Nature Energy **1**, 9 (2016). https://doi.org/10.1038/nenergy.2016.157
19. K. Nikolaidou et al., Hybrid perovskite thin films as highly efficient luminescent solar concentrators. Adv. Opt. Mater. **4**, 2126–2132 (2016). https://doi.org/10.1002/adom.201600634
20. Y. Li, P. Miao, W. Zhou, X. Gong, X. Zhao, N-doped carbon-dots for luminescent solar concentrators. J. Mater. Chem. A **5**, 21452–21459 (2017). https://doi.org/10.1039/c7ta05220k
21. F. Meinardi et al., Doped halide perovskite nanocrystals for reabsorption-free luminescent solar concentrators. Acs Energy Lett. **2**, 2368–2377 (2017). https://doi.org/10.1021/acsenergylett.7b00701
22. F. Meinardi, F. Bruni, S. Brovelli, Luminescent solar concentrators for building-integrated photovoltaics. Nature Rev. Mater. **2** (2017). Doi: https://doi.org/10.1038/natrevmats.2017.72
23. F. Meinardi et al., Highly efficient luminescent solar concentrators based on earth-abundant indirect-bandgap silicon quantum dots. Nature Photonics **11**, 177-+ (2017). Doi: https://doi.org/10.1038/nphoton.2017.5
24. M. Sharma et al., Near-unity emitting copper-doped colloidal semiconductor quantum wells for luminescent solar concentrators. Adv. Mater. **29** (2017). Doi: https://doi.org/10.1002/adma.201700821
25. L. Tan et al., Ultrasmall PbS quantum dots: a facile and greener synthetic route and their high performance in luminescent solar concentrators. J. Mater. Chem. A **5**, 10250–10260 (2017). https://doi.org/10.1039/c7ta01372h
26. H. Zhao, Y. Zhou, D. Benetti, D. Ma, F. Rosei, Perovskite quantum dots integrated in large-area luminescent solar concentrators. Nano Energy **37**, 214–223 (2017). https://doi.org/10.1016/j.nanoen.2017.05.030
27. M.R. Bergren et al., High-performance $CuInS_2$ quantum dot laminated glass luminescent solar concentrators for windows. Acs Energy Lett. **3**, 520–525 (2018). https://doi.org/10.1021/acsenergylett.7b01346
28. G Liu, H. Zhao, F. Diao, Z. Ling, Y. Wang, Stable tandem luminescent solar concentrators based on CdSe/CdS quantum dots and carbon dots. J. Mater. Chem. C **6**, 10059–10066 (2018)
29. X. Gong et al., Fabrication of high-performance luminescent solar concentrators using N-doped carbon dots/PMMA mixed matrix slab. Org. Electron. **63**, 237–243 (2018). https://doi.org/10.1016/j.orgel.2018.09.028
30. G. Liu, H. Zhao, F. Diao, Z. Ling, Y. Wang, Stable tandem luminescent solar concentrators based on CdSe/CdS quantum dots and carbon dots. J. Mater. Chem. C **6**, 10059–10066 (2018). https://doi.org/10.1039/c8tc02532k
31. J. Shu et al., Monte-Carlo simulations of optical efficiency in luminescent solar concentrators based on all-inorganic perovskite quantum dots. Phys. B **548**, 53–57 (2018). https://doi.org/10.1016/j.physb.2018.08.021
32. B.R. Sutherland, Cost competitive luminescent solar concentrators. Joule **2**, 203–204 (2018). https://doi.org/10.1016/j.joule.2018.02.004

33. K. Wu, H. Li, V.I. Klimov, Tandem luminescent solar concentrators based on engineered quantum dots. Nature Photonics **12**, 105-+ (2018). Doi: https://doi.org/10.1038/s41566-017-0070-7
34. H. Zhao et al., Efficient and stable tandem luminescent solar concentrators based on carbon dots and perovskite quantum dots. Nano Energy **50**, 756–765 (2018). https://doi.org/10.1016/j.nanoen.2018.06.025
35. Y. Zhou et al., Colloidal carbon dots based highly stable luminescent solar concentrators. Nano Energy **44**, 378–387 (2018). https://doi.org/10.1016/j.nanoen.2017.12.017
36. E. Bagherzadeh-Khajehmarjan, A. Nikniazi, B. Olyaeefar, S. Ahmadi-Kandjani, J.M. Nunzi, Bulk luminescent solar concentrators based on organic-inorganic $CH_3NH_3PbBr_3$ perovskite fluorophores. Sol. Energy Mater. Sol. Cells **192**, 44–51 (2019). https://doi.org/10.1016/j.solmat.2018.12.009
37. T.A. Cohen et al., Quantum-cutting $Yb^{3+}$-doped perovskite nanocrystals for monolithic bilayer luminescent solar concentrators. J. Mater. Chem. A **7**, 9279–9288 (2019). https://doi.org/10.1039/c9ta01261c
38. H. Zhao, R. Sun, Z. Wang, K. Fu, X. Hu, Y. Zhang, Zero-dimensional perovskite nanocrystals for efficient luminescent solar concentrators. Adv. Funct. Mater. 1902262 (2019). Doi: https://doi.org/10.1002/adfm.201902262
39. S.K.E. Hill et al., Silicon quantum dot-poly(methyl methacrylate) nanocomposites with reduced light scattering for luminescent solar concentrators. Acs Photonics **6**, 170–180 (2019). https://doi.org/10.1021/acsphotonics.8b01346
40. X. Luo, T. Ding, X. Liu, Y. Liu, K. Wu, Quantum-cutting luminescent solar concentrators using ytterbium-doped perovskite nanocrystals. Nano Lett. **19**, 338–341 (2019). https://doi.org/10.1021/acs.nanolett.8b03966
41. B. Mendewala et al., The potential of scalability in high efficiency hybrid perovskite thin film luminescent solar concentrators. Sol. Energy **183**, 392–397 (2019). https://doi.org/10.1016/j.solener.2019.03.042
42. J. Tong et al., Fabrication of highly emissive and highly stable perovskite nanocrystal-polymer slabs for luminescent solar concentrators. J. Mater. Chem. A **7**, 4872–4880 (2019). https://doi.org/10.1039/c8ta12149d
43. X. Luo, T. Ding, X. Liu, Y. Liu, K. Wu, Quantum-cutting luminescent solar concentrators using ytterbium-doped perovskite nanocrystals. Nano Lett. (2019). Doi: https://doi.org/10.1021/acs.nanolett.8b03966
44. Y. You et al., Eco-friendly colloidal quantum dot-based luminescent solar concentrators. Adv. Sci.**6** (2019). Doi: https://doi.org/10.1002/advs.201801967
45. H. Zhao, Refractive index dependent optical property of carbon dots integrated luminescent solar concentrators. J. Lumin. **211**, 150–156 (2019). https://doi.org/10.1016/j.jlumin.2019.03.039
46. C. Yan, H. Zhao, D.F. Perepichka, F. Rosei, Lanthanide ion doped upconverting nanoparticles: synthesis, structure and properties. Small **12**, 3888–3907 (2016). https://doi.org/10.1002/smll.201601565
47. H. Zhao et al., Absorption enhancement in "giant" core/alloyed-shell quantum dots for luminescent solar concentrator. Small **12**, 5354–5365 (2016). https://doi.org/10.1002/smll.201600945
48. M.J. Talite et al., Greener luminescent solar concentrators with high loading contents based on in situ cross-linked carbon nanodots for enhancing solar energy harvesting and resisting concentration induced quenching. ACS Appl. Mater. Interfaces **10**, 34184–34192 (2018). https://doi.org/10.1021/acsami.8b10618
49. L.H. Slooff et al., A luminescent solar concentrator with 7.1% power conversion efficiency. Phys. Status Solidi-Rapid Res. Lett. **2**, 257–259 (2008). Doi: https://doi.org/10.1002/pssr.200802186
50. J.A.H.P. Sol, G.H. Timmermans, A.J. van Breugel, A.P.H.J. Schenning, M.G. Debije, Multistate luminescent solar concentrator "smart" windows. Adv. Energy Mater. **8** (2018). Doi: https://doi.org/10.1002/aenm.201702922

51. H.-J. Song et al., Performance limits of luminescent solar concentrators tested with seed/quantum-well quantum dots in a selective-reflector-based optical cavity. Nano Lett. **18**, 395–404 (2018). https://doi.org/10.1021/acs.nanolett.7b04263
52. Q.A. Akkerman, A.L. Abdelhady, L. Manna, Zero-dimensional cesium lead halides: history, properties, and challenges. J. Phys. Chem. Lett. **9**, 2326–2337 (2018). https://doi.org/10.1021/acs.jpclett.8b00572
53. L. Dou et al., Atomically thin two-dimensional organic-inorganic hybrid perovskites. Science **349**, 1518–1521 (2015). https://doi.org/10.1126/science.aac7660
54. J. Yin et al., Intrinsic lead ion emissions in zero-dimensional $Cs_4PbBr_6$ nanocrystals. Acs Energy Lett. **2**, 2805–2811 (2017). https://doi.org/10.1021/acsenergylett.7b01026
55. Z. Yuan et al., One-dimensional organic lead halide perovskites with efficient bluish white-light emission. Nature Commun. **8** (2017). Doi: https://doi.org/10.1038/ncomms14051
56. Y. Zhang et al., Direct-indirect nature of the bandgap in lead-free perovskite nanocrystals. J. Phys. Chem. Lett. **8**, 3173–3177 (2017). https://doi.org/10.1021/acs.jpclett.7b01381
57. C. Zhou et al., Luminescent zero-dimensional organic metal halide hybrids with near-unity quantum efficiency. Chem. Sci. **9**, 586–593 (2018). https://doi.org/10.1039/c7sc04539e
58. S. Mirershadi, S. Ahmadi-Kandjani, Efficient thin luminescent solar concentrator based on organometal halide perovskite. Dyes Pigm. **120**, 15–21 (2015). https://doi.org/10.1016/j.dyepig.2015.03.035
59. S.C. Erwin et al., Doping semiconductor nanocrystals. Nature **436**, 91–94 (2005). https://doi.org/10.1038/nature03832
60. G.K. Grandhi, R. Viswanatha, Tunable infrared phosphors using Cu doping in semiconductor nanocrystals: surface electronic structure evaluation. J. Phys. Chem. Lett. **4**, 409–415 (2013). https://doi.org/10.1021/jz3021588
61. A.K. Guria, S.K. Dutta, S. Das Adhikari, N. Pradhan, Doping $Mn^{2+}$ in lead halide perovskite nanocrystals: successes and challenges. Acs Energy Lett. **2**, 1014–1021 (2017). Doi: https://doi.org/10.1021/acsenergylett.7b00177
62. M. He et al., Mn-doped cesium lead halide perovskite nanocrystals with dual-color emission for WLED. Dyes Pigm. **152**, 146–154 (2018). https://doi.org/10.1016/j.dyepig.2018.01.045
63. C.-H. Hsia, A. Wuttig, H. Yang, An accessible approach to preparing water-soluble $Mn^{2+}$-doped (CdSSe)ZnS (core) shell nanocrystals for ratiometric temperature sensing. ACS Nano **5**, 9511–9522 (2011). https://doi.org/10.1021/nn2025622
64. G. Huang et al., Ag- and Mn-doped ZnInS/ZnS dual-emission quantum dots with zone tunability in the color coordinate. Nanotechnology **27** (2016). Doi: https://doi.org/10.1088/0957-4484/27/18/185602
65. J.Y. Kim et al., Highly loaded PbS/Mn-doped CdS quantum dots for dual application in solar-to-electrical and solar-to-chemical energy conversion. Appl. Catal. B-Environ. **227**, 409–417 (2018). https://doi.org/10.1016/j.apcatb.2018.01.041
66. W.Y. Liu et al., $Mn^{2+}$-doped lead halide perovskite nanocrystals with dual-color emission controlled by halide content. J. Am. Chem. Soc. **138**, 14954–14961 (2016). https://doi.org/10.1021/jacs.6b08085
67. D. Mocatta et al., Heavily doped semiconductor nanocrystal quantum dots. Science **332**, 77–81 (2011). https://doi.org/10.1126/science.1196321
68. Y. Pan et al., Inherently $Eu^{2+}/Eu^{3+}$ codoped $Sc_2O_3$ nanoparticles as high-performance nanothermometers. Adv. Mater. (Deerfield Beach, Fla.) (2018). Doi: https://doi.org/10.1002/adma.201705256
69. V.A. Vlaskin, N. Janssen, J. van Rijssel, R. Beaulac, D.R. Gamelin, Tunable dual emission in doped semiconductor nanocrystals. Nano Lett. **10**, 3670–3674 (2010). https://doi.org/10.1021/nl102135k
70. Z. Wang et al., H-doped black titania with very high solar absorption and excellent photocatalysis enhanced by localized surface plasmon resonance. Adv. Func. Mater. **23**, 5444–5450 (2013). https://doi.org/10.1002/adfm.201300486
71. P.Arunkumar et al., Probing molecule-like isolated octahedra via-phase stabilization of zero-dimensional cesium lead halide nanocrystals. Nature Commun.**9** (2018). Doi: https://doi.org/10.1038/s41467-018-07097-x

72. B.M. Benin et al., Highly emissive self-trapped excitons in fully inorganic zero-dimensional tin halides. Angew. Chem. Int. Ed. **57**, 11329–11333 (2018). https://doi.org/10.1002/anie.201806452
73. T. Jun et al., Lead-free highly efficient blue-emitting $Cs_3Cu_2I_5$ with 0D electronic structure. Adv. Mater. **30**, 6 (2018). https://doi.org/10.1002/adma.201804547
74. Y. Wu et al., In situ passivation of $PbBr_6{}^{4-}$ octahedra toward blue luminescent $CsPbBr_3$ nanoplatelets with near 100% absolute quantum yield. Acs Energy Lett. **3**, 2030–2037 (2018). https://doi.org/10.1021/acsenergylett.8b01025
75. J. Yin et al., Point defects and green emission in zero-dimensional perovskites. J. Phys. Chem. Lett. **9**, 5490–5495 (2018). https://doi.org/10.1021/acs.jpclett.8b02477
76. Y. Zhang et al., Zero-dimensional $Cs_4PbBr_6$ perovskite nanocrystals. J. Phys. Chem. Lett. **8**, 961–965 (2017). https://doi.org/10.1021/acs.jpclett.7b00105
77. M. Wei et al., Ultrafast narrowband exciton routing within layered perovskite nanoplatelets enables low-loss luminescent solar concentrators. Nature Energy **4**, 197–205 (2019). https://doi.org/10.1038/s41560-018-0313-y
78. W. Ma et al., Carbon dots and AIE molecules for highly efficient tandem luminescent solar concentrators. Chem. Commun. (Camb.) **55**, 7486–7489 (2019). https://doi.org/10.1039/c9cc02676b
79. F. Liu et al., Highly luminescent phase-stable $CsPbI_3$ perovskite quantum dots achieving near 100% absolute photoluminescence quantum yield. ACS Nano **11**, 10373–10383 (2017). https://doi.org/10.1021/acsnano.7b05442

# Chapter 9
# Perovskite Quantum Dots for Photovoltaic Applications

**Xu Chen, Siyuan Huang, Yue Tian, Tingming Jiang, and Yang (Michael) Yang**

**Abstract** Perovskite solar cell with easy solution processing and high efficiency has been considered as a promising photovoltaic technology, but the material instability makes it questionable for practical applications. Perovskite quantum dots (QDs) provide new possibilities to resolve these concerns. In this chapter, we will discuss the backgrounds, fundamentals, synthesis methods and device physics of the perovskite QDs based solar cells.

## 9.1 Introduction to Perovskite Solar Cells

Photovoltaic devices, converting solar energy into electricity, have continuously been a research hotspot during the past decades. In respect to the semiconducting materials used in the devices, solar cells (SCs) can be divided to three categories: the first generation of silicon-based cells, the second generation of multi-component thin-film cells including copper indium gallium selenide (GIGS), cadmium telluride (CdTe), GaAs, and the third generation of novel solar cells that mainly include organic solar cells, quantum dots solar cells, and perovskite solar cells (PSCs). Among them, perovskite materials with $ABX_3$ crystal structure [(A) organic or inorganic cations including normally Cs, Rb, methylammonium (MA) and formamidinium (FA); (B) Pb or Sn; (C) halides I, Br and Cl], becomes one of the most outstanding candidates to rival other solar cells, (Fig. 9.1a) since it possesses promising optoelectrical properties as well as cost-effective manufacturing process, *i.e.* long carrier lifetime, high defeat tolerance, tunable bandgaps, easy depositing routes, cheap precursors, etc. To date, mesoporous negative-intrinsic-positive (n-i-p) configuration PSCs have achieved a certified power conversion efficiency (PCE) of 25.2%, which is very close to the best Si cell.

X. Chen · S. Huang · Y. Tian · T. Jiang · Y. Yang (✉)
State Key Laboratory of Modern Optical Instrumentation, College of Optical Science and Engineering, Zhejiang University, Hangzhou 310058, Zhejiang, China
e-mail: yangyang15@zju.edu.cn

 
Y. Zhou and Y. Wang (eds.), *Perovskite Quantum Dots*, Springer Series in Materials Science 303, https://doi.org/10.1007/978-981-15-6637-0_9

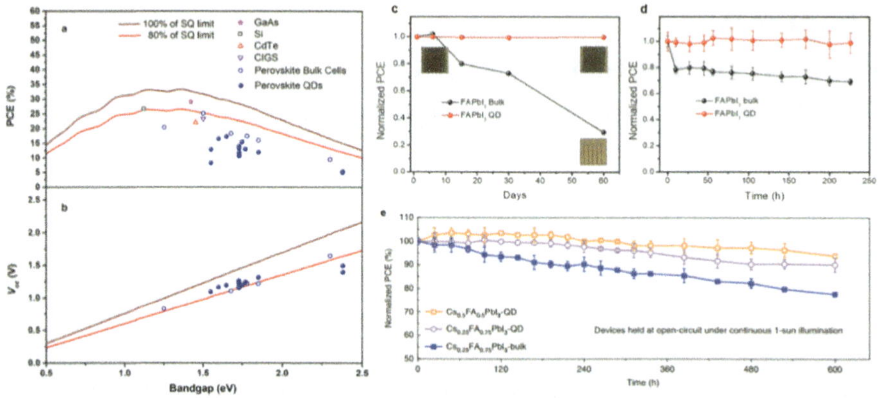


**Fig. 9.1** Comparison of theoretical PCE, open-circuit voltage ($V_{oc}$) limit and experimental performance of reported photovoltaics devices. **a** Theoretical and experimental PCE of current hot photovoltaic devices. GaAs (five-pointed star, 29.1%), Si (square, 26.7%), CdTe (triangle, 22.1%), CIGS (Inverted triangle, 23.4%) and the certified record perovskite-base solar cell (hollow circle at ~1.5 eV bandgap, 25.2%) are available in the NREL's best research-cell efficiencies chart. The literatures of perovskite bulk devices (hollow circles) are listed as Refs. [20, 39–41], while that of Pe-QD photovoltaics (solid circles) are listed as Refs. [22, 26–28, 30–32, 35–37, 42–45]. The wine and red lines represent 100, 80% of Shockley Queisser (SQ) limit, respectively. **b** Theoretical and experimental $V_{oc}$ of current hot photovoltaic devices. **c**, **d**, compared the humidity (without encapsulation) and light stability (encapsulation stored at room temperature) of $FAPbI_3$ QD with that of $FAPbI_3$ bulk devices respectively. Adapted from Ref. [31], Copyright 2018 by the Elsevier. **e** Comparison of the light stability of $Cs_{0.5}FA_{0.5}PbI_3$ QD, $Cs_{0.25}FA_{0.75}PbI_3$ OD and $Cs_{0.25}FA_{0.75}PbI_3$ bulk devices. Adapted from Ref. [36], Copyright 2020 by the Springer Nature

Despite of the rapid progress of device efficiency, the long-term stability of PSCs is still a big concern. Perovskite solar cell is often recognized as a low-cost photovoltaic technology, but this argument is true only if it can last for sufficiently long time. According to the cost analysis [1, 2], if the perovskite modules with an average 16% PCE operate steadily for more than 20 years, the levelized cost of energy (LCOE) will be less than 8.0 cents per kW. However, the cell is very sensitive to ambient air. The water molecules can react with both the organic cations [3] and halides [4] and generate hydrogen bonds and volatile acid, such as HI, to induce the decomposition of perovskite crystals. Since the volume of the halide vacancy is similar to that of molecular oxygen, the perovskite crystals with halide vacancies can absorb and transport molecular oxygens and create $O_2^-$, which is extremely easy to produce water through acid–base reaction with the A site molecules of perovskite [5]. As a result, both water and oxygen will accelerate the device degradation. In addition, perovskite is also questionable when it is under operational conditions. It is found that PSCs suffer from faster degradation associated to the ion migration under external stresses including electric field [6], heat [7, 8] and light [9, 10]. To eliminate those problems, various of approaches including encapsulation technique [11, 12], composition engineering [13–15] interface modification [16–18], and additive engineering [19–21] have been developed for improving crystal structure robustness, reducing ion migration. Despite the existing methods that continuously improve the stability of the perovskite thin film solar cells, the emergence of halide perovskite quantum dots (Pe-QD) might provide a novel and effective pathway to realize highly stable perovskite solar cells.

## 9.2 Perovskite Quantum Dot Solar Cells with Enhanced Device Stabilities

The first Pe-QD solar cell was fabricated by Swarnker and his coworkers [22], it was made of the cubic $CsPbI_3$ ($\alpha$-$CsPbI_3$) without volatile organic group [23] showing 10.77% efficiency and improved stability. Compared with the thermodynamically stable but optically undesirable orthorhombic $CsPbI_3$ ($\delta$-$CsPbI_3$) of 2.82 eV bandgap, the 1.73 eV $\alpha$-$CsPbI_3$ generally exhibits poor phase stability at room temperature [24]. Nevertheless, $\alpha$-$CsPbI_3$ QDs shows significantly lowered phase transition temperature and can be stable for months at room temperature, owing to the variation of the surface energy [25]. After that, several effective pathways were carried out for promoting the charge transport in the $\alpha$-$CsPbI_3$ QDs based optoelectronic devices, in order to enhance the conversion efficiency. Wang et al. repressed the agglomeration-induced phase transition of $\alpha$-$CsPbI_3$ QDs by crosslinking QDs with the $\mu$-graphene, and achieved a 11.4% PCE and enhanced tolerance to humidity [26]. Chen et al. utilized some short-chain ligands as the capping ligands to replace the common long insulating ligands oleic acid (OA) and oleylamine (OLA), and obtained a higher PCE of 11.87% [27]. In addition to the optimization of the perovskite QD layer, the charge carrier transport layer is also crucial. Yuan et al. substituted polymer hole transport layer (HTL) for frequently-used Spiro-OMeTAD to accelerate the charge extraction at QD/HTL interface, obtaining the increased PCE of 12.55% [28]. Element doping in QDs is also found helpful. Shi et al. doped appropriate amounts of Ytterbium during synthesis through hot-injection method and gained 13.1% efficiency, which is attributed to the decrease of trap states [22]. Sanehira et al. and Ling et al. optimized the coupling between $\alpha$-$CsPbI_3$ QDs via surface passivation treatment of cation halide salt (AX) and built corresponding champion devices with 13.43% and 14.1% PCEs, respectively [29, 30].

Obviously, the relatively wide bandgap of $\alpha$-$CsPbI_3$ QDs limits its efficiency [25] Yang et al. controlled precisely the polarity of antisolvents during the post-synthetic treatments and demonstrated low band gap $FAPbI_3$ QDs solar cells with PCE of 8.38% [31]. Later, they achieved a breakthrough in PCE up to 13% by interface engineering [32]. The organic cations of $FAPbI_3$ ODs are weaker in $PbI_6$ octahedra than the inorganic cations of $CsPbI_3$ QDs, thus they are relatively more sensitive to polar solvent, greatly restricting the applications of $FAPbI_3$ QD-based photovoltaics [31, 33]. It has been demonstrated that simply mixing colloidal $FAPbI_3$ QDs with $CsPbI_3$ ODs as a mixed photoactive layer doesn't give good results [34]. Li et al. sequentially constructed a CsPbI3-QDs/$FAPbI_3$-QDs architecture via the layer-by-layer method and the resulting devices showed better humidity stability and a improved PCE up to 15.6% [35]. Hao et al. used OA ligands controllably assisting the cation-exchange during synthesizing $Cs_{1-x}FA_xPbI3$ QDs and obtained the corresponding hero devices with a certified PCE of 16.6% [36]. Zhao et al. revealed the exiting of a heterojunction in the $Cs_{0.25}FA_{0.75}PbI_3/CsPbI_3$ (ratio = 1:3) QD interface facilitating charge separation that leads to 17.4% efficiency [37]. Figure 9.1a, b depicts the theoretical limit of single junctions and experimental results of perovskite bulk solar cells and Pe-QD photovoltaics in the literatures. It shows that the efficiency of Pe-QD photovoltaics is gradually catching up with the conventional perovskite polycrystal thin film devices.

In addition to the progress of PCE, Pe-QD photovoltaics behave more stable under external stresses [31, 36]. Figure 9.1c–e show comparison of humidity and light stability of perovskite polycrystal thin film devices and Pe-QD devices. As it is shown, $FAPbI_3$ QD-based photovoltaics without any encapsulation almost remained their PCEs under ambient condition for 200 h, in the contrast the $FAPbI_3$ polycrystal thin film devices lost nearly 70% of their efficiencies. Furthermore, the encapsulated $FAPbI_3$ QD photovoltaics operate 225 h under continuous 1-sun illumination at open-circuit condition without any efficiency decay, while the control device only remains 70% of its initial PCE. The operational stability can be further enhanced with Cs incorporations. Both $Cs_{0.25}FA_{0.75}PbI_3$ QD and $Cs_{0.5}FA_{0.5}PbI_3$ QD photovoltaics can maintain 90% of their initial PCEs after 600 h continuous 1-sun illumination. Indeed, Pe-QD photovoltaics have demonstrated promising advantages on device stability. As mentioned above, $\alpha$-$CsPbI_3$ QDs exhibit reduced phase transition temperature of bellow 100 °C compared to 320 °C for CsPbI3 bulk perovskite, indicating that the $\alpha$-$CsPbI_3$ perovskite phase is thermodynamically more stable in the forms of Quantum dots [38]. The Layer-to-Layer deposition method is employed in fabricating sufficiently thick Pe-QDs layer to ensure adequate absorption of sunlight. In the following parts, we will cover more details on the synthesis methods of Pe-QDs, as well as the device structures and physics.

## 9.3 Quantum Dots Synthesis Methods

By far, Pe-QDs based solar cells have primarily employed the systems of formamidinium lead iodide ($FAPbI_3$) and all-inorganic cesium lead halide ($CsPbX_3$, X = $Cl^-$, $Br^-$, $I^-$) as photoactive layers. Formamidinium lead iodide ($FAPbI_3$) possesses the narrowest bandgap (1.45–1.51 eV) among lead halide perovskites, and exhibits improved thermal stability relative to methylammonium lead iodide ($FAPbI_3$) due to its elevated decomposition temperature. During the rapid development of organic–inorganic hybrid perovskite solar cells, all-inorganic perovskites ($CsPbX_3$, X = $Cl^-$, $Br^-$, $I^-$) without the volatile organic components have been gaining more and more attention because they are more thermally stable and can sustain temperature exceeding 400 °C without any phase degradation.

Although all-inorganic perovskites are more thermally stable, it has been found that the α-phase of $CsPbI_3$ (black phase) could be rapidly converted into non-photovoltaic δ-phase (yellow phase) in an ambient environment with high moisture. Encouragingly, α-$CsPbI_3$ QDs could retain the cubic phase because of the large contribution of surface energy [46] in contrast to the easy phase transformation of instability of the bulk α-$CsPbI_3$. The $CsPbI_3$ QDs synthesis generally follows the synthetic protocols of hot injection method reported by Protesescu et al. [46] which is schematically shown in Fig. 9.2a. Firstly, the Cs-oleate precursor is prepared from the chemical reaction between $Cs_2CO_3$ and oleic acid (OA), then octadecene (ODE) and $PbX_2$ are loaded into a 3-neck flask and dried under vacuum. A 1:1 mixture of dry oleylamine (OAm) and dry oleic acid are added to solubilize $PbX_2$ salt and stabilize the nanocrystals. After the temperature is raised, the pre-heated Cs-oleate is quickly injected into the high boiling solvent, and the crude solution is cooled by the ice-water bath. Then the aggregated nanocrystals are separated by centrifugation. Finally the supernatant is discarded and the particles are re-dispersed in toluene or hexane forming long-term colloidally stable solutions.

In 2016, Swarnker et al. modified the synthetic route and purification approach of $CsPbI_3$ QDs in order for extensive characterization and further application in PV cells [47]. The softer basic nature of $I^-$ as compared with $Br^-$ results in weaker acid–base interactions between the halide and the oleylammonium ligand [48]. Therefore, in the isolation process, the loss of ligand during extraction could cause agglomeration and conversion of α-$CsPbI_3$ to the orthorhombic phase. Thus, in the isolation step, they used methyl acetate (MeOAc) instead of polar non-solvent to purify the QDs without full removal of the surface species or inducing agglomeration. Based on this modification, the obtained QDs were more stable in the cubic phase for months under ambient storage. The solar cells based on these $CsPbI_3$ QDs delivered a PCE of 10.77% with an open-circuit voltage of 1.23 V. To further optimize this synthesis method, Wang et. al. utilized bis-(2,2,4-trimethylpentyl) phosphinic acid (TMPPA) as an oleic acid alternative in the synthesis of α–$CsPbI_3$ to stabilize the α-$CsPbI_3$ QDs [49].

On the other hand, many groups have also found that partial replacement of I- by Br- can make α–$CsPbI_xBr_{3-x}$ QDs more phase-stable. Ghosh et al. reported the photovoltaic application of $CsPbBr_xI_{3-x}$ QDs, which could create an optimal balance between the PV device efficiency and its stability [50]. The photovoltaic devices made from $CsPbBr_{1.5}I_{1.5}$ showed power conversion efficiency (PCE) ~7.94% with open circuit voltage ~1.0 V, fill factor 0.70 and long-term stability. In addition, the other mixed-halide perovskites, such as $CsPb(Cl/Br)_3$ and $CsPb(Br/I)_3$, can be readily produced by combining appropriate ratios of $PbX_2$ salts [46].

In addition to the hot injection synthesis method, Li et al. synthesized perovskite QDs under room temperature based on the princible of supersaturation recrystallization [51], as shown in Fig. 9.2b. OA and OAm were added to stabilize the precursor solution after dissolution of $PbX_2$ and CsX n DMF or DMSO. Then the precursor solution was quickly dipped into toluene by the assist of vigorous stirring. $CsPbX_3$ Pe-QDs could be obtained immediately upon completion of the injection.

Apart from the all-inorganic Pe-QDs with vulnerable ionic bonds and large bandgap, $FAPbI_3$ Pe-QDs have also attracted attentions of the photovoltaic community because of the narrower optical bandgaps. Xue's group reported the $FAPbI_3$ Pe-QDs solar cells with power-conversion efficiency of 8.38%, realized by three cycles of surface treatment to improve the inter-dot coupling [31]. In order to further improve the low charge separation efficiency due to quantum confinement, Yang et al. raised a strategy to introduce a conjugated small molecule ITIC while removing the original insulating surface ligands of $FAPbI_3$ QDs film [32] consequently leading to a high PCE of 12.7% for the devices. The $FAPbI_3$ QDs used above were both synthesized via a modified hot-injection two-precursor method reported by Protesescu et al. [52] Unlike the synthesis of $CsPbI_3$ QDs, which required a high excess of Pb (molar ratio Pb:Cs = 3.75) and high temperatures (120–200 °C), $FAPbI_3$ QDs form exclusively under conditions with excess FA (FA:Pb = 2.7) and at 80 °C. And note that excess OLA could lead to the quick decomposition of $FAPbI_3$ QDs before the cooling process. In another three precursor method reported by Protesescu et al., $Pb(acetate)_2 \cdot 3H_2O$, FA-acetate, ODE, and OA were combined and dried under vacuum [52]. The mixture was heated to 80 °C under N2, followed by the injection of Oleylammonium Halide (OAmX, X = Br, I), before the cooling, centrifugation, and washing routine.

Nevertheless, Joseph M. Luther group firstly realized the layer-by-layer deposition of $CsPbI_3$ QDs and $Cs_xFA_{1-x}PbI_3$ QDs [37]. The $Cs_xFA_{1-x}PbI_3$ alloys were obtained by mixing the $CsPbI_3$ and $FAPbI_3$ QDs colloidal solutions in different volume ratios to yield $Cs_xFA_{1-x}PbI_3$ QDs with the desired Cs/FA stoichiometry, after the concentrations were calibrated. Furthermore, Lianzhou Wang's group also reported an effective oleic acid (OA) ligand-assisted cation-exchange strategy that allowed controllable synthesis of $Cs_{1-x}FA_xPbI_3$ QDs [36]. In an OA-rich environment, the cross-exchange of cations was facilitated, enabling rapid formation of $Cs_{1-x}FA_xPbI_3$ QDs with reduced defect density (Fig. 9.2c).

## 9.4 Device Structure and Physics

Perovskite solar cells with an architecture of "n-i-p" possess generally better device performance than their counterparts with "p-i-n" configuration [53–55]. Likewise, Pe-QDs based solar cells with n-i-p structure have also gathered much attention with a configuration of Glass/ITO(or FTO)/ETL/Pe-QD film/HTL/Electrode. Figure 9.3a shows a cross-section TEM image of a typical Pe-QDs based solar cell. The commonly used ETL are mesoporous or planar $TiO_2$ and $SnO_2$, while Spiro-OMeTAD, PTAA, P3HT, etc. [22, 26–32, 36, 42] are generally utilized as HTL. Based on the conventional structure, Yang Yang group proved that the ITIC that was mainly distributed between CQD film and ETL could produce extra charge driving force attributed to the ingredient band structure [32]. The conventional structure is adopted by all Pe-QD devices for high efficiency at present, but the selection of HTL in such structure is not varied, and these expensive HTLs literally increase device

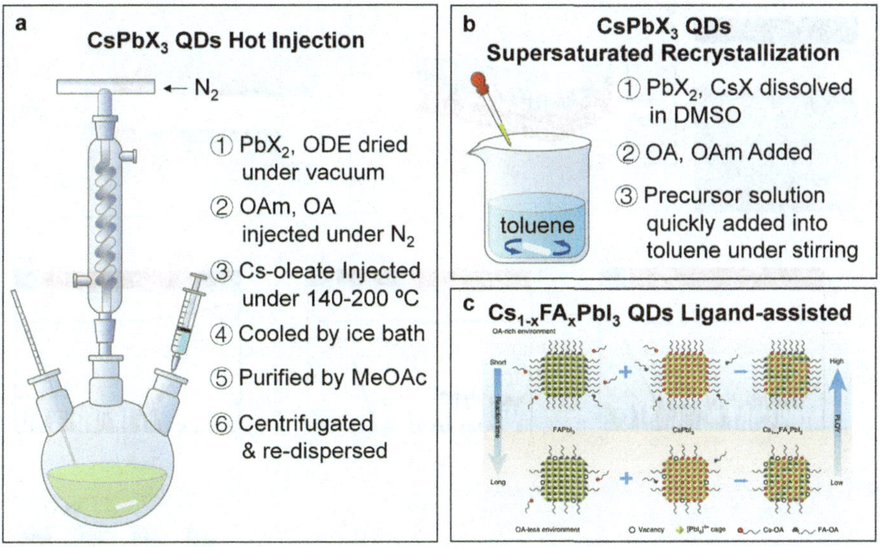


**Fig. 9.2** Schematic diagram of representative synthesis methods of Perovskite quantum dots applied in photovoltaic cells. **a** The hot injection synthetic method of $CsPbX_3$ QDs. **b** The supersaturation recrystallization synthetic method of $CsPbX_3$ QDs. **c** Ligand-assisted cation-exchange synthesis of $Cs_{1-x}FA_xPbI_3$ QDs, Copyright 2020 by Nature Energy

cost. Therefore, it's necessary to develop the inverted structure (p-i-n) Pe-QD photovoltaics in the near future [56]. Figure 9.3b delivers that schematic illustration of the perovskite solar cell with an α-$CsPbI_3$/$FAPbI_3$ QD structure. The distinct difference of the device structure of Pe-QD photovoltaics to that of perovskite-bulk solar cells is that multiple and diverse Pe-QD films can be directly deposited in sequence without destroying the underlying layers [29, 57]. Up to now, Most of highly-efficient Pe-QD SCs including the record cell are based on only one layer of Pe-QD film, but from a technical point of view, it is feasible to deposit two or multiple Pe-QD layers as light-absorping layer. Joseph M. Luther group measured the energy band positions for Pe-QD compositions of $Cs_{1-x}FA_xPbI_3$. Figure 9.3c describes the energy band positions data showing that the Fermi level and $E_c$ become deeper as FA concentration increases. From the perspective of bandgap, $FAPbI_3$ presents a lower bandgap of 1.55 eV which is closer to the ideal bandgap and much narrower than that of $CsPbI_3$ (~1.74 eV), while the cation-mixed $Cs_{1-x}FA_xPbI_3$ offers a bandgap range from 1.55 and 1.74 eV. Thus, two or multiple Pe-QD layers based on different $Cs_{1-x}FA_xPbI_3$ compositions are deposited, because the light-absorping layer can make fuller use of the solar spectrum due to the diverse and gradient bandgaps. Moreover, a heterojunction formed by two $Cs_{1-x}FA_xPbI_3$ quantum dots with different values of x exit inside Pe-QDs films, can result in extra driving force which assists in pushing the electron to the ETL while pushing the hole to the HTL. Joseph M. Luther group firstly realized the layer-by-layer deposition of Pe-QD. With the time-of-flight secondary ion mass spectrometry measurement, the compositions on different depth are given.

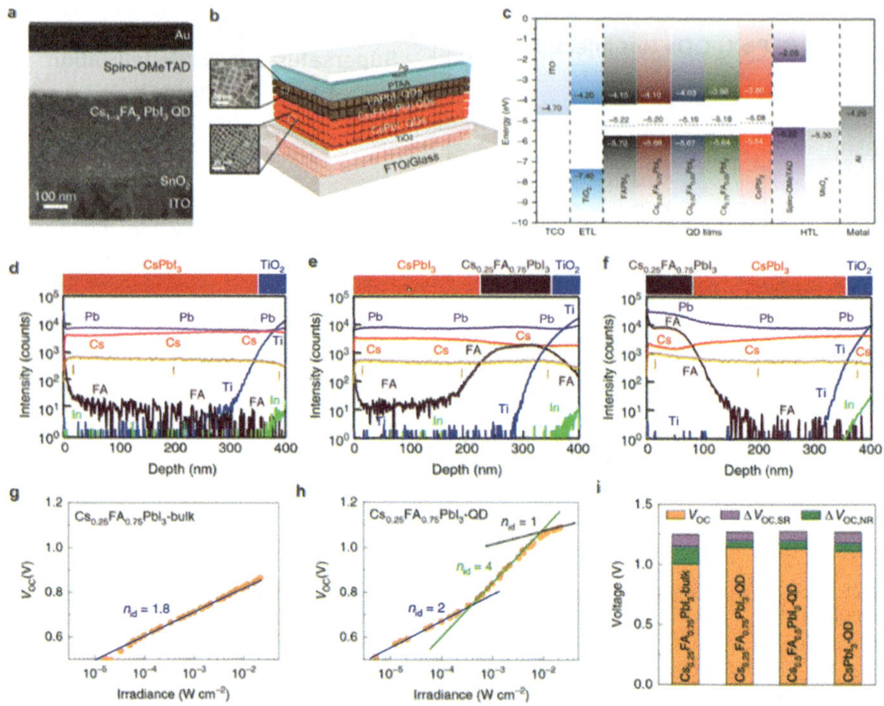


**Fig. 9.3** **a** Cross-section TEM image of the $Cs_{1-x}FA_xPbI_3$ QDSC. Adapted from Ref. [36], Copyright 2020 by the Springer Nature. **b** Schematic illustration of the perovskite solar cell with an α-$CsPbI_3$ + $FAPbI_3$ QD structure. Adapted from Ref. [35], Copyright 2019 by American Chemical Society. **c** Energy band positions for perovskite QD compositions and contact layers. Adapted from Ref. [37], Copyright 2019 by the Springer Nature. **d–f** ToF–SIMS depth profile of samples with the structure and interface location indicated in the bar above the plots of $CsPbI_3/TiO_2$. **d** $CsPbI_3/Cs_{0.25}FA_{0.75}PbI_3/TiO_2$. **e** $Cs_{0.25}FA_{0.75}PbI_3/CsPbI_3/TiO_2$. **f** All relevant elements from perovskite QD layer and $TiO_2$ are shown. Adapted from Ref. [37], Copyright 2019 by the Springer Nature. **g–h** Light intensity-dependent VOC measurements over a large dynamic range in $Cs_{0.25}FA_{0.75}PbI_3$-Bulk device. **g** $Cs_{0.25}FA_{0.75}PbI_3$-QD device. **h** The solid lines denote linear fits to the experimental data and the ideality factors ($n_{id}$) are provided in the panels. **i** Voltage loss due to the surface recombination and non-radiative recombination in bulk and QD devices. Adapted from Ref. [36], Copyright 2020 by the Springer Nature

Figure 9.3d–f, it is obvious that $CsPbI_3$ quantum dots and $Cs_{0.25}FA_{0.75}PbI_3$ quantum dots have formed a heterojunction perfectly with evidence of the change of FA concentration [37]. With the optimal thickness ratio of $Cs_{0.25}FA_{0.75}PbI_3$ and $CsPbI_3$, finally the PCE of champion devices broke the 17% threshold. Except for the heterojunction formed by $CsPbI_3$ quantum dots and $Cs_{0.25}FA_{0.75}PbI_3$ quantum dots, Wanli Ma group realized the bilayers Pe-QD films by stacking $FAPbI_3$-QD on $CsPbI_3$-QD with a graded heterojunction facilitating the photocarrier harvesting, and boosting both PCE and ambient stablity [35]. Combination of different Pe-QD films is an unique and effective strategy to enhance the advantageous property of Pe-QD SCs, which needs to be researched more.

The general solar cell model is applicable to Pe-QD solar cells to calculate the theoretical PCE limit and reveal the mechanism of recombination and electrical loss as well. At present, the highest efficiency achieved is still far from the SQ limit, and the fabrication technology still needs further development. Considering the potential of PCE and the long-term stability both, the material composition of $Cs_{1-x}FA_xPbI_3$ could be the best choice among various Pe-QDs. The detailed balanced model assumes that the material has only radiative recombination, but in fact, Auger recombination and non-radiative recombination caused by recombination centers (impurities, defects, etc.) inevitably exist in the material. The dominant recombination in perovskite-bulk solar cells is radiative recombination and non-radiative recombination while the ideal factor is between 1 and 2. Impurities and density of defects are at a low level inside Pe-QD solar cells due to the better crystallinity of Pe-QD while the ideal factor can even reach a level close to 1. Lianzhou Wang group calculated the ideal factors of corresponding devices based on $Cs_{1-x}FA_xPbI_3$ Pe-QD shown in Fig. 9.3g, h. The $Cs_{0.25}FA_{0.75}PbI_3$-bulk solar cells have an ideal factor of 1.8 while the $Cs_{0.25}FA_{0.75}PbI_3$-QD solar cells have an ideal factor of 1 under high irradiance [36]. These results indicate that the non-radiative recombination of Pe-QD solar cells based on optimized component of $Cs_{1-x}FA_xPbI_3$ QDs ($x = 0.75$) can be effectively suppressed, consequently leading to the lower voltage loss of the Pe-QD SCs. (Fig. 9.3i).

## 9.5 Conclusion

In general, Pe-QD solar cells have better long-term stability compared to perovskite-bulk solar cell, but still have great room for development. More efforts should be put on the controllable cation-exchange method balancing the trade-off, suppressing phase-seperation of $Cs_{1-x}FA_xPbI3$ QDs and ultimately achieving over 20% PCE for single junction [36]. Meanwhile, the electronic properties of relatively insulated capping ligands of Pe-QDs are still obscure, which are related to trap states, electron coupling between Pe-QDs, etc.[30]. (1) From the perspective of industrialization, inverted Pe-QD SCs should be considered and explored to further reduce cost [56]. (2) Owe to the unequivalent PCE of Pe-QD photovoltaics to that of perovskite-bulk SCs (Fig. 9.1a, b), how to apply the wide-bandgap Pe-QDs as the top subcells of tandem SCs to obtain highly-efficienct and stable two-terminal or four-terminal tandem SCs will be a research hotspot [58]. Finally, referred to the development history of perovskite-bulk SCs and PbS quantum SCs, we anticipate that Pe-QD SCs will step forward by leaps and bounds.

## References


1. Z. Song et al., A technoeconomic analysis of perovskite solar module manufacturing with low-cost materials and techniques. Energy Environ. Sci. **10**, 1297–1305 (2017). https://doi.org/10.1039/c7ee00757d
2. J. Guo, J. Min, A cost analysis of fully solution-processed ITO-free organic solar modules. Adv. Energy Mater. **9**, 1802521 (2019). https://doi.org/10.1002/aenm.201802521
3. A. Jeffrey, A.M.H. Pierre, V.K. Prashant, Transformation of the excited state and photovoltaic efficiency of CH3NH3PbI3 perovskite upon controlled exposure to humidified air. J. Am. Chem. Soc. **137**, 1530–1538 (2015)
4. J.M. Frost et al., Atomistic origins of high-performance in hybrid halide perovskite solar cells. Nano Lett. **14**, 2584–2590 (2014)
5. C.C. Boyd, R. Cheacharoen, T. Leijtens, M.D. McGehee, Understanding degradation mechanisms and improving stability of perovskite photovoltaics. Chem. Rev. **119**, 3418–3451 (2018)
6. T. Leijtens et al., Mapping electric field-induced switchable poling and structural degradation in hybrid lead halide perovskite thin films. Adv. Energy Mater. **5**, 1500962 (2015)
7. S. Bai et al., Planar perovskite solar cells with long-term stability using ionic liquid additives. Nature **571**, 245–250 (2019). https://doi.org/10.1038/s41586-019-1357-2
8. G. Divitini et al., In situ observation of heat-induced degradation of perovskite solar cells. Nat. Energy **1**, 1–6 (2016)
9. K. Domanski et al., Migration of cations induces reversible performance losses over day/night cycling in perovskite solar cells. Energy Environ. Sci. **10**, 604–613 (2017)
10. H.J. Snaith et al., Anomalous hysteresis in perovskite solar cells. J. Phys. Chem. Lett. **5**, 1511–1515 (2014)
11. L. Shi et al., Accelerated lifetime testing of organic-inorganic perovskite solar cells encapsulated by polyisobutylene. ACS Appl. Mater. Interfaces **9**, 25073–25081 (2017). https://doi.org/10.1021/acsami.7b07625
12. R. Cheacharoen et al., Design and understanding of encapsulated perovskite solar cells to withstand temperature cycling. Energy Environ. Sci. **11**, 144–150 (2018)
13. Y. Zhang, Y. Liu, Y. Li, Z. Yang, S. Liu, Perovskite CH3NH3Pb(BrxI1-x)(3) single crystals with controlled composition for fine-tuned bandgap towards optimized optoelectronic applications. J. Mater. Chem. C **4**, 9172–9178 (2016). https://doi.org/10.1039/c6tc03592b
14. B.G. Kim, W. Jang, J.S. Cho, D.H. Wang, Tailoring solubility of methylammonium lead halide with non-stoichiometry molar ratio in perovskite solar cells: morphological and electrical relationships for high current generation. Sol. Energy Mater. Sol. Cells **192**, 24–35 (2019). https://doi.org/10.1016/j.solmat.2018.12.010
15. Z. Wang et al., Efficient ambient-air-stable solar cells with 2D–3D heterostructured butylammonium-caesium-formamidinium lead halide perovskites. Nat. Energy **2**, 17135 (2017)
16. G.-W. Kim, G. Kang, M. Malekshahi Byranvand, G.-Y. Lee, T. Park, Gradated mixed hole transport layer in a perovskite solar cell: improving moisture stability and efficiency. ACS Appl. Mater. Interfaces **9**, 27720–27726 (2017)
17. F. Han et al., Cesium iodide interface modification for high efficiency, high stability and low hysteresis perovskite solar cells. Electrochim. Acta **236**, 122–130 (2017). https://doi.org/10.1016/j.electacta.2017.03.139
18. H. Zhu, et al., Tailored amphiphilic molecular mitigators for stable perovskite solar cells with 23.5% efficiency. Adv. Mater., e1907757. https://doi.org/10.1002/adma.201907757 (2020)
19. Q. Han et al., Additive engineering for high-performance room-temperature-processed perovskite absorbers with micron-size grains and microsecond-range carrier lifetimes. Energy Environ. Sci. **10**, 2365–2371 (2017). https://doi.org/10.1039/c7ee02272g
20. J. Tong et al., Carrier lifetimes of >1 mus in Sn-Pb perovskites enable efficient all-perovskite tandem solar cells. Science **364**, 475–479 (2019). https://doi.org/10.1126/science.aav7911
21. S. Yang et al., Tailoring passivation molecular structures for extremely small open-circuit voltage loss in perovskite solar cells. J. Am. Chem. Soc. **141**, 5781–5787 (2019)

22. A. Swarnkar et al., Quantum dot-induced phase stabilization of alpha-CsPbI3 perovskite for high-efficiency photovoltaics. Science **354**, 92–95 (2016). https://doi.org/10.1126/science.aag2700
23. S.S. Shin et al., Colloidally prepared La-doped BaSnO3 electrodes for efficient, photostable perovskite solar cells. Science **356**, 167–171 (2017)
24. R.E. Beal et al., Cesium lead halide perovskites with improved stability for tandem solar cells. J. Phys. Chem. Lett. **7**, 746–751 (2016)
25. G.E. Eperon et al., Inorganic caesium lead iodide perovskite solar cells. J. Mater. Chem. A **3**, 19688–19695 (2015)
26. Q. Wang et al., μ-graphene crosslinked CsPbI3 quantum dots for high efficiency solar cells with much improved stability. Adv. Energy Mater. **8**, 1800007 (2018)
27. K. Chen et al., Short-chain ligand-passivated stable α-CsPbI3 quantum dot for all-inorganic perovskite solar cells. Adv. Func. Mater. **29**, 1900991 (2019)
28. J. Yuan et al., Band-aligned polymeric hole transport materials for extremely low energy loss α-CsPbI3 perovskite nanocrystal solar cells. Joule **2**, 2450–2463 (2018)
29. E.M. Sanehira, et al., Enhanced mobility CsPbI3 quantum dot arrays for record-efficiency, high-voltage photovoltaic cells. Sci. Adv. **3**, eaao4204 (2017)
30. X. Ling, et al., 14.1% CsPbI3 perovskite quantum dot solar cells via cesium cation passivation. Adv. Energy Mater. **9**, 1900721 (2019)
31. J. Xue et al., Surface ligand management for stable FAPbI3 perovskite quantum dot solar cells. Joule **2**, 1866–1878 (2018)
32. J. Xue et al., A small-molecule, "charge driver" enables perovskite quantum dot solar cells with efficiency approaching 13%. Adv. Mater. **31**, 1900111 (2019)
33. F. Zhang et al., Colloidal synthesis of air-stable CH3NH3PbI3 quantum dots by gaining chemical insight into the solvent effects. Chem. Mater. **29**, 3793–3799 (2017)
34. A. Hazarika et al., Perovskite quantum dot photovoltaic materials beyond the reach of thin films: full-range tuning of A-site cation composition. ACS Nano **12**, 10327–10337 (2018)
35. F. Li, et al., Perovskite quantum dot solar cells with 15.6% efficiency and improved stability enabled by an α-CsPbI3/FAPbI3 bilayer structure. ACS Energy Lett. **4**, 2571–2578 (2019)
36. M. Hao et al., Ligand-assisted cation-exchange engineering for high-efficiency colloidal Cs1−xFAxPbI3 quantum dot solar cells with reduced phase segregation. Nat. Energy **5**, 79–88 (2020). https://doi.org/10.1038/s41560-019-0535-7
37. Q. Zhao et al., High efficiency perovskite quantum dot solar cells with charge separating heterostructure. Nat. Commun. **10**, 1–8 (2019)
38. S. Sharma, N. Weiden, A. Weiss, Phase diagrams of quasibinary systems of the type: ABX3—A′ BX3; ABX3—AB′ X3, and ABX3—ABX′ 3; X= halogen. Zeitschrift für Physikalische Chemie **175**, 63–80 (1992)
39. M. Abdi-Jalebi, et al., Maximizing and stabilizing luminescence from halide perovskites with potassium passivation. *Nature* **555**, 497+. https://doi.org/10.1038/nature25989 (2018)
40. Y. Wang et al., Thermodynamically stabilized beta-CsPbI3-based perovskite solar cells with efficiencies >18. Science **365**, 591–595 (2019). https://doi.org/10.1126/science.aav8680
41. W. Chen et al., Precise control of crystal growth for highly efficient CsPbI2Br perovskite solar cells. Joule **3**, 191–204 (2019). https://doi.org/10.1016/j.joule.2018.10.011
42. J. Shi et al., Efficient and stable CsPbI 3 perovskite quantum dots enabled by in situ ytterbium doping for photovoltaic applications. J. Mater. Chem. A **7**, 20936–20944 (2019)
43. Q. Zeng, et al., Polymer-passivated inorganic cesium lead mixed-halide perovskites for stable and efficient solar cells with high open-circuit voltage over 1.3 V. Adv. Mater. (Deerfield Beach, Fla.). https://doi.org/10.1002/adma.201705393 (2018)
44. J.B. Hoffman, G. Zaiats, I. Wappes, P.V. Kamat, CsPbBr 3 solar cells: controlled film growth through layer-by-layer quantum dot deposition. Chem. Mater. **29**, 9767–9774 (2017)
45. Q.A. Akkerman et al., Strongly emissive perovskite nanocrystal inks for high-voltage solar cells. Nat. Energy **2**, 1–7 (2016)
46. L. Protesescu et al., Nanocrystals of cesium lead halide perovskites (CsPbX3, X= Cl, Br, and I): novel optoelectronic materials showing bright emission with wide color gamut. Nano Lett. **15**, 3692–3696 (2015)

47. A. Swarnkar et al., Quantum dot–induced phase stabilization of α-CsPbI3 perovskite for high-efficiency photovoltaics. Science **354**, 92–95 (2016)
48. Q.A. Akkerman et al., Tuning the optical properties of cesium lead halide perovskite nanocrystals by anion exchange reactions. J. Am. Chem. Soc. **137**, 10276–10281 (2015)
49. C. Wang, A.S. Chesman, J.J. Jasieniak, Stabilizing the cubic perovskite phase of CsPbI 3 nanocrystals by using an alkyl phosphinic acid. Chem. Commun. **53**, 232–235 (2017)
50. D. Ghosh, M.Y. Ali, D.K. Chaudhary, S. Bhattacharyya, Dependence of halide composition on the stability of highly efficient all-inorganic cesium lead halide perovskite quantum dot solar cells. Sol. Energy Mater. Sol. Cells **185**, 28–35 (2018)
51. X. Li et al., CsPbX3 quantum dots for lighting and displays: room-temperature synthesis, photoluminescence superiorities, underlying origins and white light-emitting diodes. Adv. Func. Mater. **26**, 2435–2445 (2016)
52. L. Protesescu et al., Dismantling the "red wall" of colloidal perovskites: highly luminescent formamidinium and formamidinium–cesium lead iodide nanocrystals. ACS Nano **11**, 3119–3134 (2017)
53. D. Luo et al., Enhanced photovoltage for inverted planar heterojunction perovskite solar cells. Science **360**, 1442–1446 (2018)
54. T. Jiang et al., Realizing high efficiency over 20% of low-bandgap Pb–Sn-alloyed perovskite solar cells by in situ reduction of $Sn^{4+}$. Solar RRL **4**, 1900467 (2019). https://doi.org/10.1002/solr.201900467
55. Q. Jiang et al., Surface passivation of perovskite film for efficient solar cells. Nat. Photonics **13**, 460–466 (2019)
56. J. Jean, Getting high with quantum dot solar cells. Nat. Energy **5**, 10–11 (2020). https://doi.org/10.1038/s41560-019-0534-8
57. L.M. Wheeler et al., Targeted ligand-exchange chemistry on cesium lead halide perovskite quantum dots for high-efficiency photovoltaics. J. Am. Chem. Soc. **140**, 10504–10513 (2018)
58. J.A. Christians, et al., *2018 IEEE 7th World Conference on Photovoltaic Energy Conversion (WCPEC) (A Joint Conference of 45th IEEE PVSC, 28th PVSEC & 34th EU PVSEC)*, pp. 81–84 (IEEE)

# Chapter 10
# Perovskite Quantum Dots Based Phototransistors

**Xiang Liu, Yuan Tao, You Zhang, Zhi Tao, and Jianhua Chang**

**Abstract** Perovskite semiconductors have attracted intensely advancing researches in optical devices, due to the long diffusion length, high quantum efficiency and excellent light absorption coefficients and so on. Especially combined with Quantum dots (QDs), this blended semiconductor can be selected as the alternative candidate for the photosensing core of the sensors and other optoelectronics. To realize the incident light-to-electric converting and directly signal's reading-out, we have put enormous efforts to design and fabricate a versatile phototransistor. The optimal device's structure also balances the electrical and optical performances to enable enhancing the photodetecting efficiency and compatibility. In this report we reviewed realizations, models, physics and metrology applications of perovskite and perovskite QDs based phototransistor. In our research group, we have designed the phototransistors with QDs as sensing core by nano-manufacturing technology and explored the photo-generation mechanism. The corresponding protocols focus on understanding of connection between these different structures and photodecting performances. The subsequent optimized studies advance technologies for integrated photodetecting applications with potentially inorganic nano-scale optical materials.

## 10.1 Introduction

Diverse advancing research progresses have been made on novel materials, device physics and electrical structures, in order to address the practical problems existed in next-generation QDs(QDs) based photodetectors [1, 2]. The recently isolated heterostructures of phototransistors are interesting hosts to study quantum phenomena with organometal halide perovskite semiconductor material and so on. Compared with conventional photodetectors, phototransistors can act as direct IR reading-out integrated circuits (ROIC) and either convert an IR incident light signal

X. Liu (✉) · Y. Tao · Y. Zhang · Z. Tao · J. Chang
School of Electronic and Information Engineering, Nanjing University of Information Science and Technology, Nanjing 210046, China
e-mail: xjlx1906@126.com



Y. Zhou and Y. Wang (eds.), *Perovskite Quantum Dots*, Springer Series in Materials Science 303, https://doi.org/10.1007/978-981-15-6637-0_10

or amplify and output electrical signals [3]. These device structures also allow modulation of carrier's transport efficiency over many orders of magnitude, and enable enhancing the photodetecting performances. In this report we review realizations, models, physics and metrology applications of perovskite QDs based photodetectors. In our research group, we have designed the phototransistors with QDs as sensing core by nano-manufacturing technology in which the trade-off between detecting speed versus high quantum efficiency is optimized by employing this heterostructure. The corresponding protocols focus on understanding of connection between these different structures and photodecting applications. The subsequent optimized protocol produces promising for scaling up to a high-efficient system level application and related to the non-negligible influence of exchange interactions among the carriers.

## 10.2 Perovskite Based Phototransistors

### *10.2.1 Overview of Perovskite Phototransistors*

Perovskite semiconductors have long diffusion length, high quantum efficiency and excellent light absorption coefficients, remarkable electronic, thermal, optical and mechanical properties, such as high mobility (2 $cm^2/Vs$), a high on/off ratio (~$10^5$), efficient responsivity (47 A/W) and quick detection speed (~5 μs) [4]. Therefore, before we lucubrate the perovskite QDs' based phototransistors, the bulk perovskite materials' optical-electrical characters and their applications in phtotransistors were researched and developed in advance. An organic–inorganic perovskite ($CH_3NH_3PbI_3 = MAPbI_3$) based transistor has been reported to achieve high electrical performance in our group [5]. Nevertheless, when subjected to a limited bandgap, a perovskite transistor usually only has highly efficient visible lights absorption.

In addition, due to the intrinsic drawbacks derived from bulk material structure, this phototransistors' limitation is obvious apparent with low efficiency and bad response wavelengths' range. Consequently, excepted with producing pristine perovskite, we also have tried to using hybrid bulk perovskite/$MoS_2$ QDs hybrid thin film, which solves the band lattice matchup and bandgap offset problem to maintain preferable electrical performance with a high-performance IR detecting capability. In view of the result of characterization, these works are conductive to decrease the photodetector's optical-electrical conversion efficiency; to extend response time (lowering detecting speed); and to increase the noise current.

## 10.2.2 Bulk Perovskite Semiconductor Phototransistors

The light's photoluminescence (PL) characteristics are investigated and illustrated preliminarily at room temperature to reveal the optical characters, as shown in Fig. 10.1. In particular, two sharp peaks are found at 560 and 820 nm respectively. Many researchers have tried to extend the detection range of perovskite transistors to the infrared wavelengths [6]. Presently, commonly used methods are involved to blend IR QDs with superior optical properties and to synthesize the perovskite QDs directly.

The shift in the peaks can be explained by the interactions between the two components and the short wavelength photons due to the increased number of surface traps among the hybrid perovskite/$MoS_2$QDs semiconductors. Moreover, because of the carriers' quenching effect, the carriers' hybrid lifetime quickly decreases compared with pristine perovskite. However, in our Transient Resolution PL (TRPL) experiment, after multiexponential fitting at room temperature, it was found that photogenerated charges in the hybrids cannot be transferred into $MoS_2$ but are stored inside the perovskite-$MoS_2$ interface.

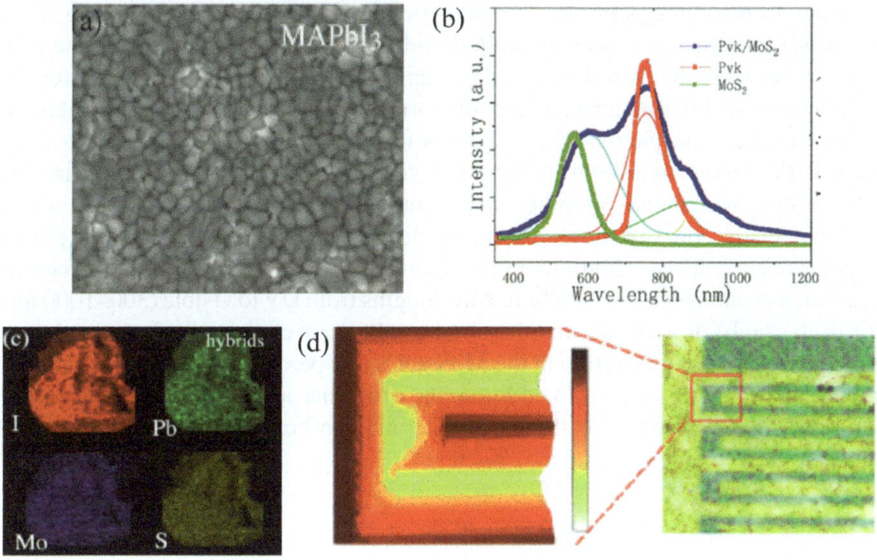


**Fig. 10.1** **a** SEM of perovskite surface on top of the transistors. **b** PL spectra of the main semiconductor components discussed in this work. **c** EDS mapping for the photo-sensing channel. **d** photocurrent mapping for the phototransistors

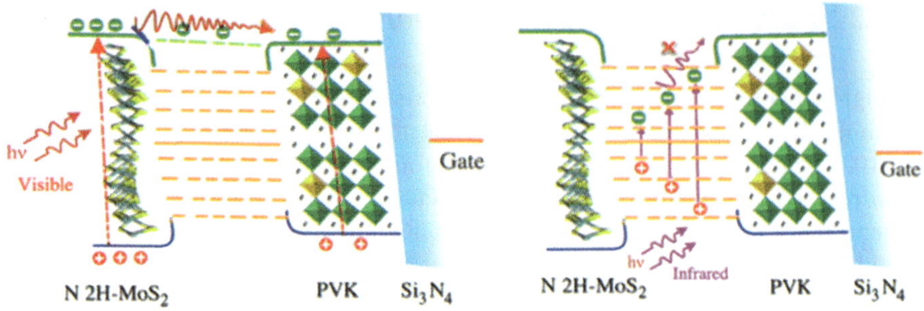


**Fig. 10.2** Bandgap alignment of the perovskite/MoS2 QD hybrids and their interfaces

### *10.2.3 Careers' Transfer Inside of Bulk Perovskite Transistor*

As shown in Fig. 10.2, many continuous energy bands exist in the interface between the $MoS_2$ QDs and $MAPI_3$ perovskite, which can generate electron–hole pairs under weak infrared illumination. However, due to the large bandgap offset between these continuous narrow energy bands and perovskite's conduction band, it is difficult for electrons to reach the perovskite's conduction band. This creates more trap states and densities, which can store these photoinduced charges. Thus, based on the relationship between the stored charges and threshold voltage, the IV curves for the phototransistor shift and generate a photocurrent via the gate voltage's modulation.

To elucidate the photocurrent's generation mechanism, we characterized the transfer IV curves, as shown in Fig. 10.2 and it was found that this mixture can achieve improved device mobility and a comparable on/off ratio property, which is a synergistic effect in terms of optical-electrical performance. Different photo-responsive currents are identified in the drain current as a function of the sweeping Vg, which occur under different light wavelengths from UV to visible (300–1000 nm, at Fig. 10.2a, b). It is shown that the obvious photocurrent increase is derived from more photoelectron generation in the $MoS_2$QDs and perovskite(the natural bandgap in these wavelengths' range).In addition, the $MoS_2$ has a similar conduction energy band and light absorption as the perovskite and can benefit the carrier's transport between these components.

### *10.2.4 Characterization of Perovskite Based Phototransistors*

By comparing the responsivity of the perovskite hybrid and pristine perovskite transistors, we found that because of the insert of $MoS_2$ QDs and its interaction with perovskite, the hybrid phototransistor exhibits much higher responsivity compared with the pristine counterpart (Fig. 10.3).

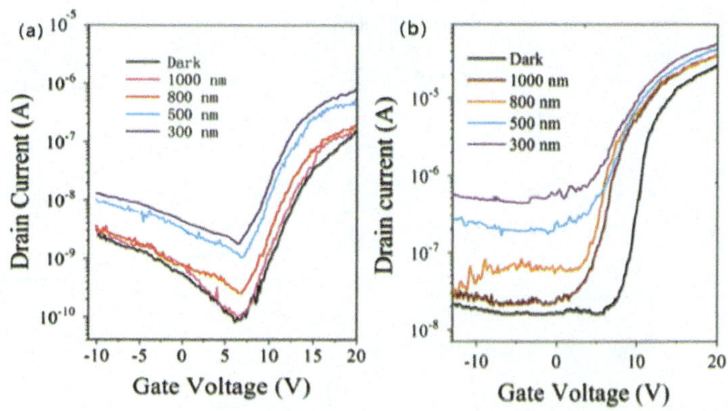


**Fig. 10.3** **a** Transfer characteristic curves of the pristine perovskite phototransistor illuminated under visible-to-near infrared wavelengths ($V_{DS}$ = 5 V). **b** Transfer characteristic curves of a perovskite hybrid phototransistor illuminated under visible-to-near infrared wavelengths ($V_{DS}$ = 5 V)

Right here, we take this perovskite phototransistors as an example to introduce our characterization approaches to characterize the device for their application in photodetection.

Figure 10.4a shows the variation of the photocurrent as a function of the light power for a 1200 nm IR wavelength. A power law can be used to fit this curve:

$$I_{PH} = \alpha P^{\beta} = 1.8 \times 10^{-9} P^{1.28},$$

where the photocurrent is given in A and the light power $P$ in μW/cm². As the responsivity is the ratio between the increase of the current under illumination and the light power (R(Responsivity) = $I_{PH}P$), the expression for responsivity as a function of light power $P$ is given by:

$$\mathrm{R} = \alpha P^{\beta-1} = 90P^{0.28}.$$

The good sensor linearity below 1 mW/cm² shows efficient electron—hole pair extraction in the IR waveband. This fitted responsivity is contrasted with the functional photoresponse obtained by:

$$\mathrm{R} = \frac{\alpha W C_{ox} \mu V_{DS}}{L} P^{\beta-1}.$$

Obviously, if we want to seek a high responsivity and sensor linearity, the most striking parameters for the transistor (mobility, drain-source voltage, geometry capacitance and channel) are beneficial to the optimized optical-electrical performance. Especially at infrared wavelengths, the responsivity (920 A/W at 1200 nm of

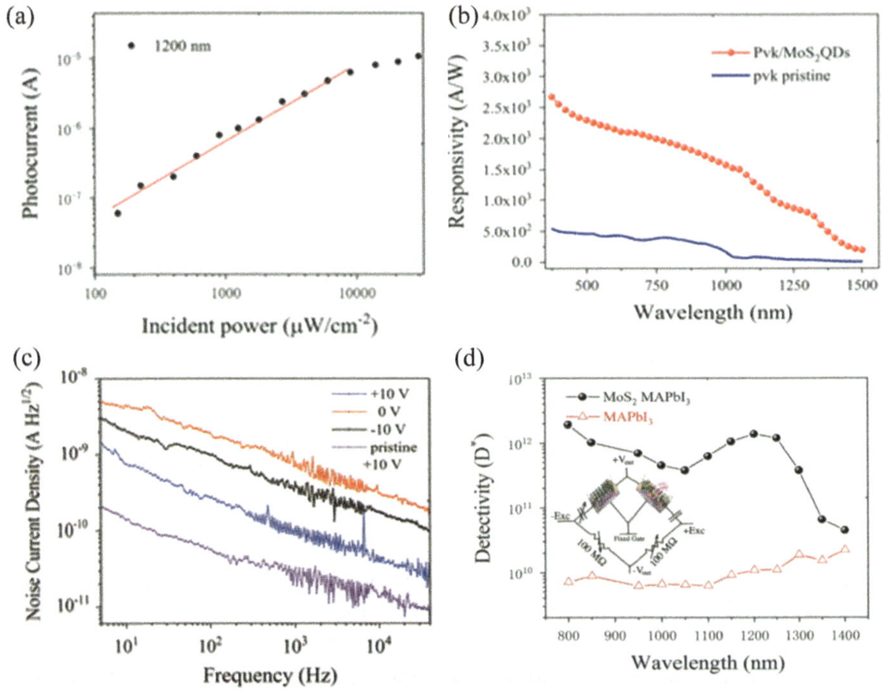


**Fig. 10.4** **a** Photocurrent versus incident light power. **b** Responsivity comparison between the PVK hybrids and pristine PVK phototransistors. **c** Noise current density for the hybrid IR phototransistor under three different biases. **d** Calculated detectivity as a function of the frequency signal for a 10 V gate bias and 100 Hz at room temperature

Fig. 10.4b) has been improved at least 100 times more than that of pristine perovskite transistor.

To further study the performance of the hybrid's phototransistor for IR detection under dynamic operations, it demonstrates this hybrid phototransistor has a quick detecting speed reaching approximately 15 μs rise time and 25 μs decay time. Notably, even at a high frequency of 22 kHz, photocurrent attenuation can be controlled within a reasonable scope (<−3 dB), which is evidence of a high speed and wide bandwidth IR detection phototransistor.

Finally, because of the greater number of granules, defects and trap states inside of the hybrids, the noise current density ($3.4 \times 10^{-9}$ A/Hz, +10 $V_G$) deteriorates compared to pristine perovskite photodetectors ($2.1 \times 10^{-10}$ A/Hz, +10 $V_G$) in Fig. 10.4c. Nevertheless, taking consideration the photo responsivity, the detection performance can be characterized by the following expression:

$$D^* = \frac{(A \cdot \Delta f)^{1/2}}{NEP} = \mathrm{R} \cdot A^{1/2}/I_N$$

where R is the responsivity, A is the illuminating area of the sensor and B is the reference bandwidth (we use 1 Hz here).

Due to a large illuminating area from the interdigital electrodes and the relatively high responsivity, this IR photodetector can achieve $3 \times 10^{12}$ Jones at IR wavebands based and high detecting frequency on Fig. 10.4d.

In summary, a phototransistor with $MoS_2$ QDs encapsulated in an $MAPbI_3$ perovskite film has been fabricated assisted by sputtering and additional solution processes. The optical-electronic measurement showed strong, effective photoinduced charges and a threshold voltage shift due to the incorporation of the $MoS_2$ QDs. With systematic characterization of the phototransistor's detection capabilities for IR wavelengths, it was possible to demonstrate a highly efficient photodetector that exhibits quick detection speed (40 μs) and a high detectivity ($3 \times 10^{12}$ Jones) under dynamic high frequency.

## 10.3 Classification of QDs Based Phototransistors

### *10.3.1 Phototransistors with QD's Hybrid Active Layer*

To improve the shortage of bulk perovskite materials such as weak optical absorbance and bad uniformity, QDs are selected as the alternative candidate for the photosensing key kernel. Especially after the emerging of the colloidal chemical synthesized QDs, it can be easily transferred and integrated in the large-scale microelectronics.

As shown in Fig. 10.5, the phototransistors only have two positions can be embedded with QDs, which are the active channel and the modulated insulator respectively. One scheme is to combine the QDs' unique optical-electrical properties and high performance transistor active semiconductors, which can both increase the photoconductivity under illumination, and guarantee the transistors' electrical performances. Another solution is to embed the QDs inside of the insulator and to generate induced charges within active layer.

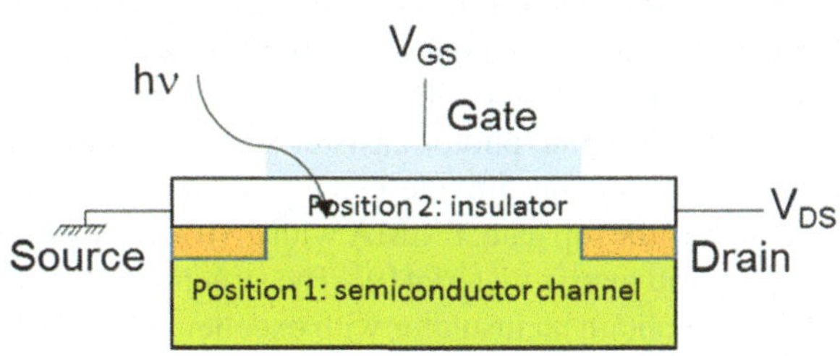


**Fig. 10.5** The structure of the phototransistor with two different photo-sensing schemes

### *10.3.2 Phototransistors with QD's Hybrid Insulator*

Hybrid QDs' photosensing insulator approach illustrated in Fig. 10.5 has a better uniformity and thick optical-electrical layers for the phototransistor than that of previous hybrid QDs' channel phototransistor. It is rarely reported to fabricate hybrid perovskite QDs' based photo-sensing insulator. Nevertheless, many researches have been presented about this type phototransistor with comparable other semiconductor QDs.

For instance, Kuo and Lai demonstrated a CMOS approach for integrating Ge QDs in the composite insulator to inject the electrons into the channel [7]. Through this fabrication approach, the distance between QDs is curtailed to transfer the carriers feasibly. The photocurrent ratio and the responsivity were $4.1 \times 10^6$ and 1.7 A/W, respectively. A superior external quantum efficiency of 240% and a very fast temporal response time of 1.4 ns have been obtained. These research results manifest the great promise for these Si-based phototransistor in the optical interconnects and telecommunications. However, the disadvantages of these QDs' synthesis methods are the complicated process, the rigid fabrication conditions, the non-uniformity and the thin as-prepared QDs layer. In particular, our group combined the results of these two types of the phototransistors and analyzed comprehensively the advantages and disadvantages of both detectors.

## 10.4 Perovskite QDs Encapsulated in Insulator Gel of Phototransistors

The combination of polymer gel and quantum dot has become a possible solution in the research of photoelectric devices [8], because polymer can be used in the packaging insulator of phototransistors with perovskite QDs. For example, in the pioneering work of phototransistors using polymethyl methacrylate (PMMA) as an insulator, a mixture of polymers and QDs is expected to play an important role in photosensitive devices.

Because the phototransistor's electrical performances can be deteriorated by the organic insulator, it is a risky scheme to adopt hybrid polymer directly as modulated insulator. On this basis, different from the single-gate phototransistor, we proposed a novel double-gate phototransistor with indium gallium zinc oxide (IGZO) as the active layer, silicon dioxide ($SiO_2$) as the bottom gate and IPQD/PMMA hybrid insulator as the top gate. PMMA with a Molecular weight of 1,10,000 was purchased as the hybrid matrix material [9]. The structure is shown in Fig. 10.6b. The aim of this study is to obtain an insulator with excellent electrical properties and to demonstrate the possibility of a high-efficiency photosensitive top-gate insulator producing high photoelectrical properties.

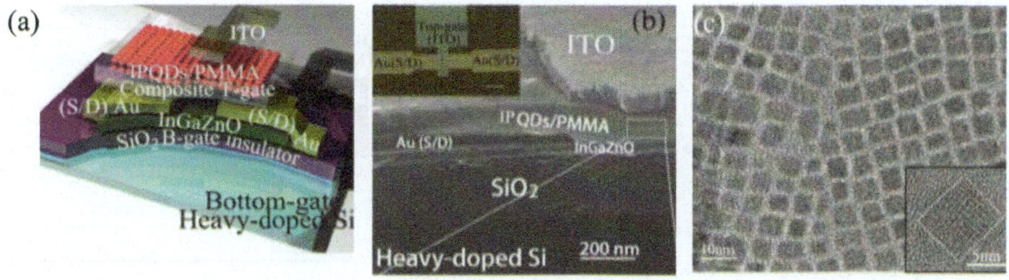


Fig. 10.6 Device schematic (**a**), optical and SEM cross sectional microscope image of dual-gate Perovskite QDs/IGZO phototransistor (**b**), and TEM picture of the perovskite QDs (**c**)

## *10.4.1 Fabrication Processes of the Hybrid Photo-Sensing Insulator*

The precursor route was used to produce tunable-size, high efficiency and stable CsPbBr3 QDs out following the chemical method. In brief, CsBr and PbBr3 (volumeratio = 1:1) were dissolved in a DMF solution. Oleylamineand oleic acid (OAM:OA = 1:2) were included to stabilize the precursor products and contribute to control the size of the crystal. Finally, polycrystalline $CsPbBr_3$ QDs were transferred to a nonpolar anti-solvent toluene solution (the nanoscale cubic morphology of QDs with W × L: 6.7 × 7.4 nm diameter illustrated in Fig. 10.6c. We have tested the characteristics of this novel phototransistor with photo-sensing insulator. The validation shows that perovskite materials can keep chemically stable with PMMA insulator and nonpolar toluene solvent. Meanwhile, PMMA can protect pervskite QDS from moisture and oxygen in the air, keeping the film for several days.

## *10.4.2 Photo-Charges' Inducing Mechanism of Phototransistors*

We have investigated the hysteresis transfer characteristics of both single topgate and bottom-gate TFT as illustrated in Fig. 10.7a. It indicates that the direct use of single-grid photosensitizing perovskite QDs/PMMA insulator leads to low electrical performance and serious stability problems, which are due to the mismatch of organic and non-organic contact surfaces (presence of more trap states). More importantly, the top-gate ITO/PMMA_Perovskite QDs/IGZO structure can form a basic metal/insulator/semiconductor (MIS) model.

However, compared with the direct blending with active layer, this type of transistor can't inject electrons into the channel. Consequently, we speculated that the photocurrent can be generated by the induced charge coupled from the MIS structures. As described in measured C–V characteristics of Fig. 10.7b (1 MHz gate

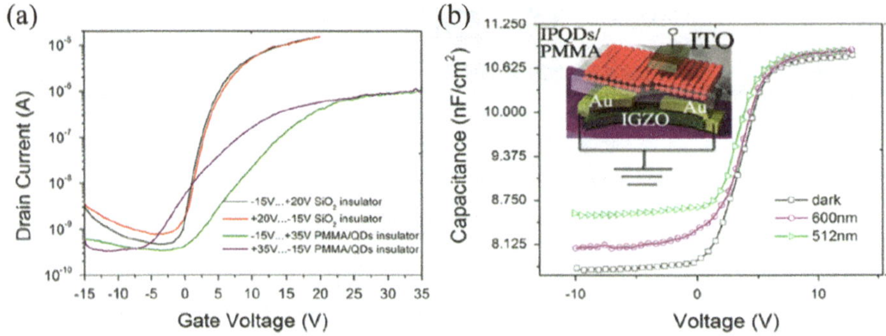


**Fig. 10.7** $I_D$-$V_{TG}$ curves of Perovskite QDs/PMMA top-gated phototransistor illuminated with same $V_D = 5$ V and different wavelengths lights and device schematic (inset)

modulation frequency), the work function offset between ITO and IGZO generates the positive flat-band voltage (positive $V_{TG}$). In addition, the incident light and negative $V_{TG}$ can decrease tunneling barriers and are beneficial for the photocarriers' transfer to IGZO. As depicted in Fig. 10.8b, it demonstrated that more interior photo-carriers are captured by the PMMA trap states.

As shown in the experiments, it can be seen that the single perovskite QDs/PMMA top-gate phototransistor has obvious threshold voltage offset under incident light irradiation. Obviously, this threshold voltage $\Delta V_T$ is determined by the accumulated photo-charges and the geometry capacitance Cox of the photo-sensing insulator. Such a large $\Delta V_T$ can phototransistor help generate high photocurrent and responsivity in the on-regime (up to 0.8 μA and 1800 A/W, extracted at $V_{TG} = 18$ V).

Based on the characteristic curves of the two-grid phototransistor shown in Fig. 10.8a, we find that the variation of the threshold voltage can be compensated by controlling the $V_{TG}$ and incident light. Through constructing the dual-gate modulation system, we realize to make use of the photo-sensing insulator but guarantee the electrical performance of the phototransistors. Especially for the dynamic

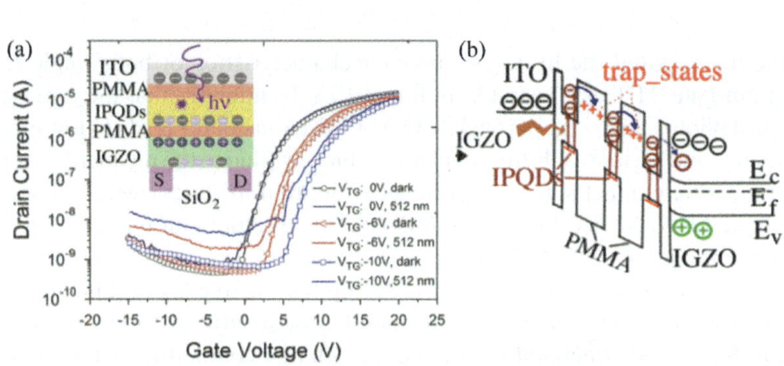


**Fig. 10.8** $I_D$-$V_{BG}$ curves of phototransistor under same light for different $V_{TG}$ and photocurrent generating mechanism's schematic (inset)

signal's response, this encapsulated polymer insulator can build up high-frequency induced photocurrent response, which has certain application value for nowadays' photodetectors.

### *10.4.3 Dynamic Analysis of the Phototransistors*

For special applications, phototransistor requires rapid extraction of photocarriers to operate in a high frequency environment. Under UV irradiation, the active layer of IGZO can be fully excited. The process of its slow rise and recovery indicates that the charge inside the oxidized semiconductor has a photoconductivity (PPC) effect. When a more negative $V_{TG}$ (−10 V, the same 512 nm light pulse) is applied (observed in Fig. 10.9a), the main mechanism of the fast photocurrent response comes from the increase in the light excitation of the capacitance (dielectric constant) of the perovskite QDS/PMMA insulator.

Another important application of this dual-gate phototransistor is the future senor-display system, so the same 20 × 100 μm channel of the phototransistor is serially connected to another IGZO TFT to form a simple 2-transistor circuit architecture (as shown in Fig. 10.9b).

The observed detailed response time can be estimated as follows: rise time = 120 ms and recovery time = 300 ms. The positive $V_{TG}$ pulse (triggered by the falling edge modulation of light pulse circuit) can reset the son-state output signal ($V_S$ terminal), achieving a simple accelerated photocurrent recovery time (less than 100 μs). This positive top-gate bias can suppress the photoelectron injection into the active layer. Notably, when a positive top-gate voltage is applied, the hybrid insulator immediately exhibits a weak increase relative to its negative voltage counterpart. Therefore, this technique can directly help the electron discharge captured in the perovskite QDs/PMMA insulator without interleaving the bottom-gate voltage (without affecting the series output TFT component).

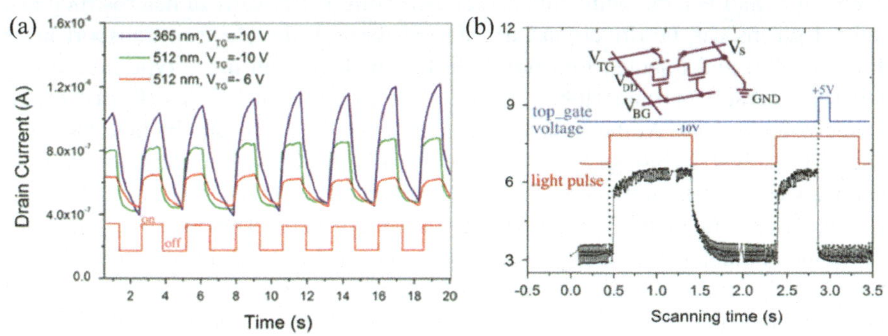


**Fig. 10.9** **a** Transient output signal as a function of time when subjected to light, and **b** $V_{TG}$ ($V_{DD}$ = 10 V, $V_{BG}$ = 7.5 V) and output signal of the phototransistor + TFT serial circuit (inset)

In summary, the dual-gate phototransistor based on $CsPbBr_3$ QDs has been realized in order to achieve excellent performance. It can be seen that this dual-gate phototransistor has high electrical properties, as well as good detection performance. In addition, its low photocurrent recovery time can even be reduced to the microsecond time scale by resetting a positive pulse. This work greatly extends the future applications of phototransistors of perovskite materials and provides a potential alternative for efficient, stable and functional optoelectronic devices.

## 10.5 Optimal Perovskite QDs Based Phototransistor and Evaluation of Figures of Merit

### *10.5.1 Optimal Heterostructures of the Perovskite Phototransistors*

Studies have shown that by using the new P-I-N vertical heterojunction with perovskite-type QDs photosensitive elements, a highly sensitive and fast photodetectors can be realized [10]. Among them, the perovskite material has a long diffusion length and high absorption efficiency. In addition, they exhibit adjustable band gap, high internal quantum yield, and many other advantages. Although vertical heterojunctions and diodes have been widely used in the past decade due to their high photosensitivity and fast detection speed, few reports have been reported on P-I-N phototransistors with vertical heterostructures to improve light detection. However, these advantages make it possible to use the similar perovskite QDs heterostructures in phototransistor. Because it can combine the high quantum yield of the quantum dot and the high carrier mobility of electrical semiconductors (such as IGZO and graphene), it has great attraction for its combination with other semiconductors.

Currently, we have proposed several methods to combine perovskite QDs with high electrical performance transistors [11]. By systematically cooperating with electrical N and P semiconductors and constructing P-I-N vertical heterostructures, photoelectrons can be driven and transferred through the electron transport layer (ETL, IGZO) and injected into the drain terminal through the P-type hole transport layer (grapheme nanoribbon, GNR in the example of Fig. 10.10) and drain-source voltage ($V_D$). At the same time, the holes will be impeded by hole blocking layer (HBL, IGZO) blocks and is retained inside the QDs' interface. In addition, an enhanced electric field along the edge of the nanoribbons can improve the electron injection efficiency of the graphene source terminals. This improves the optoelectronic characteristics, and you can easily adjust the vertical transistor. Furthermore, we also employed this optimal P-I-N channel heterostructure and its figure of merits in other application scenarios at some other regimes (such as novel fabrication process and optoelectronic devices), which will be introduced in detail at the following sections.

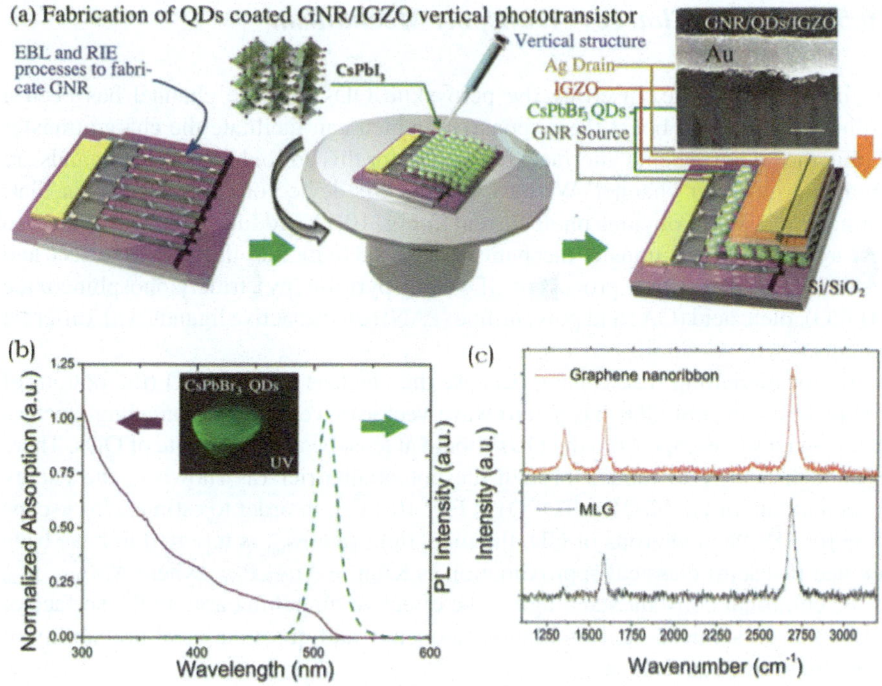


**Fig. 10.10** **a** Structure and fabrication processes of GNR/CsPbBr3 QD/IGZO phototransistors. Corresponding SEM image of the device (inset, scale bar: 1 μm). **b** Photoluminescence and absorption of CsPbBR3 QDs and photograph of QDs under UV light (inset). **c** Raman characterization of monolayer grapheme (MLG) and GNRs

As shown in the example of Fig. 10.10b, the data shows that $CsPbBr_3$ QDs have high absorption efficiency (80%) in the ultraviolet and visible light range, sharp emission peaks, and long exciton lifetimes. These characteristics are the combination of high-quality transistors with highly efficient vertical heterogeneity. The key to obtain excellent photoelectric properties are the junction integration. As shown in Fig. 10.1c, Raman characteristics of single-layer graphene (MLG) and GNRs are studied to detect the quality of graphene. Especially for GNRs, a high-intensity D-peak display the slight p-type characteristics of GNRs. Perovskite materials, especially as a light absorber in this vertical structure, have the cubic crystal structure shown (as shown in Fig. 10.1a), no impurities, and are in the photodetector Excellent intrinsic semiconductor.

### 10.5.2 Photo-Charges' Transport Mechanism

As introduced above, inserting the perovskite QDs into the channel has been a ordinary solution to boost the responsivity which can facilitate the charge transfer within the QD/grapheme interface. Especially for this colloidal QDs, the ligands are the charge transfer channel. We have exploit this device on IGZO transistor platform, which shows obvious photocurrent in Fig. 10.11. Additionally, with expect to research the charges' transfer mechanism, we tried to modify the QDs' interface and perform charge exchange process of QDs using pyridine(py), trioctylphosphineoxide (TOPO), oleic acid (OA) and polyanilline (PANI) electroactive ligands with different lengths.

As for tunnelling mechanism, because the electron on ELUMO (the bottom of conduction band) of QDs has a zero wave vector (k-vector), the tunneling electron that manages to escape from the QDs should at least be at the 1S-state of QDs. There is enough energy to tunnel through the potentialbarrier (as shown in the energy level diagram of GNM/QDs (TOPO) in Fig. 10.11b). In order to estimate the escape time for 1S-state electrons of QD, the tunnelling rate $K_{tun}$ is reported to have been applied in a semi-classical approximation as Ktun = $Af_{QDs}P_{tun.}$, where A, $f_{QDs}$, $P_{tun}$ is the empirical constant standing for the effective tunnelling area in the surface of QD, electron semiclassical oscillation frequency and quasi-classical probability of tunnelling, respectively.

For the tunneling rate $K_{tun}$, $f_{QDs}$, $P_{tun}$ can be calculated as [12]:

$$f_{QDs} = v_e/2d_{QDs}, v_e = \sqrt{\frac{2E_{1S}}{m_{eff}}}$$

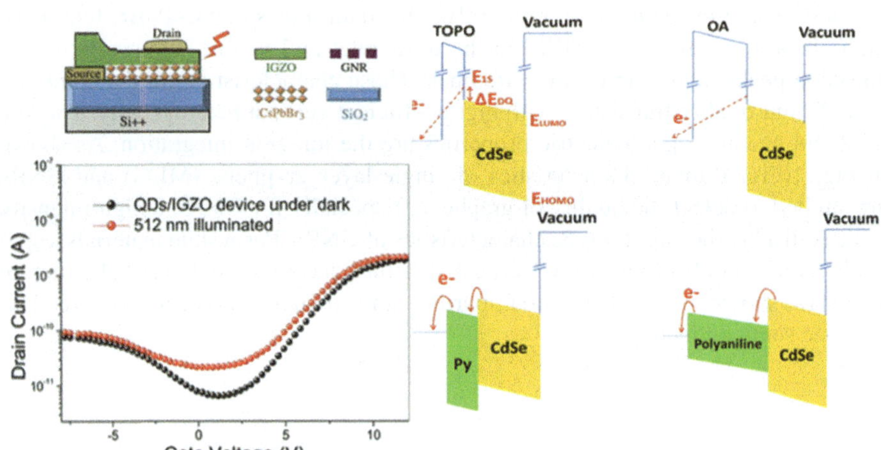


**Fig. 10.11** **a** Transfer character of the QDs/IGZO channel phototransistor under dark and illumination state. **b** Band alignment diagram of the charger transfer of QDs/ligand/GNM structure

$$P_{tun} = e^{-\frac{2}{\hbar}\int_0^{lig} dz|p(z)|},\ p(z) = \sqrt{2m_{eff}(z)(E_{1S} - U(z))}$$

where, $v_e$ is the characteristic electron velocity, while electron is in the 1S-state, dQD is the QDs diameter, $m_{eff}$ is the effective mass of the electron, $\hbar$ is plank constant, the integral is taken over the tunnel path ($L_{lig}$, ligand length) going through the potential barrier, U(z) is the electron potential along the tunneling path and meff(z) is the effective mass of the electron taken along the tunneling path. In experiment, we acquired the similar variation trend with the different lengths ligands and bandgap barriers as shown in Fig. 10.11b.

### *10.5.3 Fabrication Compatibility of the Optimal Perovskite QDs' Phototransistors*

Another importance is the fabrication compatibility of the optimal perovksite QDs' phototransistor to meet the diverse application demand. As described and illustrated at above Fig. 10.10a, perovskite is spin-coated on top conventional microelectronics fabrication processes (such as lithorgraphy and evaporation). Of course, the spin-coating is an easy-fabricating and cost-effective solution fabrication process. However, this extensive spin fabrication process can't guarantee the uniformity and be embedded in the refined device's micro or nano structures. Therefore, there is still a lot of work also for further research on the new processes. Herein, we are fully tapping combination potential, especially with large-scale fabrication processes and novel nano-structures.

Printable processes are emergying heated processes to assemble electronics at large-scale conviently, which can dramatically reduce material waste and manufacturing process steps while lowering manufacturing costs [13]. By taking into consideration of the recent progress in optoelectronics devices, combination of these two technologies can be useful in overlapping more regions of optical applications. Because the ink jet printing process usually need soluble organic semiconductor and polymer insulator, Fullerene organic thin-film transistor (OTFT) presents not only interesting mobility but also high electrical stability. Fittingly, QDs are also an infrequent inorganic semiconductors for electronics, which can offer convenience for the assembly of the phototransistors.

In particular, the QDs based P-I-N structure phototransistor with both electron and hole blockers show low dark current and high detectivity derived from its enhanced structure. Although the additional photocarriers provide substantial benefits and great potential for applications, the developments of P-I-N heterostructures combined with a phototransistor are still rarely reported.

For studying the characteristics of the P-I-N heterojunction in organic phototransistors, we focused on the perovskite QDs' functions. As illustrated in Fig. 10.12a, b, after fabricating the organic phototransistor based on the Fullerene ($C_{60}$),

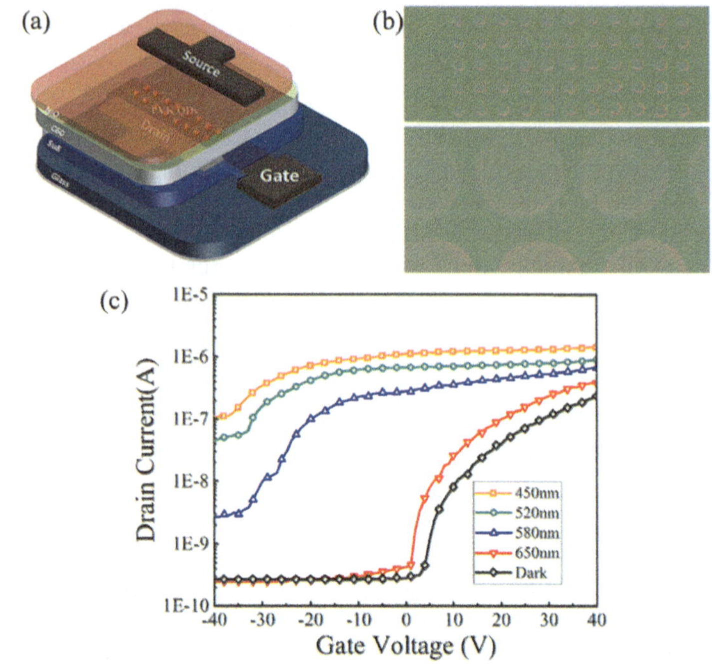


**Fig. 10.12** **a** The structure of the P-I-N organic photo-transistors. **b** Optical image of the drop-by-drop printing nozzles for IGZO and QDs' printing test. **c** The transfer character curves ($V_{DS} = 5$ V) of the printed perovskite QDs based phototransistor at different incident lights' wavelengths

perovskite QDs was printed with a thickness of 40 nm using 16 nozzles Q-class piezoelectric printhead for the silver pad electrodes, organic semiconductors and insulators. And the organic semiconductors (IGZO and perfovskite QDs in Fig. 10.12b) can be formed by using corresponding piezoelectric pressure gush printhead. Then, the QDs layer was covered by the NiO for forming the P-I-N heterojunction. Transfer characteristics are investigated to exhibit an excellent photocurrent increase and low dark current, which are rectified by the P-I-N heterostrucutre under dark and with various wavelengths incident light as presented in Fig. 10.12c.

As can be seen in Fig. 13a, With 450 nm incident light ($V_{DS} = 5$ V, $V_{GS} = 20$ V,), the responsivity of the device is calculated as $\sim 3.5 \times 10^3$ A/W. The detectivity of the P-I-N phototransistor is found at the level of the lowest noise current density due to the low dark current and longer diffusion length of the PVK QDs in the heterostructures (Fig. 10.13b). The P-I-N phototransistor's ultimate dynamic performances can be derived and are presented in Fig. 10.6. The spectrum analyzer measurement bandwidth can be estimated by the formula $\int_0^\infty NEP(f)d(f) = NEP(f)\Delta f$. Moreover, this organic PIN phototransistor simultaneously achieves $6.7 \times 10^9$ cmHz$^{1/2}$/W. Through this verification, embedded intrinsic photo-sensing perovskite QDs can adapt the novel fabrication processes and even relieve the shortages of organic electronics, such as low mobility, low efficiency and bad frequency characters.

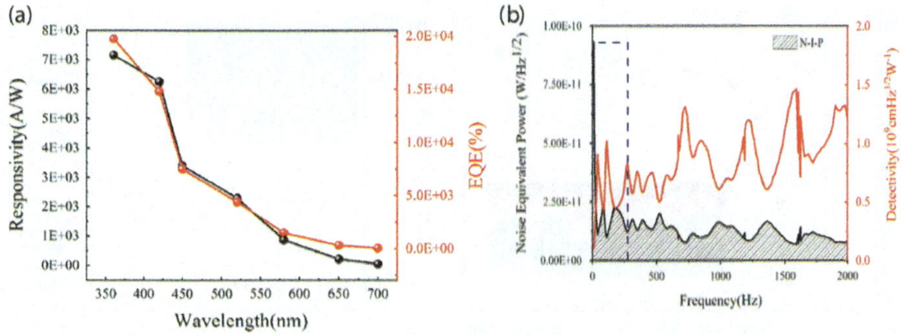


**Fig. 10.13** **a** The responsivity and EQE of the photo-transistor as function of wavelength. **b** Noise equivalent power (NEP) and detectivity spectrum under a dynamic detecting environment

So far, except for the large-scale fabrication processes, to fabricate a perovskite QDs-based phototransistors with nano-structure is also a focused research topic to boost the intrinsic electrical and optical performance. The analysis on optimizing the phototransistor mainly includes two respects: analyzing in theory and verifying by experiment (giving a detailed introduction in last two part of this chapter).

### *10.5.4 Origins' Exploration of the Optimal Phototransistors' Performances*

After above-mentioned verification, perovskite QDs have fundamentally the application potential in phototransistor and high performance detectors with optimizing. In improving the photoelectric characteristics of vertical P-I-N heterojunction phototransistors, the influence of the 2-D atomic component's nano-structure (P-type GNR) in Fig. 10.14a has been checked to verify the correlation of the averaged normalized electric field averages calculated at the edges. It can be confirmed that the wider gap results in a higher electric field at the edge of the GNR. When the gap exceeds 500 nm, the electric field tends to extreme value and enters a saturated state. However, in the experiment, the photoelectricity with 500 nm (duty cycle: width/gap = 0.6) nanoribbon exhibits the highest photosensitivity and photocurrent as shown in Fig. 10.14a.

According to previous literature on graphene/QD phototransistors, it proves that with the enhancement of the fringe electric field, the optimized 500 nm gap device shows high photocurrent with optimized device mobility. As presented in Fig. 10.14b, from the transfer characteristic curve of $V_D$, it can be known that at low $V_D = 0.3$ and 0.6 V, the on/off ratio of the transistor can reach $10^3$. Meanwhile, perovskite QDs' P-I-N nano heterostructure transistors can obtain excellent electrical performance.

It is worth noting that the operating speed of CMOS circuits depends on parasitic capacitance in digital electronics. Consequently, except for the charges' flow, the

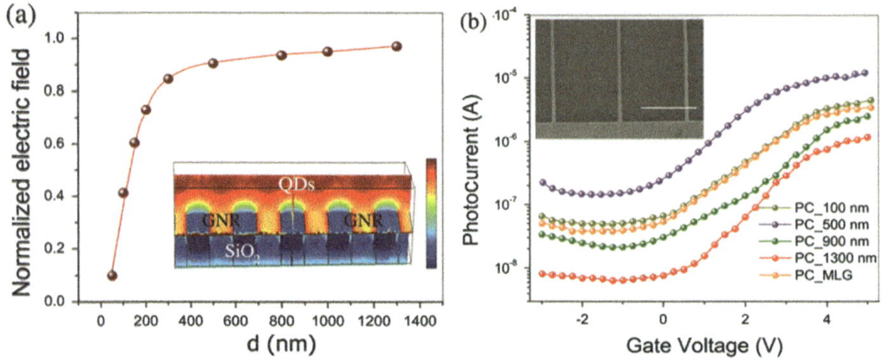


**Fig. 10.14** **a** Simulated 3D electric field of devices with different widths and the electric field distribution of the 500 nm GNR, shown in the inset. **b** Photocurrent of phototransistors illuminated by 512 nm incident light (0.1 mW cm$^{-2}$) with different GNR gaps and illustrated SEM image of GNR with 1300 nm gap (scale bar: 1 μm)

P-I-N heterostructure's junction capacitance and charges' distribution have been identified with the following model. In this semiconductor heterostructure, the area where the barrier height is a spatially varying (depleted) region can be modulated by an applied voltage bias, and carriers are extracted from the depleted region. In our drain-source structure, the model of capacitance is the abrupt junction capacitance of the IGZO n-type and ultra-thin GNR p-type devices shown in Fig. 10.15b. The width of the depletion region can be expressed as: $X_D = X_n + X_p$. The entire semiconductor heterostructure meets the following electrically neutral conditions: $QN_A\ X_p = QN_D\ X_n = Q$, $N_A\ X_p = N_D\ X_n$.

However, due to the P-I-N + abrupt junction characteristics of this heterostructure as shown in Fig. 10.15b, the barrier is close to the side of the lightly doped

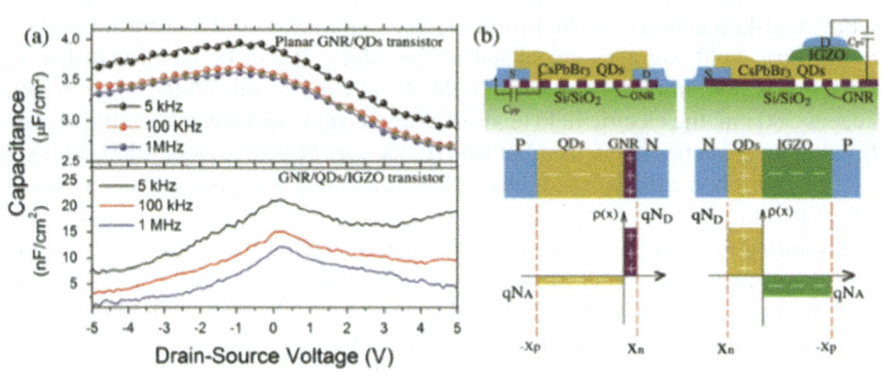


**Fig. 10.15** **a** Parasitic drain/source capacitance C–V measurements for GNR/QD and GNR/QD/IGZO transistors. **b** The structure and carrier concentration of the parasitic capacitance between the drain and source terminals

semiconductor element. The capacitance can be calculated as [14]:

$$C_T = \frac{A\varepsilon_r\varepsilon_0}{X_D},\ X_D = \sqrt{V_D\left(\frac{2\varepsilon_r\varepsilon_0}{q}\right)\left(\frac{N_A + N_D}{N_A N_D}\right)}$$

where A is the area of junction, $\varepsilon_r$ is the dielectric constant of the depletion region, $\varepsilon_0$ is the vacuum dielectric and $X_D$ is the width of the depletion region. Because $N_A \gg N_D$, $X_n \gg X_p$ and $X_D \approx X_n$, the barrier is near the side of the lightly doped semiconductor component. Therefore, the thin $CsPbBr_3$ QDs (30 nm) play more vital roles in the value of the capacitance compared with heavily n-doped IGZO n-type component.

Then, we performed a low-frequency CV test to study the relationship with the detection speed. As shown in Fig. 10.15a, the parasitic capacitance of the planar GNR/QD (peak: 4 $\mu$F $cm^{-2}$) transistor is about 170 times that of the GNR/QD/IGZO vertical transistor (peak: 23 nF $cm^{-2}$). At low frequencies, the carriers in the depletion region can follow the slow AC voltage and produce a higher excess Capacitance. Since this structure can be equivalent to an RC oscillating circuit, the theoretical time constant $\tau$ = RC for GNR/QD planar structure and GNR/QD/IGZO vertical structure can be predicted as 0.17 ms and 20 $\mu$s. This result can be examined through successive experimental findings.

### *10.5.5 Verification and Characterization for Practical Detecting Application*

As investigated in Fig. 10.16a and b, photo-generated electrons in $CsPbBr_3$ quantum dots can be driven to n-type IGZO and holes are transferred to p-type GNR. Due

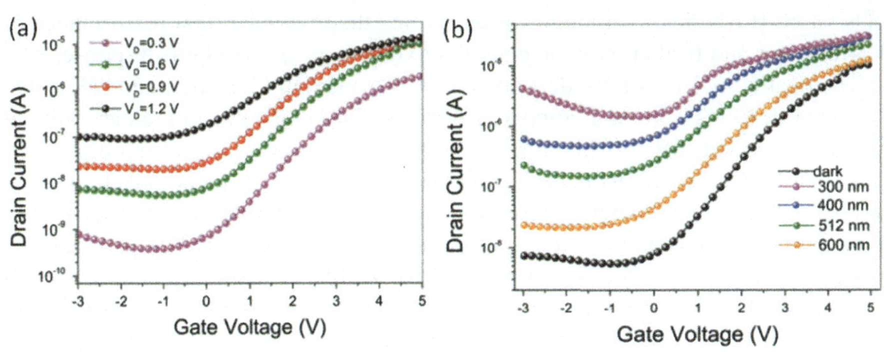


**Fig. 10.16** **a** Transfer characteristics of a phototransistor with different $V_D$. **b** Transfer characteristics of a phototransistor at $V_D$ = 0.6 V under 0.1 mW $cm^{-2}$ illumination of different wavelengths

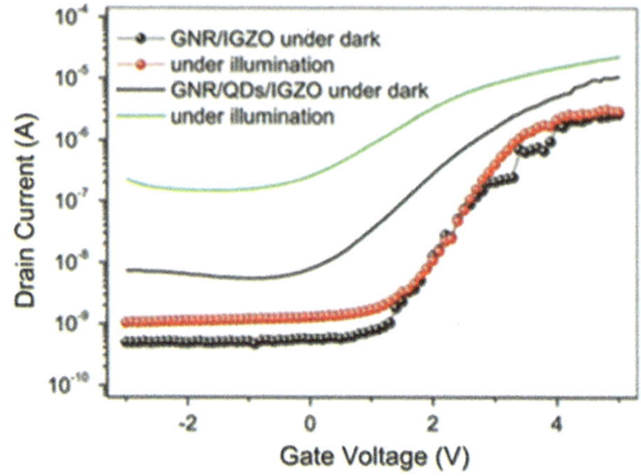


**Fig. 10.17** Transfer character curves of the GNR/IGZO vertical phototransistor with and without perovskite QDs under dark and illumination

to the hole-blocking characteristics of n-type IGZO, holes can be retained in the in QDs/GNR interface. As a result, the electron and hole separation efficiency is improved. In addition, the P-I-N (GNR/perovskite/metal oxide) heterojunction has a significant light leakage current under light. In addition, the intrinsic transistor structure causes the current to self-amplify capability. Finally, as can be seen in Fig. 10.1b, the dark current is very small, which results in high device photosensitivity and detection.

According to devices' comparison of Fig. 10.17, the following strong evidence can be convinced: (1) Since IGZO is a typical wide band gap semiconductor, it has significant absorption in the ultraviolet range. Consequently, QDs brought about an impressive increase in photocurrent under visible 512 nm incident light; (2) The GNR/IGZO or GNR/QD heterojunction can limit the current and achieve a low operating/threshold voltage. The channel between the source/drain can be considered as a photoconductor for conventional hybrid phototransistor, resulting in high current and Operating voltage. Therefore, a phototransistor with a perovskite QDs-based P-I-N heterostructure can guarantee the high electrical performance of the transistor and further gain the potential to obtain excellent photo-response.

The peak photosensitivity depends on the wavelength.Measured and calculated as shown in Fig. 10.5a, and compared with the flat GNR/QD counterpart without IGZO, it is found that this phototransistor has a photosensitivity in the ultraviolet range of 800. The sensitivity in the visible range is as high as 300. The detection rate is another important parameter representing the limit sensitivity of the phototransistor. In particular, we attribute the noise current to shot noise generated by dark current, which is usually randomly affected and hidden Effective photoelectric signal. The left vertical axis of Fig. 10.4b shows the change in detection rate at different wavelengths, which is as high as $6.9 \times 10^{14}$ cm Hz $^{1/2}$ W $^{-1}$ (Jones) in the ultraviolet band, and in the visible band up to $7.5 \times 10^{14}$ Jones. It is worth noting that this vertical heterostructure

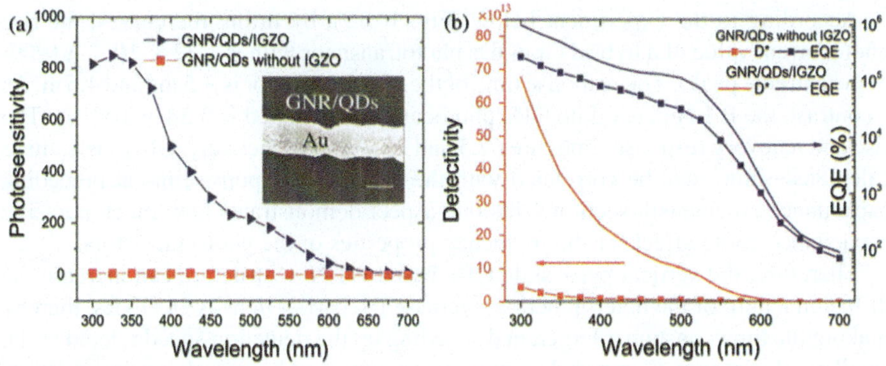


**Fig. 10.18** **a** Comparison of the peak photosensitivity for GNR/QD/IGZO and planar GNR/QD phototransistors ($V_D = 0.6$ V, light power = 0.1 mW $cm^{-2}$) and an inset of the SEM image of the GNR/QD device, scale bar: 1 μm. **b** Comparison of the detectivity and EQE for GNR/QD/IGZO and planar GNR/QD phototransistors

has extremely high sensitivity compared to planar graphene/QD phototransistors (Fig. 10.18).

Generally speaking, compared with the reported easy-fabricated composite phototransistors, this device has very high photosensitivity and detection. There is a trade-off usually between the key optical-electrical parameters for the phototransistor (phototransistors have higher efficiency and responsiveness, while photodiodes have higher sensitivity and detection speed). For graphenebased transistors (graphene is an excellent material for a high frequency transistor), the gate parasitic capacitance and graphene quantum capacitance are the two key parameters for delaying the electrical response time. We speculate that because the photocurrent is caused by the careers' generation and the drift at the source-drain. Thus, the charges' storage influence of the parasitic capacitance of the drain-source is greater for the responding time.

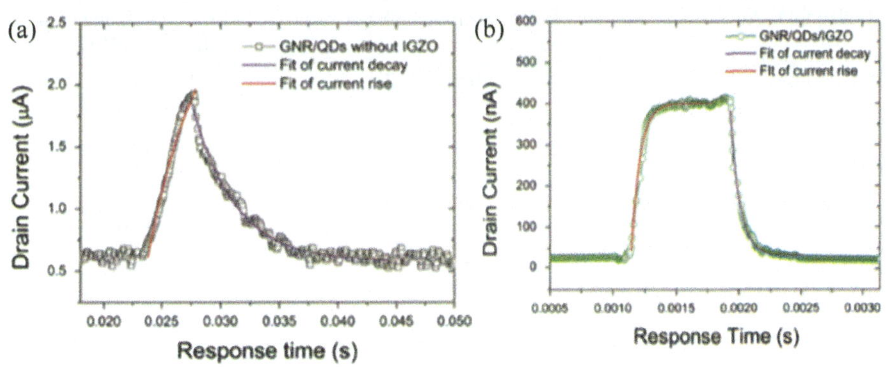


**Fig. 10.19** The drain current's response to the light (512 nm) on/off test for the GNR/QD **a** and GNR/QD/IGZO **b** transistors at $V_D = 0.6$ V, $V_G = 1$ V and mean power = 0.1 mW $cm^{-2}$

According to the experiment below (Fig. 10.19), by fitting the curves, the rise photocurrent value of a hybrid Gra/QDs phototransistor with is $1.32 \times 10^{-6}$ A (90% of the extreme peak). The response time of the phototransistor is 3.5 ms and 4.6 m. As a contrast, the P-I-N perovskite QDs photocurrent calculated is $3.58 \times 10^{-7}$ A. The rise and recovery response times are 67.5 and 73.7 μs, respectively. Obviously, these calculated values can be correlated with the estimated response time at preceding capacitance mechanism section. This one aspect demonstrates how much parasitic capacitance could affect for the frequency properties of the phototransistors.

Therefore, the vertical perovskite QDs based P-I-N + structure can increase the diffusion length of the heterojunction, increase the carrier density in GNRs, thereby making the time constant τ larger, and speeding up the detection speed related to the small overlap region. Compared to other commonly used hybrid-channel transistors, our device can benefit from this fast detection speed, with higher sensitivity and similar efficiency.

In summary, phototransistors based on the perovskite QDs heterojunction can be redesigned and optimized to possess excellent electrical properties and high optical-electrical performances, for instance photosensitivity (800), detection rate ($7.5 \times 10^{14}$ jones), efficiency ($5 \times 10^5$%) and the improved detection speed (~141 μs) is shown in Table 10.1. According to our study, this difference can be attributed to the diffusion loss region of the $CsPbBr_3$ quantum dot light sensing element and intrinsic nano-structures.

**Table 10.1** Optical–electrical performances of the different QD/graphene phototransistors and photodiodes with vertical (V) and planar transistor (P) structures

| Device type | Photo-sensitivity | Decay time | $D^*(10^{13}$ Jones) | EQE (%) | References |
|---|---|---|---|---|---|
| Optimal device | 800 | 74 s | 75 | $5 \times 10^5$ | / |
| P.GNR/Perovskite QDs | 5 | 4.6 ms | 0.5 | $2 \times 10^5$ | / |
| V. NW/QDs/Au | 2 | 80 ms | 0.7 | $3 \times 10^5$ | Zhang et al. [15] |
| P. Gra/provskite | / | 10 ms | 2 | $10^4$ | Kwak et al. [16] |
| P.GNR/perovskite QDs | 1.5 | 3.65 s | 70 | / | Zhao et al. [17] |
| Vertical perovskite diode | >300 | 560 s | 0.08 | 600 | Manga et al. [18] |
| Vertical Gra/ QDs/$TiO_2$ diode | $10^4$ | 400 ns | 6 | 140 | [19] |

## 10.6 Conclusions and Outlook

At present, due to the novel and broad application in flexible, low-cost, and light electronic applications such as flat-panel displays, radio-frequency identification tags (RFID), complementary integrated circuits array (CMOS array) and biological and medical applications, the complexity and large area applications call for an efficient, interactive and transplantable optical-electrical phototransistors. In order to realize the practical application of the perovskite phototransistors and the like, heterojunctions' mechanism and process development are crucial to design an optimal device.

The understanding of their functioning together with the encouraging experimental results conduct to propose several schemes to promote the performances. Interestingly, perovskite QDs both possess the unique material's properties (long diffusion length and quantum efficiency) and the QDs' size-dependent quantum confinement. These figures of merits are perfect to fit the demand of photodetectors, especially for phototransistors. According to our researching, it can reinforce the dynamic frequency performances simultaneously with increasing the conventional optical-electrical conversion efficiency. All these research findings further demonstrate the great potential of perovskite quantum dots in optoelectronics and provide a new candidate for commercial phototransistors.

## References

1. J. Lu, X. Sheng, G. Tong, Z. Yu, X. Sun, L. Yu, X. Xu, J. Wang, J. Xu, Y. Shi, K. Chen, Ultrafast solar-blind ultraviolet detection by inorganic perovskite $CsPbBr_3$ quantum dots radial junction architecture. Adv. Mater. **29**(23), 1700400 (2017)
2. X. Yu, T.J. Marks, A. Facchetti, Metal oxides for optoelectronic applications. Nat. Mater. **15**(4), 383–396 (2016)
3. S. Jeon, S.E. Ahn, I. Song, C.J. Kim, U.I. Chung, E. Lee, I. Yoo, A. Nathan, S. Lee, J. Robertson, K. Kim, Gated three-terminal device architecture to eliminate persistent photoconductivity in oxide semiconductor photosensor arrays. Nat. Mater. **11**(4), 301–305 (2012)
4. F. Li, C. Ma, H. Wang, W. Hu, W. Yu, A.D. Sheikh, T. Wu, Ambipolar solution-processed hybrid perovskite phototransistors. Nat. Commun. **6**, 8238 (2015)
5. X. Liu, C.-L. Li, T.-F. Dai, Z. Tao, W.-X. Zhou, W. Lei, J. Chang, Infrared phototransistor induced by MoS2 quantum dots encapsulated in lead iodide perovskite. IEEE Elec. Dev. Lett. **40**(5), 746–749 (2019)
6. L. Qian, Y. Sun, M. Wu, D. Xie, L. Ding, G. Shi, A Solution-processed high-performance phototransistor based on a perovskite composite with chemically modified graphenes. Adv. Mater. **29**(22), 1606175 (2017)
7. M.H. Kuo, W.T. Lai, T.M. Hsu, Y.C. Chen, C.W. Chang, W.H. Chang, P.W. Li, Designer germanium quantum dot phototransistor for near infrared optical detection and amplification. Nanotech **26**(5), 055203 (2015)
8. M. Meyns, M. Perálvarez, A. Heuer-Jungemann, W. Hertog, M. Ibánez, R. Nafria, A. Genç, J. Arbiol, M.V. Kovalenko, J. Carreras, Polymer-Enhanced stability of inorganic perovskite nanocrystals and their application in color conversion LEDs. Acs Appl. Mater. Inter. **8**(30), 19579–19586 (2016)

9. X. Liu, Z. Tao, W. Kuang, Q.Q. Huang, Q. Li, J. Chen, W. Lei, Dual-gate phototransistor with perovskite quantum dots-PMMA photosensing nanocomposite insulator. IEEE Elec. Dev. Lett. **38**(6), 1270–1273 (2017)
10. J.W. Lee, D.Y. Kim, F. So, Unraveling the gain mechanism in high performance solution-processed PbS infrared PIN photodiodes. Adv. Func. Mater. **25**(8), 1233–1238 (2015)
11. X. Liu, W. Kuang, H. Ni, Q.Q. Huang, J. Chen, Q. Liu, J. Chang, W. Lei, A highly sensitive and fast graphene nanoribbon/$CsPbBr_3$ quantum dot phototransistor with enhanced vertical metal oxide heterostructures. Nanoscale **10**(21), 10182–10189 (2018)
12. Controlling photoinduced electron transfer from PbS@CdS core@shell quantum dots to metal oxide nanostructured thin films. Nanoscale **6**(12), 7004 (2014)
13. L. Shi, L. Meng, F. Jiang, Y. Ge, F. Li, X.-g. Wu, H. Zhong, In situ inkjet printing strategy for fabricating perovskite quantum dot patterns. Adv. Func. Mater. **29**(37), 1903648.1–1903648.6 (2019)
14. A. Singh, M.A. Uddin, T. Sudarshan, G. Koley, Tunable reverse-biased graphene/silicon heterojunction schottky diode sensor. Small **10**(8), 1555–1565 (2014)
15. Y. Zhang, M. Cao, X. Song, J. Wang, Y. Che, H. Dai, X. Ding, G. Zhang, J. Yao, Multi-heterojunction phototransistors based on graphene-pbse quantum dot hybrids. J. Phys. Chem. C. **119**(37), 21739–21743 (2015)
16. D.-H. Kwak, D.-H. Lim, H.S. Ra, P. Ramasamy, J.S. Lee, High performance hybrid graphene–$CsPbBr_3-xIx$ perovskite nanocrystal photodetector. Rsc Adva. **6**(69), 65252–65256 (2016)
17. F. Zhao, K. Xu, X. Luo, Y. Liang, Y. Peng, F. Lu, Toward high uniformity of photoresponse broadband hybrid organic-inorganic photodiode based on PVP-modified perovskite. Adv. Opt. Mater. **6**(1), 1700509 (2018)
18. K.K. Manga, J. Wang, M. Lin, J. Zhang, M. Nesladek, V. Nalla, W. Ji, K.P. Loh, High-performance broadband photodetector using solution-processible pbse–tio2–graphene hybrids. Adv. Mater. **24**(13), 1697–1702 (2012)

# Chapter 11
# Perovskite Quantum Dots Based Lasing-Prospects and Challenges

**Yue Wang and Siyang Xia**

**Abstract** Since the first report of stimulated emission (SE) and lasing action from colloidal perovskite quantum dots (Pe-QDs) in 2015, Pe-QDs have made great strides in constructing high-performance optically pumped lasers. By virtue of the quantum confinement effect and merits of halide perovskites, the Pe-QDs hold the promise for developing practical and cost-effective lasers based on optical pumping and even electrical injection. In this Chapter, the status and prospects of Pe-QD lasers are introduced. First of all, we present the basic photophysics of Pe-QDs that benefits light emission and SE. Then, the underlying gain mechanisms are discussed. We comprehensively introduce the emerging kinds of Pe-QD lasers based on various high-quality optical resonators. At last, we point out the existing challenges toward the development of high-performance lasers utilizing the Pe-QDs and research trends of Pe-QD lasers in the future.

## Abbreviations

| | |
|---|---|
| 1 PA | One-photon absorption |
| 2 PA | Two-photon absorption |
| ASE | Amplified spontaneous emission |
| BnOH | Benzyl alcohol |
| CB | Conduction band |
| CBM | Conduction band minimum |
| CIE | Commission Internationale de l'Eclairage |
| CLC | Cholesteric liquid crystal |
| CW | Continuous-wave |
| DA | 2-hexyldecanoic acid |
| DBR | Distributed bragg reflector |
| DFB | Distributed feedback |

Y. Wang (✉) · S. Xia
College of Materials Science and Engineering, Nanjing University of Science and Technology, Nanjing 210094, China
e-mail: ywang@njust.edu.cn

 
Y. Zhou and Y. Wang (eds.), *Perovskite Quantum Dots*, Springer Series in Materials Science 303, https://doi.org/10.1007/978-981-15-6637-0_11

| | |
|---|---|
| FA | Formamidinium |
| fs | Femtosecond |
| FWHM | Full-width at half maximum |
| LC | Liquid crystal |
| LEDs | Light emitting diodes |
| MA | Methylammonium |
| NTSC | North American national television system committee |
| ns | Nanosecond |
| PA | Photoinduced absorption |
| PAN | Polyacrylonitrile |
| PB | Photobleaching |
| Pe-QDs | Perovskite quantum dots |
| PL | Photoluminescence |
| PL QY | Photoluminescence quantum yield |
| PMMA | Poly (methyl methacrylate) |
| OA | Oleic acid |
| OAm | oleylamine |
| QD | Quantum dot |
| QDs | Quantum dots |
| Q factor | Quality factor |
| SE | Stimulated emission |
| TA | Transient absorption |
| TEM | Transmission electron microscopy |
| TR PL | Time-resolved photoluminescence |
| VCSEL | Vertical cavity surface emitting laser |
| VB | Valence band |
| VBM | Valence band maximum |
| WGM | Whispering-gallery-mode |

## 11.1 Introduction

Quantum dots (QDs) or nanocrystals represent the ultimate nanostructured materials which exhibits the quantum confinement effect in three-dimension (3D). The bandgap of QDs has been tuned through size control to yield the emission spectra spanning across the entire visible and near-infrared-region [1]. Thanks to the unique electronic structure and solution processability, QDs have been recognized as the promising lasing materials, which may serve as the supplement to the technologically mature epitaxial semiconductors. In the past few years, perovskite quantum dots (Pe-QDs), endowed with excellent photonic and optoelectronic properties, have been demonstrated as promising candidate as optical gain media. These Pe-QDs own the advantages of both QDs and halide perovskites, such as the solution-processability, bandgap tunability, large exciton binding energy and unusual defect-tolerant nature [2]. Compared with the traditional Cd-chalcogenide-based QDs,

Pe-QDs were found to show much lower stimulated emission (SE) thresholds in the range of $\mu J\ cm^{-2}$ [3, 4]. Especially, the all-inorganic compositions feature enhanced photo- and thermo-stability by virtue of the intrinsic higher melting and decomposition points compared with the hybrid organic-inorganic counterparts. Since the first SE behavior observed in the year of 2015 [5], great advances have been made in developing high-performance Pe-QDs-based lasers. The gain origin accounting for lasing in Pe-QDs has been intensively investigated and different mechanisms have been put forward. Moreover, as a vital branch of nonlinear optics, multiphoton pumped SE and lasers in Pe-QDs have been developed. However, from the point view of better understanding the photophysics and improving the device performance, several issues and challenges remain which shall be addressed in order to develop the commercially available Pe-QDs-based lasers.

In this chapter, a comprehensive presentation of Pe-QDs-based lasing including the all-inorganic and hybrid perovskites will start from the introduction of photophysical processes in Pe-QDs; In Sect. 11.2, we will discuss the underlying gain mechanisms as well as multiphoton absorption and multiphoton pumped lasers based on the Pe-QDs. Then, Pe-QDs-based lasers with various high-quality optical resonators are discussed in detail in Sect. 11.3. In Sect. 11.4, we list the existing issues that hinder the development of high-performance lasers utilizing the Pe-QDs. At last, research trend of Pe-QD lasers in the future is proposed.

## 11.2 Photophysics of Perovskite Quantum Dots

### *11.2.1 Crystal and Electronic Band Structure*

Optical properties of semiconductors are in close relationship with the intrinsic crystal and electronic structures. For the Pe-QDs, the chemical structure can be represented by a general formula of $ABX_3$. Regarding the A-site cation, it can be divided into two kinds of Pe-QDs: (1) the all-inorganic cesium halide Pe-QDs; (2) the organic-inorganic hybrid halide Pe-QDs with A-site of $CH_3NH_3$ (MA) or $CH(NH_2)_2$ (FA); B is a divalent metal cation (typically $Pb^{2+}$ or $Sn^{2+}$) and X is halide anion (Cl, Br and I). It is noted that the newly emerging lead-free double Pe-QDs have recently attracted research interest, featuring the chemical formula of $A_2BB'X_6$. Generally, A is monovalent cations $Cs^+$, $MA^+$, B is +1 cations $Ag^+$, $Au^+$, $Cu^+$, B' is +3 cations $Bi^{3+}$, $Sb^{3+}$, and X is the halogen anion $Cl^-$, $Br^-$, $I^-$ (see the typical perovskite crystal structures in Fig. 11.1).

As the representative inorganic perovskite, $CsPbX_3$ perovskites maintain the pure cubic space group (*Pm3m*) at high temperatures, where $[PbX_6]^{4-}$ octahedrons are interconnected in three-dimensions by corner-sharing (Fig. 11.1). With the decreased temperature, the $[PbX_6]^{4-}$ octahedron tends to tilt and the pure cubic phase will transform into the tetragonal and further to the orthorhombic phase, showing the distorted perovskite structures. Table 11.1 summarizes the typical phase-transition temper-

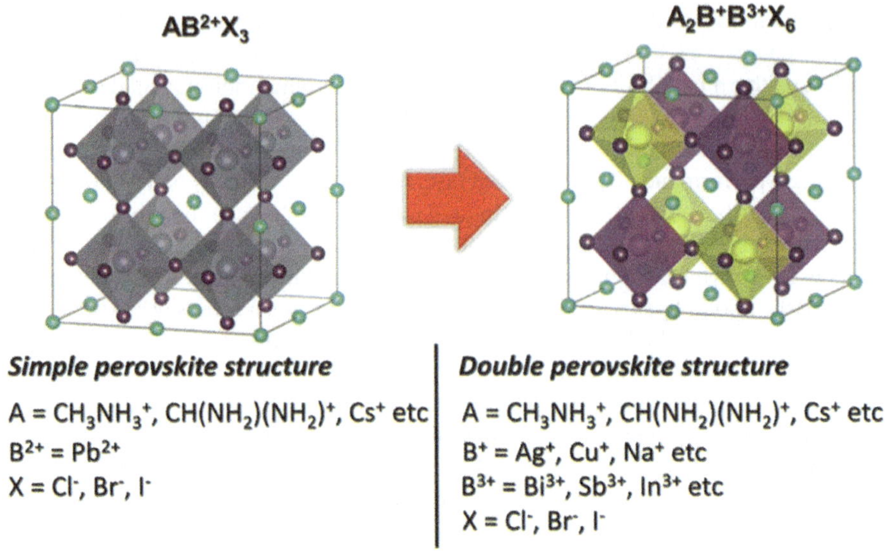


**Fig. 11.1** Representative schematic illustration of the structures of both normal and double perovskites, the difference is lying in the alternating elements arrangement in the B-site in the octahedral cavity (i.e., $2Pb^{2+} \rightarrow B^{+1} + B^{3+}$). Reprinted with permission from Ref. [6]. Copyright 2019, Published by John Wiley & Sons

**Table 11.1** The summary of the typical phase-transition temperatures for perovskite $ABX_3$ bulk crystals

| Active perovskites/References | Cubic phase (K) | Tetragonal phase (K) | Orthorhombic phase (K) |
|---|---|---|---|
| $CsPbI_3$ [18] | >578 | | |
| $CsPbBr_3$ [18] | >403 | 373–403 | <373 |
| $CsPbCl_3$ [18] | >320 | 315–320 | <315 |
| $MAPbI_3$ [9] | >327.4 | 162.2–327.4 | <162.2 |
| $MAPbBr_3$ [9] | >236.9 | 155.1–236.9 (I4/mcm)<br>149.5–155.1 (P4/mmm) | <144.5 |
| $MAPbCl_3$ [9] | >178.8 | 172.9–178.8 | <172.9 |
| $CsSnI_3$ [8] | >478 | 350–478 | <300 |
| $CsSnBr_3$ [8] | Cubic phase with no subsequent phase-transition | | |
| $CsSnCl_3$ [8] | >390 | 373–390 | <373 |

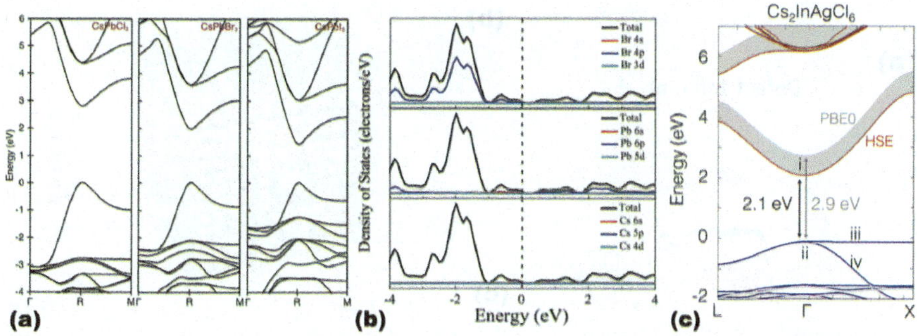


**Fig. 11.2** Electronic band structures of both normal and double perovskites. **a** Calculated electronic band structures for cubic $CsPbX_3$ perovskites based on the density functional theory and considering the relativistic corrections and spin-orbit interactions, respectively. Reprinted with permission from Ref. [2]. Copyright 2015, Published by American Chemical Society. **b** Density of states calculation in cubic-phased $CsPbBr_3$ to extract different elemental contributions to the energy band. Reprinted with permission from Ref. [11]. Copyright 2016, Published by American Chemical Society. **c** Band structures and bandgap demonstration of $Cs_2InAgCl_6$ based on the hybrid functionals. Reprinted with permission from Ref. [13]. Copyright 2017, Published by American Chemical Society

atures for perovskite $ABX_3$ bulk crystals [7–9]. The phase transition can significantly modify the emission properties. For example, the cubic or black phase $CsPbI_3$ (>578 K) exhibit bright emission with small bandgap ($E_g$), while orthorhombic $CsPbI_3$ shows a yellow phase with non-luminescent nature and larger $E_g$ [2]. When the crystal phase transforms from the tetragonal and further into the orthorhombic polymorphs, the crystal symmetry becomes lower compared with the cubic one and hence, the bandgap and the effective carrier masses are increased. The small effective carrier masses are beneficial to the high carrier mobilities (~4500 $cm^2\ V^{-1}\ s^{-1}$) [10].

The overall electronic structures of different $CsPbX_3$ (X = Cl, Br and I) are nearly independent of the X-site halogen compositions, except for the $E_g$ values as shown in Fig. 11.2a [11]. The calculated electronic structures of cubic $CsPbX_3$ are based on density functional theory and considering relativistic corrections and spin-orbit interactions influencing the conduction band minimum. All of three cubic-phase $CsPbCl_3$, $CsPbBr_3$, $CsPbI_3$ feature a direct bandgap character. The direct bandgap is a key ingredient for achieving efficient luminescence and high optical gain [2, 12]. The band structure of double perovskite $Cs_2InAgCl_6$ is shown in Fig. 11.2c, the most striking finding is that the band transition is also direct, in which both valence band maximum (VBM) and conduction band minimum (CBM) locate at the center of the Brillouin zone (Γ point) [13].

The nature of the VBM and CBM will have a deep influence on (i) the charge transfer and (ii) the optical transitions, which are related to the optoelectronic applications. The $CsPbX_3$ perovskites is known to possess an inverted band structure, as depicted in Fig. 11.3a, where the VBM is characterized by the antibonding hybridization ($\sigma^*$) between the Pb 6 s and X np orbitals (Cl: n = 3, Br: n = 4, I: n = 5). On the other hand, the CBM state is mostly governed by mutual interactions between

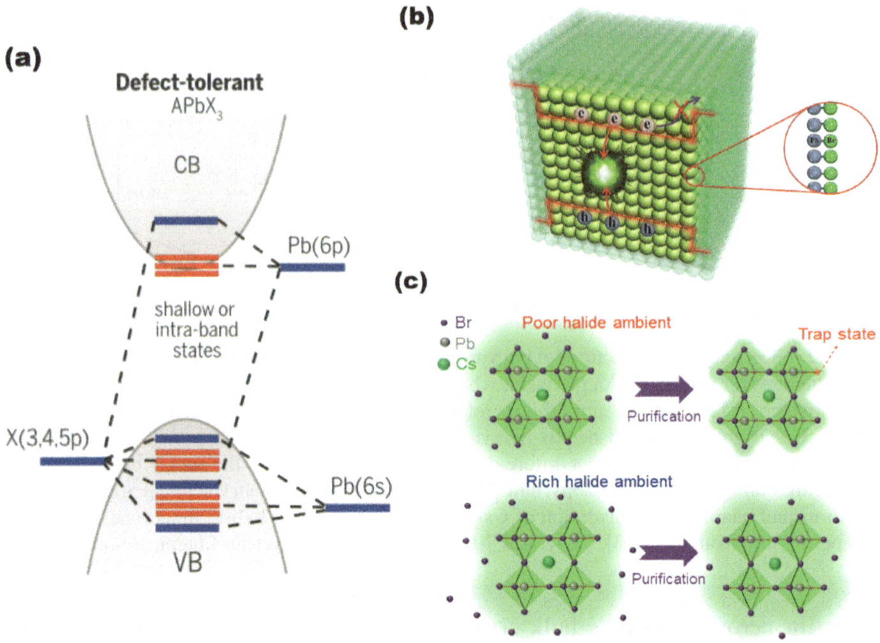


**Fig. 11.3** Defect-tolerant nature and surface passivation effect in lead halide perovskites. **a** Electronic structures of lead halide perovskites with defect-tolerant nature. Reprinted with permission from Ref. [19]. Copyright 2017, Published by Science. **b** Schematic illustration of self-passivating strategy with abundant Br atoms. Reprinted with permission from Ref. [11]. Copyright 2016, Published by John Wiley & Sons. **c** Enhanced PL QY in $CsPbBr_3$ QDs under halide rich circumstance. Reprinted with permission from Ref. [23]. Copyright 2017, Published by American Chemical Society

Pb 6p and X np orbitals confined in the $[PbX_6]^{4-}$ octahedral structures [11, 14]. See the Fig. 11.2b for the density of states of the cubic-phase $CsPbBr_3$. When the halide compositions in X-site changing from chlorine, to bromine and iodine (Fig. 11.2a), the VBM of perovskites moves upwards with the halide np orbits, while the CBM remains nearly unchanged [12]. It indicates that the emission wavelength can tuned by simply changing the stoichiometry of halogen components. Theoretically, the A-site cation is less contributive to the electronic structure of band edge because the electronic contributions of the A-site cation locate several eV (electron volts) below the VBM. With the decreased cation radius at the A-site, the perovskite lattice will shrink and resulting in an increase in the $E_g$ [15]. When replacing lead ion with tin at the B-site, the CBM of perovskites decreases more than the VBM. Hence, the tin-based perovskites have a narrower bandgap than the lead perovskite analog, indicating the promising applications in narrow-band optoelectronics [16]. For example, Xing et al. successfully demonstrated solution-processed $CsSnX_3$ with favorable optical gain properties in lasing regime with the emission wavelength extendable to the near-infrared-region [17].

## 11.2.2 Optical Properties

### 11.2.2.1 High Photoluminescence Quantum Yield

The solution-processable Pe-QDs exhibit ultrahigh photoluminescence quantum yield (PL QY) (nearly 95% for $CsPbBr_3$) with narrow line width, rendering them as promising candidates for photonic applications. In comparison to the tradition Cd-chalcogenide-based QDs, the Pe-QDs can maintain high PL QY with no deliberate surface passivation [2]. It is demonstrated that the halide perovskites feature a highly defect-tolerance thanks to its unique electronic structure, where most of the intrinsic defects residing within the valence band (VB) and conduction band (CB) [19–21]. This characteristic is also validated in the high photophysical quality of hybrid organic-inorganic perovskites ($MAPbX_3$) [5, 22].

Moreover, during the synthesis of Pe-QDs, the additive surfactants can help to improve the dispersion stability and modify the crystal growth kinetics [14]. Although the peculiar defect-tolerant nature of lead halide perovskites potentially enables the high PL QY. It is found that surface passivation is essential in order to achieve bright emission. Inspired by the intentional surface passivation of traditional QDs, Li et al. reported a self-passivating strategy with abundant Br atoms. As depicted in Fig. 11.3b, rich Br on QDs surface tends to combine with Pb atoms to form $PbBr_x$ analogues with larger bandgap. In addition to suppressing the formation of halogen vacancies, the type-I core-shell-structured band alignment can localize the electron and hole wavefunctions in the interior of the nanocrystal, thus inhibiting the trapping of excited carriers and resulting in high PL QY [11]. Self-passivating strategy with abundant halide atoms is also validated by Liu et al., who offer a halide rich circumstance in virtue of $NH_4Br$ and PbO to substitute for conventional $PbBr_2$ (Fig. 11.3c). The similar case is also valid for the organic-inorganic hybrid Pe-QDs.

## 11.2.3 Carrier Dynamics

Carrier dynamics, especially the hot carrier cooling and multi-carrier recombination, are closely related to the gain and lasing performance of QDs. Before the discussion of carrier dynamics in Pe-QDs, the fundamentals of dynamics characterization by transient absorption is introduced.

### 11.2.3.1 Transient Absorption Signal and Photoinduced Processes

The operation process of transient absorption (TA) spectrometer includes the following steps. A pump laser is used to excite the sample, and then, a time-delayed broad-band probe pulse interrogates the photoinduced dynamic changes, such as hot carrier cooling [24], carriers trapping [25], exciton and multiexciton recombination [26], etc. With the pump on $\left(I(\lambda)_{pump-on}\right)$ or off $\left(I(\lambda)_{pump-off}\right)$, the detector measures the intensity of the probe pulse, separately. There are generally three kinds of signals (marked as ***b***, ***c*** and ***d*** upon photoexcitation ***a*** in Fig. 11.4) in the obtained absorbance difference spectrum $(\Delta A(\lambda) = -\log(I(\lambda)_{pump-on}/I(\lambda)_{pump-off})$. The process of ***a*** represents the hot carriers generation by pumping, the processes of ***b***, ***c*** and ***d*** denote the photobleaching (PB, $\Delta A < 0$), stimulated emission (SE, $\Delta A < 0$) and photoinduced absorption (PA, $\Delta A > 0$), respectively. PB describes the ground state bleach due to state filling upon photoexcition. PA represents the transition from newly occupied excited states to the higher energy levels [27, 28].

The typical TA spectra of $MAPbI_3$ over different delay times are shown in Fig. 11.5b. The two concave parts with $\Delta A < 0$ are marked as PB1 (480 nm) and PB2 (760 nm) [27]. The formation of the strong photobleaching signal (PB2) has generally reached a consensus of being optical transition between band-edges. However, there are still several different claims about the cause of the weak signal (PB1): the existence of a double valence band [29], the existence of a double valence band and double conduction band [24], effect of the presence of $PbI_2$ impurities [30], and a dual nature of excited state induced by the formation of plumbate complexes formed by $Pb^{2+}$ in the presence of excess iodide ions [31]. The above four cases are represented in Fig. 11.5c to 11.5f, respectively. The wavelength range of 550 to 650 nm ($\Delta A > 0$) is recorded as PA signal explained by photoinduced reflection differences caused by refractive index changes [32].

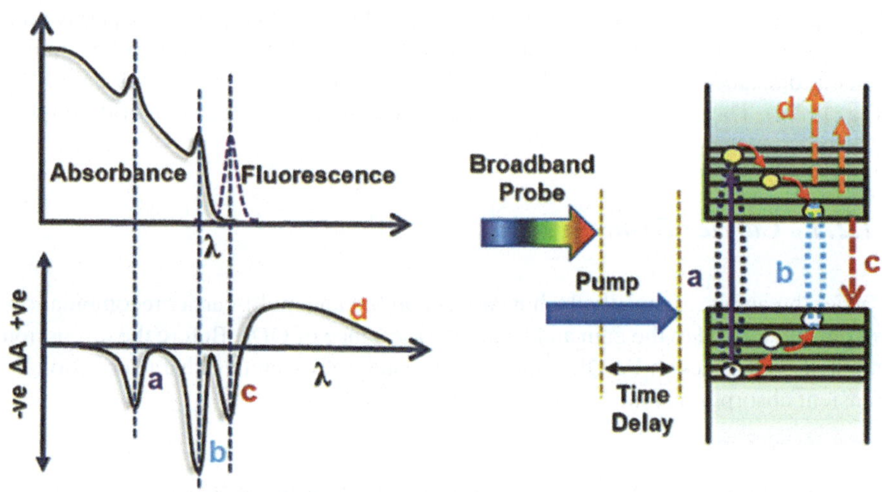


**Fig. 11.4** Steady-state absorption and Photoluminescence (PL) spectra, a fundamental TA signal (left panels). Schematic diagram of different photoinduced processes corresponding to the left TA signal (right panel), a, b, c, and d marks represent the generation of hot carriers, photobleaching, stimulated emission, and photoinduced absorption in order. Reprinted with permission from Ref. [28]. Copyright 2019, Published by Royal Society of Chemistry

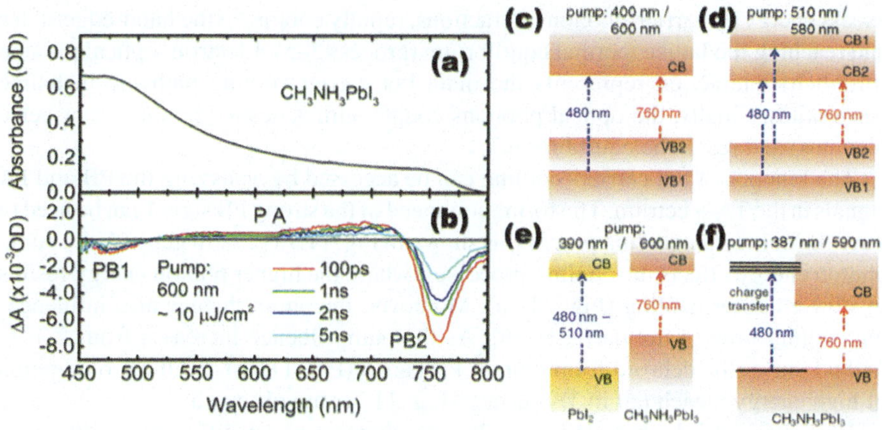


**Fig. 11.5** A representative case of $CH_3NH_3PbI_3$: **a** Linear absorption spectrum with two absorption edges at 480 and 760 nm. **b** TA spectra involving PB and PA signals at different probe delay times. **c** Double valence band model. **d** Double valence band and double conduction band model. **e** and **f** Overlapping signals induced by $PbI_2$ or plumbate complexes. Reproduced with permission from Ref. [27]. Copyright 2016, Published by American Chemical Society

#### 11.2.3.2 Carrier Dynamics of Pe-QDs

Figure 11.6 depicts the typical dynamic process of carrier generation and relaxation in semiconductors [33]. Photons with energies higher than the bandgap energy create electrons and holes that attain higher energy states (process 1). Then, a process called carrier thermalization (process 2) that results in carriers non-equilibrium distribution subject to Fermi-Dirac distribution with carrier temperature, $T_c$ [33]. It should be noted that, for carriers whose energy is far greater than the Fermi energy, it follows the Maxwell-Boltzmann (M-B) statistics [34, 35]. Next, the hot carriers consume

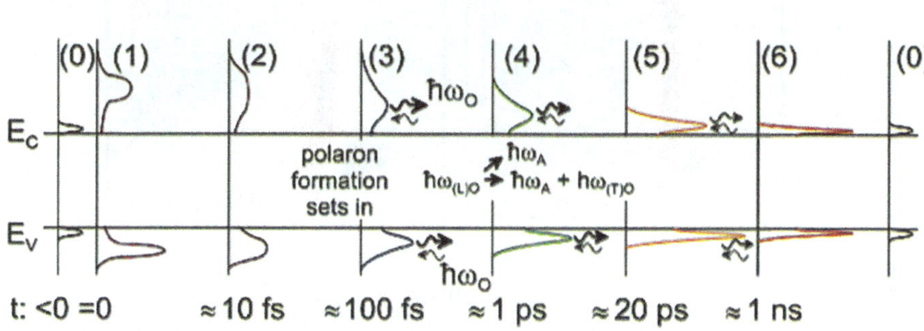


**Fig. 11.6** Schematic diagram of hot carrier relaxation kinetics. (0) thermal equilibrium; (1) optical excitation; (2) carrier thermalization; (3) carrier-optical phonon interaction (optical phonon emission); (4) optical phonons decay into acoustic phonons; (5) acoustic phonon emission; (6) initiation of carrier recombination. Reprinted with permission from Ref. [33]. Copyright 2010, Published by Elsevier

excess energy by carrier-phonon interactions, rapidly cooling to the band-edge states and reaching the lattice thermal equilibrium (process 3–5). Electron–optical phonon (Fröhlich) interaction represents the main hot carrier cooling pathway in halide perovskites. Finally, the optical phonons couple with acoustic phonons (transverse phonons) and reach the equilibrium.

The kinetics of hot carriers cooling can be accessed by analyzing the PB and PA signals in the TA spectrum. The formation speed of the strong PB signal can be used to trace the hot carrier cooling process, as shown in Fig. 11.7a [25]. In general, the pump photon energy affects the cooling processes, where the higher photon energy results in slower carrier cooling (Fig. 11.7a). Moreover, the pump fluence also influences the cooling time of hot carriers [36]. As the pump fluence increases from 2.3 to 340 μJ cm$^{-2}$, the delayed formation of PB signal (Fig. 11.7b) and the prolongation of high energy bleach tail in TA surface (Fig. 11.7c) are observed.

Understanding the hot carrier cooling mechanism under different conditions is essential to improve light emission and amplification from Pe-QDs [36]. In this

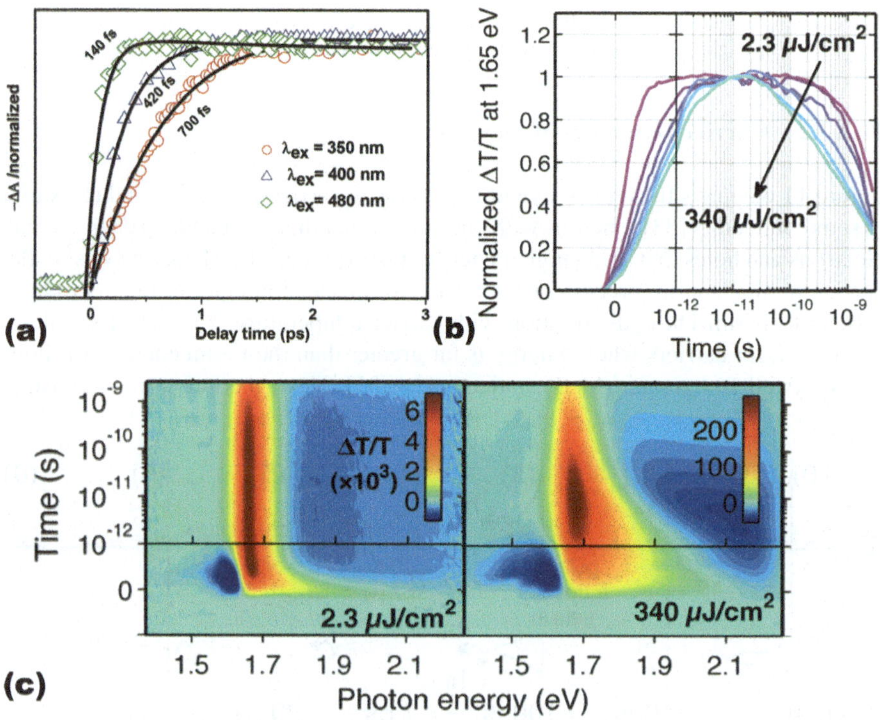


**Fig. 11.7** **a** Formation kinetics of the PB of $CsPbBr_3$ NCs at different excitation wavelengths. Reprinted with permission from Ref. [25]. Copyright 2017, Published by Royal Society of Chemistry. For $MAPbI_{3-x}Cl_x$, under 400 nm (3.1 eV) excitation, **b** PB kinetics over various pump fluences. **c** TA surfaces at two pump fluences. Reproduced with permission from ref. [36]. Copyright 2015, Published by American Chemical Society

regard, different mechanisms have been proposed to account for the observed phenomenon. For example, the hot-phonon bottleneck effect is presented to interpret the slow cooling of hot carriers in perovskites under high excitation powers [34, 37, 38], acoustical-optical phonon up-conversion [39] and large polaron screening [40] are used to rationalize the prolonging of hot carriers relaxation in perovskites.

#### 11.2.3.3 Multi-Carrier Recombination Dynamics

It can be seen from the TA spectrum that a fast decay component of PB becomes evident with the increase of excitation densities (Fig. 11.8a). By subtracting the decay process corresponding to the single exciton recombination under low excitation intensities, the fast lifetime component of ~20 ps can be derived (Fig. 11.8b) [41]. Wang et al. attributed this fast decay component to nonradiative biexciton Auger recombination [4]. Makarov et al. also demonstrated and extended this conclusion to universal $CsPbX_3$ nanocrystals [26]. However, Butkus et al. argued that the appearance of the fast decay component is not necessarily due to the formation of biexcitons because the recombination of multiple free unbound charges confined in the nanocrystals will also lead to the emergence of fast decay component [41]. On the other hand, Mondal et al. used the similar treatment to verify that the 29 ps decay constant in the exciton dynamics of $CsPb(Cl_{0.20}Br_{0.80})_3$ NCs is ascribed to the biexciton Auger recombination [42]. They also pointed out that the asymmetric derivative peak feature in the subpicosecond delay interval of TA spectrum results from the biexciton state, which explains the observation of the peak shift phenomenon (Fig. 11.9a) [26, 43]. With the increase of delay time, the hot carrier relaxes to the low energy states, resulting in bleaching signal caused by state filling. Figure 11.9b depicts the effect of exciton states on TA dynamics where a clear biexciton redshift is observed.

Moreover, Castañeda et al. systematically studied the size and composition dependence of the biexciton interaction in $CsPbX_3$ QDs [44]. They find that the biexciton

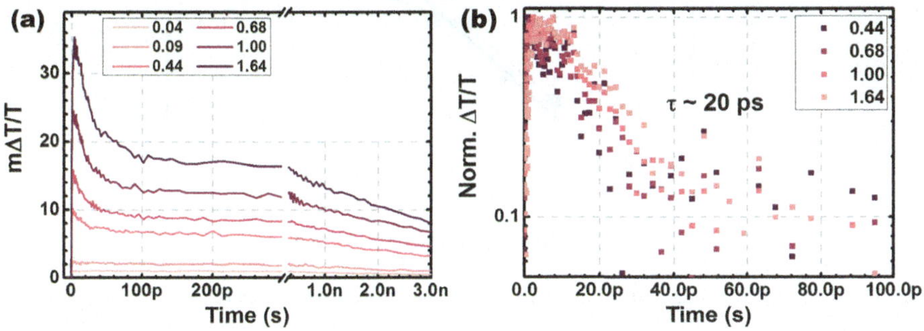


**Fig. 11.8** **a** TA kinetics at different excitation densities for NC with 4.1 nm edge length. **b** Biexciton kinetics dependence of excitation density obtained by subtracting the contribution of single-exciton for NC with 4.1 nm edge length. Reproduced with permission from ref. [41]. Copyright 2017, Published by American Chemical Society

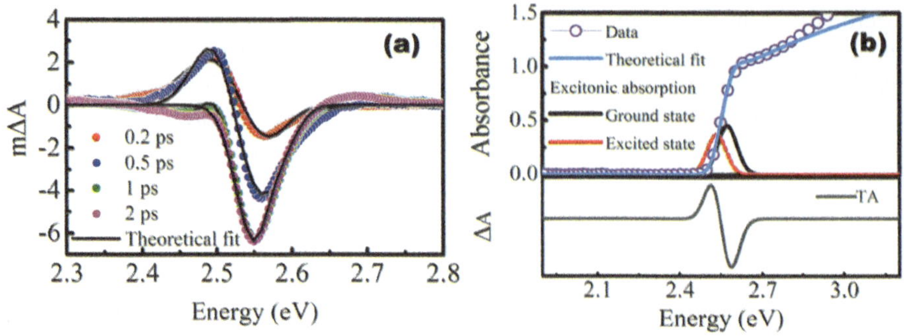


**Fig. 11.9** **a** TA spectrum of CsPb $(Cl_{0.20}Br_{0.80})_3$ NCs over different delay times at $1.8 \times 10^{16}$ $cm^{-3}$ carrier density. Black solid line is fitted by TA model formula. **b** Influence of exciton states on TA spectrum. Reproduced with permission from ref. [42]. Copyright 2018, Published by American Physical Society

lifetime depends linearly on the QD volume for $CsPbI_3$ and small-sized $CsPbBr_3$ QD (in the strong confinement regime) (Fig. 11.10). When the edge length is greater than the exciton Bohr radius ($\alpha_{B,Br} \cong 7.1$ nm), locating at the weak confinement regime, the previously linear dependence of $CsPbBr_3$ QD becomes sublinear dependence. Moreover, the Auger rate of $CsPbBr_3$ QD becomes larger than that in $CsPbI_3$ QD. That may be due to that the iodized QDs have a larger Bohr radius ($\alpha_{B,I} \cong 12$ nm), resulting in a weaker quantum confinement effect at similar crystal size with respect

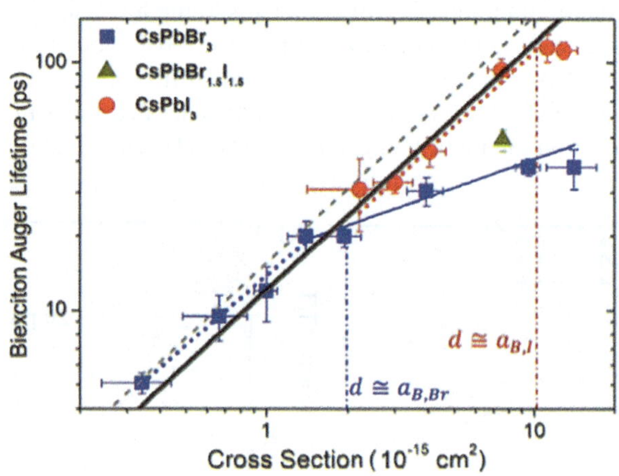


**Fig. 11.10** Biexciton Auger lifetime dependence of the absorption cross section at 3.1 eV of $CsPbX_3$ QDs. The solid black line represents linear dependence between Biexciton Auger lifetime and absorption cross section. The dashed blue and red lines respectively represent the nearly linear dependence of small size $CsPbBr_3$ and $CsPbI_3$ QDs. The solid blue line is the sublinear dependence of the $CsPbBr_3$ QDs with the cross section above $2 \times 10^{-15}$ $cm^2$. Reproduced with permission from ref. [44]. Copyright 2016, Published by American Chemical Society.

to that of $CsPbBr_3$. When comparing the biexciton lifetimes of perovskite QDs with CdSe and PbSe QDs possessing the comparable crystal sizes, it is found that the multiexciton interaction is much stronger in perovskite QDs which leads to a larger Auger recombination rate [26, 45, 46].

## *11.2.4 Optical Gain in Pe-QDs*

### 11.2.4.1 Stimulated Emission

The first observation of stimulated emission (SE) in colloidal QDs (the canonical CdSe QDs) was reported in year 2000 [47]. The phenomenon of SE represents the fundamental principle underlying the operation of lasers. SE can be viewed as a special kind of luminescence. The essential thermodynamic conditions for SE occurrence differ from the common occupancy of electronic states in luminescence processes, that is, population inversion states. In other words, photons generated by SE dominates over the photon absorption in ground state. If population inversion was achieved, then the spontaneous emission may be amplified in semiconductors with high probability of radiative recombination. Given that the gain overcame the optical loss, a sharp SE band will appear in the PL spectrum. This new emission band is also termed as amplified spontaneous emission (ASE).

Here, it is important to point out that the distinction between ASE and lasing. ASE refers to emission amplification by SE in a resonator-free device, whereas lasing action is strongly dependent of the optical cavity. Due to the categorized complexity of the Pe-QDs, we will discuss the SE from the all-inorganic Pe-QDs and organic-inorganic Pe-QDs, as well as nontoxic lead-free halide Pe-QDs, separately, in the following sections.

#### All-Inorganic Halide Perovskite Quantum Dots

The first demonstration of SE and optical gain in all-inorganic perovskites (microcrystalline $CsPbBr_3$ films prepared by crystallization from the amorphous phase) at room temperature was reported by Kondo et al. in 2005 [48]. Although the $CsPbBr_3$ films exhibited large optical gain due to giant oscillator strength effect of confined exciton, the pump threshold was extremely high because of the poor microcrystalline quality of $CsPbBr_3$ films. Soon after the advent of cesium lead halide perovskite nanocrystals in 2015, the favorable optical gain properties were reported by Kovalenko et al. and Sun et al. [3, 4] Both of the groups demonstrated low-threshold SE and lasing action under either femtosecond or nanosecond pulsed excitation, highlighting the potential of perovskite nanocrystals in laser application. Figure 11.11a shows pump intensity-dependent PL spectra and the appearance of SE band with a narrow bandwidth of 4–9 nm. Reabsorption effect in the single exciton gain mechanism and the biexciton gain origin may be the reason for the red-shifted

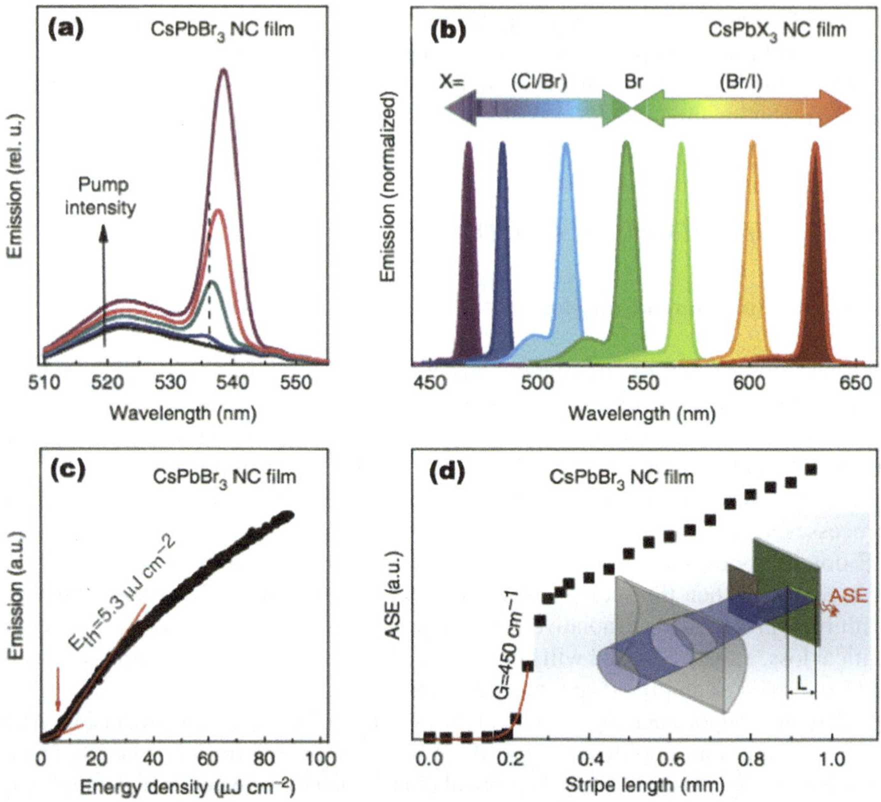


**Fig. 11.11** SE spectra from thin films of $CsPbX_3$ nanocrystals. **a** Pump intensity dependence of the emission. (pumping intensity range was 3–25 $\mu J\ cm^{-2}$). **b** Spectral emission-tunability via modulating halide constituents in a $CsPbX_3$ QDs film. **c** Low-threshold behavior for the intensity of the SE band of the $CsPbBr_3$ QDs film shown in (**a**). **d** Estimation of modal net gain of $CsPbBr_3$ QDs with the aid of the variable stripe length technique. All spectra were pumped at $\lambda = 400$ nm with 100 fs laser pulses. Reprinted with permission from Ref. [3]. Copyright 2015, Published by Nature

SE with respect to that of the spontaneous emission peak [49, 50]. By facilely modulating halide constituents and sizes of $CsPbX_3$ QDs, the SE spectrum can be freely tuned from blue, green to red region, exhibiting the spectral tunability of SE in the entire visible spectral range (440–700 nm, Fig. 11.11b). As shown in Fig. 11.11c, the threshold of SE is about $5 \pm 1\ \mu J\ cm^{-2}$ for $CsPbBr_3$ Pe-QDs.

The possible reasons that contribute to the low SE threshold of $CsPbBr_3$ can be concluded as following:

1. Extremely high PL QY (~90%, Ref [2].).
2. Large absorption cross-section (~$2.5 \times 10^{-14}\ cm^2$ for 9 nm $CsPbBr_3$ QDs, Ref [4, 51].).

3. Large biexciton binding energy restraining the reabsorption effect (20–100 meV in different nm-sized $CsPbBr_3$ QDs, Ref [4, 44, 52].)
4. Mitigated non-radiative loss (~105 ps of the Auger process in $CsPbBr_3$ QDs, Ref [4].)

In addition to the low SE threshold, the net modal gain is also a pivotal factor from a practical perspective. Through the amplification of inter-band recombination emission, the SE intensity increases exponentially with propagation length along an optical waveguide. Generally, there are two factors that determine the gain of a semiconductor: the dipole transition moment and the distribution of the states of electrons and holes near the Fermi level [26]. Semiconductors such as perovskites, possessing a large absorption coefficient, also potentially have a large optical gain. The net modal gain can be estimated by using the variable stripe length technique, as shown in the inset of Fig. 11.11d, where the excitation light is shaped as a variable-length line on the sample surface through a cylindrical lens. An established formula can be employed to fit the optical gain with the detected emission intensity [53]:

$$I(\lambda) = \frac{AP_0}{g_{tot}(\lambda)}\{\exp(g_{tot}(\lambda)L) - 1\}. \qquad (2.4.1)$$

where $I(\lambda)$ is the emission intensity, $AP_0$ indicates the SE proportional to excitation intensity, $g_{tot}$ is the total optical gain of the waveguide, and $L$ is the excitation stripe length. For $CsPbBr_3$ QDs film, the net modal gain was estimated to be ~$450 \pm 30\ cm^{-1}$, which is relatively high and beneficial to obtaining high-performance laser devices. Moreover, in addition to SE from the thin close-packed Pe-QD films, stable liquid-based green SE (a threshold value of ~$105\ \mu J\ cm^{-2}$) from $CsPbBr_3$ nanocrystals has been demonstrated by modifying the surface ligands [54]. The threshold is found to be dependent on solvent composition, where the added proportions of toluene may bring about fast nonradiative relaxation pathways. Anyway, the development of colloidal perovskites as liquid gain media may offer new possibilities in on-chip optofluidic coherent light sources.

***Other nano-structures***. Yet, besides the $CsPbX_3$ QDs, $CsPbX_3$ in form of other nanostructures have also been reported with excellent lasing performance. In year of 2016, Yang et al. demonstrated low-threshold lasing ($5\ \mu J\ cm^{-2}$) in $CsPbBr_3$ nanowires with high stability (maintain robust lasing more than 1 h or over $10^9$ excitation cycles), as shown in Fig. 11.12a [55]. Also, lasing action based on single-crystal $CsPbCl_3$ nanowires and well-faceted $CsPbBr_3$ nanoplates have also been realized (Fig. 11.12b and c). The lasing thresholds were determined to be $86\ \mu J\ cm^{-2}$ and $38\ \mu J\ cm^{-2}$, respectively. These self-formed nanostructures can serve as the gain media and the optical resonators, simultaneously. Moreover, low-defect $CsPbBr_3$ perovskite nanorods (see TEM image of nanorods with uniform surface inset Fig. 11.12d) was synthesized by Wang et al., via employing an intermediate monomer reservoir synthetic strategy [56]. The stable $CsPbBr_3$ nanorods exhibited a high net mode gain of $980\ cm^{-1}$ and low SE threshold of $7.5\ \mu J\ cm^{-2}$ pumped by a nanosecond laser, as shown in Fig. 11.12e and f.

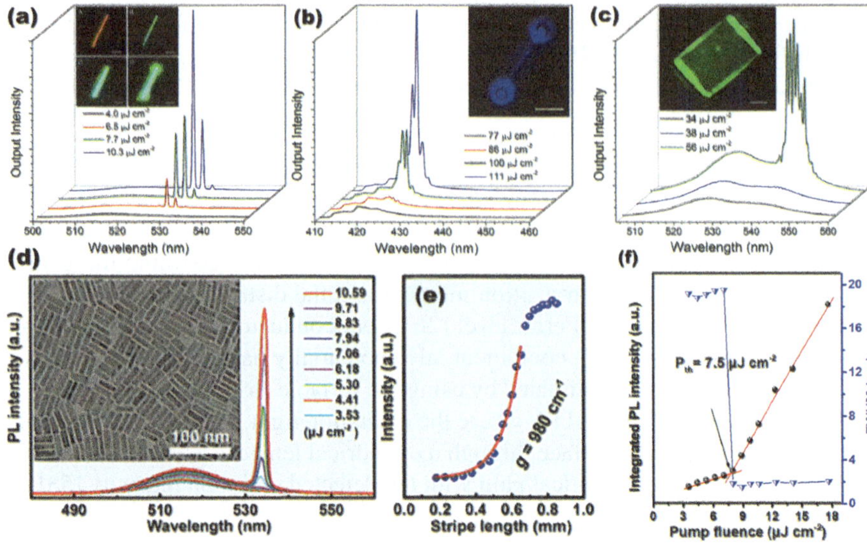


**Fig. 11.12** Lasing action in $CsPbX_3$ nano-structures besides the quantum dots. **a** Power-dependent emission spectra from the $CsPbBr_3$ nanowire, narrow emission peaks at ~530 nm indicates lasing. (Inset) Dark-field image and snaps of nanowire under excitation from a femtosecond pulsed laser with the increased excitation fluence (Scale bar, 2 μm). **b** Lasing in $CsPbCl_3$ nanowire power-dependent emission spectra at ~425 nm. (Inset) Optical image of $CsPbCl_3$ nanowire above the lasing threshold (Scale bar, 5 μm). **c** Lasing in $CsPbBr_3$ nanoplate power-dependent emission spectra at ~545 nm. (Inset) Optical image of $CsPbBr_3$ nanoplate above the lasing threshold (Scale bar, 5 μm). Reprinted with permission from Ref. [55]. Copyright 2016, Published by National Academy of Sciences. **d** Pump fluence-dependent emission of $CsPbBr_3$ perovskite nanorods with uniform surface, inset: the typical TEM images of low-defect $CsPbBr_3$ nanorods. **e** The net modal gain of $CsPbBr_3$ nanorods. **f** Integrated PL intensity and FWHM of $CsPbBr_3$ nanorods as a function of pump fluences. Reprinted with permission from Ref. [56]. Copyright 2019, Published by American Chemical Society

## Organic-Inorganic Hybrid Halide Perovskite Quantum Dots

In the year of 2014, the first SE behavior (the measured threshold fluence ~12 $\pm$ 2 μJ $cm^{-2}$) has been observed in low-temperature solution-processed organic-inorganic hybrid halide $CH_3NH_3PbX_3$ ($CH_3NH_3$, that is methylammonium, simplified by MA) thin film [5], revealing that these hybrid halide perovskites can not only serve as the excellent photovoltaic materials, but also the promising laser media. Afterwards, although the optically pumped CW lasing above threshold of ~17 kW $cm^{-2}$ for over 1 h below the tetragonal-to-orthorhombic phase transition temperature of $T \approx 160$ K in $MAPbI_3$ films has been demonstrated [57], the relatively poor stability against ambient environment of hybrid halide perovskites have always been considered as major obstacles for the realization of practically available laser devices.

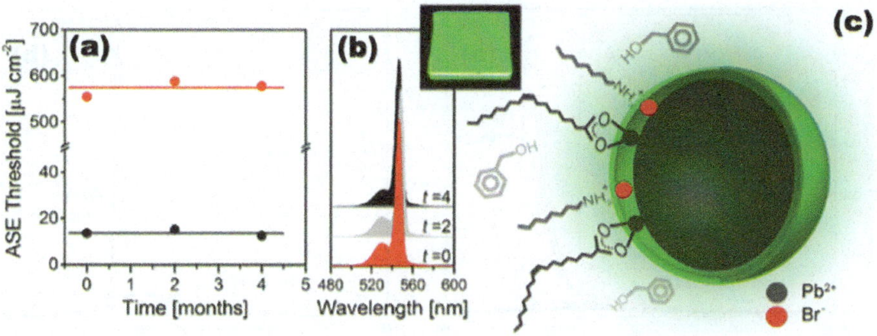


**Fig. 11.13** Highly stable SE behavior in BnOH/$MAPbBr_3$ QDs. **a** Corresponding SE threshold values for 1PA (black) and 2PA (red) excitation, respectively. **b** Consistency of steady-state PL and SE peak position above the ASE threshold of a freshly prepared (red), 2 month old (gray), and 4 month old (black) QDs stored under ambient conditions (400 nm excitation; fluence 38 μJ $cm^{-2}$). **c** Schematic illustration of surface ligands modified BnOH/$MAPbBr_3$ QDs. Reprinted with permission from Ref. [58]. Copyright 2017, Published by American Chemical Society

In contrast, by virtue of the passivating ligands and controlled ligand binding motifs on the $MAPbBr_3$ QDs surface (Fig. 11.13c), Veldhuis et al. reported the SE from benzyl alcohol (BnOH)/$MAPbBr_3$ Pe-QDs with ultralow ASE thresholds (13.9 ± 1.3 μJ $cm^{-2}$ at one-photon absorption (400 nm, 1 PA); 569.7 ± 6 μJ $cm^{-2}$ at two-photon absorption (800 nm, 2 PA), respectively) and high-stability (>4 months in air, Fig. 11.13a and b), as well as outstanding optical modal gain (520 $cm^{-1}$) signatures through the detailed ultrafast spectroscopic studies [58]. The outstanding low ASE thresholds were further corroborated by the one- and two-photon cross sections (using the Z-scan technique). The 1PA and 2PA cross sections were $\sigma_{abs}^{(1)} \sim 1.4 \times 10^{-13} cm^2$ and $\sigma_{abs}^{(2)} \sim 1.0 \times 10^5$GM (where 1 GM = $10^{-50}$ $cm^4$ s), respectively, which compares favorably to CdSe/ZnS chalcogenide and all-inorganic halide perovskite QDs systems [3, 4]. Furthermore, the biexciton binding energy was claimed to be ~70 meV (sufficiently far above the room-temperature thermal energy of $k_BT$ ~ 26 meV), which can suppress the detrimental reabsorption losses. It is highlighted that surface ligand engineering and chemical treating are effective strategies to improve the thermal and environmental stability of organic-inorganic hybrid components.

Compared to the organic-inorganic MA perovskite analogue, the pristine formamidinium (FA, $CH(NH_2)_2$) perovskites manifest attractive features including higher thermodynamic and chemical stability. It is demonstrated that the $FAPbI_3$ film exhibited excellent photostability upon sustained pumping (a 532 nm pulsed laser, 5 ns, 10 Hz), where the SE intensity retained mostly 90% of the starting value [59]. Two factors can contribute to the enhanced SE stability of $FAPbI_3$: (i) less susceptible to the heat induced by continuously laser illumination due to better thermal stability than $MAPbI_3$. (ii) moderated light-induced ion migration due to larger FA cation size and the enhanced interaction between FA cations and iodine ions than in $MAPbI_3$. Moreover, despite the major successes based on all-inorganic Pe-QDs, the

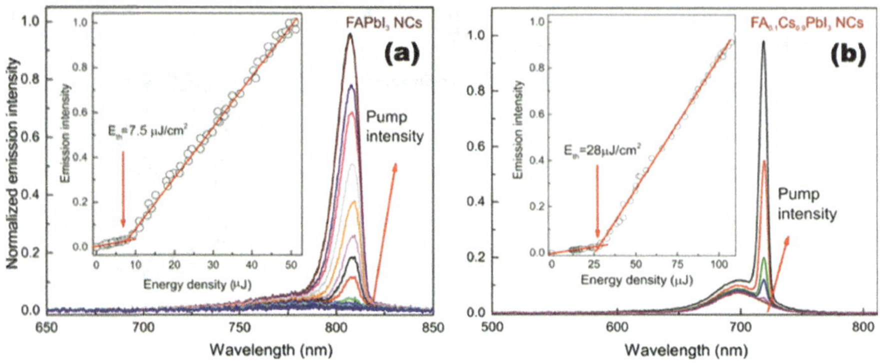


**Fig. 11.14** SE behavior of **a** $FAPbI_3$ and **b** $FA_{0.1}Cs_{0.9}PbI_3$ QDs film pumped by pulsed laser with duration of 100 fs. Insets: the integrated PL intensity as a function of energy density indicating the SE behavior with ultralow thresholds. Reprinted with permission from Ref. [60]. Copyright 2017, Published by American Chemical Society

unstable $CsPbI_3$ thin films and nanocrystals hamper such materials for laser applications in the red region, forming the "perovskite red wall". In 2017, through cationic modulation, the monodisperse and cubic-shaped $FAPbI_3$ and $FA_{0.1}Cs_{0.9}PbI_3$ (both average sizes of 10-15 nm) were reported by Protesescu et al., which breaks the "red wall" and extends the emission spectra into red and even near-infrared range (e.g., 690 nm for $FA_{0.1}Cs_{0.9}PbI_3$ and 780 nm for $FAPbI_3$ QDs) [60]. Furthermore, these Pe-QDs have been tested as qualified optical gain media under femtosecond-pulsed excitation, as shown in Fig. 11.14, low pump thresholds of 28 and 7.5 $\mu J\ cm^{-2}$ were revealed to develop SE from $FA_{0.1}Cs_{0.9}PbI_3$ and $FAPbI_3$ QDs, respectively. The SE threshold ~7.5 $\mu J\ cm^{-2}$ of $FAPbI_3$ was among the lowest values of the red to near-IR emitting perovskites (5–10 $\mu J\ cm^{-2}$). The ultralow pump threshold can be attributed to the significantly longer biexciton lifetimes ($FAPbI_3$ QDs, 0.4 ns) [61] compared to Cs-based QDs ($CsPbI_3$ QDs, 93 ps) [62]. It is noted that robust SE from close-packed films of $FAPbI_3$ QDs under excitation in the nanosecond regime has also been demonstrated. Systematic optimization of the $FAPbI_3$ QDs results in high SE net modal gain up to 604 $cm^{-1}$ [63].

Lead-Free Halide Perovskites

In the previous two sections, we have discussed the optical gain in all-inorganic and organic-inorganic hybrid halide perovskites. However, the intrinsic toxicity of lead ($Pb^{2+}$) remains a roadblock for real-life optoelectronic applications based on these Pe-QDs. To address this issue, the lead-free halide Pe-QDs have been explored as laser media.

Solution-processed lead-free tin-based halide perovskites $CsSnX_3$ have been demonstrated with exceptional optical gain properties in the near-infrared lasing.

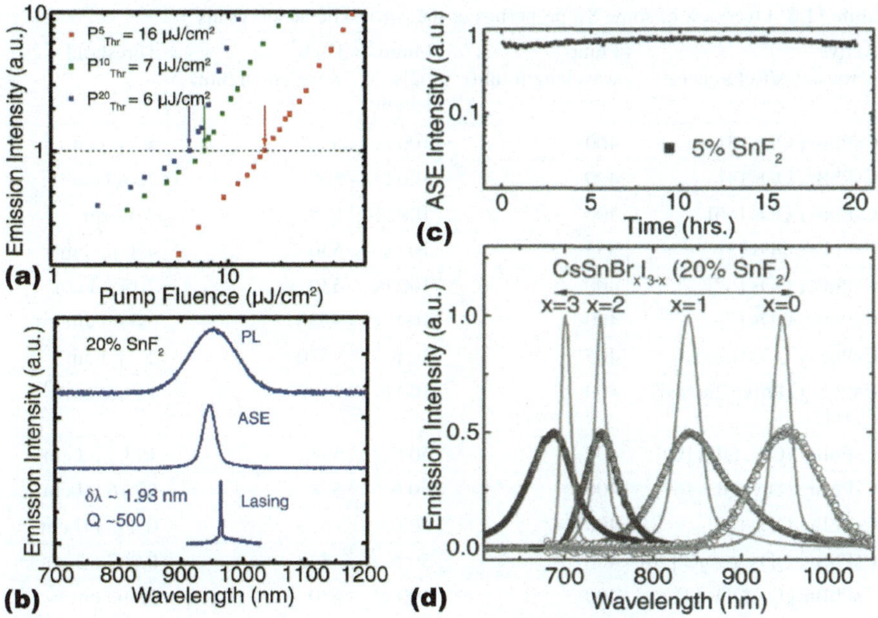


**Fig. 11.15** Superior gain properties of 20% $SnF_2$-treated $CsSnI_3$. **a** The emission intensity as a function of variable fluence, reveal the onset of SE thresholds of the different mol concentrations of $SnF_2$-treated $CsSnI_3$. **b** The PL, SE, and single mode lasing of 20% $SnF_2$-treated $CsSnI_3$. **c** High stability of the 5% $SnF_2$-treated $CsSnI_3$. (a pump fluence of 70 μJ cm$^{-2}$, 600 nm, 50 fs, 1 kHz, under excitation for ≈ 20 h conducted at room temperature). **d** Tunable PL and SE wavelength up to 1000 nm in $CsSnBr_xI_{3-x}$ films. (pumped with 500 nm pump pulses, 50 fs, 1 kHz). Reprinted with permission from Ref. [17]. Copyright 2016, Published by John Wiley & Sons

It is noted that the longest gain wavelength of up to ≈ 1000 nm achieved from the $CsSnX_3$ perovskites is unavailable in the lead based analogues. By doping the $Sn^{2+}$ ions from the $SnF_2$, the optimized 20% $SnF_2$-treated $CsSnI_3$ films exhibited the ultralow SE threshold (≈6 μJ cm$^{-2}$, Fig. 11.15a) and large gain (200 cm$^{-1}$). Leveraging on the photonic crystal corrugations in a butterfly wing, $CsSnI_3$-based lasing (Q factor ≈ 500) was obtained as shown in Fig. 11.15b. Through detailed transient PL investigation, Xing et al. determined that the free carrier recombination dominates the emission in these $CsSnI_3$ gain medium [17]. At higher pump fluence (>0.6 μJ cm$^{-2}$), the quadratic behavior of PL decay clearly indicates that the PL emanates from the bimolecular recombination between free carriers, while at low pump fluence, the linear behavior shows that the PL is dominated by trap-assisted recombination. The free-carrier mechanism of the radiative recombination is consistent with the small exciton binding energy (~0.1 meV) estimated for $CsSnI_3$ [64]. The bimolecular free-electron-hole recombination constant is calculated to be $10^{-8}$ cm$^3$ s$^{-1}$, which is 1–2 orders larger than $Pb^{2+}$-based $MAPbI_3$ counterparts (9.4 × $10^{-10}$ to 8.7 × $10^{-11}$ cm$^3$ s$^{-1}$) [65]. The large bimolecular free-electron-hole recombination constant contributes to the superior gain properties. Furthermore, in the TA

**Table 11.2** Overview of some SE properties of the perovskite nanocrystals

| Active perovskites/References | Pump wavelength (nm) | Pump pulse length | Peak wavelength (nm) | Threshold |
|---|---|---|---|---|
| $CsPbBr_3$ QDs [3] | 400 | 100 fs | 530 | 5.3 μJ $cm^{-2}$ |
| $CsPbBr_3$ QDs [4] | 400 | 100 fs | 525 | 20 μJ $cm^{-2}$ |
| $CsPbBr_3$ QDs [66] | 400 | 100 fs | 535 | 2 μJ $cm^{-2}$ |
| $CsPbBr_3$ QDs [3] | 355 | 10 ns | 530 | 450 μJ $cm^{-2}$ |
| $CsPbBr_3$ QDs [67] | 400 | 100 fs | 530 | 2.14 μJ $cm^{-2}$ |
| $CsPbBr_3$ QDs [7] | 400 | 100 fs | ~530 | 3.8 μJ $cm^{-2}$ |
| $CsPbBr_3$ QDs [68] | 405 | 50 fs | ~530 | 2.2 μJ $cm^{-2}$ |
| $CsPbBr_3$ ($PbBr_2$-treated) QDs [7] | 400 | 100 fs | ~530 | 1.2 μJ $cm^{-2}$ |
| $CsPbBr_3$ (OA) QDs [69] | 400 | 50 fs | 538 | 193.5 μJ $cm^{-2}$ |
| $CsPbBr_3$ (DA) QDs [69] | 400 | 50 fs | 538 | 89.76 μJ $cm^{-2}$ |
| $CsPbBr_3$ QDs [70] | 400 | 50 fs | ~538 | 0.292 mJ $cm^{-2}$ |
| $CsPbBr_3$ QDs @ZnO [70] | 400 | 50 fs | ~538 | 0.207 mJ $cm^{-2}$ |
| $CsPbBr_{0.8}I_{2.2}$ QDs [4] | 400 | 100 fs | 620 | 55 μJ $cm^{-2}$ |
| $CsPbCl_3$ QDs [4] | 400 | 100 fs | 460 | 10.9 μJ $cm^{-2}$ |
| $CsPbCl_{1.4}Br_{1.6}$ QDs [4] | 400 | 100 fs | 475 | 48 μJ $cm^{-2}$ |
| $CsPbI_3$ QDs [3] | 400 | 100 fs | 630 | 22 μJ $cm^{-2}$ |
| Liquid $CsPbBr_3$ [54] | 400 | 100 fs | 526 | 105 μJ $cm^{-2}$ |
| $MAPbBr_3$ QDs [58] | 400 | 100 fs | 548 | 13.9 μJ $cm^{-2}$ |
| $MAPbBr_3$ QDs [68] | 405 | 50 fs | ~550 | 4.4 μJ $cm^{-2}$ |
| $FAPbBr_3$ QDs [68] | 405 | 50 fs | ~530 | 8.1 μJ $cm^{-2}$ |
| $FAPbBr_3$ QDs [71] | 400 | 35 fs | ~550 | 57 μJ $cm^{-2}$ |
| $FAPbI_3$ QDs [60] | 400 | 100 fs | 808 | 7.5 μJ $cm^{-2}$ |
| $FAPbI_3$ QDs [63] | 532 | 4 ns | 808 | 140 μJ $cm^{-2}$ |
| $FA_{0.1}Cs_{0.9}PbI_3$ QDs [60] | 400 | 100 fs | 720 | 28 μJ $cm^{-2}$ |
| $CsSnI_3$ films [17] | 650 | 50 fs | 950 | 6 μJ $cm^{-2}$ |

measurements, a strong PB signal appears at around 950 nm, which may arise from: (i) stimulated emission of the carriers near bandgap; (ii) band-filling near the band edge states. The TA results clearly highlight the reduced trap state in 20% $SnF_2$-treated $CsSnI_3$. Also, excellent photostability and tunable PL/SE wavelength are shown in Fig. 11.15c and d. After that, $CsSnI_3$ Pe-QD based lasing action has been demonstrated. With aid of the cholesteric liquid crystal (CLC) as a microresonator, the $CsSnI_3$ QDs-CLC laser with tunable lasing wavelength and long-term stability has been realized (see details in Sect. 11.3.6). The main SE properties (wavelength and threshold) and pump laser properties (wavelength and pulse length) of all the above-discussed perovskite nanocrystals are summarized in Table 11.2.

#### 11.2.4.2 Gain Mechanism

Higher-order excitons (multiexcitons) can occur in the individual perovskite nanocrystal with the increased pumping intensity. Since the degeneracy of the band edge states in Pe-QDs is multi-fold (e.g., $CsPbBr_3$ QDs has been demonstrated to show 2-fold degeneracy due to the asymmetry in cubic-shaped crystal structure and/or electron-hole exchange interactions) [26], population inversion (SE, or gain) in Pe-QDs can be achieved when the average number of electron-hole pairs (excitons) per nanocrystal $\langle N \rangle$ is larger than 1, which implies that these QDs at least contain biexcitons (XX) and multiexcitons. The photocarriers recombination rate can be described by the kinetics equation within a certain range of excitation intensity:

$$\frac{dn(t)}{dt} = -k_1 n - k_2 n^2 - k_3 n^3 \tag{2.4.2}$$

where $n$ is the density of photocarriers, $t$ is the time, and $k_i$ is the recombination constant [7]. The right side of the kinetics equation can be referred to three carrier recombination processes: single-exciton recombination, exciton-exciton interaction, and three-body (even many-body) Auger recombination, respectively. Until now, three possible gain mechanisms have been demonstrated in Pe-QDs, they are (i) biexciton-based gain; (ii) trion-based (X*, charged excitons; for a negative trion, it comprises two electrons and a hole; and two holes and an electron for a positive trion.) gain; (iii) single-exciton (X) gain, respectively, as shown in Fig. 11.16.

***Biexciton-based gain***. Although the gain mechanism in Pe-QDs is still in a matter of debate, the biexcitons (XX) gain mechanism has been widely recognized and mostly reported. Let us first discuss the biexciton gain model based on the Poisson distribution. The Poisson distribution can be used to describe the probability of absorbing $i$ photons per nanocrystal which eventually creates $i$ excitons:

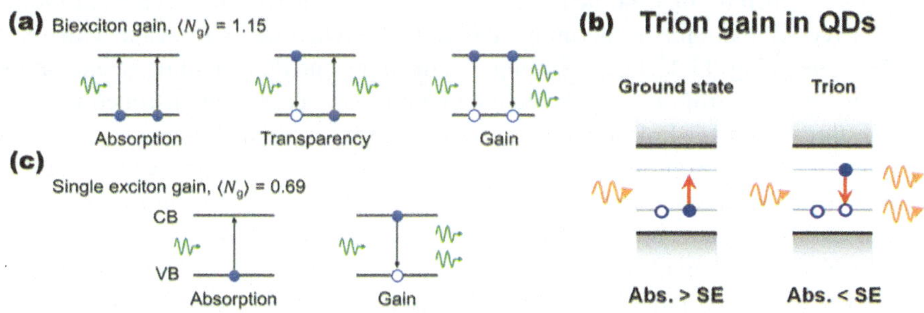


**Fig. 11.16** Three gain mechanisms in perovskite nanocrystals with doubly degenerate band edge states. **a** Biexciton-based gain. Reprinted with permission from Ref. [72]. Copyright 2019, Published by American Chemical Society. **b** Trion-based gain. Reprinted with permission from Ref. [7]. Copyright 2018, Published by American Chemical Society. **c** Single-exciton gain. Reprinted with permission from Ref. [72]. Copyright 2019, Published by American Chemical Society

$$P_{\langle N\rangle}(i) = \left(\langle N\rangle^i / i!\right)\exp(-\langle N\rangle) \tag{2.4.3}$$

where the ⟨N⟩ is average number of excitons per nanocrystal.

In a two-fold degeneracy system, $g_e = g_h = 2$, such as the case of $CsPbX_3$, three possible states may appear in a QD, expressed as $P_0$ (no exciton), $P_X$ (one exciton) and $P_{XX}$ (biexciton). The mechanism for a biexciton-based gain can be written as:

$$\alpha_0 \propto 2\gamma \tag{2.4.4}$$

$$-\Delta\alpha \propto 2\gamma + 2\gamma(P_{XX} - P_0) \tag{2.4.5}$$

$$-\Delta\alpha/\alpha_0 = 2 - 2P(0) - P(1) \tag{2.4.6}$$

where the $\gamma$ indicates the rate of photon emission and absorption by a QD.

The properties of multiexciton states can be assessed through transient-PL signatures. Figure 11.17a displays fluence-dependent PL dynamics obtained for $CsPbI_3$ nanocrystals. At low pump fluences ($\langle N\rangle \ll 1$), the measured PL shows tens-of-ns decay typical of a single-exciton recombination. Note that a much faster decay (tens-of-ps scale, superlinearly grows with excitation fluence) than that of single-exciton recombination emerges with the increased excitation fluences (to ⟨N⟩ of ~1 and higher), corresponding to nonradiative multiexciton Auger recombination.

Similar phenomena are also observed for $CsPbBr_3$ nanocrystals in other literatures [4, 26, 44, 73]. Auger recombination can be considered as a dominant decay channel of multiexciton states, where the energy from the recombination of an exciton transferred to the third carrier through nonradiative recombination. The measured late-time PL decay dependences can be accurately fitted by the Poisson distribution (for single exciton, $I_{PL} \propto 1 - e^{-\sigma j_p}$; for biexciton, $I_{PL} \propto 1 - e^{-\sigma j_p} - ej_p e^{-\sigma j_p}$, lines in Fig. 11.17b, σ is the QD absorption cross-section and $j_p$ is per-pulse photon fluence), confirming that the fast and early-time PL component arises from multi-exciton recombination. Furthermore, the fast PL dynamics is in scale with $\langle N\rangle^2$ (dashed line in Fig. 11.17d), confirming that the transient-PL signatures at excitation fluences originate from biexciton recombination. Xiao et al. demonstrated that the low SE threshold in $CsPbBr_3$ nanocrystals is largely determined by the competition between SE from biexcitons and stimulated absorption from single-exciton to biexciton states through two-dimensional electronic spectroscopy, confirming the optical gain in $CsPbBr_3$ nanocrystals indeed originates from biexcitons [72].

However, a serious obstacle to the realization of practical Pe-QDs-based laser occurs when ultrafast nonradiative Auger decay dominates the carrier recombination. In fact, the biexciton lifetimes calculated for $CsPbX_3$ QDs are shorter than those in CdSe and PbSe QDs of similar sizes, which is in a time scale of tens to hundreds of picoseconds [26]. The biexciton lifetime $\tau_{XX} = 93$ ps in 12 nm-sized $CsPbI_3$ can be fitted based on the single-exponential decay (dashed lines in Fig. 11.17c), similar value (~100 ps) for ~11 nm-sized $CsPbI_3$ QDs reported by Park et al. [62] $\tau_{XX}$ determined in literatures differs for various sized $CsPbBr_3$ QDs: e.g., 105 ps for

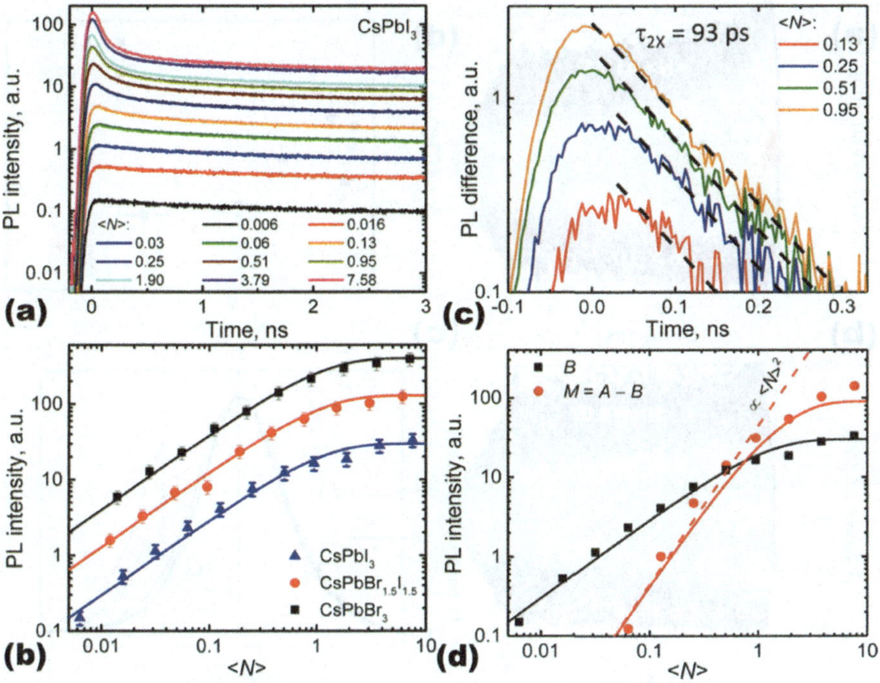


**Fig. 11.17** Pump-fluence dependence of PL dynamics and biexciton lifetimes in $CsPbI_3$ Pe-QDs. **a** Fluence-dependent time-resolved PL of the $CsPbI_3$. (the sample is excited at 3.6 eV with 220 fs pulses). **b** The fluence dependence of the late-time PL signal for perovskite QDs of different compositions. Lines were fits to calculated for the Poisson distribution of initial QDs occupancies. **c** The biexciton lifetimes is isolated from the single-exciton decay and fitted to single-exponential decay (dashed lines). **d** The quadratic scaling of PL signal suggest the biexciton recombination. Reprinted with permission from Ref. [26]. Copyright 2016, Published by American Chemical Society

9 nm; 85 ps for 8.6 ± 1.5 nm; [73] 47 ps for 9.3 nm; [26] 44 ps for 7 nm; [74] and 38 ps for 8.1 nm [26]. However, much slower $\tau_{XX}$ reported in the phase of α–FAPbI3 (*Pm3m*) with average size ~13 ± 3 nm is 0.4 ns, highlighting these hybrid QDs are promising candidates in the red-emitting lasing applications [60, 61].

The precise determination of $\tau_{XX}$ in $CsPbX_3$ nanocrystals is not easy because $\tau_{XX}$ is short-lived and dependent on the fabrication procedure, size, shape, excitation intensity and even surface nature. It is well-known the $\tau_{XX}$ increased with decreased QDs volume, and in chalcogenide QDs, the $\tau_{XX}$ dependence follows a linear volume scaling. However, in $CsPbBr_3$ QDs, the $\tau_{XX}$ was shown to be less dependent on QDs volume, following a $\tau_{XX}$ proportional to $V^{0.5}$ relation, indicating the different multi-carrier interactions in the Pe-QDs and chalcogenide QDs [26].

On the other hand, the biexciton binding energy has strong effect on the gain performance. According to the strong biexciton interaction in Pe-QDs by PL dynamics investigation, the strength of the biexciton interaction can be calculated in virtue of the temporally and spectrally resolved photoluminescence. Figure 11.18a

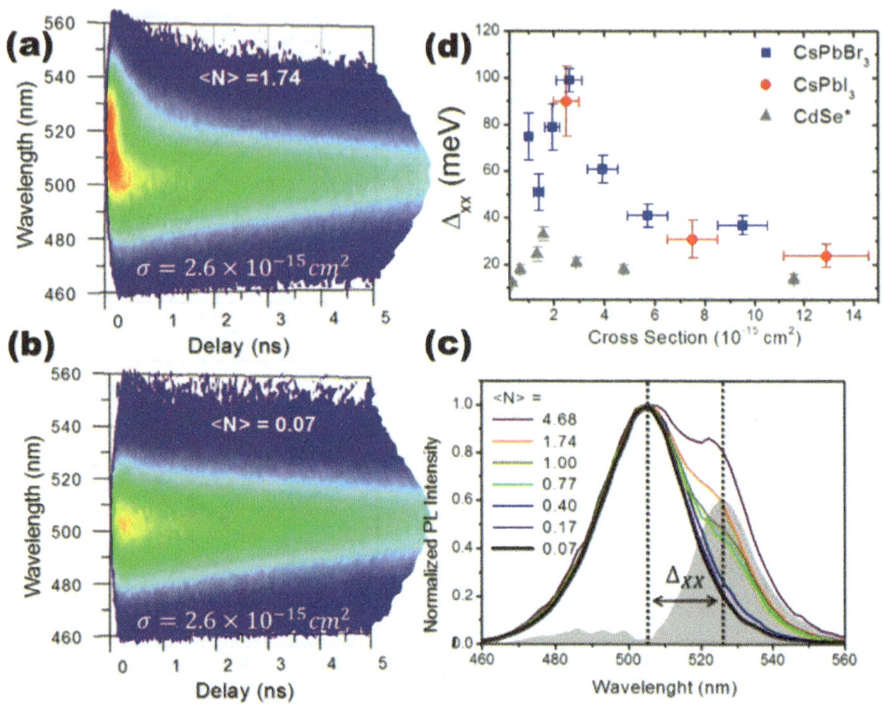


**Fig. 11.18** Temporally and spectrally resolved photoluminescence of 7 nm-sized $CsPbBr_3$ QDs. **a** High excitation density. **b** Low excitation density. **c** Extraction of $E_b^{XX}$ according to the difference between the high and the low excitation intensity at the short delay spectra (≈100 ps) (shaded area). **d** The size-dependence of $E_b^{XX}$ in $CsPbBr_3$, $CsPbI_3$ and CdSe QDs. Reprinted with permission from Ref. [44]. Copyright 2016, Published by American Chemical Society

and b separately show the temporally and spectrally resolved photoluminescence of 7 nm-sized $CsPbBr_3$ QDs samples at low excitation levels (only single-excitons, $\langle N \rangle = 0.07$) and at high excitation levels (typically, $\langle N \rangle \cong 2$), which confirm the presence of biexcitons [44]. Meanwhile, it is found that the biexciton interaction becomes strong with the decreased QDs size due to the strong quantum confinement [44, 75]. The extremely large biexciton binding energy ($E_b^{XX}$) have both been reported in $CsPbBr_3$ (100 meV, the value is three times larger than the maximum observed for CdSe QDs, Fig. 11.18c) and $CsPbI_3$ QDs (nearly 100 meV) [44]. The red-shifted SE peak again suggests the biexciton-based optical gain and large $E_b^{XX}$ could effectively reduce the reabsorption loss in an inhomogeneous QD ensemble (Fig. 11.18d), which can contribute to the low pump threshold [4].

***Trion State in Optical Gain***. Trions (X*), which are charged excitons, have been reported as one promising alternative to the photogeneration of XX (or multiexcitons) to achieve SE (optical gain). X* in QDs generally appears during carrier trapping or unbalanced charge injection by virtue of electrical doping or photochemical doping.

Previous literatures have shown that the formation of X* in $CsPbX_3$ QDs occurs spontaneously under pulsed excitation at low pump fluence, where the theoretical threshold of trion gain is calculated to be $\langle N \rangle_{th} = 0.58$ [7, 26, 44]. As depicted in Fig. 11.16b, the optical gain condition can also be realized by a trion state apart from the neutral QDs similar to the case of biexciton (or higher-order) states. Makarov et al. has demonstrated that the Auger lifetime of X* is expected to be 4 times longer than that of XX ($\tau_{X*} = 4\tau_{XX}$) [26], which may result in a longer gain lifetime. Indeed, the $\tau_{X*}$ was calculated to be ~235 ps (the $\tau_{XX}$ was 47 ps under the same condition), highlighting the trion state may be more favorable for the SE. By virtue of time-resolved PL (TR PL) spectroscopy, the recombination dynamics of various excitonic states for X, XX and X* can be characterized. Figure 11.19a shows the fluence-dependent TRPL of $PbBr_2$-treated $CsPbBr_3$ QDs. Similar to the Fig. 11.17a, under lowest pump fluence of 0.02 μJ cm$^{-2}$, the TR PL traces feature single-exciton recombination. By removing the contribution from single exciton decay kinetics, the

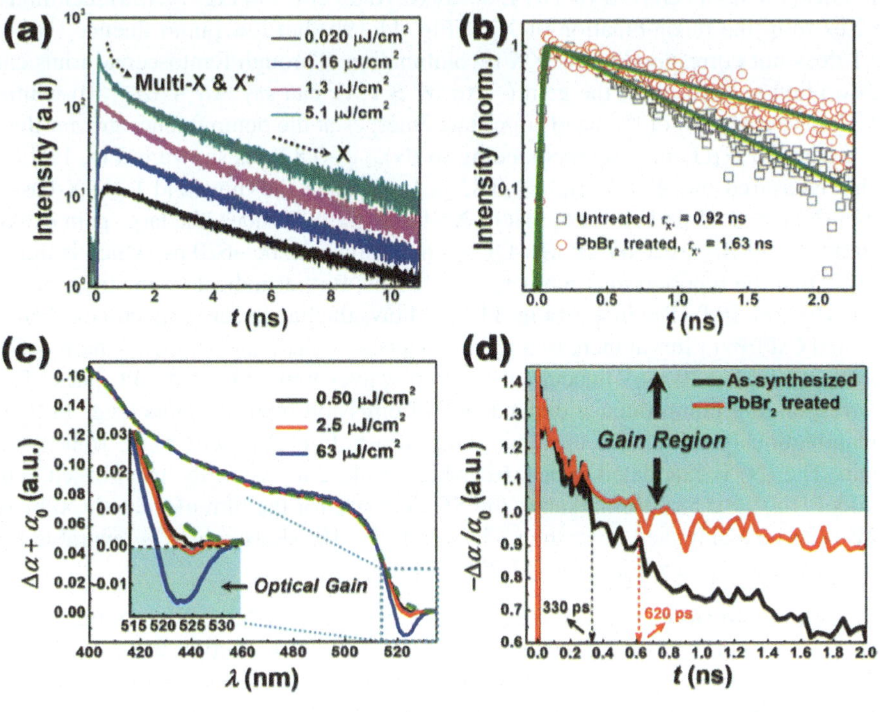


**Fig. 11.19** Trion-based gain in $PbBr_2$-treated $CsPbBr_3$. **a** Pump-fluence-dependent TRPL decay profiles of $PbBr_2$-treated $CsPbBr_3$. **b** Respective PL recombination kinetics of trions and biexcitons fitted by monoexponential functions. **c** TA spectra of treated $CsPbBr_3$ QDs measured at various pump fluences. The inset displays signature of the optical gain region. **d** Decay kinetics of the photobleaching signal in untreated/$PbBr_2$-treated $CsPbBr_3$ QDs, probed at 525 nm with a pump fluence of 2.5 μJ cm$^{-2}$. Reprinted with permission from Ref. [7]. Copyright 2018, Published by American Chemical Society

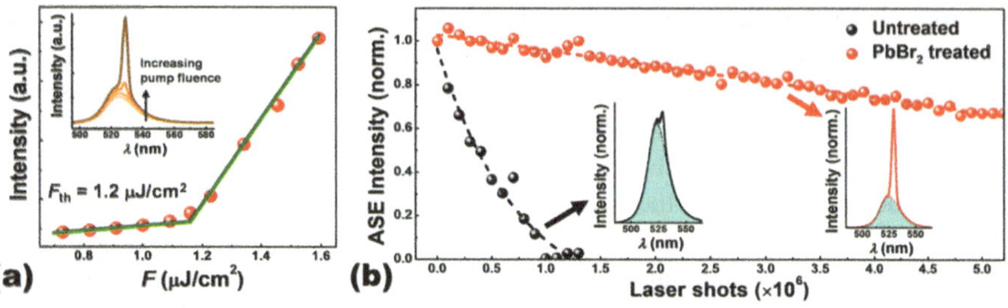


**Fig. 11.20** Ultralow SE threshold and stable SE behavior based on trion state in $CsPbBr_3$ QDs. **a** Emission intensity versus pump fluence plots show a low pump threshold. **b** Highly stable SE intensity under ambient conditions as a function of the number of 100 fs pulses from 1 kHz Ti:sapphire laser at 400 nm. Reprinted with permission from Ref. [7]. Copyright 2018, Published by American Chemical Society

monoexponential behavior of TRPL decay (0.16 $\mu J\ cm^{-2}$ in Fig. 11.19a) seemingly arises from the recombination of XX, (Fig. 11.19b) but low pump fluence $\langle N \rangle < 0.1$ does not correspond to the XX recombination. Through femtosecond transient absorption spectroscopy, the gain ($-\Delta\alpha/\alpha_0 > 1$, or can say $\Delta\alpha + \alpha_0 \leq 0$) within the long-energy tail of the band edge state emerges at the pump fluence greater than or equal to 1.6 $\mu J\ cm^{-2}$ (corresponding to $\langle N \rangle_{th} \sim 0.82$), as shown in Fig. 11.19c. This measured threshold $\langle N \rangle_{th} \sim 0.82$ is lower than the threshold for XX-based gain $\langle N \rangle_{th} = 1.15$ and closer to that of X*-based gain. Besides, the large gain decay lifetime in $PbBr_2$-treated $CsPbBr_3$ QDs is determined to be ~620 ps, which is much larger that the reported maximum $\tau_{XX}$ of 105 ps, further indicating the trion-based gain (Fig. 11.19d). The inset of Fig. 11.20a shows the fluorescence spectra of $PbBr_2$-treated $CsPbBr_3$ films at increased pump fluences. A sharp and spectrally narrow SE peak (FWHM ~ 20 meV) occurs at the long-wavelength side of the PL peak. The corresponding pump fluence dependence of emission intensity plots (Fig. 11.20a) unambiguously exhibit the ultralow pump threshold (1.2 $\mu J\ cm^{-2}$) of trion-based gain. The $\langle N \rangle$ is calculated to be 0.62, which is close to that of the trion-based gain model ($\langle N \rangle = 0.58$). Remarkably, the SE intensity for the film of treated $CsPbBr_3$ QDs can be maintained more than 3 h over $1.2 \times 10^7$ shots, which is favorable for practical laser applications (Fig. 11.20b) [7].

***Single-Exciton Gain***. The 2-fold degeneracy of the band edge states requires that at least a part of the excited QDs must contain biexcitons (or, multiple excitons, that is, $\langle N \rangle > 1$) to obtain optical gain. However, non-radiative Auger recombination will have a detrimental effect on lasing performance due to the multiexcitonic nature of optical amplification [76]. Through nonradiative recombination channel, the exciton will transfer its energy to a third particle and then re-excite to a higher energy state hampering the PL efficiency. In this regard, single-exciton gain mechanism will be a favorable approach for achieving low optical gain that can circumvent the problem of Auger decay. The radiative single-exciton lifetime is typically orders of magnitude

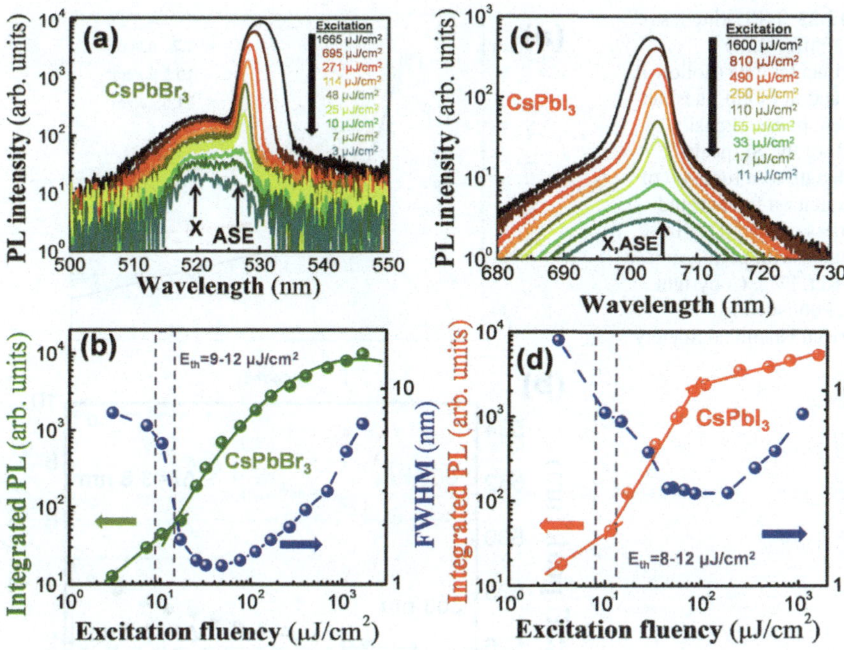


**Fig. 11.21** Single-exciton gain in $CsPbX_3$ QDs film. The pump fluence-dependence of emission in **a** $CsPbBr_3$ and **c** $CsPbI_3$ QDs film. The integrated PL intensity and FWHM as a function of excitation fluence in **b** $CsPbBr_3$ and **d** $CsPbI_3$ QDs film. All measurements were performed at 15 K. Reprinted with permission from Ref. [77]. Copyright 2019, Published by American Chemical Society

longer than the Auger-decay time constants. Klimov et al. has reported the carrier-induced Stark effect (strong X-X repulsion, $\Delta_{XX} \sim 100$ meV) will completely eliminate absorption losses and hence, the single-exciton-based gain can be realized in the type-II CdS/ZnSe core-shell QDs [76]. Figure 11.21a and c shows the SE behavior of thin films of $CsPbBr_3$ and $CsPbI_3$ QDs excited at 355 nm with increasing pump intensity, respectively. Energy difference between the PL and the SE peak (around 20.8 meV, Fig. 11.21a) seemingly conforms with the biexciton-based gain [26, 44]. Besides, the quadratic dependence of the integral SE intensity upon the excitation fluence (see Fig. 11.21b and d) for both $CsPbBr_3$ and $CsPbI_3$ QDs can be an additional evidence of biexcitonic origin of gain. However, there are existing several reasons to support the single-exciton gain instead of the biexciton-based gain mechanism in $CsPbX_3$ perovskite system:

1. If in the case of biexcitonic gain, the dependence of the PL QY on the excitation intensity (corresponding $\langle N \rangle = 1 \sim 2$) should not drop in Fig. 11.22a, and this phenomenon can be ascribed to the single-exciton annihilation due to Auger recombination [75].
2. In $CsPbI_3$ QDs film (Fig. 11.21c), the SE band emerges exactly at the maximum wavelength of the single-excitonic PL band. It implies that the biexciton binding

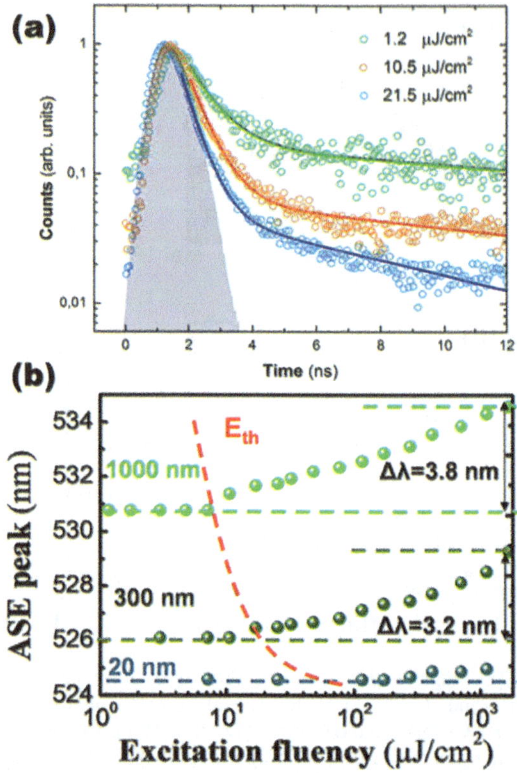


**Fig. 11.22** **a** PL kinetics of the $CsPbBr_3$ films experimental data (hollow dots) and fit (solid lines) at different pump intensity. **b** Shift of the SE peak wavelength as a function of the excitation fluence and thickness of the QDs film. Reprinted with permission from Ref. [77]. Copyright 2019, Published by American Chemical Society

energy $E_b^{XX} \sim 0\,\text{meV}$, which is not consistent with the biexcitonic nature of SE emission [44, 51].

3. If the gain mechanism is based on the biexciton, the $E_b^{XX}$ ($E_b^{XX} = E_x - E_{xx}$, or X-X Coulomb coupling) will get larger with film thickness reduction because of the strong quantum confinement in the decreased nanocrystal volume [76]. However, as shown in Fig. 11.22b, the energy difference $E_x$–$E_{xx}$ is nearly 0 meV for a single-monolayer (20 nm) QDs film, which indicates the single-exciton gain mechanism. Moreover. The threshold value of ⟨N⟩ is calculated to be 0.7, which is close to the 2/3 calculated based on single-exciton gainmechanism.

Although the dominant gain mechanism in the stimulated emission process is still a matter of debate, the demonstration of single-exciton-based gain offers us a new strategy to reducing Auger loss for optimization of optical gain. The theoretically reduced lasing threshold in single-exciton-regime may allow the realization of Pe-QDs-based lasing under CW excitation, which is the prerequisite towards the electrically-pumped laser.

#### 11.2.4.3 Multiphoton Absorption and Multiphoton Pumped Lasers

Multiphoton absorption has been considered as a vital branch of nonlinear optics. This process features long excitation wavelength and nonlinear excitation intensity dependence of the luminescence. Hence, it brings about the advantages of deeper penetration depth, higher damage threshold, higher image contrast and lesser scattering effects, which is beneficial to study high-lying electronic states and many other photonic application fields. In this section, we will briefly introduce the multiphoton absorption and multiphoton pumped lasers based on the Pe-QDs.

All-Inorganic Halide Perovskite Quantum Dots

Figure 11.23a illustrates the process of one-, two-, and three-photon excited PL for $CsPbBr_3$ QDs schematically. Multiphoton absorption can be described as the simultaneous absorption of multiple photons with same energy [78]. Pumped by a 800 nm femtosecond laser beam, the $CsPbBr_3$ QDs show bright emission and integrated PL

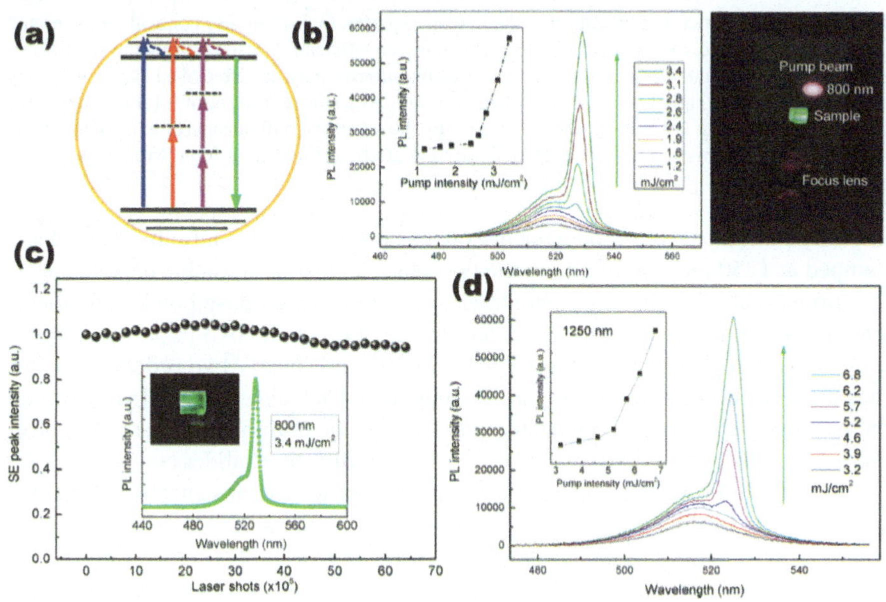


**Fig. 11.23** **a** The whole process of one-, two-, and three-photon excited PL through virtual states for $CsPbBr_3$ QDs. **b** Two-photon pumped PL spectra from $CsPbBr_3$ QDs with increased intensities, inset right (**b**): photograph of the stripe pumping configuration employed to pump the samples. **c** Stable SE behavior pumped by two-photon absorption with pumping intensity above the threshold. (5.2 mJ $cm^{-2}$) **d** SE induced by three-photon absorption and the inset shows the plot of integrated PL intensity dependence of pump intensity. Reprinted with permission from Ref. [51]. Copyright 2015, Published by American Chemical Society

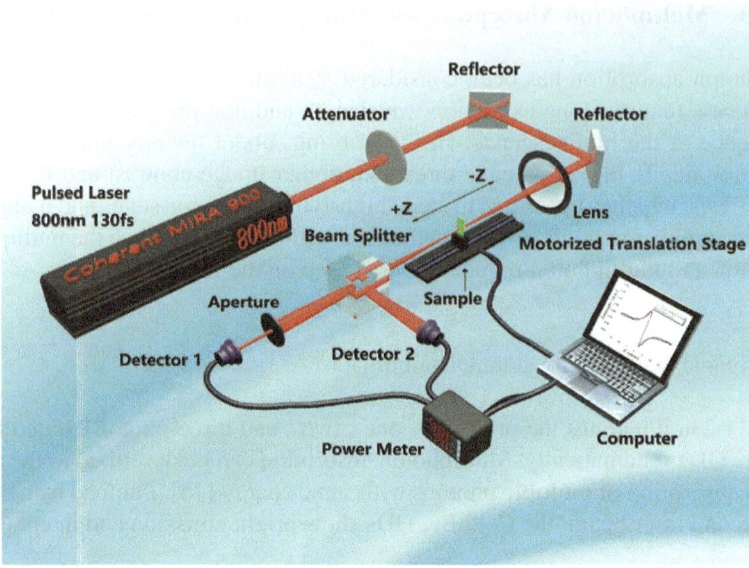


**Fig. 11.24** Schematically illustration of the Z-scan experimental set-up. By controlling the motorized translation stage, the samples injected into quartz cuvette can be moved along the Z-axis back and forth. The attenuated excitation laser beam is split into two parts in virtue of a beam splitter. The nonlinear refraction signal is probed by a detector with an aperture in front of it (closed aperture), whereas the nonlinear absorption signal is probed by the other one without aperture (open aperture). Reprinted with permission from Ref. [81]. Copyright 2016, Published by John Wiley & Sons

intensity exhibits quadratic dependence on the pump intensity (and cubic dependence pumped at 1250 nm), highlighting the two-photon absorption (cubic dependence of PL corresponding to the three-photon absorption) process. Based on Z-scan theory (see the set-up of the Z-scan experiment in Fig. 11.24), the two-photon absorption cross section ($\sigma_{abs}^{(2)}$) is derived to be ~ $1.2 \times 10^5$ GM, nearly two orders of magnitude higher than conventional green-emitting metal chalcogenide QDs [51]. Through increasing the Pauling electronegativity of halides, the multiphoton absorption properties for $CsPbX_3$ QDs can be tuned by bandgap of different halides components (see details in Table 11.3) [78]. Chen et al. adopted femtosecond transient absorption (TA) spectroscopy and found that the $\sigma_{abs}^{(2)}$ can be tuned by the size of the $CsPbBr_3$ QDs, where proportionality to the 3.3th power of the QDs size [79]. In addition, $\sigma_{abs}^{(2)}$ is found to be proportional to the linear one photon absorption. The power-law dependence and better understanding the photophysical process of two-photon absorption will pave the way for developing high-performance Pe-QDs-based nonlinear optical devices.

Moreover, multiphoton pumped SE from $CsPbBr_3$ QDs has been demonstrated inspired by the strong two-photon absorption and two-photon excited emission. As shown in Fig. 11.23b, by virtue of the stripe pumping configuration, the SE behavior under 800 nm laser beam with pulse width of 100 fs has been manifested. The pump

**Table 11.3** Bandgap of different halides components in $CsPbX_3$ QDs, as well as the tunable multiphoton absorption properties

| Active perovskites | Bandgap energy (eV) | 1 PA cross Section ($10^{-14}$ $cm^2$) | 2 PA cross Section ($10^{-50}$ $cm^4$ s $photon^{-1}$) | 3 PA cross Section ($10^{-80}$ $cm^6$ $s^2$ $photon^{-2}$) |
|---|---|---|---|---|
| $CsPbCl_3$ | 3.02 | 3.8 | $3.8\times10^4$ | $1.2\times10^4$ |
| $CsPbCl_{1.5}Br_{1.5}$ | 2.61 | 6.1 | $8.8\times10^4$ | $6.9\times10^4$ |
| $CsPbBr_3$ | 2.30 | 9.3 | $1.8\times10^5$ | $3.1\times10^5$ |
| $CsPbBr_{1.5}I_{1.5}$ | 2.11 | 16.4 | $3.2\times10^5$ | $9.1\times10^5$ |
| $CsPbBr_{0.5}I_{2.5}$ | 1.94 | 25.2 | $6.7\times10^5$ | $6.2\times10^6$ |
| $CsPbI_3$ | 1.73 | 36.6 | $2.1\times10^6$ | $1.1\times10^7$ |

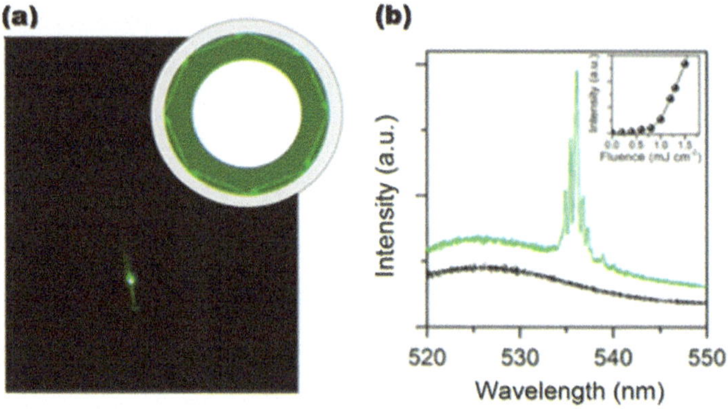


**Fig. 11.25** Two-photon pumped WGM lasing from $CsPbBr_3$ QDs. **a** Snap of fluorescence of a lasing cylindrical microcapillary tube embedded with $CsPbBr_3$ QDs. Inset: the WGM lasing action within the micro-ring resonator. **b** Two-photon pumped PL spectra plotted with excitation fluence below (0.7 mJ $cm^{-2}$, black) and above threshold (1.0 mJ $cm^{-2}$, green), respectively. Inset: the emission intensity dependence of pump fluence. Reprinted with permission from Ref. [82]. Copyright 2016, Published by American Chemical Society

threshold is derived to be ~2.5 mJ $cm^{-2}$ and the $\langle N \rangle$ is calculated to be 1.2 according to the equation $\langle N \rangle = \frac{f^2 \sigma_{abs}^{(2)}}{\tau}$, where $f$ is the pump fluence (photons/$cm^2$) and $\tau$ is the pulse width, agreeing with the biexcitonic gain mechanism. As depicted in Fig. 11.23c, the SE pumped by two-photon absorption shows the excellent photostability, withstanding more than 1.7 h under $6 \times 10^6$ uninterrupted laser shots. In the same way, three-photon pumped SE behavior has been displayed in Fig. 11.23d, the pump threshold (~5.2 mJ $cm^{-2}$) is demonstrated to be much lower than that of CdSe/CdS/ZnS core/double-shells QDs emitting in red regiem [80].

Thanks to the excellent multiphoton absorption pumped SE behavior, two-photon pumped micro-ring lasing can be obtained from inorganic Pe-QDs. As shown in Fig. 11.25a, frequency up-converted lasers have been demonstrated by incorporating the $CsPbBr_3$ QDs into resonators of glass microcapillary tubes. The whispering-gallery-mode (WGM) lasing with high quality factor and uniformly distributed lasing peaks can be attributed to the total internal reflections between the inner tubular surfaces and QDs solid (details about WGM lasers will be elaborated in Sect. 11.3.3). The pump intensity dependence of the emission (Inset, Fig. 11.25b) shows a threshold of ~0.9 mJ $cm^{-2}$, which is two orders of magnitude smaller than that of red-emitting CdSe/ZnS QDs (~21 mJ $cm^{-2}$) [82]. The subsequent work shows the magnitude of SE threshold pumped under 2PA exhibiting clear Pe-QDs size dependence, which has a strong correlation with the increasing on the biexciton binding energy [83]. Strong biexcitonic Coulomb interactions can effectively reduce the absorptive losses and offer favorable optical gain in Pe-QDs at low threshold. These results indicate that two-photon-pumped lasers based on Pe-QDs can serve as a promising strategy to achieve frequency up-conversion.

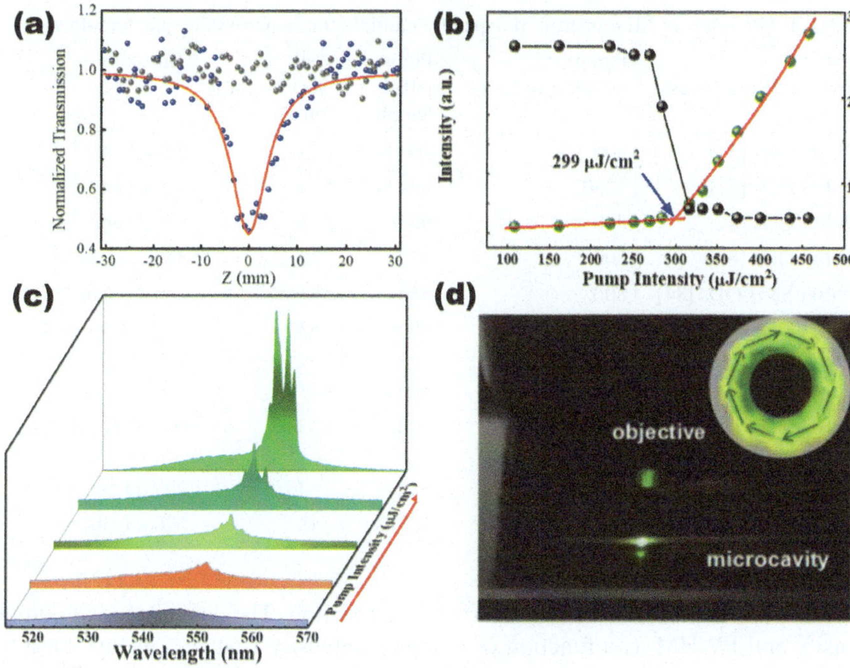


**Fig. 11.26** **a** Open aperture Z-scan measurement for $FAPbBr_3$ QDs. The red line shows the fitting based on Z-scan theory. **b** The SE threshold pumped under two-photon (800 nm) absorption. **c** Output spectral dependence of excitation intensity from $FAPbBr_3$ QDs in a microcapillary tube. **d** Optical image of two-photon-pumped cylindrical microcapillary lasing above the lasing threshold. Reprinted with permission from Ref. [71]. Copyright 2019, Published by American Chemical Society

### Organic-Inorganic Hybrid Halide Perovskite Quantum Dots

We have seen in Section "Organic-Inorganic Hybrid Halide Perovskites Quantum Dots" that solution-processed organic-inorganic hybrid halide Pe-QDs can extend the emission spectra into red and near-infrared range and serve as the promising laser media under one-photon excitation. In this section, we show that the two-photon pumped gain and lasing are also feasible, indicating them as a promising candidate in nonlinearly optical applications.

With the two-photon absorption measurement by Z-scan technique (Fig. 11.26a), the two-photon absorption coefficient of $FAPbBr_3$ QDs is derived to be ~0.76 cm $GW^{-1}$ at excitation intensity of 797 GW $cm^{-2}$ [71]. The net modal gain of $FAPbBr_3$ QDs under two-photon excitation was estimated to be ~480 $cm^{-1}$ by the variable stripe length method (fitting with the Eq. (2.4.1)). This high value is beneficial to the low threshold SE and lasing. Here, 800 nm femtosecond laser (35 fs, 1 kHz) was employed to study the SE behavior of $FAPbBr_3$ QDs under two-photon excitation. With the increased pump intensity from 108 to 458 μJ $cm^{-2}$, the red-shift

**Table 11.4** Overview of SE properties pumped by multiphoton in perovskite quantum dots

| Active perovskites/References | Pump wavelength (nm) | Pump pulse length | Peak wavelength (nm) | Threshold |
|---|---|---|---|---|
| $CsPbBr_3$ QDs [51] | 800 | 100 fs | 525 | 2.5 mJ $cm^{-2}$ |
| $CsPbBr_3$ QDs [51] | 1250 | 100 fs | 525 | 5.2 mJ $cm^{-2}$ |
| $CsPbBr_3$ QDs [82] | 800 | 90 fs | 533 | 0.8 mJ $cm^{-2}$ |
| $CsPbBr_3$ QDs [84] | 800 | 35 fs | 533 | 480.8 μJ $cm^{-2}$ |
| $CsPbBr_3/SiO_2$ QDs [84] | 800 | 35 fs | 533 | 230.8 μJ $cm^{-2}$ |
| $CsPbBr_3$ QDs (lateral sizes 7.4–12.5 nm) [83] | 800 | 50 fs | ~530 | 1.6~2.4 mJ $cm^{-2}$ |
| $CsPbBr_3$ QDs [70] | 800 | 50 fs | ~538 | 0.569 mJ $cm^{-2}$ |
| $CsPbBr_3$ QDs @ZnO [70] | 800 | 50 fs | ~538 | 0.679 mJ $cm^{-2}$ |
| $FAPbBr_3$ QDs [71] | 800 | 35 fs | ~546 | 299 μJ $cm^{-2}$ |
| $MAPbBr_3$ QDs [58] | 800 | 100 fs | 548 | 570 μJ $cm^{-2}$ |

PL peak with linewidth of 6 nm conforms the SE action. The dependence of output intensity and FWHM as a function of pumping intensity is shown in Fig. 11.26b, the SE threshold induced by two-photon absorption is extracted as 299 μJ $cm^{-2}$, which is comparable to that of the $CsPbX_3$ QDs (see Table 11.4 for comparison). Moreover, these $FAPbBr_3$ QDs shows strong temperature-tolerent SE action with the temperature increasing from 120 to 340 K, based on the following equation:

$$P_{\text{th}}(T) = P_{\text{th},0}\exp(T/T_0) \tag{2.4.7}$$

where $T_0$ is the characteristic temperature (extracted as 308 K), signifying the capacity of the temperature-tolerance of SE. The large value of $T_0$ (typical $T_0$ ~ 100 K of bulk or quantum well-based lasers and perovskite microcrystal) indicates that the temperature-insensitive SE action of $FAPbBr_3$ QDs [71].

Leveraging on the efficient nonlinear absorption and emission in the $FAPbBr_3$ QDs, low-threshold (310 μJ $cm^{-2}$) and high quality (Q) factor (~1150) two-photon-pumped WGM lasing is achieved by incorporating the perovskite $FAPbBr_3$ QDs into optical resonators of microtubes (40 μm diameter). Figure 11.26c displays that with the increased pump intensity, the multiple narrow PL peaks appears from the microcavity embedded with $FAPbBr_3$ QDs, indicating the development of lasing action. In Fig. 11.26d, the optical image of cylindrical microcapillary tube under lasing action was shown, indicating the bright lasing output.

In order to obtain a more stable multiphoton absorption in the air, benzyl alcohol (BnOH)-treated 11 nm sized $MAPbBr_3$ QDs were chosen for nonlinear investigation [58]. Here, we mainly focus on the SE action pumped under two-photon (800 nm) absorption. Excited by 800 nm pump pulses (150 fs, 1 kHz), the evolution of output intensity from spontaneous emission to SE (Fig. 11.27) confirms the nonlinear two-

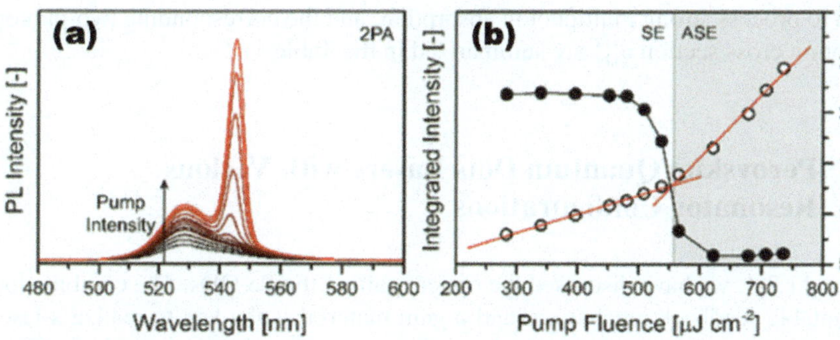


**Fig. 11.27** Excited steady-state PL emission spectra of BnOH-modified $MAPbBr_3$ QDs under two-photon absorption. **a** Pump intensity-dependent the emission spectra from the BnOH/$MAPbBr_3$ QDs film under 800 nm 100 fs laser excitation. **b** The corresponding integrated PL intensity and FWHM as a function of the pump fluence under two-photon excitation. Reprinted with permission from Ref. [58]. Copyright 2017, Published by American Chemical Society

photon pumped absorption and emission process. The SE threshold ~570 $\mu J\ cm^{-2}$ is among the lowest threshold recorded for halide perovskites (see Table 11.4 for comparison). The benzyl alcohol can passivate the QDs surface (Fig. 11.13c) and enhance the robustness toward chemical or photo-chemical changes of the $MAPbBr_3$ QDs under ambient conditions. Thereby, after 4 months storage in air, the untreated samples' two-photon pumped absorption threshold increased about 1.5 time to approximately ~1.17 $mJ\ cm^{-2}$, where the SE thresholds of BnOH/$MAPbBr_3$ QDs remain unchanged (Fig. 11.13a). It highlighted that the fairly low SE threshold and photostability can be improved by surface ligand engineering and chemical treating. By virtue of detailed single-beam Z-scan technique, the $\sigma_{abs}^{(2)}$ of $MAPbBr_3$ QDs is calculated to be $5.23 \times 10^6$ GM (and $\beta \sim 4.18\ cm\ GW^{-1}$) [81], both values are orders of magnitude higher compared with $CsPbBr_3$ QDs/toluene sample in Ref [51]. This difference may be attributed to the larger organic group methylamine (MA) has more flexible and changeable spatial polarization orientation than Cs cation under light exposure [81].

Lead-Free Halide Perovskites

Previous literatures have indicated that Sn, Ge, Cu, Bi, and Sb ions could be alternative ions and form a new environmentally-friendly lead-free perovskite structure. Compared to their lead-based analogues, the lead-free halide Pe-QDs show quite low PL QY (<1%) under ambient conditions and hence showing disappointed lasing performances. To the best of our knowledge, multiphoton-pumped SE based on the lead-free halide QDs has not yet been achieved. On the other hand, the less lead-containing Pe-QDs, such as the transition metal ions-doped $CsPbX_3$ QDs, have been shown to process strong multiphoton absorption, and the corresponding two-photon absorption cross section $\sigma_{abs}^{(2)}$ are summarized in the Table 11.5.

## 11.3 Perovskite Quantum Dots Lasers with Various Resonator Configurations

In Sect. 11.2.4, we have discussed the optical gain in the Pe-QDs. The combination of a suitable feedback mechanism and a gain material is the key to realize a laser device. In this regard, a variety of high-quality optical resonators are used to achieve favorable coherent light output based on Pe-QDs.

### *11.3.1 Random Lasers*

Random lasers are unconventional lasers where the lasing action comes from the multiple scattering in disordered materials. They are also the simplest laser configuration which avoids the extra resonator cavity and manifests low cost fabrication. The optical feedback is offered by the constructive interference of the scattered light in the disordered system. It has been demonstrated that random lasers can be engineered to provide low spatial coherence and fulfill the speckle-free full-field imaging [88].

The first perovskite-based random lasing was reported in $MAPbI_3$ planar microcrystal networks in 2014 [89]. Random lasers can be employed as a new class of light source for imaging with high photon degeneracy/spectral radiance and low spatial coherence, as displayed in Fig. 11.28a [88]. The perfect crystalline structure of the Pe-QDs makes them as the favorable candidates in the random lasers because the grains in the Pe-QDs film can serve as the multiple scattering centers for the optical gain. Combined with the monodispersed silica spheres, Li et al. developed an amination-mediated nucleation and growth strategy to anchor the $CsPbBr_3$ QDs onto the surfaces of monodisperse silica, which can avoid the aggregation of the Pe-QDs [90]. Hence, the stable random lasing (Fig. 11.28b, inset: the SE threshold ~40 $\mu J\ cm^{-2}$) based on the $CsPbBr_3$ QDs/aminated-$SiO_2$ spheres can be realized (inset the Fig. 11.28c: schematic description of random lasing in the $CsPbBr_3$ QDs/aminated-$SiO_2$ spheres film). Moreover, the random lasing emission can be collected in different angles, as shown in Fig. 11.28c. Different lasing spectra at different collecting angles under fixed pump intensity is consistent with the characteristic of a coherent random laser. Similarly, by incorporating $CsPbBr_3$ QDs into inorganic $TeO_2$-based glass ($CsPbBr_3$ QDs@glass), angle-unlimited random up-conversion lasing (the SE threshold ~200 $\mu J\ cm^{-2}$) has been demonstrated, as shown in Fig. 11.28d [91]. The snap of the robust up-conversion lasing above threshold is inset the Fig. 11.28d. Also, the random laser based on the $CsPbBr_3$ QDs@ZnO under

**Table 11.5** Summary of the two-photon absorption cross section of transition metal ions-doped perovskite quantum dots. N/A evidences non available data

| Active perovskites/References | Pump wavelength (nm) | Pump pulse length | PL wavelength (nm) | Two-photon absorption coefficient $\beta$ (cm $GW^{-1}$) | Two-photon absorption cross section $\sigma^{(2)}_{abs}$ (GM) |
|---|---|---|---|---|---|
| $Mn^{2+}$-$CsPbCl_3$ QDs [85] | 720 | fs pulse | 410 and 600 | N/A | $3.18 \times 10^5$ |
| $Fe^{3+}$-$CsPb(Cl/Br)_3$ QDs [86] | 800 | 140 fs | 514 | 210 | N/A |
| $Ni^{2+}$-$CsPbCl_3$ QDs [87] | 800 | 50 fs | 566 | 0.1 | $4.12 \times 10^4$ |

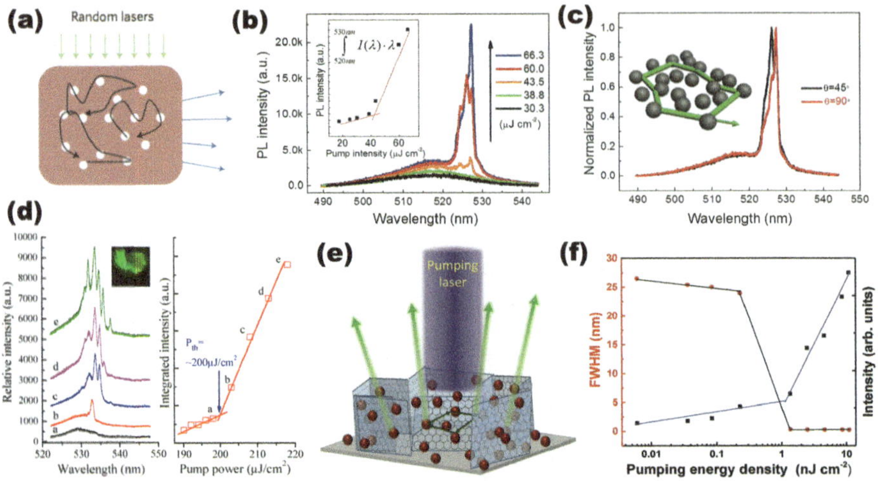


**Fig. 11.28** Random lasing based on the metal halide Pe-QDs. **a** Schematic illustration of random lasing in a gain media. Reprinted with permission from Ref. [88]. Copyright 2012, Published by Nature. **b** Excitation intensity dependence of PL spectra from the $CsPbBr_3$ QDs/aminated-$SiO_2$ spheres composite film. Inset: the plot of the integrated lasing intensity versus pump intensity. **c** Lasing spectra collected at various angles under fixed pump intensity. Inset: the schematic illustration of random lasing in $CsPbBr_3$ QDs/aminated-$SiO_2$ spheres. Reprinted with permission from Ref. [90]. Copyright 2017, Published by John Wiley & Sons. **d** Up-conversion random lasing spectra of $CsPbBr_3$ QDs@glass. Reprinted with permission from Ref. [91]. Copyright 2018, Published by American Chemical Society. **e** Schematic illustration of random lasing by virtue of vertical graphene nano-walls. **f** The plot of the integrated lasing intensity and FWHM versus pump intensity in $MAPbBr_3$ QDs/Ag/$SiO_2$ composition. Reprinted with permission from Ref. [92]. Copyright 2018, Published by John Wiley & Sons

400 nm and 800 nm fs excitation has been reported [70]. All the $SiO_2$ spheres, glass and ZnO nanoparticles can exhibit strong scattering effects and provide the extra scattering loop, which are pivotal to realize a random up-conversion lasing. By depositing $MAPbBr_3$ QDs on a heterostructures of porous nest-like 3D graphene nano-walls (see the Fig. 11.28e for the schematic illustration), low threshold random lasing (Fig. 11.28f) can be achieved by virtue of strong scattering of emitted photons and trapped photons by leachy vertical SiC networks [92]. The highly confined photons can overcome the reflection loss and provide a coherent emission of photons. The lasing threshold can be further lowered by the combined effect of the improved scattering cross section and plasmonic field enhancement by introducing extra Ag/$SiO_2$ particles.

### 11.3.2 Distributed Feedback Lasers

An advantage of the distributed feedback (DFB) laser is the low cost and high reproducibility in device manufacturing, which can realize the pre-set and tunable lasing wavelength. By virtue of DFB cavity, single-mode operation (with single frequency and narrow emission) can be achieved in macroscopic laser devices. The single-mode output of lasers can be applied in sensing, optical communication and spectroscopy [93]. A DFB laser structure is basically made by an active waveguide (gain medium), where the optical feedback comes from the Bragg scattering within the partially reflected media at the wavelengths satisfying the Bragg condition by periodical modulation of the refractive index. The schematic illustration of the DFB lasers is shown in the Fig. 3.2.1a. The Bragg condition is given by:

$$2n_{eff}\Lambda = m\lambda_{Bragg} \tag{3.2.1}$$

where $n_{eff}$ is effective refractive index of the corrugated structure, $\Lambda$ is the periodicity, $m$ is diffraction order, $\lambda_{Bragg}$ is the Bragg wavelength. The first- ($m = 1$, 1st) and second-order ($m = 2$, 2nd) DFB cavities corresponding edge- and surface-emitting lasers, respectively. This characteristic indicates the laser based on DFB cavity only works for a precisely patterned active medium (corrugated structure, Fig. 3.2.1a, processed by micro-fabrications mainly include lithography and molding) [12]. High-performance DFB lasers have been realized with an amount of active gain materials including semiconductor polymers, dyes, organic and inorganic media [93]. Especially, the metal halide perovskites with "soft crystalline structure" can be readily combined with DFB cavity geometry through micro-fabrication technique. The first perovskite-based DFB structure was demonstrated in year of 2015 in virtue of nanoimprinting lithography. The SE emission can be tuned between 770 and 793 nm simply by changing the grating periodicity $\Lambda$ [93]. Treated as such, continuous-wave (CW) lasing in 2nd DFB $MAPbI_3$ perovskite film on a silicon substrate at room temperature has been realized, the ultralow lasing threshold ~13 W $cm^{-2}$ [94]. Furthermore, the short pulsed current (durations $\leq$25 ns, repetition rates exceeding 2 MHz) drive to minimize threshold has been reported by virtue of diode-pumped DFB lasers based on $MAPbI_3$ perovskites film [95]. It represents a key step towards the electrically-driven perovskite lasers. Inspired by these previous researches, multi-color room-temperature-operated CW laser has been realized by depositing Pe-QDs-embedded polyacrylonitrile (PAN) films into a DFB cavity [96]. Excited by a 405 nm semiconductor CW laser (schematic setup in Fig. 3.2.1b), the CW lasing threshold power is as low as 15 W $cm^{-2}$ (Fig. 3.2.1d). The laser peak centered at 538.7 nm with a FWHM (full width at half-maximum) of 0.45 nm, corresponding to a Q value of ~1200 (Fig 3.2.1c). The emission color can be tuned into red and blue by simply varying the $PbX_2$ salts in the precursors, highlighting the potential to integrate the laser source into on-chip photonics based on Pe-QDs (Fig. 11.29).

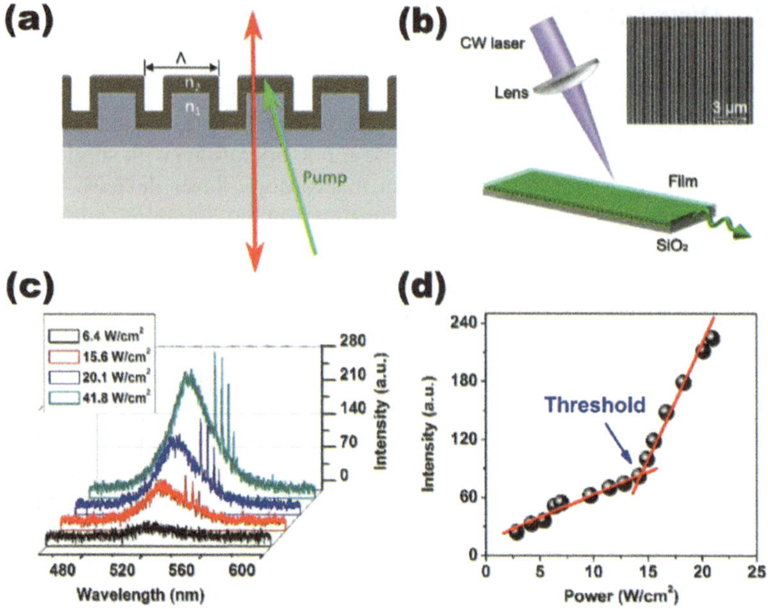


**Fig. 11.29** DFB lasing based on the metal halide Pe-QDs. **a** The schematic illustration of a generic DFB cavity. Reprinted with permission from Ref. [93]. Copyright 2016, Published by John Wiley & Sons. **b** The schematic setup of a 405 nm CW laser at room-temperature. **c** Emission spectra of the $MAPbBr_3$ QDs/PAN film on the DFB cavity. **d** The plots of emission intensity centered at 538.7 nm as a function of excitation power. Reprinted with permission from Ref. [96]. Copyright 2019, Published by American Chemical Society

### *11.3.3 Whispering-Gallery-Mode Lasers*

Most reported perovskite-based lasers are made from whispering-gallery-mode (WGM) cavities, in which feedback is provided by optical modes around the circumference of a circular or polygonal resonator by total internal reflection at the cavity boundaries [1]. WGM is often observed in dielectric spheres of index $n$ and radius $a$ with large size parameters to allow for multi-wavelengths in the sphere perimeter ($k$ is the wavenumber and $\lambda$ the wavelength in vacuum).

$$X = ka = 2\pi\, na/\lambda \tag{3.3.1}$$

Standing waves at discrete wavelengths will form when the emitted waves-induced oscillation in any closed loop. Multi-mode laser light will emit when pumping the gain media above gain threshold. But obvious disadvantages of WGM lasing is the lack of directional lasing output and multi-mode operation will have problems of false signaling and random fluctuation, exasperating the color purity and device stability due to mode oscillation.

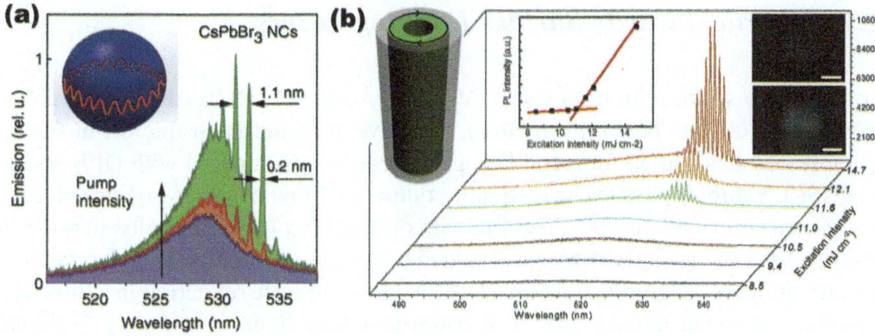


**Fig. 11.30** WGM lasing based on $CsPbBr_3$ QDs. **a** Evolved WGM lasing action with increased pump intensity in a silica microsphere covered by a film of $CsPbBr_3$ QDs. Reprinted with permission from Ref. [3]. Copyright 2015, Published by Nature. **b** WGM lasing action realized by infiltrating $CsPbBr_3$ nanocrystals into a capillary tube. The right insets show the $CsPbBr_3$ QDs infiltrated capillary tube below (up) and above (down) the lasing threshold. Scale bar is 20 μm. Reprinted with permission from Ref. [4]. Copyright 2015, Published by John Wiley & Sons

The precursory observation on WGM lasing based on the colloidal CdSe QDs [97] and CdSe/ZnS quantum rods [98] has been demonstrated around year of 2000, nearly the same time with the most revealing report on the favorable optical gain and SE in nanocrystal QDs [47]. By virtue of both colloidal nature and structural softness, the Pe-QDs can be well-assigned to the WGM cavities. Cylindrical or spherical microcavities, which have a diameter of a few microns, can naturally serve as WGM cavities which have high quality factor (Q factor) and extremely low leakage of the emitted light. Soon after the advent of $CsPbX_3$ QDs in 2014, the $CsPbX_3$ QDs-based WGM lasing was reported by Kovalenko et al. and Sun et al. nearly simultaneously in 2015 [3, 4]. Kovalenko et al. coated the $CsPbBr_3$ QDs onto silica microspheres (radius $a$ ~15μm), as shown inset Fig. 11.30a, where the emitted light orbits around the circumference of silica spheres. Excited at the wavelength 400 nm with 100 fs laser pulses, sharp SE peak with the narrow linewidths of lasing modes (0.15–0.20 nm), (Fig. 11.30a), indicating the excellent WGM lasing performance. By infiltrating $CsPbBr_3$ nanocrystals into a capillary tube, (radius a ~25μm, Fig. 11.30b) Wang et al. realized the high-quality Pe-QDs-based WGM lasers with a high Q factor of ~2000, the Q factor is calculated based on the equation

$$Q = \lambda/\Delta\lambda, \quad (3.3.2)$$

where $\lambda$ is lasing spike wavelength and $\Delta\lambda$ is the lasing linewidth. As shown in Fig. 11.30b, both uniformly-spaced lasing spectra and low lasing threshold indicate the occurrence of excellent WGM lasing action at the interface of solid $CsPbBr_3$ QDs film and capillary tube [4].

### 11.3.4 Vertical Cavity Surface Emitting Lasers

Vertical cavity surface emitting lasers (VCSEL) feature low lasing threshold, unidirectional output (low beam divergence), and have attracted great interest in fundamental researches and commercial lasing applications. Compared with DFB lasers, the VCSEL exhibiting better beam quality output and low beam divergence characteristics can contribute to easy coupling with optical fiber [99]. Basically, in order to achieve a VCSEL, an active film was cladded between two in-plane highly-reflecting dielectric mirrors. The optical feedback is given by multiple reflection in a direction perpendicular to the mirrors, and well-controlled gain layer morphology is pivotal to realize VCSEL. If active layers are thin enough, single-mode emission with high color purity and temporal stability can be achieved [100, 101]. The first demonstration of metal halide perovskite-based VCSEL was reported by Deschler et al. in 2014 [22] and first realization of the high-performance VCSEL from solution-processed $CsPbX_3$ QDs was reported by Wang et al. in 2017 [101]. The schematic diagram of VCSEL employed in Wang's work is shown inset the Fig. 11.31a. As shown in Fig. 11.31a, a sharp and robust lasing peak around 510 nm, narrow spectral (FWHM ~0.6 nm) and nonlinear PL intensity versus pump intensity (the threshold ~11 μJ $cm^{-2}$) indicate the achievement of lasing action in $CsPbX_3$ QDs-based VCSEL. By tuning the halide composition, blue-emitting $CsPb(Br/Cl)_3$ QDs and red-emitting $CsPb(I/Br)_3$ QDs VCSEL can be obtained (Fig. 11.31b), paving the way towards single-source pumped white lasers. To date, the lowest reported threshold in $CsPbBr_3$ QD lasers is 0.39 μJ $cm^{-2}$ [102], the cavity type of which is also VCSELs, as depicted in Fig. 11.31c. Note that this VCSEL manifests the superior long-term stability under both femtosecond (short-pulse) and nanosecond (quasi-continuous-wave) pumping regimes (Fig. 11.31d).

### 11.3.5 Multicolored Laser Arrays

The laser arrays are initially raised for semiconductor lasers power scaling and have become an important branch of on-chip integration of photonic and electronic circuits [12]. Multicolored laser arrays with emissions across the whole visible spectrum have great potential in the applications of full-color laser display, laser lighting, and sensing. Actually, the fabrication of well-aligned perovskite-based laser arrays is the most essential to realize high-performance laser arrays. So far, the miniaturized perovskite-based laser arrays have been demonstrated in crystal structures like nanowire [103], microwire [104], microdisk [105, 106], and mircoplates [107, 108] in virtue of fabrication techniques based on liquid-phase self-assembly and vapor-phase deposition (bottom-up strategy). One common feature of these high-performance laser arrays is the controllable growth of single-crystalline perovskite with uniform sizes, precise positioning, and high quality.

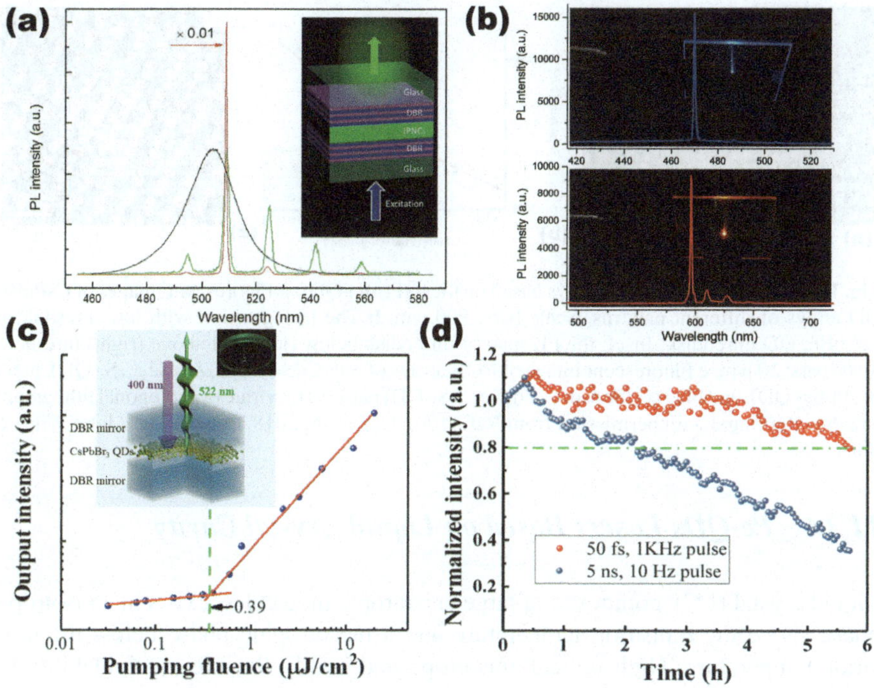

**Fig. 11.31** VCSEL based on $CsPbX_3$ QDs. **a** High-performance VCSEL in virtue of the $CsPbBr_3$ QDs sandwiched between two DBRs and corresponding schematic illustration of adopted VCSEL. **b** The robust lasing spectral of blue- and red-emitting VCSEL. Reprinted with permission from Ref. [101]. Copyright 2017, Published by John Wiley & Sons. **c** The ultralow pumping threshold of $CsPbBr_3$ QDs-based VCSEL and the schematic illustration of the cavity configuration. **d** The temporal stability of $CsPbBr_3$ QDs-based VCSEL under both fs and ns pulsed excitation with the pump fluence above threshold. Reprinted with permission from Ref. [102]. Copyright 2017, Published by American Chemical Society

By virtue of high-resolution lithographic methods, the Cd-based QDs have been successfully integrated into micro- and nanoscale devices. But the intrinsic ionic nature of $CsPbX_3$ QDs yields them not compatible with the traditional lithographic methods. In order to overcome this obstacle, Lin et al. proposed a novel top-down orthogonal lithography strategy to form large-area high-resolution $CsPbBr_3$ QDs laser arrays [109]. With the assistance of fluorinated polymer as a sacrificial layer and standard photolithography, the $CsPbBr_3$ QDs attached to the substrate remain strong emissive, the synthesized large-area (2 cm × 2 cm) $CsPbBr_3$ QDs laser arrays are depicted in Fig. 11.32a. The performance of $CsPbBr_3$-based laser in the cavity of QD microdisk (WGM, total internal reflection at the circumference, inset the Fig. 11.32b) is investigated, showing a low lasing threshold of ~200 $\mu J\,cm^{-2}$. Furthermore, leveraging on the orthogonal lithography strategy, the RGB QDs arrays with high-resolution ~1000 pixels per inch are realized, the fluorescent image of QDs patterns is shown in the Fig. 11.32c.

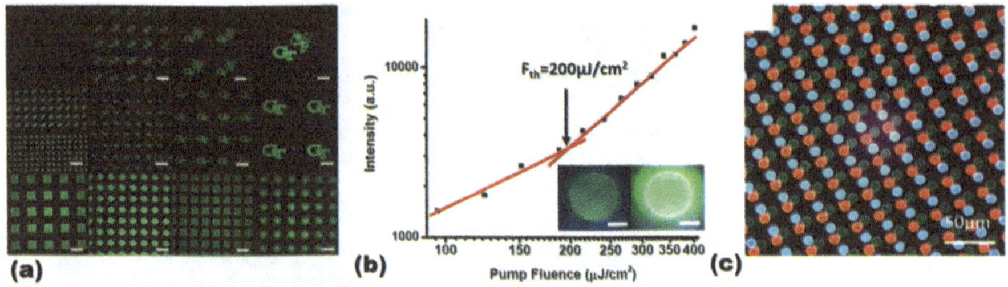


**Fig. 11.32** Multicolored laser arrays based on the $CsPbBr_3$ QDs. **a** Fluorescent images of $CsPbBr_3$ QD arrays of different patterns. Scale bars: 200 μm. **b** The lasing action with low threshold in $CsPbBr_3$ QD mircodisk. Inset: the PL snap of mircodisk below (left) and above (right) threshold. Scale bars: 10 μm. **c** Fluorescent image of RGB arrays of red ($CdSe/Cd_{1-x}Zn_xSe_{1-y}S_y$ QD), green ($CsPbBr_3$ QD), and blue ($CdSe/Cd_{1-x}Zn_xSe_{1-y}S_y$ QD) patterns by virtue of orthogonal lithography strategy. Reprinted with permission from Ref. [109]. Copyright 2018, Published by John Wiley & Sons

## *11.3.6 Pe-QDs Lasers Based on Liquid Crystal Cavity*

Liquid crystal (LC), composed of large anisotropic molecules, exists in an isotropic phase above the transition temperature and a mesotrophic phase below the transition temperature. High optical anisotropy and highly flexible manipulability by external sources (e.g. temperature, electric field) yield LC as a high-quality microresonator instead of traditional inorganic Bragg reflectors. Liquid crystal-based lasers feature wavelength-tunability, large area of coherence and potential for multidirectional outputs have become a burgeoning area in the field of soft-materials photonics [110]. Note that in a LC cavity, the lasing can be suppressed inside the bandgap and boosted at the band-edges (lasing action occurs when one edge of bandgap overlaps the emission spectrum with maximize coupling efficiency), which is known as the LC photonic bandgap, the corresponding laser is termed as band-edge LC laser [110].

The cholesteric phase, (cholesteric liquid crystal, short for CLC) as a type of the mesotrophic phase, exhibits strong nonlinear optical effects caused by highly correlated orientation of adjacent molecules inside. Soon after the successful demonstration of lasing action in $MAPbX_3$ (X = Cl, Br and I) perovskite thin films, Stranks et al. reported robust and stable SE behavior from the $MAPbI_3$ perovskite films within a glass/CLC/$Al_2O_3$/$MAPbI_3$/PMMA/Au sandwiched structure [111]. This work indicated the CLC reflectors offer a new promising strategy to realize single-mode lasers on flexible substrates. Combined with the Pe-QDs as high gain media, Chen et al. fabricated the $CsSnI_3$ QDs-doped lasers with planar CLC cavities and demonstrated the tunable lasing-wavelength and long-term stability under ambient condition at room temperature [112]. The microscopy image of the planar texture of $CsSnI_3$ QDs-CLC cell was observed under polarizing optical microscopy, as shown in Fig. 11.33a. The well-dispersed $CsSnI_3$ QDs-CLC mixtures (perfect planar texture

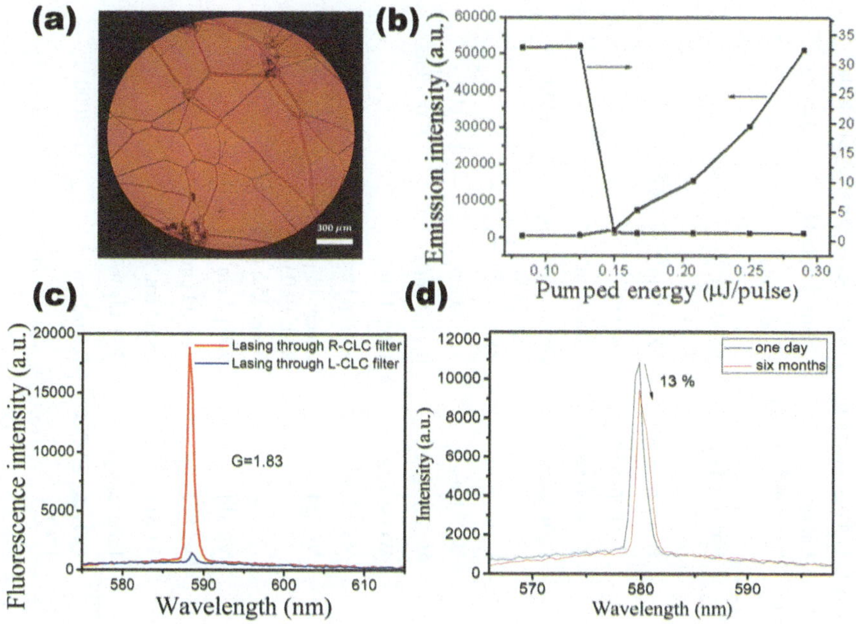


**Fig. 11.33** Stable lasing action in $CsSnI_3$-doped CLC cavities. **a** The microscopy image of the planar mixtures of $CsSnI_3$ QDs-CLC observed under polarizing optical microscopy. **b** The emission intensity and FWHM dependence of pumped energy with a threshold of 0.15 μJ/pulse. **c** The measured lasing spectra of $CsSnI_3$ QDs-CLC through right- or left-handed CLC filter inserted between the cell and the spectrometer. **d** The long-term stability of $CsSnI_3$ QDs-CLC laser experiment. Reprinted with permission from Ref. [112]. Copyright 2018, Published by American Chemical Society

domains in Fig. 11.33a) ensure the occurrence of excellent lasing action. Pumped at the wavelength of 532 nm with pulse width of 8 ns, the emission intensity and FWHM as a function of pumped energy are plotted in Fig. 11.33b. The typical band-edge lasing mechanism can be confirmed with the measurement of emission intensity through right- and left-handed CLC filter (Fig. 11.33c), indicating the perfect left-circular property for the lasing emission. The long-term stability of $CsSnI_3$ QDs-CLC-based laser is shown in Fig. 11.33d, and the lasing action can be tuned by external stimuli like temperature and electric field, in accordance with the previous report on the favorable wavelength-tunability of LC-based laser [110]. The lasing properties based on the Pe-QDs with above-mentioned resonator configurations are summarized in Table 11.6.

**Table 11.6** Summary of lasing properties based on the perovskite quantum dots with various resonator configurations. N/A evidences non available data

| Active perovskites/References | Pump wavelength (nm) | Pump pulse length | Peak wavelength (nm) | Lasing configuration | Lasing threshold | Q factor |
|---|---|---|---|---|---|---|
| $CsPbBr_3$ QDs/aminated-$SiO_2$ spheres film [90] | 400 | 100 fs | ~530 | Random laser | 40 μJ $cm^{-2}$ | |
| $CsPbBr_3$ QDs@glass [91] | 800 | fs pulse | ~530 | Random laser | 200 μJ $cm^{-2}$ (77 K) | |
| $MAPbBr_3$ QDs/Ag/$SiO_2$ [92] | 374 | 55 fs | ~540 | Random laser | 1 nJ $cm^{-2}$ | |
| $CsPbBr_3$ QDs-$SiO_2$ [113] | 800 | 35 fs | 537 | Random laser | 665 μJ $cm^{-2}$ | |
| $MAPbBr_3$ QDs/PAN film [96] | 405 | N/A | Green: 539<br>Red: 614<br>Blue: 482 | DFB laser | Green: 15 W $cm^{-2}$<br>Red: 24 W $cm^{-2}$<br>Blue: 58 W $cm^{-2}$ | ~1200 |
| $CsPbBr_3$ QDs [4] | 400 | 100 fs | ~530 | WGM laser | ~11 μJ $cm^{-2}$ | ~1600 |
| $CsPbBr_3$ QDs [82] | 800 | 90 fs | ~536 | WGM laser | 0.9 mJ $cm^{-2}$ | 1700–3500 |
| $CsPbBr_3$ QDs-$SiO_2$ [113] | 800 | 35 fs | 530 | WGM laser | 430 μJ $cm^{-2}$ | ~1532 |
| $FAPbBr_3$ QDs [71] | 800 | 35 fs | 541 | WGM laser | ~310 μJ $cm^{-2}$ | ~1150 |
| $CsPbBr_3$ QDs [101] | 400 | 100 fs | Green: ~510<br>Red: ~590<br>Blue: ~470 | VCSEL | Green: 11 μJ $cm^{-2}$<br>Red: 19.0 μJ $cm^{-2}$<br>Blue: 25.5 μJ $cm^{-2}$ | |
| $CsPbBr_3$ QDs [101] | 400 | 5 ns | ~505 | VCSEL | 900 μJ $cm^{-2}$ | |
| $CsPbBr_3$ QDs [102] | 400 | 50 fs | 522 | VCSEL | 0.39 μJ $cm^{-2}$ | |
| $CsPbBr_3$ QDs [102] | 355 | 5 ns | 522 | VCSEL | 98 μJ $cm^{-2}$ | |
| $CsPbBr_3$ QDs [109] | 450 | 7 ns | 534 | Laser arrays | 200 μJ $cm^{-2}$ | 500–700 |
| $CsSnI_3$ QDs-CLC | 532 | 8 ns | 579 | QDs-doped CLC laser | 0.8 mJ $cm^{-2}$ | ~2000 |

## 11.4 Issues and Challenges

Although these Pe-QDs have been demonstrated as the outstanding candidates for developing solution-processed lasers, there remain issues or challenges with regard to developing practical and commercially available lasers utilizing the Pe-QDs.

### *11.4.1 Stability*

Practical lasing applications are severely hampered by the relatively poor stability of Pe-QDs affected by chemical and environmental sources (e.g., oxygen, moisture, heat, and continuous light illumination). Both photo- and thermally induced aggregation of Pe-QDs will generate rich defects and leads to the degradation of PL QY. The intrinsic characteristics of ionic-crystal nature and low crystal lattice energy render that Pe-QDs are prone to dissolve in all polar solvents. To resolve these issues, encapsulated Pe-QDs with other optoelectronic materials (host matrix like silica [84, 113], polymers [96], and nanoparticles [70, 92]) will be a possible strategy to avoid the degradation caused by atmospheric environment. Frequency up-converted WGM and random lasing at room temperature have been reported in virtue of the $CsPbBr_3$ QDs evenly distributed in silica sphere [113]. Have mentioned in Sect. 11.3.1, Li et al. developed an amination-mediated nucleation and growth strategy, anchoring $CsPbBr_3$ QDs onto the surfaces of monodisperse silica, effectively depressed the optical degradation and hence, achieving highly stable and low-threshold lasing (Fig. 11.28b and c) [90]. In another case, Yuan et al. fabricated the $CsPbBr_3$ QDs in glass matrix through in situ crystallization synthesis, which can not only protect the nanocrystals from the ambient conditions, but also greatly hinder the aggregation of these Pe-QDs. As such, stable lasing emission was achieved (Fig. 11.28d) [91]. Huang et al. synthesized $SiO_2$-encapsulated $MAPbBr_3$ QDs based on the high hydrolysis rate of tetramethyl orthosilicate in “waterless” toluene to avoid the addition of water and catalyst. As a result, the PL of the encapsulated QDs powders was maintained at 94.10% after 470 nm LED illumination (21 mW $cm^{-2}$) for 7 h [114]. It is noted that by inserting $CsPbBr_3$ QDs into a wider bandgap $Cs_4PbBr_6$ matrix, the thermal stability of Pe-QDs is greatly improved and eventually realized robust high-temperature perovskite lasers [115].

Surface ligands treatment is an another simple and effective strategy to improve the stability of Pe-QDs. Organic ligands attached to the surface of $CsPbBr_3$ QDs are vulnerable and thermal degradable. By introducing 2-hexyldecanoic acid (DA) with larger binding energy during the synthesis process, DA-modified $CsPbBr_3$ QDs exhibit no aggregation indication when kept in air for over 70 days [69]. Besides, the SE thresholds of DA-modified $CsPbBr_3$ QDs for both under one- and two- photon excitation are significantly reduced and the SE photostability are improved, compared with the pristine $CsPbBr_3$ QDs. New ligand capping strategy was proposed by Krieg et al. who took ordinary long-chain zwitterionic molecules to replace the conventional surfactants (oleylamine (OAm) and oleic acid (OA)) [66]. Long-ligands can contribute to the isolation of clean $CsPbBr_3$ QDs and allow the ultralow SE threshold under femtosecond optical excitation. Also, the benzyl alcohol (BnOH)/$MAPbBr_3$ QDs manifest ultralow ASE thresholds in virtue of the enhanced surface-trap passivation on the $MAPbBr_3$ QDs surface (see Fig. 11.13) [58]. It has been validated with the use of benzoyl halides as a provider for halides-rich environment [68], the PL QY of as-synthesized perovskite $APbBr_3$ (A = Cs, MA, FA) is around 90%, 55% for $APbI_3$ systems. The record PL QY value (~65%) in $CsPbI_3$ QDs has a potential to dismantle the "perovskite red wall" as robust red-emitting media.

In addition, other methods such as doping and X-ray radiation have also been demonstrated as effective strategies to improve the Pe-QD stability.

### *11.4.2 Toxicity*

Most of the existing above-mentioned research progresses are mainly made in developing Pe-QDs lasers based on lead-containing ones. The use of toxic heavy-metal lead (Pb) in their chemical composition is harmful to the ecosystem and human health, which is counter to constructing an environment-friendly society. As a result, it will probably become an indispensable trend to study non-toxic QDs and developing heavy metal-free Pe-QDs laser. Up to now, several Pb-substituted halide perovskites has been reported with nontoxic candidate elements as follow: (1) tin; (2) germanium; (3) bismuth; (4) indium; (5) antimony; (6) copper; (7) double perovskites, etc. Note that solution-processed tin-based halide perovskites $CsSnX_3$ have been demonstrated with exceptional optical gain properties in the near-infrared lasing [17]. (Section "Lead-Free Halide Perovskites") Previous literatures have demonstrated that these low-toxic perovskites feature higher optical absorption, higher charge mobilities ($10^2$ to $10^3$ $cm^2\ V^{-1}\ s^{-1}$) and narrower bandgap, which yield them as promising photosensitive materials in the application of photodetectors and solar cells [116]. However, in the applications of laser regimes, there are barely no reports due to its instabilities under ambient conditions compared to their lead-based analogues. More efforts need to be put into the photophysical study, wet-chemistry fabrication and characterization of these lead-free perovskites. From the perspective of lead-containing perovskites, stability enhancement, better encapsulation technologies, and integrated recycling utilization improvement will reduce and avoid the toxicity of Pb ions.

### *11.4.3 Towards Electrically Pumped Lasing*

Theoretically, multiple excitons ($\langle N \rangle > 1$) are needed to obtain gain due to intrinsic band-edge state degeneracy in Pe-QDs. However, non-radiative Auger recombination raised from multiexcitonic nature will increase the lasing threshold and limit the gain lifetime. In this regard, single-exciton-regime will be a favorable approach for achieving low gain threshold which can eliminate the problem of Auger decay and prolong the optical gain lifetime [77]. Low gain threshold can contribute to the CW operated lasers and the optically pumped CW lasing is an critical step before moving forward to realizing the electrically pumped laser diodes. High-quality crystalline Pe-QDs and optimized low-cost cavity architectures are favorable in order to further decrease the threshold for CW lasing. Recently, room-temperature-operated CW laser has been realized by depositing Pe-QDs-embedded polyacrylonitrile (PAN) films into a DFB cavity [96]. The achieved ultralow threshold values for green, red, and blue lasers are one order lower than the existing values for the Cd-chalcogenide-based QDs CW lasers, considered as the milestone in achieving the goal of on-chip color-tunable CW lasers at room temperature.

In all, more work needs to be done to drive perovskite-based LEDs into the more increased current density regimes for lasing. Schematic illustration of electrically injected laser is depicted in Fig. 11.34a. There are mainly three challenges ahead: (a) the heat management; (b) the balanced charges; (c) integration between electrodes and the active material [1, 100]. Moreover, the carrier injection and transport are hindered from the organic ligands attached to the surface of Pe-QDs, which generally exhibit poor electric conductivity. As shown in Fig. 11.34b, compared with the commercial injection lasers, such as GaAs and InP, perovskites exhibit small thermal conductivity and carriers mobilities, where detrimental heat will generate under current injection. Non-radiative Auger recombination will also result in the extra heat. Further obtaining the low gain threshold and realizing efficient current injection into the gain layer would be the all-important task towards electrically-pumped Pe-QDs lasers.

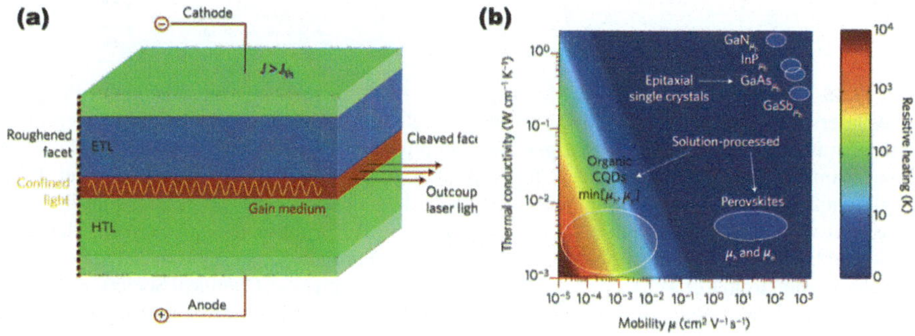


**Fig. 11.34** Electrically driven laser and the resistive heating under current of existing semiconductors. **a** Schematic illustration of electrically injected laser. Lasing action occurs when the injected current is greater than the stimulated emission threshold current $J > J_{th}$. **b** The resistive heating under typical current injection of 1 kA $cm^{-2}$ of the existing compound semiconductor injection lasers. Reprinted with permission from Ref. [1]. Copyright 2016, Published by Nature

## 11.5 Summary and Outlook

Pe-QDs have widely attracted attention in the development of practical and cost-effective lasers and manifest advantages over the conventional metal-chalcogenide nanocrystals. Figure 11.35 shows the timeline of tremendous progress made in developing Pe-QDs-based lasers since the first demonstration of low threshold SE and lasing in Pe-QDs. Featuring the advantages of both the QDs and halide perovskites, such as the quantum confinement effect, large absorption cross-section, wavelength-tunability and defect-tolerant nature, these Pe-QDs are competitive candidates for optical gain media. Better understanding of gain mechanism will offer a new path to construct high-performance lasers. In this regard, lasing action in single-exciton-regime will be a favorable approach to eliminate the non-radiative Auger recombination. Developing nontoxic lead-free halide perovskites and improving the stability of Pe-QDs will further help to improve the performance of Pe-QDs-based lasers for commercial utilization.

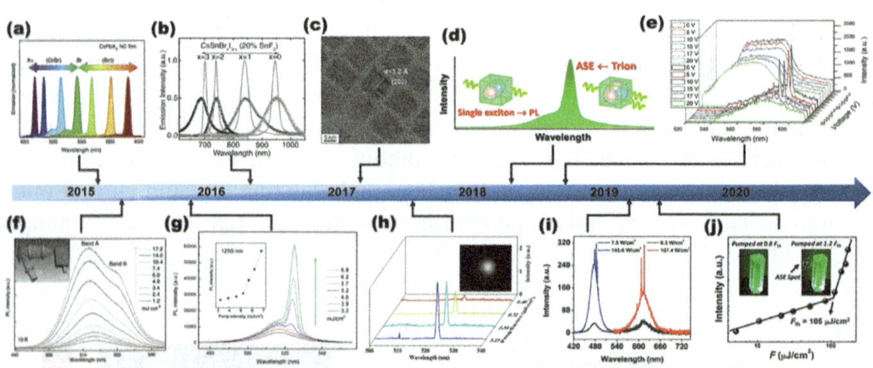


**Fig. 11.35** Timeline shows the tremendous progress made in achieving the state-of-art Pe-QDs-based lasers. **a** First achievement of tunable lasing wavelength in $CsPbX_3$ Pe-QDs in 2015. Reprinted with permission from Ref. [3]. Copyright 2015, Published by Nature. **b** Robust and stable SE in the near-infrared range in $CsSnBr_xI_{3-x}$ films in 2016. Reprinted with permission from Ref. [17]. Copyright 2016, Published by John Wiley & Sons. **c** Lasing action in $FAPbI_3$ QDs film in 2017. Reprinted with permission from Ref. [60]. Copyright 2017, Published by American Chemical Society. **d** Demonstration of trion-states-based optical gain in $CsPbBr_3$ Pe-QDs in 2018. Reprinted with permission from Ref. [7]. Copyright 2018, Published by American Chemical Society. **e** Tunable lasing action in the $CsSnI_3$ QDs-doped CLC cavities. Reprinted with permission from Ref. [112]. Copyright 2018, Published by American Chemical Society. **f** Biexcitons gain mechanism reported in $CsPbBr_3$ Pe-QDs in 2015. Reprinted with permission from Ref. [4]. Copyright 2015, Published by John Wiley & Sons. **g** Three-photon pumped SE in $CsPbBr_3$ Pe-QDs in 2015. Reprinted with permission from Ref. [51]. Copyright 2015, Published by American Chemical Society. **h** Lasing action in the $CsPbBr_3$ QDs VCSEL in 2017. Reprinted with permission from Ref. [102]. Copyright 2017, Published by American Chemical Society. **i** Room-temperature-operated CW lasers based on the Pe-QDs in 2019. Reprinted with permission from Ref. [96]. Copyright 2019, Published by American Chemical Society. **j** Lasing action with low threshold demonstrated in $CsPbBr_3$ QDs solution in 2019. Reprinted with permission from Ref. [54]. Copyright 2019, Published by American Institute of Physics

Opportunities and challenges co-exist and we envisage that the colloidal Pe-QDs will play a pivotal role in the real-word applications of photonics and optoelectronic. Stable and robust integrated lasers based on the Pe-QDs will become an promising complement to traditional epitaxial semiconductor lasers in the near future.

## References


1. B.R. Sutherland, E.H. Sargent, Perovskite photonic sources. Nat. Photon **10**, 295 (2016)
2. L. Protesescu, S. Yakunin, M.I. Bodnarchuk, F. Krieg, R. Caputo, C.H. Hendon, R.X. Yang, A. Walsh, M.V. Kovalenko, Nanocrystals of cesium lead halide perovskites ($CsPbX_3$, X = Cl, Br, and I): novel optoelectronic materials showing bright emission with wide color gamut. Nano Lett. **15**, 3692 (2015)
3. S. Yakunin, L. Protesescu, F. Krieg, M.I. Bodnarchuk, G. Nedelcu, M. Humer, G. De Luca, M. Fiebig, W. Heiss, M.V. Kovalenko, Low-threshold amplified spontaneous emission and lasing from colloidal nanocrystals of caesium lead halide perovskites. Nat. Commun. **6**, 8056 (2015)
4. Y. Wang, X. Li, J. Song, L. Xiao, H. Zeng, H. Sun, All-inorganic colloidal perovskite quantum dots: a new class of lasing materials with favorable characteristics. Adv. Mater. **27**, 7101 (2015)
5. G. Xing, N. Mathews, S.S. Lim, N. Yantara, X. Liu, D. Sabba, M. Grätzel, S. Mhaisalkar, T.C. Sum, Low-temperature solution-processed wavelength-tunable perovskites for lasing. Nat. Mater. **13**, 476 (2014)
6. F. Igbari, Z.K. Wang, L.S. Liao, Progress of lead-free halide double perovskites. Adv. Energy Mater. **9**, 1803150 (2019)
7. Y. Wang, M. Zhi, Y.-Q. Chang, J.-P. Zhang, Y. Chan, Stable, ultralow threshold amplified spontaneous emission from $CsPbBr_3$ nanoparticles exhibiting trion gain. Nano Lett. **18**, 4976 (2018)
8. T. Leijtens, G.E. Eperon, N.K. Noel, S.N. Habisreutinger, A. Petrozza, H.J. Snaith, Stability of metal halide perovskite solar cells. Adv. Energy Mater. **5**, 1500963 (2015)
9. T. Baikie, Y. Fang, J.M. Kadro, M. Schreyer, F. Wei, S.G. Mhaisalkar, M. Graetzel, T.J. White, Synthesis and crystal chemistry of the hybrid perovskite ($CH_3$ $NH_3$) PbI 3 for solid-state sensitised solar cell applications. J. Mater. Chem. A **1**, 5628 (2013)
10. G.R. Yettapu, D. Talukdar, S. Sarkar, A. Swarnkar, A. Nag, P. Ghosh, P. Mandal, Terahertz conductivity within colloidal $CsPbBr_3$ perovskite nanocrystals: remarkably high carrier mobilities and large diffusion lengths. Nano Lett. **16**, 4838 (2016)
11. X. Li, Y. Wu, S. Zhang, B. Cai, Y. Gu, J. Song, H. Zeng, $CsPbX_3$ quantum dots for lighting and displays: room-temperature synthesis, photoluminescence superiorities, underlying origins and white light-emitting diodes. Adv. Funct. Mater. **26**, 2435 (2016)
12. H. Dong, C. Zhang, X. Liu, J. Yao, Y.S. Zhao, Materials chemistry and engineering in metal halide perovskite lasers. Chem. Soc. Rev. **49**, 951 (2020)
13. G. Volonakis, A.A. Haghighirad, R.L. Milot, W.H. Sio, M.R. Filip, B. Wenger, M.B. Johnston, L.M. Herz, H.J. Snaith, F. Giustino, $Cs_2InAgC_{l6}$: a new lead-free halide double perovskite with direct band gap. J. Phys. Chem. Lett. **8**, 772 (2017)
14. X. Li, F. Cao, D. Yu, J. Chen, Z. Sun, Y. Shen, Y. Zhu, L. Wang, Y. Wei, Y. Wu, All inorganic halide perovskites nanosystem: synthesis, structural features, optical properties and optoelectronic applications. Small **13**, 1603996 (2017)
15. H. Zhu, M.T. Trinh, J. Wang, Y. Fu, P.P. Joshi, K. Miyata, S. Jin, X.Y. Zhu, Organic cations might not be essential to the remarkable properties of band edge carriers in lead halide perovskites. Adv. Mater. **29**, 1603072 (2017)
16. T.C. Jellicoe, J.M. Richter, H.F. Glass, M. Tabachnyk, R. Brady, S.E. Dutton, A. Rao, R.H. Friend, D. Credgington, N.C. Greenham, M.L. Bohm, Synthesis and optical properties of lead-free cesium tin halide perovskite nanocrystals. J. Am. Chem. Soc. **138**, 2941 (2016)

17. G. Xing, M.H. Kumar, W.K. Chong, X. Liu, Y. Cai, H. Ding, M. Asta, M. Grätzel, S. Mhaisalkar, N. Mathews, Solution-processed tin-based perovskite for near-infrared lasing. Adv. Mater. **28**, 8191 (2016)
18. Y. Wang, H. Sun, All-inorganic metal halide perovskite nanostructures: from photophysics to light-emitting applications. Small Methods **2**, 1700252 (2018)
19. R.E. Brandt, J.R. Poindexter, P. Gorai, R.C. Kurchin, R.L.Z. Hoye, L. Nienhaus, M.W.B. Wilson, J.A. Polizzotti, R. Sereika, R. Žaltauskas, L.C. Lee, J.L. MacManus-Driscoll, M. Bawendi, V. Stevanović, T. Buonassisi, Searching for "Defect-Tolerant" photovoltaic materials: combined theoretical and experimental screening. Chem. Mater. **29**, 4667 (2017)
20. J. Kang, L.W. Wang, High defect tolerance in lead halide perovskite $CsPbBr_3$. J. Phys. Chem. Lett. **8**, 489 (2017)
21. M.V. Kovalenko, L. Protesescu, M.I. Bodnarchuk, Properties and potential optoelectronic applications of lead halide perovskite nanocrystals. Science **358**, 745 (2017)
22. F. Deschler, M. Price, S. Pathak, L.E. Klintberg, D.D. Jarausch, R. Higler, S. Huttner, T. Leijtens, S.D. Stranks, H.J. Snaith, M. Atature, R.T. Phillips, R.H. Friend, High photoluminescence efficiency and optically pumped lasing in solution-processed mixed halide perovskite semiconductors. J. Phys. Chem. Lett. **5**, 1421 (2014)
23. P. Liu, W. Chen, W. Wang, B. Xu, D. Wu, J. Hao, W. Cao, F. Fang, Y. Li, Y. Zeng, R. Pan, S. Chen, W. Cao, X.W. Sun, K. Wang, Halide-rich synthesized cesium lead bromide perovskite nanocrystals for light-emitting diodes with improved performance. Chem. Mater. **29**, 5168 (2017)
24. M. Li, J. Fu, Q. Xu, T.C. Sum, Slow hot-carrier cooling in halide perovskites: prospects for hot-carrier solar cells. Adv. Mater. **31**, e1802486 (2019)
25. N. Mondal, A. Samanta, Complete ultrafast charge carrier dynamics in photo-excited all-inorganic perovskite nanocrystals ($CsPbX_3$). Nanoscale **9**, 1878 (2017)
26. N.S. Makarov, S. Guo, O. Isaienko, W. Liu, I. Robel, V.I. Klimov, Spectral and dynamical properties of single excitons, biexcitons, and trions in cesium-lead-halide perovskite quantum dots. Nano Lett. **16**, 2349 (2016)
27. T.C. Sum, N. Mathews, G. Xing, S.S. Lim, W.K. Chong, D. Giovanni, H.A. Dewi, Spectral features and charge dynamics of lead halide perovskites: origins and interpretations. Acc. Chem. Res. **49**, 294 (2016)
28. N. Mondal, A. De, S. Das, S. Paul, A. Samanta, Ultrafast carrier dynamics of metal halide perovskite nanocrystals and perovskite-composites. Nanoscale **11**, 9796 (2019)
29. G. Xing, N. Mathews, S. Sun, S.S. Lim, Y.M. Lam, M. Grätzel, S. Mhaisalkar, T.C. Sum, Long-range balanced electron-and hole-transport lengths in organic-inorganic $CH_3NH_3PbI_3$. Science **342**, 344 (2013)
30. L. Wang, C. McCleese, A. Kovalsky, Y. Zhao, C. Burda, Femtosecond time-resolved transient absorption spectroscopy of $CH_3NH_3PbI_3$ perovskite films: evidence for passivation effect of $PbI_2$. J. Am. Chem. Soc. **136**, 12205 (2014)
31. K.G. Stamplecoskie, J.S. Manser, P.V. Kamat, Dual nature of the excited state in organic–inorganic lead halide perovskites. Energy Environ. Sci. **8**, 208 (2015)
32. M.B. Price, J. Butkus, T.C. Jellicoe, A. Sadhanala, A. Briane, J.E. Halpert, K. Broch, J.M. Hodgkiss, R.H. Friend, F. Deschler, Hot-carrier cooling and photoinduced refractive index changes in organic-inorganic lead halide perovskites. Nat. Commun. **6**, 8420 (2015)
33. D. König, K. Casalenuovo, Y. Takeda, G. Conibeer, J. Guillemoles, R. Patterson, L. Huang, M. Green, Hot carrier solar cells: principles, materials and design. Phys. E **42**, 2862 (2010)
34. Y. Yang, D.P. Ostrowski, R.M. France, K. Zhu, J. Van De Lagemaat, J.M. Luther, M.C. Beard, Observation of a hot-phonon bottleneck in lead-iodide perovskites. Nat. Photon **10**, 53 (2016)
35. M. Li, S. Bhaumik, T.W. Goh, M.S. Kumar, N. Yantara, M. Gratzel, S. Mhaisalkar, N. Mathews, T.C. Sum, Slow cooling and highly efficient extraction of hot carriers in colloidal perovskite nanocrystals. Nat. Commun. **8**, 14350 (2017)
36. K. Chen, A.J. Barker, F.L. Morgan, J.E. Halpert, J.M. Hodgkiss, Effect of carrier thermalization dynamics on light emission and amplification in organometal halide perovskites. J. Phys. Chem. Lett. **6**, 153 (2015)

37. J. Fu, Q. Xu, G. Han, B. Wu, C.H.A. Huan, M.L. Leek, T.C. Sum, Hot carrier cooling mechanisms in halide perovskites. Nat. Commun. **8**, 1300 (2017)
38. J.M. Richter, F. Branchi, Valduga de Almeida, F. Camargo, B. Zhao, R.H. Friend, G. Cerullo, F. Deschler, Ultrafast carrier thermalization in lead iodide perovskite probed with two-dimensional electronic spectroscopy. Nat. Commun. **8**, 376 (2017)
39. J. Yang, X. Wen, H. Xia, R. Sheng, Q. Ma, J. Kim, P. Tapping, T. Harada, T.W. Kee, F. Huang, Acoustic-optical phonon up-conversion and hot-phonon bottleneck in lead-halide perovskites. Nat. Commun. **8**, 1 (2017)
40. H. Zhu, K. Miyata, Y. Fu, J. Wang, P.P. Joshi, D. Niesner, K.W. Williams, S. Jin, X.-Y. Zhu, Screening in crystalline liquids protects energetic carriers in hybrid perovskites. Science **353**, 1409 (2016)
41. J. Butkus, P. Vashishtha, K. Chen, J.K. Gallaher, S.K. Prasad, D.Z. Metin, G. Laufersky, N. Gaston, J.E. Halpert, J.M. Hodgkiss, The evolution of quantum confinement in $CsPbBr_3$ perovskite nanocrystals. Chem. Mater. **29**, 3644 (2017)
42. A. Mondal, J. Aneesh, V.K. Ravi, R. Sharma, W.J. Mir, M.C. Beard, A. Nag, K. Adarsh, Ultrafast exciton many-body interactions and hot-phonon bottleneck in colloidal cesium lead halide perovskite nanocrystals. Phys. Rev. B **98**, 115418 (2018)
43. J. Aneesh, A. Swarnkar, V. Kumar Ravi, R. Sharma, A. Nag, K. Adarsh, Ultrafast exciton dynamics in colloidal CsPbBr 3 perovskite nanocrystals: biexciton effect and Auger recombination. J. Phys. Chem. C **121**, 4734 (2017)
44. J.A. Castañeda, G. Nagamine, E. Yassitepe, L.G. Bonato, O. Voznyy, S. Hoogland, A.F. Nogueira, E.H. Sargent, C.H.B. Cruz, L.A. Padilha, Efficient biexciton interaction in perovskite quantum dots under weak and strong confinement. ACS Nano **10**, 8603 (2016)
45. V.I. Klimov, Spectral and dynamical properties of multiexcitons in semiconductor nanocrystals. Annu. Rev. Phys. Chem. **58**, 635 (2007)
46. H. Zhu, Y. Yang, T. Lian, Multiexciton annihilation and dissociation in quantum confined semiconductor nanocrystals. Acc. Chem. Res. **46**, 1270 (2013)
47. V. Klimov, A. Mikhailovsky, S. Xu, A. Malko, J. Hollingsworth, aC Leatherdale, H.J. Eisler, M. Bawendi, Optical gain and stimulated emission in nanocrystal quantum dots. Science **290**, 314 (2000)
48. S. Kondo, K. Takahashi, T. Nakanish, T. Saito, H. Asada, H. Nakagawa, High intensity photoluminescence of microcrystalline $CsPbBr_3$ films: evidence for enhanced stimulated emission at room temperature. Curr. Appl. Phys. **7**, 1 (2007)
49. C. Dang, J. Lee, C. Breen, J.S. Steckel, S. Coe-Sullivan, A. Nurmikko, Red, green and blue lasing enabled by single-exciton gain in colloidal quantum dot films. Nat. Nanotech. **7**, 335 (2012)
50. J.Q. Grim, S. Christodoulou, F. Di Stasio, R. Krahne, R. Cingolani, L. Manna, I. Moreels, Continuous-wave biexciton lasing at room temperature using solution-processed quantum wells. Nat. Nanotech. **9**, 891 (2014)
51. Y. Wang, X. Li, X. Zhao, L. Xiao, H. Zeng, H. Sun, Nonlinear absorption and low-threshold multiphoton pumped stimulated emission from all-inorganic perovskite nanocrystals. Nano Lett. **16**, 448 (2015)
52. K.E. Shulenberger, M.N. Ashner, S.K. Ha, F. Krieg, M.V. Kovalenko, W.A. Tisdale, M.G. Bawendi, Setting an upper bound to the biexciton binding energy in $CsPbBr_3$ perovskite nanocrystals. J. Phys. Chem. Lett. **10**, 5680 (2019)
53. A. Dey, P. Rathod, D. Kabra, Role of localized states in photoluminescence dynamics of high optical gain $CsPbBr_3$ nanocrystals. Adv. Opt. Mater. **6**, 1800109 (2018)
54. M.J. Tan, Y. Wang, Y. Chan, Solution-based green amplified spontaneous emission from colloidal perovskite nanocrystals exhibiting high stability. Appl. Phys. Lett. **114**, 183101 (2019)
55. S.W. Eaton, M. Lai, N.A. Gibson, A.B. Wong, L. Dou, J. Ma, L.-W. Wang, S.R. Leone, P. Yang, Lasing in robust cesium lead halide perovskite nanowires. P. Nati. Acad. Sci. USA **113**, 1993 (2016)

56. S. Wang, J. Yu, M. Zhang, D. Chen, C. Li, R. Chen, G. Jia, A.L. Rogach, X. Yang, Stable, strongly emitting cesium lead bromide perovskite nanorods with high optical gain enabled by an intermediate monomer reservoir synthetic strategy. Nano Lett. **19**, 6315 (2019)
57. Y. Jia, R.A. Kerner, A.J. Grede, B.P. Rand, N.C. Giebink, Continuous-wave lasing in an organic–inorganic lead halide perovskite semiconductor. Nat. Photon. **11**, 784 (2017)
58. S.A. Veldhuis, Y.K.E. Tay, A. Bruno, S.S. Dintakurti, S. Bhaumik, S.K. Muduli, M. Li, N. Mathews, T.C. Sum, S.G. Mhaisalkar, Benzyl alcohol-treated $CH_3NH_3PbBr_3$ nanocrystals exhibiting high luminescence, stability, and ultralow amplified spontaneous emission thresholds. Nano Lett. **17**, 7424 (2017)
59. F. Yuan, Z. Wu, H. Dong, J. Xi, K. Xi, G. Divitini, B. Jiao, X. Hou, S. Wang, Q. Gong, High stability and ultralow threshold amplified spontaneous emission from formamidinium lead halide perovskite films. J. Phys. Chem. C **121**, 15318 (2017)
60. L. Protesescu, S. Yakunin, S. Kumar, J. Bär, F. Bertolotti, N. Masciocchi, A. Guagliardi, M. Grotevent, I. Shorubalko, M.I. Bodnarchuk, Dismantling the "red wall" of colloidal perovskites: highly luminescent formamidinium and formamidinium–cesium lead iodide nanocrystals. ACS Nano **11**, 3119 (2017)
61. H.H. Fang, L. Protesescu, D.M. Balazs, S. Adjokatse, M.V. Kovalenko, M.A. Loi, Exciton recombination in formamidinium lead triiodide: nanocrystals versus thin films. Small **13**, 1700673 (2017)
62. Y.-S. Park, S. Guo, N.S. Makarov, V.I. Klimov, Room temperature single-photon emission from individual perovskite quantum dots. ACS Nano **9**, 10386 (2015)
63. P. Papagiorgis, A. Manoli, L. Protesescu, C. Achilleos, M. Violaris, K. Nicolaides, T. Trypiniotis, M.I. Bodnarchuk, M.V. Kovalenko, A. Othonos, G. Itskos, Efficient optical amplification in the nanosecond regime from formamidinium lead iodide nanocrystals. ACS Photonics **5**, 907 (2018)
64. L-y Huang, W.R. Lambrecht, Electronic band structure, phonons, and exciton binding energies of halide perovskites $CsSnCl_3$, $CsSnBr_3$, and $CsSnI_3$. Phys. Rev. B **88**, 165203 (2013)
65. C. Wehrenfennig, G.E. Eperon, M.B. Johnston, H.J. Snaith, L.M. Herz, High charge carrier mobilities and lifetimes in organolead trihalide perovskites. Adv. Mater. **26**, 1584 (2014)
66. F. Krieg, S.T. Ochsenbein, S. Yakunin, S. Ten Brinck, P. Aellen, A. Suess, B. Clerc, D. Guggisberg, O. Nazarenko, Y. Shynkarenko, S. Kumar, C.J. Shih, I. Infante, M.V. Kovalenko, Colloidal $CsPbX_3$ (X = Cl, Br, I) Nanocrystals 2.0: Zwitterionic Capping Ligands for Improved Durability and Stability. ACS Energy Lett. **3**, 641 (2018)
67. Y. Tong, E. Bladt, M.F. Ayguler, A. Manzi, K.Z. Milowska, V.A. Hintermayr, P. Docampo, S. Bals, A.S. Urban, L. Polavarapu, J. Feldmann, Highly luminescent cesium lead halide perovskite nanocrystals with tunable composition and thickness by ultrasonication. Angew. Chem. Int. Ed. Engl. **55**, 13887 (2016)
68. M. Imran, V. Caligiuri, M. Wang, L. Goldoni, M. Prato, R. Krahne, L. De Trizio, L. Manna, Benzoyl halides as alternative precursors for the colloidal synthesis of lead-based halide perovskite nanocrystals. J. Am. Chem. Soc. **140**, 2656 (2018)
69. D. Yan, T. Shi, Z. Zang, T. Zhou, Z. Liu, Z. Zhang, J. Du, Y. Leng, X. Tang, Ultrastable $CsPbBr_3$ perovskite quantum dot and their enhanced amplified spontaneous emission by surface ligand modification. Small **15**, e1901173 (2019)
70. C. Li, Z. Zang, C. Han, Z. Hu, X. Tang, J. Du, Y. Leng, K. Sun, Highly compact $CsPbBr_3$ perovskite thin films decorated by ZnO nanoparticles for enhanced random lasing. Nano Energy **40**, 195 (2017)
71. Z. Liu, Z. Hu, Z. Zhang, J. Du, J. Yang, X. Tang, W. Liu, Y. Leng, Two-photon pumped amplified spontaneous emission and lasing from formamidinium lead bromine nanocrystals. ACS Photonics **6**, 3150 (2019)
72. W. Zhao, Z. Qin, C. Zhang, G. Wang, X. Huang, B. Li, X. Dai, M. Xiao, Optical gain from biexcitons in $CsPbBr_3$ nanocrystals revealed by two-dimensional electronic spectroscopy. J. Phys. Chem. Lett. **10**, 1251 (2019)
73. E.M.L.D. de Jong, G. Yamashita, L. Gomez, M. Ashida, Y. Fujiwara, T. Gregorkiewicz, Multiexciton lifetime in all-inorganic $CsPbBr_3$ perovskite nanocrystals. J. Phys. Chem. C **121**, 1941 (2017)

74. K. Wu, G. Liang, Q. Shang, Y. Ren, D. Kong, T. Lian, Ultrafast interfacial electron and hole transfer from $CsPbBr_3$ perovskite quantum dots. J. Am. Chem. Soc. **137**, 12792 (2015)
75. G.E. Eperon, E. Jedlicka, D.S. Ginger, Biexciton auger recombination differs in hybrid and inorganic halide perovskite quantum dots. J. Phys. Chem. Lett. **9**, 104 (2018)
76. V.I. Klimov, S.A. Ivanov, J. Nanda, M. Achermann, I. Bezel, J.A. McGuire, A. Piryatinski, Single-exciton optical gain in semiconductor nanocrystals. Nature **447**, 441 (2007)
77. J. Navarro-Arenas, I. Suárez, V.S. Chirvony, A.F. Gualdrón-Reyes, I. Mora-Seró, J. Martínez-Pastor, Single-exciton amplified spontaneous emission in thin films of $CsPbX_3$ (X = Br, I) perovskite nanocrystals. J. Phys. Chem. Lett. **10**, 6389 (2019)
78. A. Pramanik, K. Gates, Y. Gao, S. Begum, P. Chandra Ray, Several orders-of-magnitude enhancement of multiphoton absorption property for $CsPbX_3$ perovskite quantum dots by manipulating halide stoichiometry. J. Phys. Chem. C **123**, 5150 (2019)
79. J. Chen, K. Zidek, P. Chabera, D. Liu, P. Cheng, L. Nuuttila, M.J. Al-Marri, H. Lehtivuori, M.E. Messing, K. Han, K. Zheng, T. Pullerits, Size- and wavelength-dependent two-photon absorption cross-section of $CsPbBr_3$ perovskite quantum dots. J. Phys. Chem. Lett. **8**, 2316 (2017)
80. Y. Wang, V.D. Ta, Y. Gao, T.C. He, R. Chen, E. Mutlugun, H.V. Demir, H.D. Sun, Stimulated emission and lasing from CdSe/CdS/ZnS core-multi-shell quantum dots by simultaneous three-photon absorption. Adv. Mater. **26**, 2954 (2014)
81. W.-G. Lu, C. Chen, D. Han, L. Yao, J. Han, H. Zhong, Y. Wang, Nonlinear optical properties of colloidal $CH_3NH_3PbBr_3$ and $CsPbBr_3$ quantum dots: a comparison study using Z-scan technique. Adv. Opt. Mater. **4**, 1732 (2016)
82. Y. Xu, Q. Chen, C. Zhang, R. Wang, H. Wu, X. Zhang, G. Xing, W.W. Yu, X. Wang, Y. Zhang, Two-photon-pumped perovskite semiconductor nanocrystal lasers. J. Am. Chem. Soc. **138**, 3761 (2016)
83. G. Nagamine, J.O. Rocha, L.G. Bonato, A.F. Nogueira, Z. Zaharieva, A.A. Watt, C.H. de Brito Cruz, L.A. Padilha, Two-photon absorption and two-photon-induced gain in perovskite quantum dots. J. Phys. Chem. Lett. **9**, 3478 (2018)
84. Z. Hu, Z. Liu, Y. Bian, S. Li, X. Tang, J. Du, Z. Zang, M. Zhou, W. Hu, Y. Tian, Enhanced two-photon-pumped emission from in situ synthesized nonblinking $CsPbBr_3/SiO_2$ nanocrystals with excellent stability. Adv. Opt. Mater. **6**, 1700997 (2018)
85. T. He, J. Li, C. Ren, S. Xiao, Y. Li, R. Chen, X. Lin, Strong two-photon absorption of Mn-doped $CsPbCl_3$ perovskite nanocrystals. Appl. Phys. Lett. **111**, 211105 (2017)
86. S. Zou, G. Yang, T. Yang, D. Zhao, Z. Gan, W. Chen, H. Zhong, X. Wen, B. Jia, B. Zou, Template-free synthesis of high-yield Fe-doped cesium lead halide perovskite ultralong microwires with enhanced two-photon absorption. J. Phys. Chem. Lett. **9**, 4878 (2018)
87. R. Ketavath, N.K. Katturi, S.G. Ghugal, H.K. Kolli, T. Swetha, V.R. Soma, B. Murali, Deciphering the ultrafast nonlinear optical properties and dynamics of pristine and Ni-doped $CsPbBr_3$ colloidal two-dimensional nanocrystals. J. Phys. Chem. Lett. **10**, 5577 (2019)
88. B. Redding, M.A. Choma, H. Cao, Speckle-free laser imaging using random laser illumination. Nat. Photonics **6**, 355 (2012)
89. R. Dhanker, A. Brigeman, A. Larsen, R. Stewart, J.B. Asbury, N.C. Giebink, Random lasing in organo-lead halide perovskite microcrystal networks. Appl. Phys. Lett. **105**, 151112 (2014)
90. X. Li, Y. Wang, H. Sun, H. Zeng, Amino-mediated anchoring perovskite quantum dots for stable and low-threshold random lasing. Adv. Mater. **29**, 1701185 (2017)
91. S. Yuan, D. Chen, X. Li, J. Zhong, X. Xu, In situ crystallization synthesis of $CsPbBr_3$ perovskite quantum dot-embedded glasses with improved stability for solid-state lighting and random upconverted lasing. ACS Appl. Mater. Inter. **10**, 18918 (2018)
92. P.K. Roy, G. Haider, H.I. Lin, Y.M. Liao, C.H. Lu, K.H. Chen, L.C. Chen, W.H. Shih, C.T. Liang, Y.F. Chen, Multicolor ultralow-threshold random laser assisted by vertical-graphene network. Adv. Opt. Mater. **6**, 1800382 (2018)
93. M. Saliba, S.M. Wood, J.B. Patel, P.K. Nayak, J. Huang, J.A. Alexander-Webber, B. Wenger, S.D. Stranks, M.T. Horantner, J.T. Wang, R.J. Nicholas, L.M. Herz, M.B. Johnston, S.M. Morris, H.J. Snaith, M.K. Riede, Structured organic-inorganic perovskite toward a distributed feedback laser. Adv. Mater. **28**, 923 (2016)

94. Z. Li, J. Moon, A. Gharajeh, R. Haroldson, R. Hawkins, W. Hu, A. Zakhidov, Q. Gu, Room-temperature continuous-wave operation of organometal halide perovskite lasers. ACS Nano **12**, 10968 (2018)
95. Y. Jia, R.A. Kerner, A.J. Grede, A.N. Brigeman, B.P. Rand, N.C. Giebink, Diode-pumped organo-lead halide perovskite lasing in a metal-clad distributed feedback resonator. Nano Lett. **16**, 4624 (2016)
96. L. Wang, L. Meng, L. Chen, S. Huang, X. Wu, G. Dai, L. Deng, J. Han, B. Zou, C. Zhang, Ultralow-threshold and color-tunable continuous-wave lasing at room-temperature from in situ fabricated perovskite quantum dots. J. Phys. Chem. Lett. **10**, 3248 (2019)
97. M. Artemyev, U. Woggon, Quantum dots in photonic dots. Appl. Phys. Lett. **76**, 1353 (2000)
98. M. Kazes, D.Y. Lewis, Y. Ebenstein, T. Mokari, U. Banin, Lasing from semiconductor quantum rods in a cylindrical microcavity. Adv. Mater. **14**, 317 (2002)
99. K. Wang, S. Wang, S. Xiao, Q. Song, Recent advances in perovskite micro-and nanolasers. Adv. Opt Mater. **6**, 1800278 (2018)
100. M.L. De Giorgi, M. Anni, Amplified spontaneous emission and lasing in lead halide perovskites: state of the art and perspectives. Appl. Sci. **9**, 4591 (2019)
101. Y. Wang, X. Li, V. Nalla, H. Zeng, H. Sun, Solution-processed low threshold vertical cavity surface emitting lasers from all-inorganic perovskite nanocrystals. Adv. Funct. Mater. **27**, 1605088 (2017)
102. C.-Y. Huang, C. Zou, C. Mao, K.L. Corp, Y.-C. Yao, Y.-J. Lee, C.W. Schlenker, A.K. Jen, L.Y. Lin, $CsPbBr_3$ perovskite quantum dot vertical cavity lasers with low threshold and high stability. Acs Photonics **4**, 2281 (2017)
103. P. Liu, X. He, J. Ren, Q. Liao, J. Yao, H. Fu, Organic-inorganic hybrid perovskite nanowire laser arrays. ACS Nano **11**, 5766 (2017)
104. Z. Yang, J. Lu, M. ZhuGe, Y. Cheng, J. Hu, F. Li, S. Qiao, Y. Zhang, G. Hu, Q. Yang, D. Peng, K. Liu, C. Pan, Controllable growth of aligned monocrystalline $CsPbBr_3$ microwire arrays for piezoelectric-induced dynamic modulation of single-mode lasing. Adv. Mater. **31**, e1900647 (2019)
105. Z. Duan, Y. Wang, G. Li, S. Wang, N. Yi, S. Liu, S. Xiao, Q. Song, Chip-scale fabrication of uniform lead halide perovskites microlaser array and photodetector array. Laser Photonics Rev. **12**, 1700234 (2018)
106. X. He, P. Liu, H. Zhang, Q. Liao, J. Yao, H. Fu, Patterning multicolored microdisk laser arrays of cesium lead halide perovskite. Adv. Mater. **29**, 1604510 (2017)
107. J. Feng, X. Yan, Y. Zhang, X. Wang, Y. Wu, B. Su, H. Fu, L. Jiang, "Liquid Knife" to fabricate patterning single-crystalline perovskite microplates toward high-performance laser arrays. Adv. Mater. **28**, 3732 (2016)
108. X. Liu, L. Niu, C. Wu, C. Cong, H. Wang, Q. Zeng, H. He, Q. Fu, W. Fu, T. Yu, C. Jin, Z. Liu, T.C. Sum, Periodic organic-inorganic halide perovskite microplatelet arrays on silicon substrates for room-temperature lasing. Adv. Sci. (Weinh) **3**, 1600137 (2016)
109. C.H. Lin, Q. Zeng, E. Lafalce, S. Yu, M.J. Smith, Y.J. Yoon, Y. Chang, Y. Jiang, Z. Lin, Z.V. Vardeny, Large-area lasing and multicolor perovskite quantum dot patterns. Adv. Opt. Mater. **6**, 1800474 (2018)
110. H. Coles, S. Morris, Liquid-crystal lasers. Nat Photon **4**, 676 (2010)
111. S.D. Stranks, S.M. Wood, K. Wojciechowski, F. Deschler, M. Saliba, H. Khandelwal, J.B. Patel, S.J. Elston, L.M. Herz, M.B. Johnston, Enhanced amplified spontaneous emission in perovskites using a flexible cholesteric liquid crystal reflector. Nano Lett. **15**, 4935 (2015)
112. L.-J. Chen, J.-H. Dai, J.-D. Lin, T.-S. Mo, H.-P. Lin, H.-C. Yeh, Y.-C. Chuang, S.-A. Jiang, C.-R. Lee, Wavelength-tunable and highly stable perovskite-quantum-dot-doped lasers with liquid crystal lasing cavities. ACS Appl. Mater. Inter. **10**, 33307 (2018)
113. Z. Liu, Z. Hu, T. Shi, J. Du, J. Yang, Z. Zhang, X. Tang, Y. Leng, Stable and enhanced frequency up-converted lasing from $CsPbBr_3$ quantum dots embedded in silica sphere. Opt. Express **27**, 9459 (2019)
114. H. Huang, B. Chen, Z. Wang, T.F. Hung, A.S. Susha, H. Zhong, A.L. Rogach, Water resistant $CsPbX_3$ nanocrystals coated with polyhedral oligomeric silsesquioxane and their use as solid state luminophores in all-perovskite white light-emitting devices. Chem. Sci. **7**, 5699 (2016)

115. Y. Wang, D. Yu, Z. Wang, X. Li, X. Chen, V. Nalla, H. Zeng, H. Sun, Solution-Grown $CsPbBr_3/Cs_4PbBr_6$ perovskite nanocomposites: toward temperature-insensitive optical gain. Small **13**, 1701587 (2017)
116. H. Hu, B. Dong, W. Zhang, Low-toxic metal halide perovskites: opportunities and future challenges. J. Mater. Chem. A **5**, 11436 (2017)

# Chapter 12
# Electrospun Nanofibers Embedded with Perovskite Quantum Dots

**Manikandan Venkatesan, Loganathan Veeramuthu, Fang-Cheng Liang, Chia-Jung Cho, and Chi-Ching Kuo**

**Abstract** Because of their tunable bandgap, wide visible emission wavelength, and high photoluminescence quantum yield and mobility, perovskite quantum dots (PQDs) have potential in various applications, such as in light emitting diodes, sensors, and photodetectors. However, the poor phase stability of PQDs in the environmental atmosphere impedes their further application. Therefore, encapsulation is important because it enhances PQD stability, with electrospinning (ES) being a versatile encapsulation method. This study investigated recent developments in ES-encapsulation size confinement for organic/inorganic hybrid OIPQDs ($CH_3NH_3PbX_3$; X = Cl, Br, I) and inorganic IPQDs ($CsPbX_3$; X = Cl, Br, I). ES fibrous systems greatly decrease perovskite size while retaining their shape and without forming aggregates; this facilitates superior photoluminescence (PL) emissiveness and stability.

## 12.1 Introduction

Graphene was introduced in 2004, and its use and synthesis has solved existing problems in the expanding field of atomic layering. When used in various high-speed nanoelectronic devices, graphene 2D nanosheets exhibit excellent optoelectronic and photo electronic properties, such as superior optical transparency, thermal conductivity, and charge carrier mobility at ambient temperature. Despite these advantages, graphene 2D nanosheets cannot be used in optoelectronic device fabrication due to its zero bandgap energy [1, 2]. To address this problem, silicene [3], hexagonal boron nitride (hBN) [4], transition-metal dichalcogenides [5], and black phosphorus [6] have been widely applied due to the ease of tuning their band gap [7]. In particular, the use of quantum dot (QD) zero-dimensional structures has completely solved the aforementioned optoelectronic and photo electronic problems to achieve higher quantum efficiency. Specifically, through alterations to their size and the shape, the

M. Venkatesan · L. Veeramuthu · F.-C. Liang · C.-J. Cho · C.-C. Kuo (✉)
Institute of Organic and Polymeric Materials, Research and Development Center of Smart Textile Technology, National Taipei University of Technology, 10608 Taipei, Taiwan
e-mail: kuocc@mail.ntut.edu.tw



Y. Zhou and Y. Wang (eds.), *Perovskite Quantum Dots*, Springer Series in Materials Science 303, https://doi.org/10.1007/978-981-15-6637-0_12

electronic structure of such QD structures can be adjusted without changing their chemical composition, potentially yielding the favorable properties of broad absorption, excellent photo stability, high quantum yield, and narrow emission peak. These unique and favorable properties allow for the design of energy-efficient electronics, photonics, and biomedical and other sensors.

The core/shell structure of CdSe/ZnS QDs and PbS have been widely exploited in many optical and electrical applications, such as in fluorescent biomarkers (for CdSe/ZnS QDs) and semiconductors (for PbS). These materials are favored because of their photochemical stability, higher photoluminescence quantum yield (PLQY), and higher bandgap energy (ZnS = 3.54 eV) [8]; their emission wavelengths can also be changed into any visible color by altering the size of the particles. However, despite these advantages, the achievement of high PLQY and stability requires surface passivation and a high reaction temperature. The use of perovskites resolves this problem. In addition to being inexpensive, perovskites deliver up to 95% PLQY and allow for the easy tuning of their emission wavelength in the visible region (410–700 nm) [9]. Such tuning occurs through the facile synthesis method (at near room temperature) and through alterations to the size of their anion composition. A breakthrough preparation of QDs based on $CsPbX_3$ and $CH_3NH_3PbX_3$ has potential in many applications, such as in light emitting diodes (LEDs) [10], sensors [11], and photodetectors [12].

The size confinement processes changes the perovskite photoluminescence by altering its bandgap. A PL blue shift occurs when the sizes of the QDs decrease (in the 10-nm scale). QDs that have a composition-tuned bandgap undergone size confinement [13] can cover the entire visible light spectrum. Perovskite QDs (PQDs) have attracted much research interest. In particular, Zhihai and co-workers synthesized PQDs using the classical hot-injection method [14] for use in white light emitting diodes (WLEDs), and Zhang et al. [15] prepared PQDs using the sol–gel method. Li et al. [16] used a two-phase emulsion extraction process to produce PQDs for use in a memory storage device, and Li et al. [17] was the first to achieve room-temperature synthesis in 2016. Because of light sensitivity, heat sensitivity, hydrophilicity, and anion exchange, perovskites are unstable when used in device fabrication and storage. Therefore, PQD stability must be improved, and many research groups have been working to prolong the material's lifetime.

Many attempts have been made to enhance PQD stability. For example, the ligand present on perovskite has been altered using the anion exchange reaction to adjust the band gap until higher luminescence is achieved [18], PQDs have been crosslinked by incorporating organic ligands [19], and hybrid organometal halide PQDs have been prepared, in addition to the doping of another stable divalent metal cation [20], use of X-ray irradiation [21], and encapsulation by means of plastic [22], silica wrapper [23], or glass matrix [24]. However, these added processes increase the cost and complexity of perovskite device fabrication. Among these methods, the use of a polymeric electrospinning platform for encapsulation is popular in commercial applications.

This book chapter summarizes the current advances and future prospects of perovskite-embedded fiber membranes (FMs) methods in rectifying PQD instability.

The direct contact between PQDs and the atmosphere greatly affects PL emission, which indicates the degradation of PQDs in the matrix. Upon encapsulation, the quantum yield of PQDs has at most a minor decrease, as with the solution phase, and remains stable in a polar solvent with favorable chemical inertness. Recently, polyvinylpyrrolidone (PVP), polyacrylonitrile (PAN), poly(methyl methacrylate) (PMMA), poly(vinylidene fluoride) (PVDF), polystyrene (PS), poly(styrene–butadiene–styrene) (SBS) and silicon resin matrices have been used as electrospun (ES) encapsulants to prevent direct contact with atmospheric stimulates, thereby improving flexibility and thermal stability.

## 12.2 Applications of PQD-Embedded Electrospun Nanofibers

### *12.2.1 Light Emitting Application*

To solve the problem of perovskite instability, Chen et al. [25] synthesized stable $CsPbMnCl_3@SiO_2$ for use in WLEDs. They introduced $Mn^{2+}$ ion to reduce the toxic Pb content and improve stability with respect to humidity and temperature. A rigid $SiO_2$ wrapper was used as an encapsulating layer to prevent anion exchange, which granted stability over 2 weeks. However, the device's rigidity limits its applicability in wearable/stretchable appliances. To enhance the flexibility and tailorability of the materials, electrospinning is the optimal technique. Hai et al. [26] recently fabricated a flexible and color-tunable device using PVP-encapsulated, silicon resin-covered PQDs. A hydrophobic silicon wafer prevented water permittivity. Encapsulated silicon resin-PVP/PQDs were able to withstand 50% humidity for 5 days, and PLQY retained its properties at 100 °C. Such moisture and thermal stability indicate the applicability of silica-coated PVP/$CsPbX_3$ in high temperature appliances. According to a photo stability test, that device's color stability was unchanged for 120 h under UV light exposure. The white light produced by the porous PVP nanofiber regulates the energy transfer and anion exchange between the $CsPbX_3$ ($X = I, Cl, Br$) and the fluorophore ultimately control the proportions of blue, green, and red respectively.

Subsequently, this method for producing white light was adopted by Tsai et al. [27] yielding improved stability lifetime of up to 30 days under 85% humidity when used in tunable white light emitting applications. To increase PLQY (to 30.9%), they used methyl ammonium cation as the emissive component. The material was encapsulated with a PAN polymer instead of traditional ligand and surfactants through the core–shell ES technique. The PAN polymeric shell effectively covers the perovskite core material. The polymeric shell PAN is a transparent semicrystalline material and has a non-interfering PLQY. Its hydrophobicity also protects the QDs from humidity (of up to 80%) and preserves QD confinement. In practical applications, a shell precursor blended with a blue light emitting polymer, such as poly[(9,9-bis(3′-(*N*,*N*-dimethylamino)propyl)-2,7-fluorene)-alt-2,7-(9,9–dioctylfluorene)] (PFN), can cohesively encapsulate the core OIPQDs with an optimized ratio of 3:1 to obtain a white light emitting device.

Kuo research group [28] developed highly stable IPQDs through biaxial electrospinning technique adopted to create core/shell nanostructure of encapsulated IPQDs using three different polymers (SBS, PMMA, PS). IPQDs@SBS nanofiber shows less aggregation and better dispersion because of polybutadiene hydrocarbon segment existing in SBS matrix, which strongly interact with ligands present on the IPQDs surface. IPQDs@PMMA and IPQDs@PS encapsulated porous nanofiber provide better thermal stability and UV stability respectively. PL of these encapsulated IPQDs examine over the water for 90 days it strongly suggesting that hydrophobic PS encapsulated IPQDs highly stable under polar solvents. Time resolved PL spectra suggesting IPQDs@SBS with less non-radioactive recombination thereby generating the higher PLQY. Utilizing the red IPQDs as core and green IPQDs as shell, the energy transfer is highly efficient producing greater enhancement with the white light emission efficiency as compared to normal multilayered NFMs developed onto the commercial blue LEDs.

Lin et al. [29] was the first to devise a SBS encapsulating method, for use in WLEDs, which produces stretchable FMs made from IPQDs (Fig. 12.1). The ES

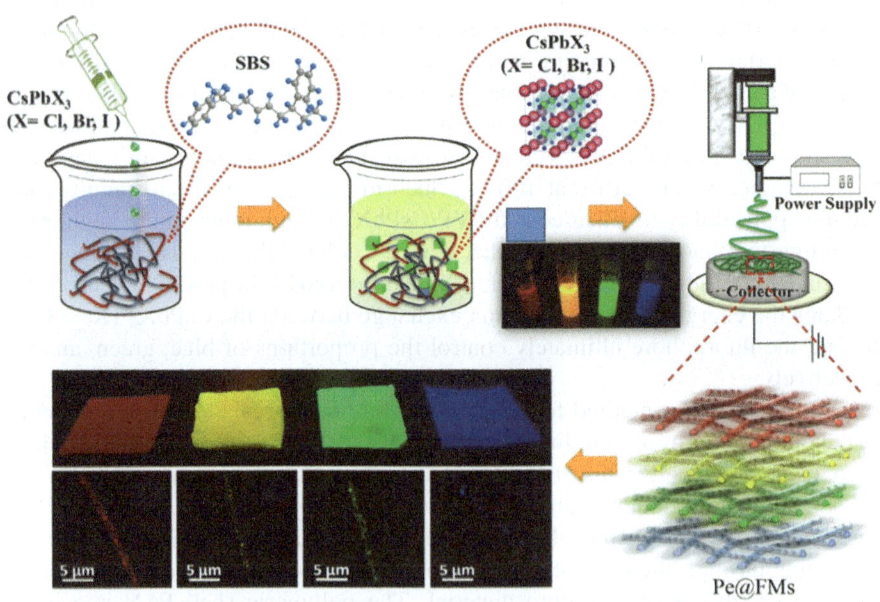


**Fig. 12.1** Schematic of the flexible $CsPbX_3$ (X = Cl, Br, I)@SBS FM fabrication process. Optical image under UV light (365 nm) excitation and fluorescence microscopic images. Reprinted with permission from Ref. [29]. Copyright 2018, American Chemical Society

method was further developed for the production of continuous IPQD-impregnated nanofibers (NFs). In general, dielectric solvent dichloromethane (DCM) is preferred for electrospinning. This is because SBS completely dissolves in DCM and does not alter the structure of IPQD nanocrystals (NCs) when mixed with $CsPbX_3$ (X = Cl, Br, and I)/toluene at an appropriate ratio. The precursor was subjected to the uniaxial electrospinning technique for generating the continuous IPQD-impregnated nanofiber. In the nanofiber, the NCs were randomly placed within the fiber matrix without aggregation; their sizes were examined using fluorescence microscopic images, transmission electron microscopy (TEM) analysis, and scanning electron microscopy (SEM). WLEDs was fabricated by combining blue LED chip with green ($CsPb(Br_{0.8}I_{0.2})_3$) and red ($CsPb(Br_{0.4}I_{0.6})_3$).

The silicon encapsulating method is typically preferred for improving in situ crystal formation. However, perovskite aggregation occurs due to its hydrophilic character, and its crystallization is greatly influenced by polar solvents through hydrolysis [30]. Conversely, hydrophobic polymers are the best choice for the commercial synthesis of functionalized nanofibers. Poly (vinylidene fluoride) (PVDF) is used as an encapsulating material because of its hydrophobic nature. Therefore, Gu et al. preferred PVDF as an encapsulating layer in uniaxial electrospinning for the preparation of QD-impregnated NFs with finer OIPQDs encapsulation [31]. PVDF controls in situ aggregation through the interaction between methyl ammonium cation of perovskite ($CH_3NH_3$) and–$CF_2$–of PVDF. A surface passivated nanofiber was then flawlessly prepared after solvent evaporation. Luminescence also increased 4.8-fold after interweaving with a stabilizer prepared by plasma treated Ag/PVP nanofiber. Confined OIPQDs were obtained by altering the ratio of Br to Cl in the perovskite to form $MAPbCl_3$, $MAPbBr_{1.2}Cl_{1.8}$, and $MAPbBr_3$. Among them, the PL spectrum of $MAPbBr_{1.2}Cl_{1.8}$ coincides with the resonance band peak of stabilizing plasma treated Ag/PVP nanofiber, thus generating highly emissive blue fluorescent membranes.

Kumar et al. [32] investigated OIPQDs NCs growth in situ in electrospun NFs. PMMA was used as a fiber precursor because of its favorable dielectric property and high refractive index value of 1.489, which approximates that of glass. The band gap was optimized by varying the concentration (wt%) of the OIPQDs core precursor to 0.16, 3.0, and 5.0 wt%, where the 0.16 and 3.0 wt% concentrations resulted in PLQYs of 20% and 80%, respectively. The PLQY increase was attributed to NCs growth, although OIPQDs began to aggregate in the case of the 5 wt% concentration. PL life time ($\tau$) also exhibited a monotonic increase from 0.5 ns (0.16 wt%) to 16 ns (5.0 wt%), through which radiative and non-radiative decay rates were determined. The increase in the ratio of radiative to non-radiative decay rates clearly indicates a significant decrease in non-radiative channels. Due to the decrease in the non-radiative channel, surface passivation and encapsulation successfully prolonged stability approximately 18 months. The ES fibers of PMMA/OIPQDs exhibited a light yellow/orange color under white light and very high green light luminescence.

### 12.2.2 Sensing Application

Wang et al. [33] synthesized a sensor to detect rhodamine 6G (R6G) pigment at very low concentrations. The detection mechanism was examined through fluorescence resonance energy transfer (FRET). The sensor consisted of IPQDs, and its core was protected from the humid environment through fiber encapsulation. Hydrophobic PS was selected as the nanofiber material due to its well-defined properties. The quantum yield of IPQDs was higher in toluene at 93%, and it decreased to 91% upon blending with PS; this negligible decrease is due to aggregation quenching effects during processing. X-ray diffraction (XRD), TEM, and X-ray photoelectron spectroscopy (XPS) were systematically used to affirm the impregnation of PQDs into the nanofiber. Water and photo stability were indicated by direct immersion in the water phase and the exposure of IPQD/PS to UV. Electrospinning appears to not affect the optical property of IPQDs. The photoluminescence and absorption spectra depicting the conformal overlap between the synthesized encapsulated perovskite nanofiber and the R6G analyte indicated high FRET performance. As evidence, the sensitive limit of detection was 0.01 ppm, with high stability in aqueous and alcoholic media.

In addition to being used as an R6G sensor, PQDs have also been used in multisensory applications. Wang et al. [34] used PMMA for the encapsulating layer due to the material's stability and receptiveness to surface modification. They enhanced the subsurface concentration of IPQDs through functionalization with optically active chemicals (Fig. 12.2). Hexamethylenediamine was immobilized on an IPQDs/PMMA surface to create—$NH_2$, which was further treated through acylation with the R6G-labeled constricted peptide CF6 (cys-Pro-Arg-Gly-R6G). This allowed for the capture of trypsin, which is responsive in cardiovascular diseases.

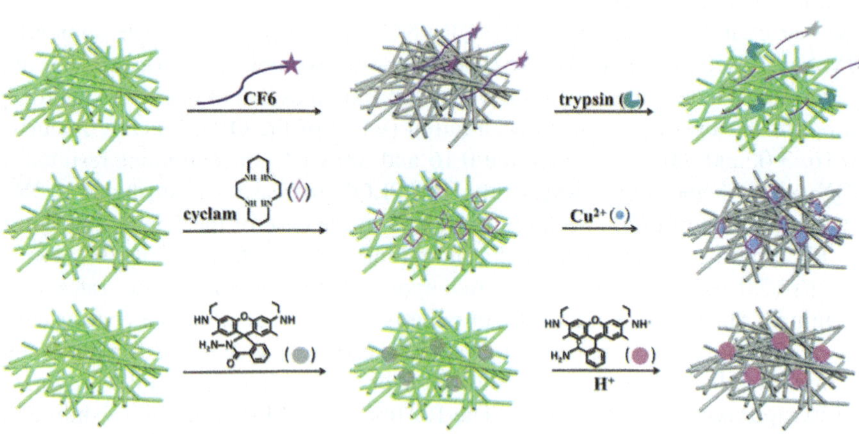


**Fig. 12.2** Schematic illustration of trypsin, Cu and pH fluorescence detection, based on the CPBQDs/PMMA FM reprinted with permission from Ref. [34]. Copyright 2016, American Chemical Society

When trypsin disrupted the peptide linkage, separation was triggered between the donor (IPQDs) and acceptor (R6G). Therefore, fluorescent intensity increased with respect to the donor, and trypsin concentration increased as well. The ester function of PMMA changed with respect to the carboxyl unit, and PMMA was subsequently treated with cyclam. Fluorescent quench was absorbed when the cyclam-grafted IPQDs/PMMA nanofibers were placed into $Cu^{2+}$ solution, and pH was measured by grafting hydrazided R6G dye on the PMMA NFs. The spirolactum ring of R6G opened when exposed to a low pH, and R6G turned from being colorless to having a red–orange color. When the pH increased from 1 to 14, the PL intensity of IPQDs at 515 nm decreased whereas that of R6G at 560 nm indirectly increased, thus indicating the occurrence of the FRET donor–acceptor process. The sensing applications with IPQDs shows relevantly promising performance because of its higher PLQY, response towards polar solvents. IPQDs can further extended into human interactive smart garment [35] applications after which it might possibly impregnated with any fabrics giving out optical and electrical signals depending on the analyte.

### *12.2.3 Photodetector Application*

Perovskite can also be used in photo sensor devices due to its excellent charge transfer properties and size tunable bandgap (3.1–1.7 eV). Unlike in LED applications, the encapsulant polymeric insulators are replaced with conductive metal electrodes and carbon composites to achieve photovoltaic suitability.

Zheng et al. [36] used $TiO_2$ conductive metal, which was drawn as a nanowire (NW) and nanotube (NT) through a simple coaxial electrospinning technique. The spin coating method was used to fabricate the perovskite (OIPQDs) onto the $TiO_2$/NTs and $TiO_2$/NWs. These integrated transparent heterostructures could cover the entire ultraviolet to visible wavelength (300–800 nm), yielding a 3 order-magnitude increase in its photo sensing property as well as the excellent photo responsivity of 0.2 $A \bullet W^{-1}$ at 700 nm. Stability tests were performed in humid air (RH: 55%) for up to 72 h, thus demonstrating the device's real-time applicability. The device retained its properties after undergoing 200 bending cycles at 90°. However, device stability was inadequate for commercial use due to the volatilization of organic compounds in ambient temperature. Zeng et al. [37] improved stability by changing organic cation through Cs cation, resulting in complete encapsulation through the use of a favorable charge transport layer carbon nanofiber (CNF) in free liquid surface (nozzle-less) electrospinning technology prior to subsequent thermal treatment (Fig. 12.3). CNF not only assisted crystal growth in situ but also participated in charge transport. The composite exhibited a wide band gap (2.71 eV) and an intense narrow full width half-maximum peak at 29 nm. This study thus demonstrated the easy control of all inorganic perovskite crystals without compromise to its emissive characteristics.

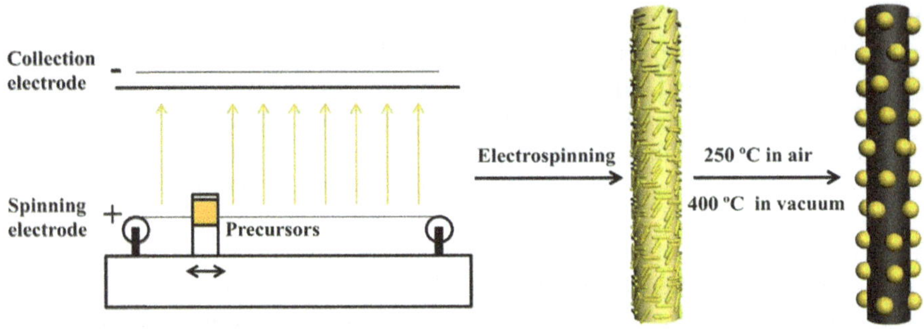


**Fig. 12.3** Schematic of synthesis process of IPQDs/carbon composite nanofiber. Reprinted with permission from Ref. [37]. Copyright 2019, Elsevier

## 12.3 Conclusion

This chapter deals with recent advances in resolving the urgent and difficult issue of perovskite quantum dots stability. ES fiber–embedded PQDs exhibited relatively higher stability and excellent emissive characteristics. Despite these advantages, the materials have not been directly employed in LED fabrication. Several studies have mounted a fibrous membrane onto a commercial blue LED to harvest WLED. The major challenge is in using these fibrous membranes as an emissive layer component in an actual LED device instead of as color convertors. In terms of sensing and photodetectors, the stability factor function as a major barrier, several efforts with improved stability can possibly give rise to good reversibility and prolonged lifetime to avoid material wastage. Stability with respect to moisture, air, and UV light must be further improved to enable commercially viable perovskite-based LEDs, sensors, solar cells, photodetectors, and other optoelectronic devices.

## References

1. Z. Sun, H. Chang, Graphene and graphene-like two-dimensional materials in photodetection: mechanisms and methodology. ACS Nano **8**(5), 4133–4156 (2014)
2. M.J. Jimenez, R.F. de Oliveira, F.M. Shimizu, C.C.B. Bufon, V. Rodrigues, ÂL. Gobbi, M.H. Piazzetta, A. Riul Jr., Poole-frenkel emission on functionalized, multilayered-packed reduced graphene oxide nanoplatelets. Nanotechnology **29**(50), 505703 (2018)
3. C. Qian, Z. Li, Multilayer silicene: structure, electronics, and mechanical property. Comput. Mater. Sci. 109354 (2019)
4. A. Razmdideh, M.T. Ahmadi, Band gap modulation by two-dimensional h-BN nanostructure. Phys. Solid State **61**(11), 2194–2199 (2019)
5. F. Chen, W. Su, S. Ding, L. Fu, The fabrication and tunable optical properties of 2D transition metal dichalcogenides heterostructures by adjusting the thickness of Mo/W films. Appl. Surf. Sci. 144192 (2019)

6. A. Sidike, G. Guo, X. Li, D. Li, Y. Nie, B. Cao, H. Duan, M. Long, Spin dependent electronic transport properties of zigzag black phosphorene nanojunctions induced by H, Li, O, Co asymmetric edge saturations. Phys. Lett. A. 126123 (2019)
7. T.S. Hartman, Z.K. Sofer, Beyond graphene: chemistry of group 14 graphene analogues: silicene, germanene, and stanene. ACS Nano. **13**(8), 8566–8576 (2019)
8. G. Ramalingam, K.V. Saravanan, T.K. Vizhi, M. Rajkumar, K. Baskar, Synthesis of water-soluble and bio-taggable CdSe@ ZnS quantum dots. RSC Adv. **8**(16), 8516–8527 (2018)
9. Y. Hu, J. Shu, X. Zhang, A. Zhao, Y. Liu, R. Li, Y. Di, H. Xu, Z. Gan, Encapsulation of colloid perovskite nanocrystals into solid polymer matrices: impact on electronic transition and photoluminescence. J. Lumin. 116938 (2019)
10. Y.-F. Li, J. Feng, H.-B. Sun, Perovskite quantum dots for light-emitting devices. Nanoscale **11**(41), 19119–19139 (2019)
11. N. Ding, D. Zhou, G. Pan, W. Xu, X. Chen, D. Li, X. Zhang, J. Zhu, Y. Ji, H. Song, Europium-doped lead-free Cs3Bi2Br 9 perovskite quantum dots and ultrasensitive Cu2+ detection. ACS Sustain. Chem. Eng. **7**(9), 8397–8404 (2019)
12. Z. Dai, Q. Ou, C. Wang, G. Si, B. Shabbir, C. Zheng, Z. Wang, Y. Zhang, Y. Huang, Y. Dong, Capillary-bridge mediated assembly of aligned perovskite quantum dots for high-performance photodetectors. J. Mater. Chem. C (2019)
13. P. Kumar, N. Ganesh, K.S. Narayan, Electrospun fibers containing emissive hybrid perovskite quantum dots. ACS Appl. Mater. Interf. (2019)
14. W. Zhihai, W. Jiao, S. Yanni, W. Jun, H. Yafei, W. Pan, W. Nengping, Z. Zhenfu, Air-stable all-inorganic perovskite quantum dot inks for multicolor patterns and white LEDs. J. Mater. Sci. **54**(9), 6917–6929 (2019). https://doi.org/10.1007/s10853-019-03382-2
15. Y.-W. Zhang, G. Wu, H. Dang, K. Ma, S. Chen, Multicolored mixed-organic-cation perovskite quantum dots (FAxMA1–xPbX3, X = Br and I) for white light-emitting diodes. Ind. Eng. Chem. Res. **56**(36), 10053–10059 (2017). https://doi.org/10.1021/acs.iecr.7b02309
16. F. Li, L. Luo, M. Wang, Y. Peng, X. Wu, Z. Su, S. Wang, F. Wang, Size controlled synthesis of $CsPbBr_3$ QDs by two-phase emulsion extraction. Funct. Mater. Lett. **12**(02), (2019) https://doi.org/10.1142/s1793604719500127
17. X. Li, Y. Wu, S. Zhang, B. Cai, Y. Gu, J. Song, H. Zeng, $CsPbX_3$ quantum dots for lighting and displays: room-temperature synthesis, photoluminescence superiorities, underlying origins and white light-emitting diodes. Adv. Func. Mater. **26**(15), 2435–2445 (2016). https://doi.org/10.1002/adfm.201600109
18. K. Abdel-Latif, R.W. Epps, C.B. Kerr, C.M. Papa, F.N. Castellano, M. Abolhasani, Facile room-temperature anion exchange reactions of inorganic perovskite quantum dots enabled by a modular microfluidic platform. Adv. Func. Mater. **29**(23), 1900712 (2019)
19. Y. Li, X. Wang, W. Xue, W. Wang, W. Zhu, L. Zhao, Highly luminescent and stable $CsPbBr_3$ perovskite quantum dots modified by phosphine ligands. Nano Res. **12**(4), 785–789 (2019)
20. H.C. Wang, Z. Bao, H.Y. Tsai, A.C. Tang, R.S. Liu, Perovskite quantum dots and their application in light-emitting diodes. Small **14**(1), (2018) https://doi.org/10.1002/smll.201702433
21. F. Palazon, Q.A. Akkerman, M. Prato, L. Manna, X-ray lithography on perovskite nanocrystals films: from patterning with anion-exchange reactions to enhanced stability in air and water. ACS Nano **10**(1), 1224–1230 (2015)
22. S.N. Raja, Y. Bekenstein, M.A. Koc, S. Fischer, D. Zhang, L. Lin, R.O. Ritchie, P. Yang, A.P. Alivisatos, Encapsulation of perovskite nanocrystals into macroscale polymer matrices: enhanced stability and polarization. ACS Appl. Mater. Interf. **8**(51), 35523–35533 (2016). https://doi.org/10.1021/acsami.6b09443
23. S.D. Liu, T.M. Chen, Synthesis and luminescent properties of polymer-silica multilayer-encapsulated perovskite quantum dots for optoelectronics. J. Chin. Chem. Soc. (2019)
24. S. Yuan, D. Chen, X. Li, J. Zhong, X. Xu, In Situ Crystallization Synthesis of $CsPbBr_3$ perovskite quantum dot-embedded glasses with improved stability for solid-state lighting and random upconverted lasing. ACS Appl. Mater. Interf. **10**(22), 18918–18926 (2018). https://doi.org/10.1021/acsami.8b05155

25. W. Chen, T. Shi, J. Du, Z. Zang, Z. Yao, M. Li, K. Sun, W. Hu, Y. Leng, X. Tang, Highly stable silica-wrapped mn-doped $CsPbCl_3$ quantum dots for bright white light-emitting devices. ACS Appl. Mater. Interf. **10**(50), 43978–43986 (2018). https://doi.org/10.1021/acsami.8b14046
26. J. Hai, H. Li, Y. Zhao, F. Chen, Y. Peng, B. Wang, Designing of blue, green, and red $CsPbX_3$ perovskite-codoped flexible films with water resistant property and elimination of anion-exchange for tunable white light emission. Chem. Commun. (Camb) **53**(39), 5400–5403 (2017). https://doi.org/10.1039/c7cc01152k
27. P.C. Tsai, J.Y. Chen, E. Ercan, C.C. Chueh, S.H. Tung, W.C. Chen, Uniform luminous perovskite nanofibers with color-tunability and improved stability prepared by one-step core/shell electrospinning. Small **14**(22), e1704379 (2018). https://doi.org/10.1002/smll.201704379
28. D.-H. Jiang, Y.-H. Tsai, L. Veeramuthu, F.-C. Liang, L.-C. Chen, C.C. Lin, T. Satoh, S.-H. Tung, C.-C. Kuo, Novel ultra-stable and highly luminescent white light-emitting diodes from perovskite quantum dots—polymer nanofibers through biaxial electrospinning. APL Mater. **7**(11), 111105 (2019)
29. C.C. Lin, D.H. Jiang, C.C. Kuo, C.J. Cho, Y.H. Tsai, T. Satoh, C. Su, Water-resistant efficient stretchable perovskite-embedded fiber membranes for light-emitting diodes. ACS Appl. Mater. Interf. **10**(3), 2210–2215 (2018). https://doi.org/10.1021/acsami.7b15989
30. Y. Huang, F. Li, L. Qiu, F. Lin, Z. Lai, S. Wang, L. Lin, Y. Zhu, Y. Wang, Y. Jiang, X. Chen, Enhancing the stability of $CH3NH3PbBr_3$ nanoparticles using double hydrophobic shells of $SiO_2$ and poly (vinylidene fluoride). ACS Appl. Mater. Interf. **11**(29), 26384–26391 (2019). https://doi.org/10.1021/acsami.9b07841
31. K. Gu, H. Peng, S. Hua, Y. Qu, D. Yang, Plasmon-enhanced blue-light emission of stable perovskite quantum dot membranes. Nanomater. (Basel) **9**(5), (2019) https://doi.org/10.3390/nano9050770
32. P. Kumar, N. Ganesh, K.S. Narayan, Electrospun fibers containing emissive hybrid perovskite quantum dots. ACS Appl. Mater. Interf. **11**(27), 24468–24477 (2019). https://doi.org/10.1021/acsami.9b08409
33. Y. Wang, Y. Zhu, J. Huang, J. Cai, J. Zhu, X. Yang, J. Shen, H. Jiang, C. Li, $CsPbBr_3$ perovskite quantum dots-based monolithic electrospun fiber membrane as an ultrastable and ultrasensitive fluorescent sensor in aqueous medium. J. Phys. Chem. Lett. **7**(21), 4253–4258 (2016). https://doi.org/10.1021/acs.jpclett.6b02045
34. Y. Wang, Y. Zhu, J. Huang, J. Cai, J. Zhu, X. Yang, J. Shen, C. Li, Perovskite quantum dots encapsulated in electrospun fiber membranes as multifunctional supersensitive sensors for biomolecules, metal ions and pH. Nanoscale Horiz. **2**(4), 225–232 (2017). https://doi.org/10.1039/c7nh00057j
35. L. Veeramuthu, W.-L. Li, F.-C. Liang, C.-J. Cho, C.-C. Kuo, W.-C. Chen, J.-H. Lin, W.-Y. Lee, C.-T. Wang, W.-Y. Lin, Smart garment energy generators fabricated using stretchable electrospun nanofibers. React. Funct. Polym. **142**, 96–103 (2019)
36. Z. Zheng, F. Zhuge, Y. Wang, J. Zhang, L. Gan, X. Zhou, H. Li, T. Zhai, Decorating perovskite quantum dots in $TiO_2$ nanotubes array for broadband response photodetector. Adv. Funct. Mater. **27**(43), (2017) https://doi.org/10.1002/adfm.201703115
37. Y. Zeng, P. Yang, L. Yang, Y. Hao, T. Shi, J. Wang, X. Zhao, In situ growth of $CsPbI_3$ on carbon nanofibers via electrospinning. Chem. Phys. Lett. **723**, 11–15 (2019). https://doi.org/10.1016/j.cplett.2019.02.052

# Chapter 13
# Strategies Towards Improving the Stability of All-Inorganic Perovskite Quantum Dots

**Kai Gu, Mu Yang, and Hongshang Peng**

**Abstract** Perovskite quantum dots (PQDs) have attracted significant interests in the past few years because of their unique optical properties. Both all-inorganic and organic–inorganic perovskite quantum dots have shown great potential in optoelectronic devices such as light-emitting diodes (LEDs) for lighting and display technology. However, these commercial applications are severely impeded by their instability towards temperature, oxygen and moisture. Recently numerous strategies towards enhancing the stability of PQDs have been developed. This chapter reviews the strategies of improving the stability of all-inorganic perovskite quantum dots in detail, which are classified into two directions: (i) compositional adjustment and (ii) surface engineering.

All-inorganic perovskite quantum dots (AIPQDs) have been regarded as promising optoelectronic materials for solar cells, [1–3] lasers, [4–6] light-emitting diodes (LED) [7–9], photodetectors, [10, 11] and bioimaging [12, 13], owing to the unique characteristics including high photoluminescence (PL) quantum yield, precise tunable bandgap, narrow emission wavelength and large absorption coefficient. Despite great achievements have been made in the past few years, they still face the issue of poor stability against air, high temperature, light or polar solvents [14–16]. Basically, the instability is attributed to the low formation energy and inherent ionic nature of halide perovskite [17]. When exposed to polar solvents or water, PQDs usually lose optical property and even structural integrity; if different halide perovskites are mixed, PL emission are significantly shifted as a consequence of fast ion-exchange reaction [18–20]. All these instabilities have inevitably complicated the synthesis, storage and device fabrication of AIPQDs. Therefore tremendous efforts have been dedicated to improve PQDs stability. In this chapter we review the recent studies in improving the stability of AIPQDs, with a focus on the advances in the past five years. These strategies are roughly classified into two categories, compositional

K. Gu · M. Yang · H. Peng (✉)
College of Science, Minzu University of China, Beijing 10081, China
e-mail: hshpeng@bjtu.edu.cn



Y. Zhou and Y. Wang (eds.), *Perovskite Quantum Dots*, Springer Series in Materials Science 303, https://doi.org/10.1007/978-981-15-6637-0_13

adjustment and surface engineering. It should be noted that there are many other reviews about strategies of improving pervoskites stability so far [14, 15]. Readers may refer to them for different perspectives.

## 13.1 Compositional Adjustment

AIPQDs feature the typical chemical composition of $ABX_3$, where A represents cesium (Cs); B represents lead (Pb); and X is halide anion (Cl, Br and I). Compositional adjustment means A-site or B-site doping with other ions, which can effectively enhance the stability of AIPQDs besides adjustment of optical property.

### *13.1.1 A-Site Doping*

Organic–inorganic hybrid perovskites generally suffer from poor thermal stability owing to the volatile organic A-site cation such as $MA^+$ ($CH_3NH_3^+$) and $FA^+$ ($[NH_2)_2CH]^+$) [21]. For AIPQDs, on the other hand, $Cs^+$ cation with a small ionic radius (1.81 Å; 2.70 and 2.79 Å for $MA^+$ and $FA^+$ cations, respectively) could lead to tilting of $PbX_6$ octahedra, which affects the structural stability of perovskite. For instance, $MAPbI_3$ and $CsPbI_3$ perovskites with 3D structure rapidly transform into tetragonal and orthorhombic structure at room temperature [22, 23]. $FAPbI_3$ perovskites have a pseudocubic structure with relatively better stability, but still transform into the wide-bandgap hexagonal phase in several months [24]. Therefore, strategies of using organic–inorganic mixed A-site cations have been adopted to enhance the stability of lead iodide perovskites [25]. Protesescu et al. reported a facile colloidal synthesis method to acquire air-stable FA-doped $FA_{0.1}Cs_{0.9}PbI_3$ PQDs with 3D orthorhombic structure [26]. The resultant $FA_{0.1}Cs_{0.9}PbI_3$ colloidal solutions had bright photoluminescence (PL) and high PL quantum yield (PLQY > 70%), which was stable for more than 6 months (Fig. 13.1a). Similarly, Rb-doped $Rb_xCs_{1-x}PbX_3$ ($x = 0$ to $x = 0.8$ and X = Br or Cl) PQDs had been reported, exhibited tunable fluorescence (395–525 nm) and PLQY up to 60% [27]. The tunable optical properties in $Rb_xCs_{1-x}PbX_3$ systems were attributed to increase of octahedral tilting due to the addition of small $Rb^+$ cations (Fig. 13.1b). Further stability tests indicated that (i) the fresh-prepared $Rb_xCs_{1-x}PbX_3$ PQDs presented a red shift of absorption in the first week and then no more changes one month later and (ii) the chloride analogues were less stable than bromide analogues. Later, Peng et al. found that PLQY of $CsPbCl_3$ PQDs could be improved from 5.7% to 13% through optimizing the doping concentration of $Rb^+$ cations (Fig. 13.1c) [28]. Ghosh et al. investigated the mechanism of Rb-doping in PQDs from the aspects of local structures and dynamics of mixed A-cation compositions: as smaller Rb cations were introduced into perovskite lattice, N–H hydrogen bonding was strengthened, which locked the octahedral rotation and resulted in the reduction of lattice dynamics and increase of phase stability accordingly [29].

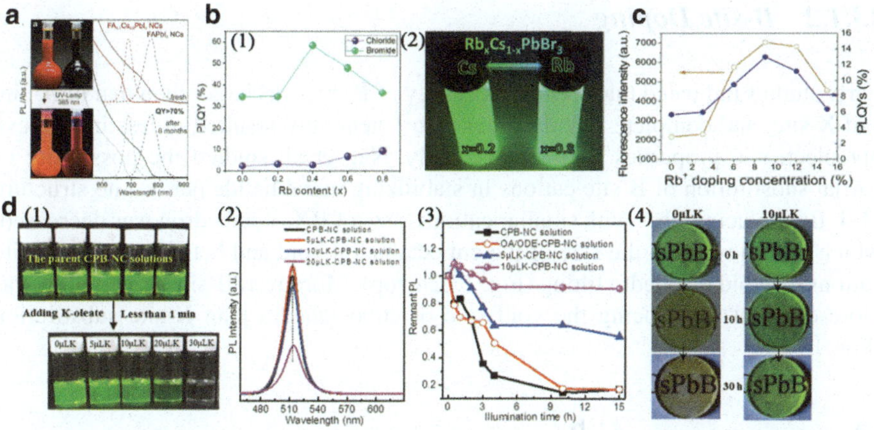


**Fig. 13.1** **a** Optical absorption and PL spectra of $FAPbI_3$ PQDs and $FA_{0.1}Cs_{0.9}PbI_3$ PQDs before and after 6 months of storage. The insets contain photographs of the $FAPbI_3$ and $FA_{0.1}Cs_{0.9}PbI_3$ PQDs colloidal solutions in toluene under daylight (upper image) and under a UV lamp ($\lambda$ = 365 nm; lower image) excitation. Reprinted from Ref. [26], copyright 2017, with permission from american chemical society. **b** Rb-doped RbxCs1-xPbX$_3$ (x = 0 to x = 0.8 and X = Br or Cl) PQDs: (1) PLQY of the PQDs with chlorine (purple) and bromine (green). (2) Photographs of the colloidal PQDs of x = 0.2 and x = 0.8 (left and right vials respectively) samples under UV light ($\lambda$ = 365 nm) showing the florescence for the extreme molar ratios of Br-based PQDs. Reprinted from Ref. [27], copyright 2018, with permission from the royal society of chemistry. **c** PL intensity and PLQY CsPbCl3: Rb PQDs (x = 0, 3, 6, 9, 12 and 15%). Reprinted from Ref. [28], copyright 2018, with permission from Elsevier. **d** KxCs1-xPbBr$_3$: (1) the optical images of the different addition amount of K-oleate precursor into the as-prepared $CsPbBr_3$ toluene solutions. (2) PL spectra of the $CsPbBr_3$ and K-modified CsPbBr3 PQDs. (3) The remnant PL emissions of the $CsPbBr_3$ and K-modified CsPbBr3 PQDs after 15 h of illumination (450 nm, 30 mW $cm^{-2}$). (4) Optical images of the $CsPbBr_3$ and K-modified $CsPbBr_3$ films as a function of the treatment time in the dark environment (50 °C, relative humidity 60%). Reprinted from Ref. [30], copyright 2018, with permission from Wiley–VCH

Another strategy of A-site doping is based on $K^+$ ion. Huang S. et al. had doped $K^+$ ions into $CsPbBr_3$ PQDs by post-modification method, which not only increased the PLQY from 65 to 83%, but also considerably improved their stability (Fig. 13.1d) [30]. Under 450-nm irradiation, the remnant PL emission was 91 and 100%, respectively, for K-modified $CsPbBr_3$ PQDs solution (15 h) and film (153 h). When heated at 50 °C for 30 h, the green emission of K-modified $CsPbBr_3$ PQDs films could maintain 97.9% of initial intensity. The improved photo- and thermo-stability was ascribed to a K-rich phase at the interface, which effectively inhibited the halide migration and suppressed the non-radiative decay. $K^+$ ions were also introduced to $CsPbCl_3$ PQDs, leading to the increase of PLQY from initial 3.2 to 10.3% [31]. It should be noted that multiple A-site cations ($Na^+$, $K^+$, $Rb^+$, $Cs^+$, $MA^+$ and $FA^+$) were also helpful to improve the stability of perovskite films, and beneficial to related optoelectronic devices [32–34].

### 13.1.2 B-site Doping

Early studies indicated that formation energy of B-site was larger than that of A-site and X-site, and complete substitution of $Pb^{2+}$ generally resulted in less impressive optoelectronic properties [35–38]. Recently, Nag et al. studied the possibility of partial substitution of B-site cations in stabilizing metal halide perovskite structure [39]. By replacing $Pb^{2+}$ with smaller cations, size of $BX_6$ octahedron was decreased, which in turn reduced the size of the cuboctahedral void and better matched A-site cations to avoid octahedra tilting (Fig. 13.2a, top). Moreover, B-site substitution was more efficient in reducing the void size of cuboctahedra than X-site substitution (Fig. 13.2a).

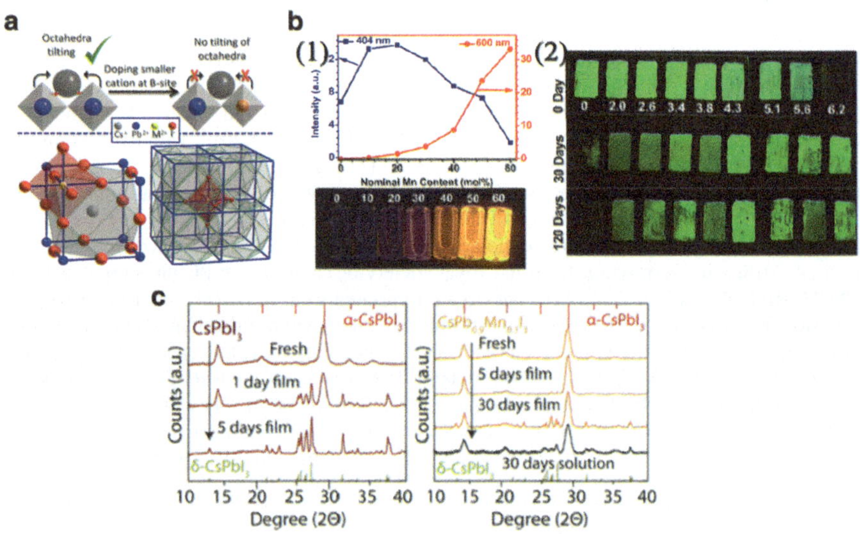


**Fig. 13.2** **a** Top: rotational distortion of $PbX_6$ octahedra can be restricted when $Pb^{2+}$ is partially substituted with smaller B-site cations. Smaller B-site cations reduce the size of the $BX_6$ octahedron, which in turn decreases the size of the cuboctahedral void for the A-site cation. Bottom: the B-site cation is located at the corner position of a perovskite unit cell and also at the center of the $BX_6$ octahedron (left panel), and the $BX_6$ octahedron is shared by eight cuboctahedra (right panel). Therefore, the substitution of one smaller B-site cation can influence the size of eight cuboctahedral voids by reducing the size of the centrally placed $BX_6$ octahedron. Reprinted from Ref. [39], copyright 2018, with permission from american chemical society. **b** Pure $CsPbCl_3$ and $CsPbCl_3$:Mn PQDs: (1) PL intensities for excitonic or $Mn^{2+}$ doping emissions $CsPbCl_3$ and $CsPbCl_3$:Mn PQDs centered at 404 and 600 nm as a function of the doping concentration of $Mn^{2+}$ ion ranging from 0 to 60 mol%, and their corresponding photographs in cyclohexane solution under 362-nm UV lamp irradiation. (2) PL emission photographs for $CsPbBr_3$:Mn PQDs coated on the surface of a glass slide with different $Mn^{2+}$ contents from 0 to 6.2 mol% taken under UV irradiation at indicated time period. Reprinted from Ref. [40], copyright 2017, with permission from American Chemical Society. **c** Stability of $CsPbI_3$ and $CsPb_xMn_{1-x}I_3$ PQDs stored under ambient conditions before and after 5/30 days. Reprinted from Ref. [44]. copyright 2017, with permission from american chemical society.

Zou and coworkers have synthesized Mn-doped $CsPbCl_3$:Mn PQDs featuring a second broad emission band peaking at ~600 nm, originated from the $^4T_1 \rightarrow {}^6A_1$ transition of $Mn^{2+}$ [40] . Interestingly, $Mn^{2+}$ dopant played a key role in enhancing the air stability of $CsPbX_3$ PQDs, which retained about 60% of the initial PL intensity under ambient air conditions within 120 days, while the counterpart pure PQDs became nonluminescent within 30 days (Fig. 13.2b). Liu et al. studied the effect of halide on the relative intensity of intrinsic band-gap versus Mn emission in $Mn^{2+}$ doped AIPQDS ($CsPbX_3$, where X = Cl, Br, or I) [41]. The maximum Mn substitution ratio was also studied, and a value up to 46% still maintained the tetragonal crystalline structure of $CsPbCl_3$ host [42]. At this doping level, Mn substitution not only increased PLQYs of $CsPbCl_3$ from 5 to 54% but also preserved the crystal structures under ambient light and atmospheres for over 3 months. Unfortunately, these $CsPb_xMn_{1-x}Cl_3$ PQDs showed severer photobleaching under strong UV irradiation (2000 W), similar to that of $CsPbCl_3$ PQDs. Mn-doped $CsPbCl_3$ nanoplatelets with quasi two-dimensional structure were also prepared, but only exhibited weak stability of 4–5 day under ambient conditions [43]. The above works mainly focused on Mn-doped stabilizing effect in $CsPbCl_3$ PQDs. Akkerman et al. studied the influence of Mn-doping in $CsPbI_3$ PQDs, and found that the highly unstable $\alpha$-$CsPbI_3$ phase was stabilized after doping due to increased Goldsmith tolerance factor and cohesive energy [44]. The alloyed $CsPb_xMn_{1-x}I_3$ PQDs showed stable and bright luminescence for up to one month, whether stored in solution or in air (Fig. 13.2c).

B-site doping with Sn(II) ions have been popularly studied. Complete replacement of Pb with Sn(II) were proved to impair the performance of PQDs. For example, $CsSnX_3$ (X = Cl, Br, I) PQDs suffered from lowered PL and instability under ambient conditions because Sn(II) ions were easily oxidized to Sn(IV) [45]. On the contrary, partial substitution with Sn(II) not only reduced the toxicity of lead content, but also improved the optical properties [46] and related devices performance [47]. Stam and coworkers had reported a partial cation exchange method to synthesize colloidal $CsPb_{1-x}Sn_xBr_3$ PQDs ($0 < x \leq 0.1$) [48]. With the partial exchange from $Pb^{2+}$ to $Sn^{2+}$, the doped PQDs still maintained the size, shape, high PLQYs and colloidal stability of $CsPbBr_3$ PQDs, except for a blue-shift in emission presumably due to lattice contraction. Deng et al. also synthesized $CsPb_{1-x}Sn_xBr_3$ PQDs at room temperature via a modified ligand-assisted reprecipitation method, which showed high PLQY (91%) and more than 80% remanent PL after 120 days storage in air (Fig. 13.3a) [49]. In comparison to Sn(II), Sn(IV) is more stable towards oxidation than Sn(II), [50] and the instability of Sn-doped AIPQDs was expected to be solved by using the more conductive Sn(IV) ions. Wang et al. synthesized Sn(IV)-doped $CsPb_{1-x}Sn_xBr_3$ PQDs by hot-injection, which not only increased the PLQY from 45 to 83%, but exhibited high environmental stability, unlike the Sn(II)-substitution [51].

B-site doping with other cations also have been performed to improve the stability of AIPQDs. Yong and coworkers reported divalent-nickel (Ni) doping in $CsPbCl_3$ PQDs [52]. The Ni-doping was thought to increase defect formation energy and local structural order, resulting in near-unity PLQY with violet luminescence and improved stability in morphology and PLQY(Fig. 13.3b). Meng et al. found that using $AlBr_3$ could afford $CsPbBr_3$ PQDs with stable blue PL [53]. The blue-emitting

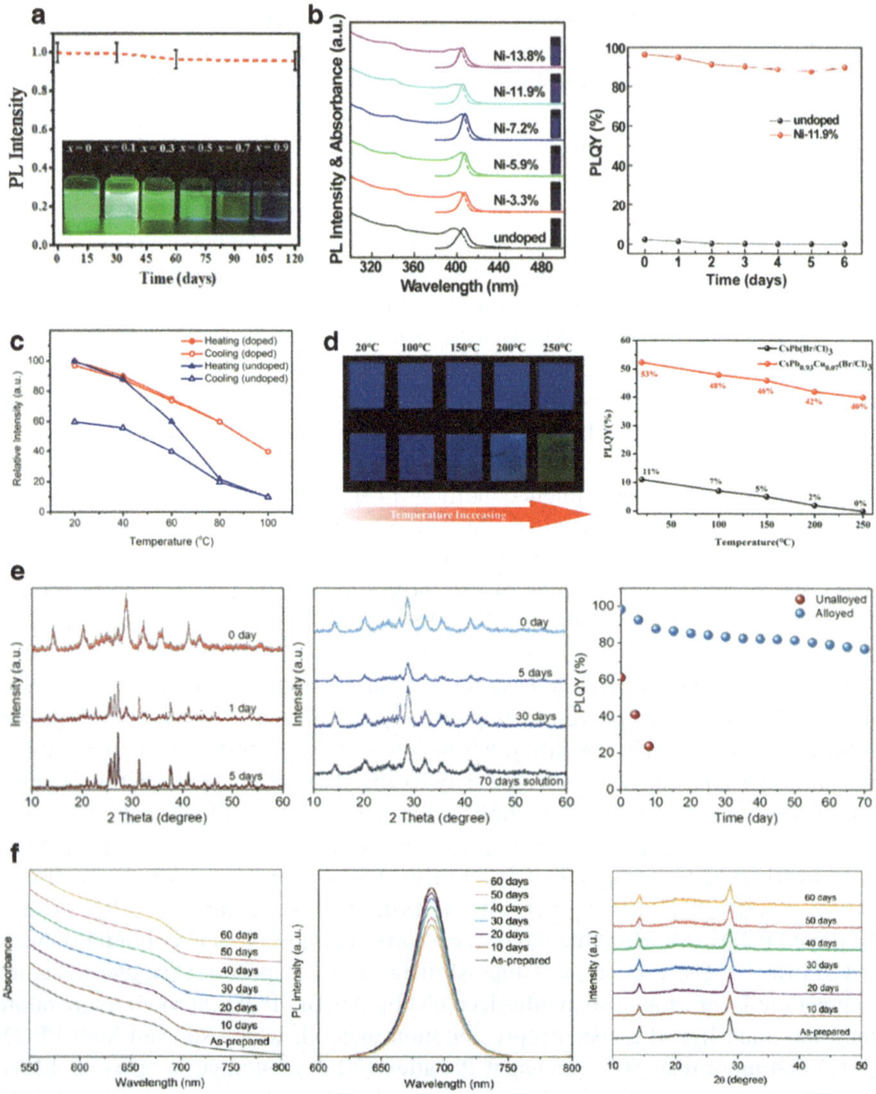
a
PL Intensity
Time (days)
b
PL Intensity & Absorbance (a.u.)
Ni-13.8%
Ni-11.9%
Ni-7.2%
Ni-5.9%
Ni-3.3%
undoped
Wavelength (nm)
PLQY (%)
undoped
Ni-11.9%
Time (days)
c
Heating (doped)
Cooling (doped)
Heating (undoped)
Cooling (undoped)
Relative Intensity (a.u.)
Temperature (°C)
d
20°C
100°C
150°C
200°C
250°C
Temperature Increasing
PLQY(%)
Temperature(°C)
e
Intensity (a.u.)
0 day
1 day
5 days
2 Theta (degree)
0 day
5 days
30 days
70 days solution
Unalloyed
Alloyed
PLQY (%)
Time (day)
f
Absorbance
60 days
50 days
40 days
30 days
20 days
10 days
As-prepared
Wavelength (nm)
PL intensity (a.u.)
Intensity (a.u.)
2θ (degree)

◂**Fig. 13.3** **a** Linear graph of the corresponding PL intensity ($\lambda_{em}$ = 519 nm) of $CsPb_{0.9}Sn_{0.1}Br_3$ PQDs in hexane solution as a function of time. The inset is photographs of $CsPb_{1-x}Sn_xBr_3$ PQDs in hexane solution under 365 nm UV lamp illumination. Reprinted from Ref. [49], copyright 2019, with permission from Elsevier. **b** Left: absorption and PL spectra of undoped and doped $CsPbCl_3$ PQDs. The inset shows the photographs of a PQD solution under UV illumination (365 nm). Right: PLQYs of Ni-doped and undoped $CsPbCl_3$ PQDs in ambient environment (humidity: ~60%; temperature: ~19.5 °C) for 6 days. Reprinted from Ref. [52], copyright 2017, with permission from american chemical society. **c** Thermal cycling of $CsPbBr_3$ and Al:$CsPbBr_3$ PQDs films. The heating ramp is 10 °C $min^{-1}$. Reprinted from Ref. [53], copyright 2017, with permission from Wiley–VCH. **d** Photographs showing different emission colors and corresponding PLQYs of $CsPb_{0.93}Cu_{0.07}(Br/Cl)_3$ PQDs (in red) and $CsPb(Br/Cl)_3$ PQDs (in black) deposited as thin films on glass slides as a function of the annealing temperature. Reprinted from Ref. [54], copyright 2019, with permission from American Chemical Society. **e** XRD patterns of $CsPbI_3$ (in red) and $Zn^{2+}$-alloyed $CsPb_{0.64}Zn_{0.36}I_3$ PQDs (in blue) dispersed in toluene solution and stored in ambient conditions (humidity: 40–50%, temperature: 20–25 °C). Change of the PL QY of the $CsPbI_3$ and $Zn^{2+}$-alloyed $CsPb_{0.64}Zn_{0.36}I_3$ PQDs solution versus storage time in the same conditions as above. Reprinted from Ref. [55], copyright 2019, with permission from american chemical society. **f** Absorption, PL spectra and XRD patterns of the $Sr^{2+}$-doped $CsPbI_3$ PQDs in toluene, sealed in glass bottle under ambient conditions for 60 days. Reprinted from Ref. [56], copyright 2018, with permission from Wiley–VCH

Al:$CsPbBr_3$ PQDs film exhibited higher thermal stability and thermal recycling than pure $CsPbBr_3$ PQDs film, and the PL intensity could be fully recovered after thermal treatment (Fig. 13.3c). Enlightened by the successful doping in traditional II–VI and II–V quantum dots, $Cu^{2+}$ with a small ionic radius (73 pm as compared to 119 pm for $Pb^{2+}$) could be another potential dopant for AIPQDs. Bi et al. demonstrated that doped $CsPb_{1-x}Cu_xX_3$ (X = Br, Br/Cl) PQDs were of impressive thermal stability with PLQY of over 80%, which were attributed to strengthened lattice binding energy and improved short-range order of the lattice (Fig. 13.3d) [54]. Shen and coworkers prepared a series of alloyed $CsPb_{1-x}Zn_xI_3$ PQDs with near-unity PLQY (Fig. 13.3e) [55]. The $CsPb_{0.64}Zn_{0.36}I_3$ PQDs could retain the α-phase for 70 days, while the α-$CsPbI_3$ was completely transformed into δ-$CsPbI_3$ within 5 days. The greatly improved stability was attributed to the progressive lattice contraction and consequent enhanced effective tolerance factors. Recently Lu et al.introduced $Sr^{2+}$ into $CsPbI_3$ PQDs by using $SrCl_2$ as co-precursor [56]. The doping of $Sr^{2+}$ significantly enhanced the PLQY (from 65 to 84%) and stability of $CsPbI_3$ PQDs. Specifically, $Sr^{2+}$-doped $CsPbI_3$ PQDs could maintain their cubic phase and optical properties for 60 days' storage under ambient conditions, while the pristine $CsPbI_3$ PQDs were transformed from α-phase to δ-phase within 3 days in solution (Fig. 13.3f).

As another strong candidates for dopants, rare earth (RE) ions exhibit rich and unique energy levels, and thus can be used to achieve multiple color emissions through energy transfer from PQDs host to RE impurities [31]. Pan and coworkers reported the doping of a serials of RE ions ($Ce^{3+}$, $Sm^{3+}$, $Eu^{3+}$, $Tb^{3+}$, $Dy^{3+}$, $Er^{3+}$, and $Yb^{3+}$) into $CsPbCl_3$ PQDs through a hot-injectionmethod [57]. The resultant RE-doped PQDs acquired not only widely tunable multicolor emissions spanning from visible to near-infrared regions, but also ameliorated stability (Fig. 13.4a).

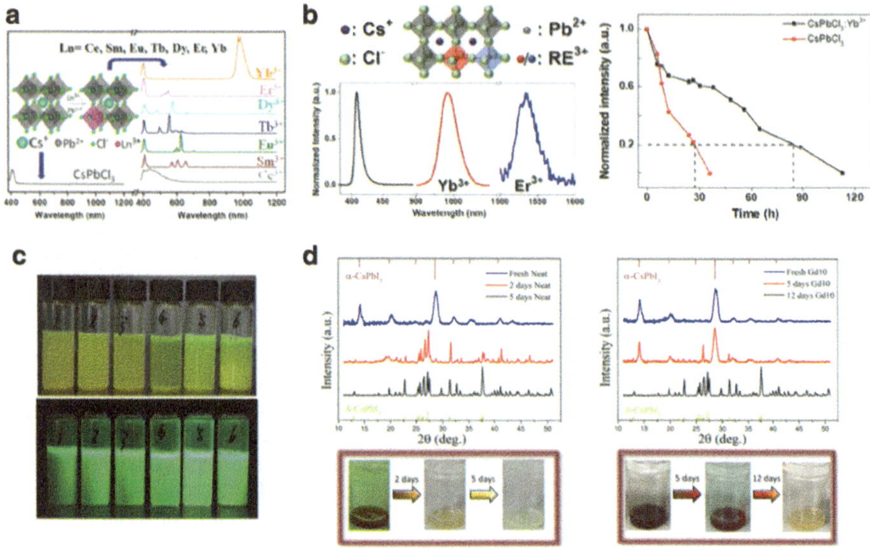


**Fig. 13.4** **a** PL spectra of the $CsPbCl_3$ PQDs doped with different lanthanide ions. Reprinted from Ref. [57], copyright 2017, with permission from American chemical society. **b** The expanding PL spectra to infrared regions of $Yb^{3+}$-doped and $Er^{3+}$-doped $CsPbCl_3$ PQDs and PL decreases of $CsPbCl_3$ and $Yb^{3+}$-doped $CsPbCl_3$ PQDs under 365 nm UV (6 W) light irradiation. Reprinted from Ref. [61], copyright 2018, with permission from the royal society of chemistry. **c** Photographs of colloidal $Ce^{3+}$-doped $CsPbBr_3$ PQDs in toluene solution after 30-day storage under ambient conditions and the corresponding UV-light irradiation (365 nm), respectively. Reprinted from Ref. [62], copyright 2018, with permission from american chemical society. **d** Time-dependent XRD patterns of $CsPbI_3$ (left) and $CsPb_xGd_{1-x}I_3$ (right) PQDs and their time-dependent photographs under day light. Reprinted from Ref. [63], copyright 2019, with permission from american chemical society

Remarkably, the incorporation of slightly smaller $Eu^{3+}$ ions (1.17 Å) into $CsPbI_3$ perovskite lattice were deemed to increase the tolerance factor and consequently stabilize the α-phase of $CsPbI_3$, which was of great importance for photovoltaic and LED applications [39, 58]. Jena et al. explored the inclusion of certain amount of $Eu^{3+}$ into precursor of $CsPbI_3$ film, and found that the α-phase Eu-doped $CsPbI_3$ was stable under ambient conditions for 30 days [59]. Similarly, stability of Eu-doped $CsPbI_2Br$ film was also greatly improved in comparison to $CsPbI_2Br$ film [60]. The former remained dark (α-phase) over the course of 6 months, whereas the latter became yellow-white (δ-phase) in less than 50 h.

In addition to Eu-doping, $Yb^{3+}/Er^{3+}$ co-doped $CsPbCl_3$ PQDs were prepared [61]. Besides a 1533 nm NIR emission, photostability of the doped PQD was significantly improved. Under the irradiation of a 6-W UV (365 nm) lamp, PL intensity of $CsPbCl_3$:Yb PQDs decreased by 35% in 27 h, in contrast to the 80% decrease of undoped PQDs (Fig. 13.4b). Yao and coworkers reported the doping of heterovalent

$Ce^{3+}$ ions into colloidal $CsPbBr_3$ PQDs through a hot-injection method [62]. Ce-doping enhanced the stability of $CsPbBr_3$ PQDs, which could preserve 60% of PLQY after being aged in toluene for 30 days under ambient environments (Fig. 13.4c). Similar to Eu-doping, $Gd^{3+}$ ions were introduced into $CsPbI_3$ PQDs with the aim of stabilizing the α-phase of $CsPbI_3$ [63]. This strategy is resultful that the phase transform from α-phase to δ-phase took 5 days for $CsPbI_3$ PQD, but spent 12 days for $CsPb_xGd_{1-x}I_3$ PQDs (Fig. 13.4d).

## 13.2 Surface Engineering

### *13.2.1 Surface Passivation*

In terms of colloidal PQDs, surface ligands play a critical role not only in reducing surface defects, but also in suppressing surface decomposition induced by oxygen and moisture. Numerous theoretical and experimental studies indicated that PQDs exhibit a high defect tolerance, but are highly susceptible to moisture and oxygen when suffering from poor surface passivation [64, 65]. Therefore, ligand-based surface passivation is important for strengthening the stability of PQDs.

Oleic acid (OA) and oleylamine (OLA) are conventional suface ligands for the synthesis of PQDs. Recently, Wang et al. replaced ligand OA with bis-(2,2,4-trimethylpentyl)phosphinic acid (TMPPA) in the synthesis of $CsPbI_3$ PQDs in order to enhance the α-phase stability [66]. Surface capping by TMPPA yielded highly stable $CsPbI_3$-TMPPA PQDs,whose α-phase was retained during the seven-day testing period. Furthermore, $CsPbI_3$-TMPPA PQDs perfectly preserved their PL property after 20 days' storage under ambient conditions, whereas $CsPbI_3$-OA PQDs decomposed within 3 days (Fig. 13.5a).

During the synthesis of $CsPbI_3$ PQDs, ammonium ligands (i.e. OLA) were prone to be lost in the necessary procedure of purification, resulting in fast agglomeration and undesired phase transformation from cubic to orthorhombic [67]. As a kind of alkylphosphine, trioctylphosphine (TOP) was generally considered to effectively coordinate to the surface of PQD to afford more complete surface passivation [68]. Liu and coworkers prepared TOP modified $CsPbI_3$ PQDs by using highly reactive TOP-$PbI_2$ as precursor [69]. After being kept for one moth, the resultant TOP-$CsPbI_3$ PQDs retained ~85% of its initial unity PLQY, while PLQY of non-TOP capped PQDs decreased from 86–60% (Fig. 13.5b). The excellent chemical stability of TOP-$CsPbI_3$ PQDs was attributed to their improved crystalline quality with less quenching defects. TOP-capping also could significantly enhanced stability of PQDs against high temperature and UV irradiation, and more impressively, made the "dead" (lost most or all of its fluorescence) PQDs "reborn" (recovered its fluorescence) (Fig. 13.5c) [70].

Koscher and coworkers proposed thiocyanate-based surface treatment for PQDs with lead-rich surface [71]. Thiocyanate salt such as ammonium thiocyanate

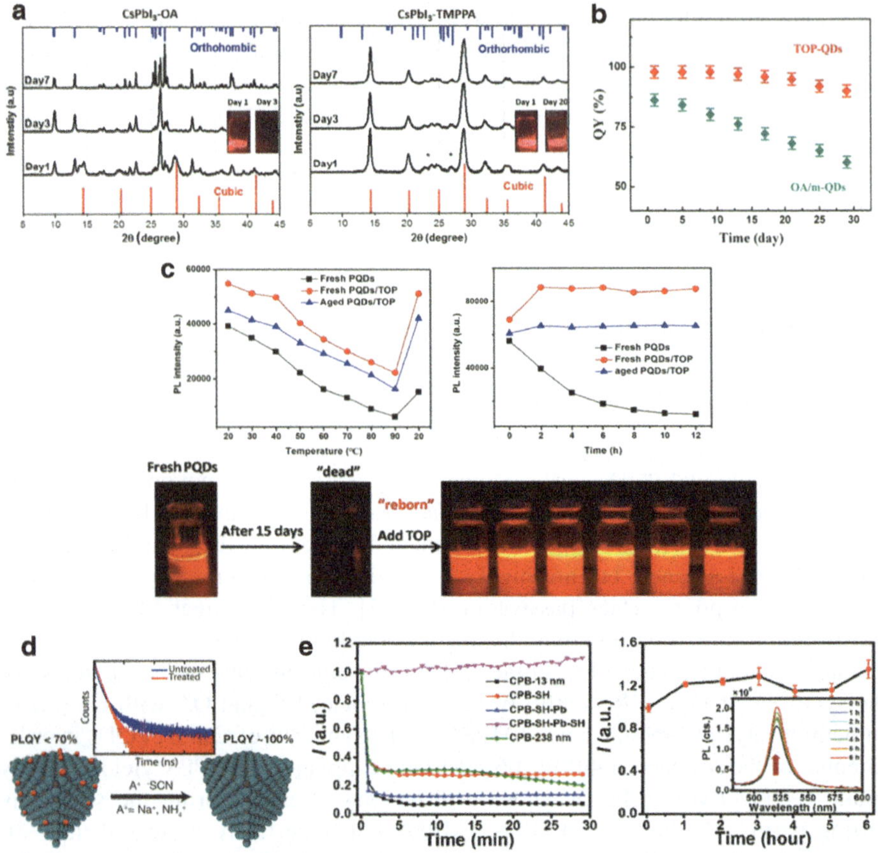


**Fig. 13.5** **a** XRD patterns of $CsPbI_3$-OA (right) and $CsPbI_3$-TMPPA (left). Standard cubic phase is shown in red, orthorhombic phase in blue. The insets are solutions of the respective washed PQDs under UV light (365 nm) at different times following synthesis. Reprinted from Ref. [66], copyright 2017, with permission from the royal society of chemistry. **b** Change of the PLQY of the OA-OLA capped and TOP capped $CsPbI_3$ PQDs versus storage time, where PQD solutions synthesized at 140 °C were stored in a sealed bottle under ambient conditions. All these PQD samples were washed twice with methyl acetate before PLQY measurement. Reprinted from Ref. [69], copyright 2017, with permission from american chemical society. **c** Thermal and UV irradiation (365 nm) stabilities of fresh PQDs and fresh/aged PQDs/TOP (top). Photos of fresh/aged $CsPbBr_{1.2}I_{1.8}$ PQDs and aged $CsPbBr_{1.2}I_{1.8}$ PQDs with adding different amounts of TOP under UV light (bottom). Reprinted from Ref. [70], copyright 2018, with permission from american chemical society. **d** Schematic of thiocyanate surface treatment on $CsPbBr_3$ PQDs and time resolved photoluminescence lifetimes of untreated and treated $CsPbBr_3$ PQDs. Reprinted from Ref. [71], copyright 2017, with permission from american chemical society. **e** PL intensity monitoring stabilities of $CsPbBr_3$ (CPB) with average size of 13 nm, a thiol capped CPB (CPB-SH), CPB-SH-Pb, CPB-SH-Pb-SH, and CPB with average size of 238 nm in toluene and ethanol (v/v = 9: 1) mixed solvent; and PL intensity monitoring the stabilities of CPB-SH-Pb-SH in ethanol; the inset shows the raw spectra. Reprinted from Ref. [72], copyright 2017, with permission from the royal society of chemistry

($NH_4SCN$) or sodium thiocyanate (NaSCN) were utilized to repair the surface trapping of $CsPbBr_3$ PQDs by removing excessive lead, left the entire PQDs untouched. Because of the surface passivation, non-radiative pathways $^{-}$SCN-treated $CsPbBr_3$ PQDs were effectively reduced, corresponding to the luminescence decay kinetics transform from multi-exponential decays to a monoexponential decay (Fig. 13.5d). Ruan et al. developed a novel surface-passivation method by using thiol molecules, which formed stable Pb–S bonds like typical X-type ligand to deeply decrease surface energy and enhance chemical stability of $CsPbBr_3$ PQDs [72]. A compact layer of thiol molecules was developed around PQDs by multi-step of ligand exchange and ion filling, and the resultant thiol capped PQDs were extremely stable even submerged in ethanol for 12 h (Fig. 13.5e).

Colloidal stability and structural integrity of PQDs were substantially impaired by the facile desorption of surface capping molecules during isolation and purification. Highly dynamic binding existed between surface capping ligands was believed as the reason of rapid desorption of protective ligand shell, and also experimentally observed to weaken the colloidal stability [23, 67]. To overcome the shortcoming of dynamic binding, Krieg et al. utilized an inexpensive long-chain sulfobetaine, 3-(N,N-dimethyloctadecylammonio)-propanesulfonate, as surface ligand to modify $CsPbX_3$ PQDs [73]. Due to the formation of tight binding between sulfobetaine ligands and PQDs, the zwitterionic surface capped $CsPbBr_3$ were kinetically stable against several cycles of washing, different solvents and aging period (Fig. 13.6a). Similarly, Pan et al. employed a bidentate ligand, namely 2,2′-iminodibenzoic acid (IDA), to passivate the surface of $CsPbI_3$ PQDs [74]. Stability of IDA-treated PQDs was greatly enhanced, which could maintain 90% of PL emission after 15 days, while PL of untreated PQDs was almost completely quenched after 5 days (Fig. 13.6b). Bidentate ligands were also used to stabilize cubic phase PQDs [75]. Palazon and coworkers deposited OA and OLA capped $CsPbX_3$ PQDs on a silicon substrate, and linked the adjacent ligands with intermolecular $C = C$ bonding formed by a low flux of X-rays. Such "graphitization" created a network of PQDs interconnected with each other, which afforded the resultant film with high stability (Fig. 13.6c).

### *13.2.2 Surface Coating*

For PQDs-based optoelectronic devices, the inevitable thermal effect significantly undermine the performance of PQDs through particle aggregation, surface ligand detachment or irreversible PL quenching. To solve this problem, inorganic surface coating with mechanically robust and highly thermal-resistant materials serves as the most effective method.

Silica is a chemically stable and transparent material that does not affect optical properties in the whole visible region, and have been widely used for surface coating of various nanoparticles, such as iron oxide, QDs and golden. Sun et al. used (3-aminopropyl) triethoxysilane (APTES) as both capping agent and silica precursor for AIPQDs [76]. The three silyl ether groups in APTES hydrolyzed to form a

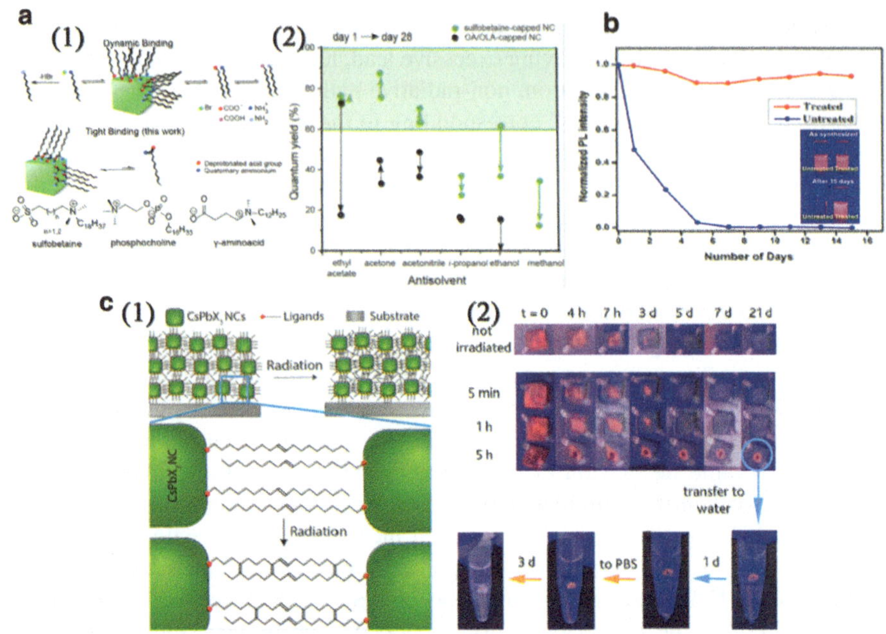


**Fig. 13.6** **a** (1) Depiction of (top) conventional ligand capping of PQD using long-chain molecules with single head groups (OA and OLA) and (bottom) a novel strategy wherein cationic and anionic groups are combined in a single zwitterionic molecule. (2) PLQY of PQD covered with the 3-(N,N-dimethyloctadecylammonio)propanesulfonate and OA/OLA after two steps of purification on day 1 and after storage for 28 days. Reprinted from Ref. [73], copyright 2018, with permission from american chemical society. **b** Normalized PLQY intensities as a function of days for untreated (blue) and IDA-treated $CsPbI_3$ PQDs (red). The insets are the corresponding photographs of untreated and treated PQDs as-synthesized and aged for 15 days. Reprinted from Ref. [74], copyright 2017, with permission from american chemical society. **c** (1) Schematic representation of the irradiation-induced graphitization process. $CsPbX_3$ PQDs (green cubes) deposited on a silicon substrate are capped with oleylamine (red dot = NH2) and oleic acid (red dot = COOH) ligands. Intermolecular C=C bonding appears as a consequence of irradiation, linking adjacent PQDs in the film. (2) Photographs of $CsPbI_3$ PQD films (ca. 5 × 5 mm2) under UV lamp (365 nm). Films irradiated at low dose (1 h or less) are completely degraded after 5 days exposure to air (room temperature and ambient humidity) and were therefore not immersed in water or biological buffer (PBS). Instead, the sample irradiated for 5 h retains its PL properties in air, water and PBS 1×. Reprinted from Ref. [75], copyright 2016, with permission from american chemical society

cross-linked matrix to encapsulate PQDs, resulting in a PQD/silica monolith which exhibited no decay and 5% decrease of PLQY after 3 months, respectively, for $CsPbBr_3$/silica and $CsPb(Br/I)_3$/silica (Fig. 13.7a). Mesoporous silica particles were also adopted to encapsulate pre-synthesized $CsPbBr_3$ PQDs [20]. The as-synthesized $CsPbBr_3$ PQD nanocomposites showed enhanced thermal stability and photostability compared with pristine $CsPbBr_3$ PQDs (Fig. 13.7b).

Hu et al. reported a water-triggered transformation process (WT) combined with a sol-gel method to prepare monodisperse $CsPbX_3/SiO_2$ and $CsPbBr_3/Ta_2O_5$ Janus nanoparticles [77]. Stability of the products was dramatically improved against air, water and light irradiation. After being treated with water for 7 days, $CsPbBr_3/SiO_2$ PQDs could maintain 80% of their original PL intensity, while WT-$CsPbBr_3$ PQDs and PQDs with hot-injection method declined to 17.8 and 9.2%. When stored in humid air (40 °C and humidity of 75%) for 4 days, a film of $CsPbBr_3/SiO_2$ PQDs only showed a slight drop in PL intensity, but the control samples of WT-$CsPbBr_3$ and HI-$CsPbBr_3$ displayed significant drop (Fig. 13.7c). Park et al. employed perhydropolysilazane (PHPS) as silica precursor to prepare silica modified PQDs [78]. On account of the robust PHPS-derived $SiO_2$ passivation, the $CsPbBr_3$-$SiO_2$ composites exhibited enhanced thermal stability and water resistance compared with $CsPbBr_3$ PQDs (Fig. 13.7d). Li and coworkers reported that anchoring PQDs on silica particle surface could also improve their stability [79]. The anchoring effects prevented the PQDs from contacting each other, which limited the regrowth of PQDs (Fig. 13.7e). After being kept for 40 days in ambient air, PL intensity of as-obtained PQDs-$SiO_2$ nearly did not change.

In addition to silica, other stable inorganic materials have been adopted to tentatively coat PQDs. Loiudice et al. encapsulated $CsPbX_3$ PQDs with amorphous alumina ($AlO_x$) via a low temperature atomic layer deposition (ALD) technique [80]. The $AlO_x$ matrix prevented the PQDs from sintering, oxygen and moisture in air, endowed them exceptional stability against water, temperature and light irradiation (Fig. 13.8a). Hierarchical $CaF_2$ nanospheres have been used to encapsulate PQDs [81].

The imperviousness of $CaF_2$-PQDs was significantly improved because the gap between small-sized $CaF_2$ nanocrystals were filled with PQDs. Therefore, the $CaF_2$-PQDs composites exhibited better water-resistant property, which preserved 60% of initial PL intensity within a day storage in a moist atmosphere, in comparison to the 95% PL decay in pristine $CsPbBr_3$ PQDs (Fig. 13.8b). Li et al. synthesized $CsPbBr_3/TiO_2$ core/shell nanocrystals through encapsulating colloidal PQDs with titanium precursor [82]. The as-obtained $CsPbBr_3/TiO_2$ were ultrastable, exhibiting excellent stability against water that their size, morphology and crystallinity were unchanged for more than 12 weeks (Fig. 13.8c). Lou and coworkers reported a $CsPbBr_3$@$NH_4Br$ (CPBr-NB-AE) nanocomposite formed via ion-exchange reaction between the nuleus $NH_4Br$ and supported $CsPbCl_3$ PQDs [83]. When placed in water, $NH_4Br$ was dissolved from CPBr-NB-AE to release ammonium ions, which capped $CsPbBr_3$ PQDs to give better performance of water resistance accordingly (Fig. 13.8d).

### *13.2.3 Polymer Encapsulation*

Polymer encapsulation has been regarded as a powerful strategy to protect PQDs from surrounding environments. Hence the perovskite-polymer composites may exhibit high stability especially superior water resistance. On the other hand, some polymers could passivate PQDs and in turn enhance their PL performances. Last, due to the

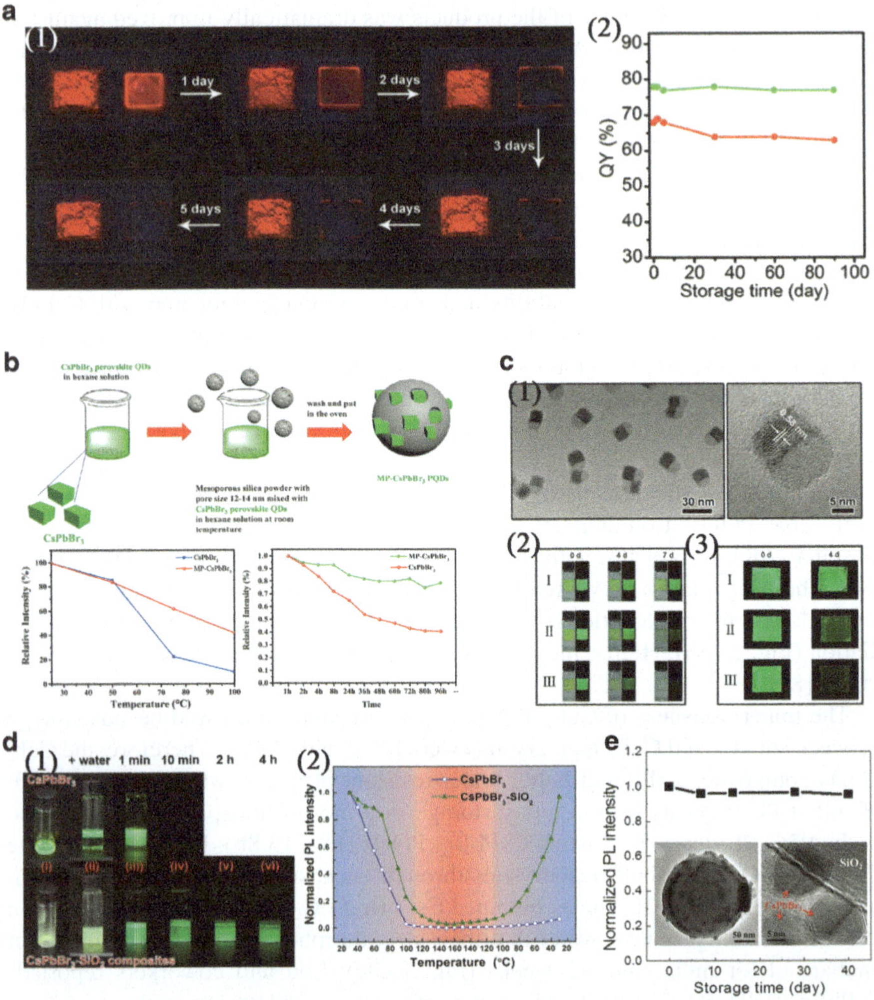
a
(1)
1 day
2 days
3 days
4 days
5 days
(2)
QY (%)
Storage time (day)
b
CsPbBr3 perovskite QDs in hexane solution
wash and put in the oven
MP-CsPbBr3 PQDs
Mesoporous silica powder with pore size 12-14 nm mixed with CsPbBr3 perovskite QDs in hexane solution at room temperature
CsPbBr3
Relative Intensity (%)
Temperature (°C)
Time
c
(1)
30 nm
5 nm
(2)
(3)
0 d
4 d
7 d
I
II
III
d
(1)
+ water
1 min
10 min
2 h
4 h
CsPbBr3
(i)
(ii)
(iii)
(iv)
(v)
(vi)
CsPbBr3-SiO2 composites
(2)
CsPbBr3
CsPbBr3-SiO2
Normalized PL intensity
Temperature (°C)
e
Normalized PL intensity
Storage time (day)
SiO2
50 nm
5 nm

◂**Fig. 13.7** **a** The photographs of red $CsPb(Br/I)_3$/silica powder and the red OLA-$CsPb(Br/I)_3$ film (dropped onto a piece of quartz glass) under UV light (365 nm) in air at room temperature; the pictures were taken once a day. (2) The PLQY stability of the green $CsPbBr_3$/silica and red $CsPb(Br/I)_3$/silica composites after 90 days of storage in air at room temperature. Reprinted from Ref. [76], copyright 2016, with permission from Wiley–VCH. **b** The synthesis process of mesoporous silica green $CsPbBr_3$ PQD nanocomposite (MP-PQDs). Thermal and photostability (UV-light, 365 nm, 6 W) stability test of MP-$CsPbBr_3$ and $CsPbBr_3$ in hexane solution. Reprinted from Ref. [20], copyright 2016, with permission from Wiley–VCH. **c** TEM image of the obtained $CsPbBr_3/SiO_2$ Janus PQDs and HRTEM image of a single $CsPbBr_3/SiO_2$ PQDs. (2) Photographs showing the stability against water of (I) $CsPbBr_3/SiO_2$ PQDs, (II) water-triggered transformation $CsPbBr_3$ PQDs (WT-$CsPbBr_3$) and (III) hot-injection $CsPbBr_3$ PQDs (HI-$CsPbBr_3$). The top layer is hexane, and the bottom layer is deionized water. Photographs were taken under daylight (left) and UV light (right, $\lambda = 365$ nm). (3) Photographs of (I) $CsPbBr_3/SiO_2$ PQDs, (II) WT-$CsPbBr_3$ PQDs and (III) HI-$CsPbBr_3$ PQDs thin film stored in humid air (40 °C and humidity of 75%). Reprinted from Ref. [77], copyright 2017, with permission from american chemical society. **d** Photographs showing water stability of $CsPbBr_3$ PQDs and $CsPbBr_3$-$SiO_2$ composites in water. The photographs were taken under illumination with (i and ii) indoor room light and (iii–vi) UV lamp ($\lambda_{ex} = 365$ nm). (2) Temperature dependent normalized PL intensities of $CsPbBr_3$ (blue open circle) and $CsPbBr_3$-$SiO_2$ composites (green solid triangle). Reprinted from Ref.[78], copyright 2017, with permission from Elsevier. **e** Improved photostability of $CsPbBr_3$/A-$SiO_2$ PQDs after 40 days of storage under ambient condition (the temperature changed from 15 to 25 °C and the humidity varied from 40 to 60%.) without any inert atmosphere protections. (No additional protection was applied that light illumination and oxygen existed during the whole measurement process.). The insets are TEM and HRTEM images of a typical $CsPbBr_3$/A-$SiO_2$ sphere. Reprinted from Ref. [79], copyright 2017, with permission from Wiley–VCH

demand of large-area backlight sources for liquid crystal displays (LCDs), PQD-polymer films have received special attention.

Lin et al. embedded $CsPbX_3$ PQDs into poly(styrene–butadiene–styrene) (SBS) fibers through one-step electrospinning strategy [84]. This flexible PQDs@SBS fiber membranes (Pe@SBS FMs) showed high photostability in aqueous solution owing to encapsulation of hydrophobic SBS polymer (Fig. 13.9a). Similarly, poly(methyl)methacrylate (PMMA) [85] and Poly(maleicanhydride-alt-1-octadecene) (PMA) [86] were both used to encapsulate $CsPbBr_3$ PQDs to improve their stability. In particular, the latter polymer could offer addition stability through its carbon chain shell surrounding PQDs (Fig. 13.9b). As regards interaction between polymers and PQDs, poly(styrene-ethylene-butylene-styrene) (SEBS) containing alkyl chain similar to the native alkyl chain ligands (OA and OLA) created better PQD-polymer interface, which minimized ligand loss and diffusion of water/ oxygen molecules towards PQD surface [87]. The as-synthesized $CsPbBr_3$-SEBS exhibited ultrahigh stability against water, PLQY of which was nearly constant when soaked in water for 60 days, while pure PQDs completely lost their fluorescence ~60 min later (Fig. 13.9c).

With the rapid development of wearable electronics, flexible LEDs and flexible displays receive growing attentions. Li et al. reported long-term stable and flexible $CsPbBr_3$ PQDs/ethylene vinyl acetate copolymer (EVA) films [88]. The films were fabricated by drying a mixture of perovskite precursor and EVA toluene solution.

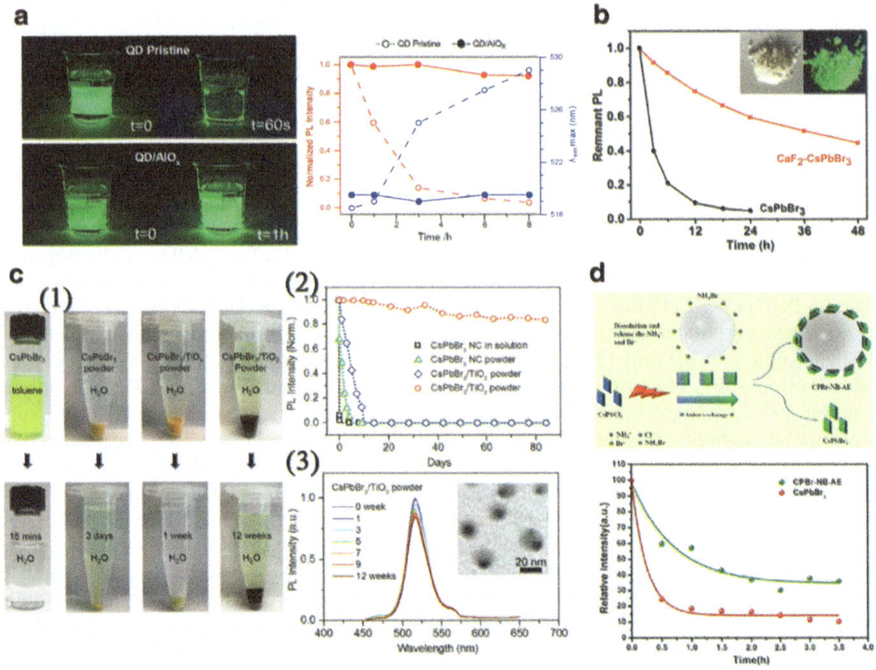


**Fig. 13.8** **a** Photographs of $CsPbBr_3$ PQDs film (top) and $CsPbBr_3/AlO_x$ nanocomposites film taken under UV illumination ($\lambda = 365$ nm) after 1 h of soaking in water and their PL properties after 8 h photosoaking under solar spectrum irradiation at 10 mW/cm$^2$. Reprinted from Ref. [80], copyright 2017, with permission from Wiley–VCH. **b** Humidity test of $CaF_2$-$CsPbBr_3$ composites and pristine $CsPbBr_3$ PQDs. This test is conducted in the sealed container with 100% humidity at room temperature. The insets are photographs of $CaF_2$-$CsPbBr_3$ powders under daylight (left) and UV light (right) illumination. Reprinted from Ref. [81], copyright 2017, with permission from Wiley–VCH. **c** Optical images of the $CsPbBr_3$ PQDs in toluene, $CsPbBr_3$ PQDs powder in water, $CsPbBr_3/TiO_x$ composite in water, and $CsPbBr_3/TiO_2$ nanocrystals in water (top) and those samples in water with different extended time under ambient conditions (bottom). (2) The relative PL intensity of $CsPbBr_3$ PQDs (without precipitation), dried $CsPbBr_3$ PQD powder, $CsPbBr_3/TiO_x$ powder, and $CsPbBr_3/TiO_2$ powder after immersing in water. (3) The relative PL intensity of $CsPbBr_3/TiO_2$ after immersing in water (0–12 weeks), the inset shows a TEM image of $CsPbBr_3/TiO_2$ after immersing in water for 12 weeks. Reprinted from Ref. [82], copyright 2017, with permission from Wiley–VCH. **d** Scheme for the synthesis of the CPBr-NB-AE nanocomposite (top) and water resistance test of the CPBr-NB-AE nanocomposite and $CsPbBr_3$ PQDs (bottom). Both samples were dispersed in 5 mL deionized water during the water resistance test. Reprinted from Ref. [83], copyright 2017, with permission from the royal society of chemistry

The resultant $CsPbBr_3$ PQDs/EVA film displayed high flexibility, PL intensity of which did not change after over 1000 cycles' bending, and outstanding stability when exposed to water vapor or even directly soaked in water (Fig. 13.9d).

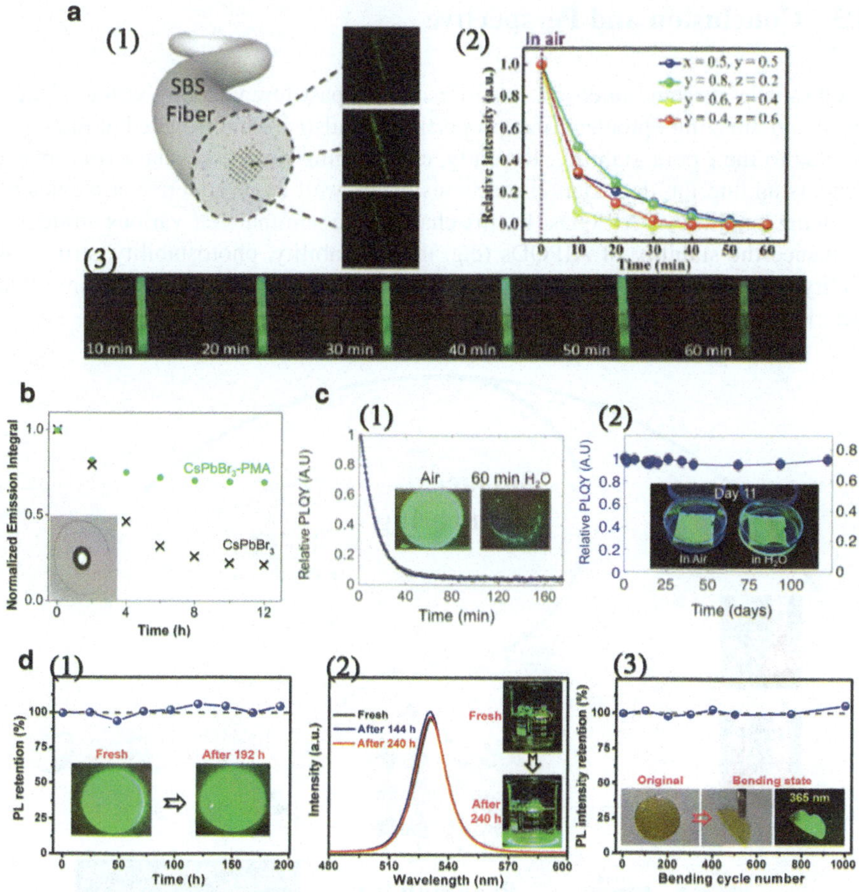


**Fig. 13.9** **a** $CsPb(Cl_x, Br_y, I_z)_3$@SBS composite: (1) Schematic of $CsPb(Br_{0.8}I_{0.2})_3$ PQD distribution in the SBS fibers. (2) Relative PL intensity of $CsPb(Cl_x, Br_y, I_z)_3$@SBS composite FMs after immersion in an aqueous solution. (3) Photographs of the $CsPb(Br_{0.8}I_{0.2})_3$@SBS fiber membrane under UV-light (365 nm) excitation when immersed in an aqueous solution. Reprinted from Ref. [84], copyright 2018, with permission from american chemical society. **b** Normalized integrals of the emission peaks between 460 and 600 nm of $CsPbBr_3$ and purified $CsPbBr_3$-PMA on a UV—LED constant irradiation (100 mW, 365 nm). The inset shows a green filter in the sample position of the integrating sphere. Reprinted from Ref. [86], copyright 2016, with permission from american chemical society. **c** Relative quantum yield of as-synthesized $CsPbBr_3$ PQDs in water as a function of soaking time in mins (black). The insets are the degradation of PQDs excited with UV ($\lambda = 365$ nm) before and after 60 min of water soaking. (2) Relative and absolute quantum yield of $CsPbBr_3$-SEBS composite films after >4 months of water-soaking. The insets are images of this composite films after 11 days of soaking in water and in air. Reprinted from Ref. [87], copyright 2016, with permission from American Chemical Society. **d** Long-term stability tests of $CsPbBr_3$ PQDs/EVA films: (1) PL intensity retention and photographs under UV (365 nm) light after keeping in the air for 0–192 h. (2) PL emission spectra and photographs under UV (365 nm) light after soaking in water (the film is pointed out by dotted line circles) for 0 ~240 h. (3) PL intensity retention of $CsPbBr_3$ PQDs/EVA film after repeatedly bending for 0 ~1000 cycles. The inset is photographs of $CsPbBr_3$ PQDs/EVA film at original and bending states. Reprinted from Ref. [88], copyright 2018, with permission from american chemical society

## 13.3 Conclusion and Perspective

AIPQDs have attracted intensive attentions in the past few years by virtue of their unique and amazing optoelectronic properties, but also are interrogated at the same time due to their poor stability. Recently, encouraging progresses have been made in understanding the degradation mechanisms, as well as in effective strategies of improving stability of AIPQDs. In this chapter, we summarized various strategies to enhance the stability of AIPQDs (e.g. thermostability, photostability, structural stability, environmental stability), and roughly categorize these strategies into two types (Fig. 13.10).

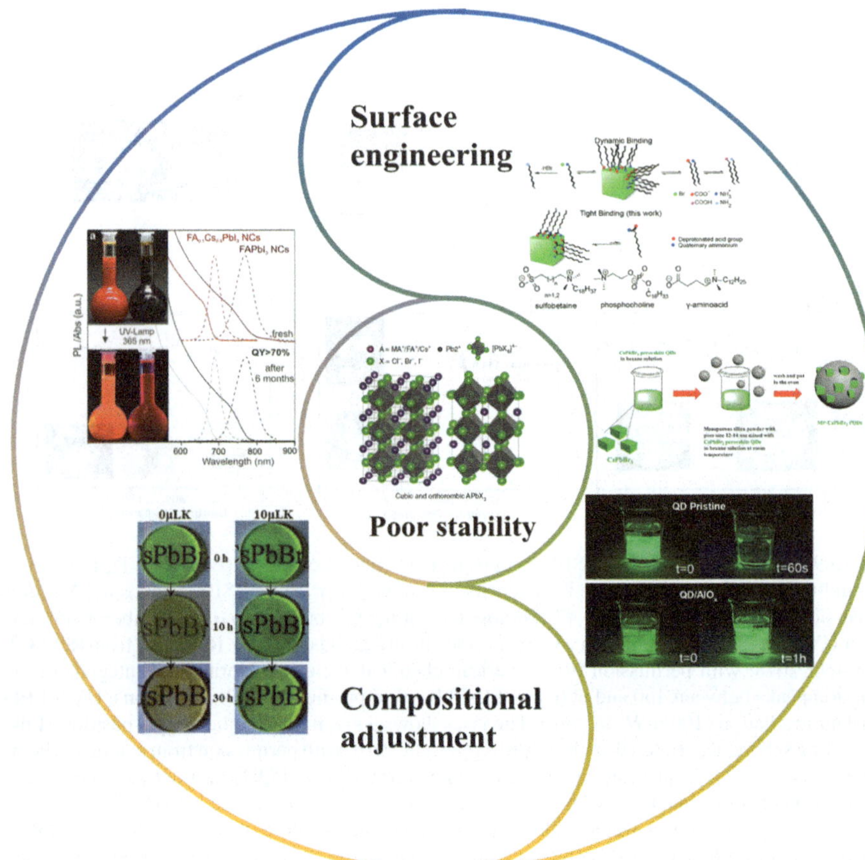


**Fig. 13.10** Schematic illustration describing two strategies towards improving the stability of AIPQDs. Figure in the section of poor stability is reprinted from Ref. [89], copyright 2018, with permission from Springer Nature. Figures in the section of compositional adjustment are reprinted from Ref. [26 and 30], copyright 2017 and 2018, with permission from american chemical society and Wiley–VCH, respectively. Figures in the section of surface engineering are reprinted from Ref. [20, 73, 80] copyright 2018, 2016 and 2017, with permission from american chemical society and Wiley–VCH, respectively

Ionic radius plays a critical role in structural stability of 3D-$ABX_3$ perovskite. As for AIPQDs, doping larger A-site cations (e.g. $MA^+$ and $FA^+$) or mixed multiple cations has been confirmed to effectively stabilizes the 3D perovskite structure. On the other hand, B-site doping not only can adjust optoelectronic property of AIPQDs, but also contribute to their stability. Additionally, substitution of $Pb^{2+}$ with some metal ions can reduce the toxicity of PQDs.

High density of surface trap states results in high non-radiative recombination rate as well as susceptivity to water, oxygen and light irradiation. OA and OLA molecules are conventional ligand for surface passivation, but suffer from highly dynamic binding and missing in the course of ageing and purification. Therefore, branch-structured ligands with large hindrance or capping ligands with strong interaction with surface of PQDs can increase the PLQY and stability of PQDs. In addition, inorganic or organic outer layer can endow PQDs with excellent environmental stability and water resistance. Most inorganic coating materials (e.g. $SiO_2$, $AlO_x$ and $TiO_2$) are dense and thermally stable, but the synthesis usually need high temperature and relatively complicated synthetic procedures. Some polymers with carboxyl or amine groups can well passivate PQD surface and result in enhanced PL properties and stability, but their thermal stability is rather limited. Hence more effective coating materials are needed to perfectly stabilize PQDs.

Overall, we summarized the optical properties and stability of optimized AIPQDs in Table 13.1. We hope that this chapter could encourage more researchers to devote sustained efforts to luminescent PQDs and promote the development of perovskite optoelectronic devices.

**Table 13.1** Summary of PL properties and stability of AIPQDs

| AIPQDs | Emission peak (nm) | FWHM (nm) | PLQYs (%) | Stability (phase, PL or PLQY) | References |
|---|---|---|---|---|---|
| $FA_{0.1}Cs_{0.9}PbI_3$ | 685 | | >70 | 95% (several months, air) | Protesescu et al. [26] |
| $CsPbBr_3$:K | 513 | ~20 | 83 | 91.1% (15 h irradiation, 450 nm, 30 mW $cm^{-2}$) | Huang et al. [30] |
| $CsPbBr_3$:Mn | 515 | 20 | 90 | 60% (120 d, air) | Zou et al. [40] |
| $CsPb_{0.76}Mn_{0.27}Cl_3$ | 580 | | 54 | 40% (1 h irradiation, 2000 W) tetragonal phase (3 months, air) | Liu et al. [42] |
| $CsPb_{0.9}Mn_{0.1}I_3$ | 680 | 40 | 82 | α-phase (30 d, air) | Akkerman et al. [44] |
| $CsPb_{0.9}Sn_{0.1}Br_3$ | 519 | | 91 | 80% (120 d, air) | Deng et al. [49] |

(continued)

**Table 13.1** (continued)

| AIPQDs | Emission peak (nm) | FWHM (nm) | PLQYs (%) | Stability (phase, PL or PLQY) | References |
|---|---|---|---|---|---|
| $CsPbCl_3$:Ni | 407 | | 96.5 | >90% (6 d, air) | Yong et al. [52] |
| $CsPbBr_3$:Al | 456 | 16 | 42 | 40% (100 °C, air) | Liu et al. [53] |
| $CsPb_{0.93}Cu_{0.07}(Br/Cl)_3$ | 453 | | 53 | 75% (250 °C, air) | Bi et al. [54] |
| $CsPb_{0.64}Zn_{0.36}I_3$ | 682 | | 98.5 | 80% (70 d, air) α-phase (70 d, air) | Shen et al. [55] |
| $CsPbI_3$:Sr | ~690 | | 84 | α-phase (60 d, air) | Lu et al. [56] |
| $CsPbCl_3$:Yb | 408 | | | 20% (85 h irradiation, 365 nm, 6 W) | Zhang et al. [61] |
| $CsPbBr_3$:Ce | ~512 | | 89 | 60% (30 d, air) | Yao et al. [62] |
| $CsPbI_3$:Gd | 688 | 34 | 80 | α-phase (4 d, air) | Guvenc et al. [63] |
| $CsPbI_3$-TMPPA | ~690 | 34 | | α-phase (7 d, air) | Wang et al. [66] |
| $CsPbI_3$-TOP | 685 | 33 | ~100 | 85% (30 d, air) | Liu et al. [69] |
| $CsPbI_3$-IDA | 680 | | ~95 | 90% (15d, air) | Pan et al. [74] |
| $CsPb(Br/I)_3$-APTES | 624 | | 68 | 95% (90 d, air) | Sun et al. [76] |
| $CsPbBr_3$-mesoporous silica | 515 | 20 | 55 | 40% (100 °C, air) 80% (96 h irradiation, 365 nm, 6 W) | Wang et al. [20] |
| $CsPbBr_3$-PHPS | ~525 | | 71.8 | 63.5% (80 °C, air) | Park et al. [78] |
| $CsPbBr_3$-$CaF_2$ | 524 | | 82 | 60% (1 d, 100% humidity) | Wei et al. [81] |
| $CsPbBr_3$-$TiO_2$ | 520 | ~33 | | 85% (90 d, in water) 75% (24 h irradiation) | Li et al. [82] |
| $CsPbBr_3$-PMA | 528 | 17 | ~53 | 60% (12 h irradiation, 365 nm, 100 mW) | Meyns et al. [86] |

(continued)

**Table 13.1** (continued)

| AIPQDs | Emission peak (nm) | FWHM (nm) | PLQYs (%) | Stability (phase, PL or PLQY) | References |
|---|---|---|---|---|---|
| $CsPbBr_3$-SEBS | 514 | 17 | ~78 | ~100% (120 d, in water) | Raja et al. [87] |
| $CsPbBr_3$-EVA | 520 | 18.6 | 40.5 | ~100% (192 h, air)<br>~100% (240 h, in water) | Li et al. [88] |

## References


1. X. Lan, O. Voznyy, F.P. García de Arquer, M. Liu, J. Xu, A.H. Proppe, G. Walters, F. Fan, H. Tan, M. Liu, Z. Yang, S. Hoogland, E.H. Sargent, 10.6% Certified colloidal quantum dot solar cells via solvent-polarity-engineered halide passivation. Nano Lett. **16**(7), 4630–4634 (2016). https://doi.org/10.1021/acs.nanolett.6b01957
2. G.H. Carey, A.L. Abdelhady, Z. Ning, S.M. Thon, O.M. Bakr, E.H. Sargent, Colloidal quantum dot solar cells. Chem. Rev. **115**(23), 12732–12763 (2015). https://doi.org/10.1021/acs.chemrev.5b00063
3. N.J. Jeon, H. Na, E.H. Jung, T.-Y. Yang, Y.G. Lee, G. Kim, H.-W. Shin, S. Il Seok, J. Lee, J. Seo, A fluorene-terminated hole-transporting material for highly efficient and stable perovskite solar cells. Nat. Energy **3**(8), 682–689 (2018). https://doi.org/10.1038/s41560-018-0200-6
4. M.M. Adachi, F. Fan, D.P. Sellan, S. Hoogland, O. Voznyy, A.J. Houtepen, K.D. Parrish, P. Kanjanaboos, J.A. Malen, E.H. Sargent, Microsecond-sustained lasing from colloidal quantum dot solids. Nat. Commun. **6**(1), 8694 (2015). https://doi.org/10.1038/ncomms9694
5. Y. Wang, X. Li, X. Zhao, L. Xiao, H. Zeng, H. Sun, Nonlinear absorption and low-threshold multiphoton pumped stimulated emission from all-inorganic perovskite nanocrystals. Nano Lett. **16**(1), 448–453 (2016). https://doi.org/10.1021/acs.nanolett.5b04110
6. Y. Wang, X. Li, V. Nalla, H. Zeng, H. Sun, Solution-processed low threshold vertical cavity surface emitting lasers from all-inorganic perovskite nanocrystals. Adv. Func. Mater. **27**(13), 1605088 (2017). https://doi.org/10.1002/adfm.201605088
7. H. Cho, S.-H. Jeong, M.-H. Park, Y.-H. Kim, C. Wolf, C.-L. Lee, J.H. Heo, A. Sadhanala, N. Myoung, S. Yoo, Overcoming the electroluminescence efficiency limitations of perovskite light-emitting diodes. Science **350**(6265), 1222–1225 (2015)
8. Y.-H. Kim, C. Wolf, Y.-T. Kim, H. Cho, W. Kwon, S. Do, A. Sadhanala, C.G. Park, S.-W. Rhee, S.H. Im, R.H. Friend, T.-W. Lee, Highly efficient light-emitting diodes of colloidal metal-halide perovskite nanocrystals beyond quantum size. ACS Nano **11**(7), 6586–6593 (2017). https://doi.org/10.1021/acsnano.6b07617
9. L.N. Quan, F.P. García de Arquer, R.P. Sabatini, E.H. Sargent, Perovskites for light emission. Adv. Mater. **30**(45), 1801996 (2018). https://doi.org/10.1002/adma.201801996
10. H. Wang, D.H. Kim, Perovskite-based photodetectors: materials and devices. Chem. Soc. Rev. **46**(17), 5204–5236 (2017). https://doi.org/10.1039/C6CS00896H
11. Y. Dong, Y. Zou, J. Song, X. Song, H. Zeng, Recent progress of metal halide perovskite photodetectors. J. Mater. Chem. C **5**(44), 11369–11394 (2017). https://doi.org/10.1039/C7TC03612D
12. H. Zhang, X. Wang, Q. Liao, Z. Xu, H. Li, L. Zheng, H. Fu, Embedding perovskite nanocrystals into a polymer matrix for tunable luminescence probes in cell imaging. Adv. Func. Mater. **27**(7), 1604382 (2017). https://doi.org/10.1002/adfm.201604382

13. S. Lou, Z. Zhou, T. Xuan, H. Li, J. Jiao, H. Zhang, R. Gautier, J. Wang, Chemical transformation of lead halide perovskite into insoluble, less cytotoxic, and brightly luminescent cspbbr 3/cspb2br5 composite nanocrystals for cell imaging. ACS Appl. Mater. Interf. **11**(27), 24241–24246 (2019). https://doi.org/10.1021/acsami.9b05484
14. F. Lang, O. Shargaieva, V.V. Brus, H.C. Neitzert, J. Rappich, N.H. Nickel, Influence of radiation on the properties and the stability of hybrid perovskites. Adv. Mater. **30**(3), 1702905 (2018). https://doi.org/10.1002/adma.201702905
15. H. Cho, Y.-H. Kim, C. Wolf, H.-D. Lee, T.-W. Lee, Improving the stability of metal halide perovskite materials and light-emitting diodes. Adv. Mater. **30**(42), 1704587 (2018). https://doi.org/10.1002/adma.201704587
16. L. Gomez, C. de Weerd, J.L. Hueso, T. Gregorkiewicz, Color-stable water-dispersed cesium lead halide perovskite nanocrystals. Nanoscale **9**(2), 631–636 (2017)
17. X. Li, F. Cao, D. Yu, J. Chen, Z. Sun, Y. Shen, Y. Zhu, L. Wang, Y. Wei, Y. Wu, H. Zeng, All inorganic halide perovskites nanosystem: synthesis, structural features. Opt. Prop. Optoelectron. Appl. Small **13**(9), 1603996 (2017). https://doi.org/10.1002/smll.201603996
18. V.K. Ravi, R.A. Scheidt, A. Nag, M. Kuno, P.V. Kamat, To exchange or not to exchange. Suppressing anion exchange in cesium lead halide perovskites with PbSO4–Oleate capping. ACS Energy Lett. **3**(4), 1049–1055 (2018). https://doi.org/10.1021/acsenergylett.8b00380
19. Q.A. Akkerman, V. D'Innocenzo, S. Accornero, A. Scarpellini, A. Petrozza, M. Prato, L. Manna, Tuning the optical properties of cesium lead halide perovskite nanocrystals by anion exchange reactions. J. Am. Chem. Soc. **137**(32), 10276–10281 (2015). https://doi.org/10.1021/jacs.5b05602
20. H.-C. Wang, S.-Y. Lin, A.-C. Tang, B.P. Singh, H.-C. Tong, C.-Y. Chen, Y.-C. Lee, T.-L. Tsai, R.-S. Liu, Mesoporous silica particles integrated with all-inorganic $CsPbBr_3$ perovskite quantum-dot nanocomposites (MP-PQDs) with high stability and wide color gamut used for backlight display. Angew. Chem. Int. Ed. **55**(28), 7924–7929 (2016). https://doi.org/10.1002/anie.201603698
21. E.J. Juarez-Perez, Z. Hawash, S.R. Raga, L.K. Ono, Y. Qi, Thermal degradation of CH 3 NH 3 PbI 3 perovskite into NH 3 and CH 3 I gases observed by coupled thermogravimetry–mass spectrometry analysis. Energy Environ. Sci. **9**(11), 3406–3410 (2016)
22. G.P. Nagabhushana, R. Shivaramaiah, A. Navrotsky, Direct calorimetric verification of thermodynamic instability of lead halide hybrid perovskites. Proc. Natl. Acad. Sci. **113**(28), 7717–7721 (2016). https://doi.org/10.1073/pnas.1607850113
23. L. Protesescu, S. Yakunin, M.I. Bodnarchuk, F. Krieg, R. Caputo, C.H. Hendon, R.X. Yang, A. Walsh, M.V. Kovalenko, Nanocrystals of cesium lead halide perovskites ($CsPbX_3$, X = Cl, Br, and I): novel optoelectronic materials showing bright emission with wide color gamut. Nano Lett. **15**(6), 3692–3696 (2015). https://doi.org/10.1021/nl5048779
24. H. Huang, M.I. Bodnarchuk, S.V. Kershaw, M.V. Kovalenko, A.L. Rogach, Lead halide perovskite nanocrystals in the research spotlight: stability and defect tolerance. ACS Energy Lett. **2**(9), 2071–2083 (2017). https://doi.org/10.1021/acsenergylett.7b00547
25. Z. Li, M. Yang, J.-S. Park, S.-H. Wei, J.J. Berry, K. Zhu, Stabilizing perovskite structures by tuning tolerance factor: formation of formamidinium and cesium lead iodide solid-state alloys. Chem. Mater. **28**(1), 284–292 (2016). https://doi.org/10.1021/acs.chemmater.5b04107
26. L. Protesescu, S. Yakunin, S. Kumar, J. Bär, F. Bertolotti, N. Masciocchi, A. Guagliardi, M. Grotevent, I. Shorubalko, M.I. Bodnarchuk, C.-J. Shih, M.V. Kovalenko, Dismantling the "Red Wall" of colloidal perovskites: highly luminescent formamidinium and formamidinium-cesium lead iodide nanocrystals. ACS Nano **11**(3), 3119–3134 (2017). https://doi.org/10.1021/acsnano.7b00116
27. D. Amgar, T. Binyamin, V. Uvarov, L. Etgar, Near ultra-violet to mid-visible band gap tuning of mixed cation RbxCs1−xPbX3 (X = Cl or Br) perovskite nanoparticles. Nanoscale **10**(13), 6060–6068 (2018). https://doi.org/10.1039/C7NR09607K
28. Z. Zhao, W. Xu, G. Pan, Y. Liu, M. Yang, S. Hua, X. Chen, H. Peng, H. Song, Enhancing the exciton emission of $CsPbCl_3$ perovskite quantum dots by incorporation of $Rb^+$ ions. Mater. Res. Bull. **112**, 142–146 (2019). https://doi.org/10.1016/j.materresbull.2018.12.004

29. D. Ghosh, P. Walsh Atkins, M.S. Islam, A.B. Walker, C. Eames, Good vibrations: locking of octahedral tilting in mixed-cation iodide perovskites for solar cells. ACS Energy Lett. **2**(10), 2424–2429 (2017). https://doi.org/10.1021/acsenergylett.7b00729
30. S. Huang, B. Wang, Q. Zhang, Z. Li, A. Shan, L. Li, Postsynthesis potassium-modification method to improve stability of $CsPbBr_3$ perovskite nanocrystals. Adv. Opt. Mater. **6**(6), 1701106 (2018). https://doi.org/10.1002/adom.201701106
31. Y. Liu, G. Pan, R. Wang, H. Shao, H. Wang, W. Xu, H. Cui, H. Song, Considerably enhanced exciton emission of $CsPbCl_3$ perovskite quantum dots by the introduction of potassium and lanthanide ions. Nanoscale **10**(29), 14067–14072 (2018). https://doi.org/10.1039/C8NR03581D
32. B. Philippe, M. Saliba, J.-P. Correa-Baena, U.B. Cappel, S.-H. Turren-Cruz, M. Grätzel, A. Hagfeldt, H. Rensmo, Chemical distribution of multiple cation (Rb+, Cs+, MA+, and FA+) perovskite materials by photoelectron spectroscopy. Chem. Mater. **29**(8), 3589–3596 (2017). https://doi.org/10.1021/acs.chemmater.7b00126
33. T.J. Jacobsson, S. Svanström, V. Andrei, J.P.H. Rivett, N. Kornienko, B. Philippe, U.B. Cappel, H. Rensmo, F. Deschler, G. Boschloo, Extending the compositional space of mixed lead halide perovskites by Cs, Rb, K, and Na doping. J. Phys. Chem. C **122**(25), 13548–13557 (2018). https://doi.org/10.1021/acs.jpcc.7b12464
34. M. Saliba, T. Matsui, K. Domanski, J.-Y. Seo, A. Ummadisingu, S.M. Zakeeruddin, J.-P. Correa-Baena, W.R. Tress, A. Abate, A. Hagfeldt, M. Grätzel, Incorporation of rubidium cations into perovskite solar cells improves photovoltaic performance. Science **354**(6309), 206–209 (2016). https://doi.org/10.1126/science.aah5557
35. C. Eames, J.M. Frost, P.R.F. Barnes, B.C. O'Regan, A. Walsh, M.S. Islam, Ionic transport in hybrid lead iodide perovskite solar cells. Nat. Commun. **6**(1), 7497 (2015). https://doi.org/10.1038/ncomms8497
36. J. Pal, S. Manna, A. Mondal, S. Das, K.V. Adarsh, A. Nag, Colloidal synthesis and photophysics of M3Sb2I9 (M=Cs and Rb) nanocrystals: lead-free perovskites. Angew. Chem. Int. Ed. **56**(45), 14187–14191 (2017). https://doi.org/10.1002/anie.201709040
37. W.J. Mir, A. Warankar, A. Acharya, S. Das, P. Mandal, A. Nag, Colloidal thallium halide nanocrystals with reasonable luminescence, carrier mobility and diffusion length. Chem. Sci. **8**(6), 4602–4611 (2017)
38. A. Swarnkar, V.K. Ravi, A. Nag, Beyond colloidal cesium lead halide perovskite nanocrystals: analogous metal halides and doping. ACS Energy Lett. **2**(5), 1089–1098 (2017). https://doi.org/10.1021/acsenergylett.7b00191
39. A. Swarnkar, W.J. Mir, A. Nag, Can B-site doping or alloying improve thermal—and phase-stability of all-inorganic $CsPbX_3$ ($X = Cl, Br, I$) Perovskites? ACS Energy Lett. **3**(2), 286–289 (2018). https://doi.org/10.1021/acsenergylett.7b01197
40. S. Zou, Y. Liu, J. Li, C. Liu, R. Feng, F. Jiang, Y. Li, J. Song, H. Zeng, M. Hong, X. Chen, Stabilizing cesium lead halide perovskite lattice through Mn(II) substitution for air-stable light-emitting diodes. J. Am. Chem. Soc. **139**(33), 11443–11450 (2017). https://doi.org/10.1021/jacs.7b04000
41. W. Liu, Q. Lin, H. Li, K. Wu, I. Robel, J.M. Pietryga, V.I. Klimov, Mn2+-doped lead halide perovskite nanocrystals with dual-color emission controlled by halide content. J. Am. Chem. Soc. **138**(45), 14954–14961 (2016). https://doi.org/10.1021/jacs.6b08085
42. H. Liu, Z. Wu, J. Shao, D. Yao, H. Gao, Y. Liu, W. Yu, H. Zhang, B. Yang, CsPbxMn1−xCl3 perovskite quantum dots with high mn substitution ratio. ACS Nano **11**(2), 2239–2247 (2017). https://doi.org/10.1021/acsnano.6b08747
43. W.J. Mir, M. Jagadeeswararao, S. Das, A. Nag, Colloidal Mn-doped cesium lead halide perovskite nanoplatelets. ACS Energy Lett. **2**(3), 537–543 (2017). https://doi.org/10.1021/acsenergylett.6b00741
44. Q.A. Akkerman, D. Meggiolaro, Z. Dang, F. De Angelis, L. Manna, Fluorescent alloy CsPbxMn1−xI3 perovskite nanocrystals with high structural and optical stability. ACS Energy Lett. **2**(9), 2183–2186 (2017). https://doi.org/10.1021/acsenergylett.7b00707

45. T.C. Jellicoe, J.M. Richter, H.F.J. Glass, M. Tabachnyk, R. Brady, S.E. Dutton, A. Rao, R.H. Friend, D. Crdgington, N.C. Greenham, M.L. Böhm, Synthesis and optical properties of lead-free cesium tin halide perovskite nanocrystals. J. Am. Chem. Soc. **138**(9), 2941–2944 (2016). https://doi.org/10.1021/jacs.5b13470
46. M. Li, X. Zhang, K. Matras-Postolek, H.-S. Chen, P. Yang, An anion-driven Sn2+ exchange reaction in CsPbBr 3 nanocrystals towards tunable and high photoluminescence. J. Mater. Chem. C **6**(20), 5506–5513 (2018). https://doi.org/10.1039/C8TC00990B
47. X. Zhang, W. Cao, W. Wang, B. Xu, S. Liu, H. Dai, S. Chen, K. Wang, X.W. Sun, Efficient light-emitting diodes based on green perovskite nanocrystals with mixed-metal cations. Nano Energy **30**, 511–516 (2016). https://doi.org/10.1016/j.nanoen.2016.10.039
48. W. van der Stam, J.J. Geuchies, T. Altantzis, K.H.W. van den Bos, J.D. Meeldijk, S. Van Aert, S. Bals, D. Vanmaekelbergh, D.C. de Mello, Highly emissive divalent-ion-doped colloidal CsPb1–xMxBr$_3$ perovskite nanocrystals through cation exchange. J. Am. Chem. Soc. **139**(11), 4087–4097 (2017). https://doi.org/10.1021/jacs.6b13079
49. J. Deng, H. Wang, J. Xun, J. Wang, X. Yang, W. Shen, M. Li, R. He, Room-temperature synthesis of excellent-performance CsPb1-xSnxBr$_3$ perovskite quantum dots and application in light emitting diodes. Mater. Des. **185**, 108246 (2020). https://doi.org/10.1016/j.matdes.2019.108246
50. B. Saparov, J.-P. Sun, W. Meng, Z. Xiao, H.-S. Duan, O. Gunawan, D. Shin, I.G. Hill, Y. Yan, D.B. Mitzi, Thin-film deposition and characterization of a Sn-deficient perovskite derivative Cs2SnI$_6$. Chem. Mater. **28**(7), 2315–2322 (2016). https://doi.org/10.1021/acs.chemmater.6b00433
51. H.-C. Wang, W. Wang, A.-C. Tang, H.-Y. Tsai, Z. Bao, T. Ihara, N. Yarita, H. Tahara, Y. Kanemitsu, S. Chen, R.-S. Liu, High-performance CsPb1−xSnxBr$_3$ perovskite quantum dots for light-emitting diodes. Angew. Chem. Int. Ed. **56**(44), 13650–13654 (2017). https://doi.org/10.1002/anie.201706860
52. Z.-J. Yong, S.-Q. Guo, J.-P. Ma, J.-Y. Zhang, Z.-Y. Li, Y.-M. Chen, B.-B. Zhang, Y. Zhou, J. Shu, J.-L. Gu, L.-R. Zheng, O.M. Bakr, H.-T. Sun, Doping-enhanced short-range order of perovskite nanocrystals for near-unity violet luminescence quantum yield. J. Am. Chem. Soc. **140**(31), 9942–9951 (2018). https://doi.org/10.1021/jacs.8b04763
53. M. Liu, G. Zhong, Y. Yin, J. Miao, K. Li, C. Wang, X. Xu, C. Shen, H. Meng, Aluminum-doped cesium lead bromide perovskite nanocrystals with stable blue photoluminescence used for display backlight. Adv. Sci. **4**(11), 1700335 (2017). https://doi.org/10.1002/advs.201700335
54. C. Bi, S. Wang, Q. Li, S.V. Kershaw, J. Tian, A.L. Rogach, Thermally stable copper (II)-doped cesium lead halide perovskite quantum dots with strong blue emission. J. Phys. Chem. Lett. **10**(5), 943–952 (2019). https://doi.org/10.1021/acs.jpclett.9b00290
55. X. Shen, Y. Zhang, S.V. Kershaw, T. Li, C. Wang, X. Zhang, W. Wang, D. Li, Y. Wang, M. Lu, L. Zhang, C. Sun, D. Zhao, G. Qin, X. Bai, W.W. Yu, A.L. Rogach, Zn-alloyed CsPbI$_3$ nanocrystals for highly efficient perovskite light-emitting devices. Nano Lett. **19**(3), 1552–1559 (2019). https://doi.org/10.1021/acs.nanolett.8b04339
56. M. Lu, X. Zhang, Y. Zhang, J. Guo, X. Shen, W.W. Yu, A.L. Rogach, Simultaneous strontium doping and chlorine surface passivation improve luminescence intensity and stability of CsPbI$_3$ nanocrystals enabling efficient light-emitting devices. Adv. Mater. **30**(50), 1804691 (2018). https://doi.org/10.1002/adma.201804691
57. G. Pan, X. Bai, D. Yang, X. Chen, P. Jing, S. Qu, L. Zhang, D. Zhou, J. Zhu, W. Xu, B. Dong, H. Song, Doping lanthanide into perovskite nanocrystals: highly improved and expanded optical properties. Nano Lett. **17**(12), 8005–8011 (2017). https://doi.org/10.1021/acs.nanolett.7b04575
58. G. Lefevre, A. Herfurth, H. Kohlmann, A. Sayede, T. Wylezich, S. Welinski, P. Duarte Vaz, S.F. Parker, J.F. Blach, P. Goldner, N. Kunkel, Electron-phonon coupling in luminescent europium-doped hydride perovskites studied by luminescence spectroscopy, inelastic neutron scattering, and first-principles calculations. J. Phys. Chem. C **122**(19), 10501–10509 (2018). https://doi.org/10.1021/acs.jpcc.8b01011

59. A.K. Jena, A. Kulkarni, Y. Sanehira, M. Ikegami, T. Miyasaka, Stabilization of α-$CsPbI_3$ in ambient room temperature conditions by incorporating Eu into $CsPbI_3$. Chem. Mater. **30**(19), 6668–6674 (2018). https://doi.org/10.1021/acs.chemmater.8b01808
60. W. Xiang, Z. Wang, D.J. Kubicki, W. Tress, J. Luo, D. Prochowicz, S. Akin, L. Emsley, J. Zhou, G. Dietler, M. Grätzel, A. Hagfeldt, Europium-doped CsPbI2Br for stable and highly efficient inorganic perovskite solar cells. Joule **3**(1), 205–214 (2019). https://doi.org/10.1016/j.joule.2018.10.008
61. X. Zhang, Y. Zhang, X. Zhang, W. Yin, Y. Wang, H. Wang, M. Lu, Z. Li, Z. Gu, W.W. Yu, Yb3+ and Yb3+/Er3+ doping for near-infrared emission and improved stability of $CsPbCl_3$ nanocrystals. J. Mater. Chem. C **6**(37), 10101–10105 (2018). https://doi.org/10.1039/C8TC03957G
62. J.-S. Yao, J. Ge, B.-N. Han, K.-H. Wang, H.-B. Yao, H.-L. Yu, J.-H. Li, B.-S. Zhu, J.-Z. Song, C. Chen, Q. Zhang, H.-B. Zeng, Y. Luo, S.-H. Yu, Ce3+-doping to modulate photoluminescence kinetics for efficient $CsPbBr_3$ nanocrystals based light-emitting diodes. J. Am. Chem. Soc. **140**(10), 3626–3634 (2018). https://doi.org/10.1021/jacs.7b11955
63. C.M. Guvenc, Y. Yalcinkaya, S. Ozen, H. Sahin, M.M. Demir, Gd3+-doped α-$CsPbI_3$ nanocrystals with better phase stability and optical properties. J. Phys. Chem. C **123**(40), 24865–24872 (2019). https://doi.org/10.1021/acs.jpcc.9b05969
64. J. Kang, L.-W. Wang, High defect tolerance in lead halide perovskite $CsPbBr_3$. J. Phys. Chem. Lett. **8**(2), 489–493 (2017). https://doi.org/10.1021/acs.jpclett.6b02800
65. J. Pan, L.N. Quan, Y. Zhao, W. Peng, B. Murali, S.P. Sarmah, M. Yuan, L. Sinatra, N.M. Alyami, J. Liu, E. Yassitepe, Z. Yang, O. Voznyy, R. Comin, M.N. Hedhili, O.F. Mohammed, Z.H. Lu, D.H. Kim, E.H. Sargent, O.M. Bakr, Highly efficient perovskite-quantum-dot light-emitting diodes by surface engineering. Adv. Mater. **28**(39), 8718–8725 (2016). https://doi.org/10.1002/adma.201600784
66. C. Wang, A.S.R. Chesman, J.J. Jasieniak, Stabilizing the cubic perovskite phase of CsPbI3 nanocrystals by using an alkyl phosphinic acid. Chem. Commun. **53**(1), 232–235 (2017). https://doi.org/10.1039/C6CC08282C
67. J. De Roo, M. Ibáñez, P. Geiregat, G. Nedelcu, W. Walravens, J. Maes, J.C. Martins, I. Van Driessche, M.V. Kovalenko, Z. Hens, Highly dynamic ligand binding and light absorption coefficient of cesium lead bromide perovskite nanocrystals. ACS Nano **10**(2), 2071–2081 (2016). https://doi.org/10.1021/acsnano.5b06295
68. M. Green, The nature of quantum dot capping ligands. J. Mater. Chem. **20**(28), 5797–5809 (2010). https://doi.org/10.1039/C0JM00007H
69. F. Liu, Y. Zhang, C. Ding, S. Kobayashi, T. Izuishi, N. Nakazawa, T. Toyoda, T. Ohta, S. Hayase, T. Minemoto, K. Yoshino, S. Dai, Q. Shen, Highly luminescent phase-stable cspbi3 perovskite quantum dots achieving near 100% absolute photoluminescence quantum yield. ACS Nano **11**(10), 10373–10383 (2017). https://doi.org/10.1021/acsnano.7b05442
70. H. Wang, N. Sui, X. Bai, Y. Zhang, Q. Rice, F.J. Seo, Q. Zhang, V.L. Colvin, W.W. Yu, Emission recovery and stability enhancement of inorganic perovskite quantum dots. J. Phys. Chem. Lett. **9**(15), 4166–4173 (2018). https://doi.org/10.1021/acs.jpclett.8b01752
71. B.A. Koscher, J.K. Swabeck, N.D. Bronstein, A.P. Alivisatos, Essentially trap-free $CsPbBr_3$ colloidal nanocrystals by postsynthetic thiocyanate surface treatment. J. Am. Chem. Soc. **139**(19), 6566–6569 (2017). https://doi.org/10.1021/jacs.7b02817
72. L. Ruan, W. Shen, A. Wang, Q. Zhou, H. Zhang, Z. Deng, Stable and conductive lead halide perovskites facilitated by X-type ligands. Nanoscale **9**(21), 7252–7259 (2017). https://doi.org/10.1039/C7NR02125A
73. F. Krieg, S.T. Ochsenbein, S. Yakunin, S. ten Brinck, P. Aellen, A. Süess, B. Clerc, D. Guggisberg, O. Nazarenko, Y. Shynkarenko, S. Kumar, C.-J. Shih, I. Infante, M.V. Kovalenko, Nanocrystals 2.0: zwitterionic capping ligands for improved durability and stability. ACS Energy Lett. **3**(3), 641–646 (2018). https://doi.org/10.1021/acsenergylett.8b00035

74. J. Pan, Y. Shang, J. Yin, M. De Bastiani, W. Peng, I. Dursun, L. Sinatra, A.M. El-Zohry, M.N. Hedhili, A.-H. Emwas, O.F. Mohammed, Z. Ning, O.M. Bakr, Bidentate ligand-passivated $CsPbI_3$ perovskite nanocrystals for stable near-unity photoluminescence quantum yield and efficient red light-emitting diodes. J. Am. Chem. Soc. **140**(2), 562–565 (2018). https://doi.org/10.1021/jacs.7b10647
75. F. Palazon, Q.A. Akkerman, M. Prato, L. Manna, X-ray lithography on perovskite nanocrystals films: from patterning with anion-exchange reactions to enhanced stability in air and water. ACS Nano **10**(1), 1224–1230 (2016). https://doi.org/10.1021/acsnano.5b06536
76. C. Sun, Y. Zhang, C. Ruan, C. Yin, X. Wang, Y. Wang, W.W. Yu, Efficient and stable white leds with silica-coated inorganic perovskite quantum dots. Adv. Mater. **28**(45), 10088–10094 (2016). https://doi.org/10.1002/adma.201603081
77. H. Hu, L. Wu, Y. Tan, Q. Zhong, M. Chen, Y. Qiu, D. Yang, B. Sun, Q. Zhang, Y. Yin, Interfacial synthesis of highly stable $CsPbX_3$/oxide janus nanoparticles. J. Am. Chem. Soc. **140**(1), 406–412 (2018). https://doi.org/10.1021/jacs.7b11003
78. D.H. Park, J.S. Han, W. Kim, H.S. Jang, Facile synthesis of thermally stable $CsPbBr_3$ perovskite quantum dot-inorganic $SiO_2$ composites and their application to white light-emitting diodes with wide color gamut. Dyes Pigm. **149**, 246–252 (2018). https://doi.org/10.1016/j.dyepig.2017.10.003
79. X. Li, Y. Wang, H. Sun, H. Zeng, Amino-mediated anchoring perovskite quantum dots for stable and low-threshold random lasing. Adv. Mater. **29**(36), 1701185 (2017). https://doi.org/10.1002/adma.201701185
80. A. Loiudice, S. Saris, E. Oveisi, D.T.L. Alexander, R. Buonsanti, $CsPbBr_3$ QD/AlOx inorganic nanocomposites with exceptional stability in water, light, and heat. Angew. Chem. Int. Ed. **56**(36), 10696–10701 (2017). https://doi.org/10.1002/anie.201703703
81. Y. Wei, H. Xiao, Z. Xie, S. Liang, S. Liang, X. Cai, S. Huang, A.A. Al Kheraif, H.S. Jang, Z. Cheng, J. Lin, Highly luminescent lead halide perovskite quantum dots in hierarchical CaF2 Matrices with enhanced stability as phosphors for white light-emitting diodes. Adv. Opt. Mater. **6**(11), 1701343 (2018). https://doi.org/10.1002/adom.201701343
82. Z.-J. Li, E. Hofman, J. Li, A.H. Davis, C.-H. Tung, L.-Z. Wu, W. Zheng, Photoelectrochemically active and environmentally stable $CsPbBr_3/TiO_2$ core/shell nanocrystals. Adv. Func. Mater. **28**(1), 1704288 (2018). https://doi.org/10.1002/adfm.201704288
83. S. Lou, T. Xuan, C. Yu, M. Cao, C. Xia, J. Wang, H. Li, Nanocomposites of $CsPbBr_3$ perovskite nanocrystals in an ammonium bromide framework with enhanced stability. J. Mater. Chem. C **5**(30), 7431–7435 (2017). https://doi.org/10.1039/C7TC01174A
84. C.C. Lin, D.-H. Jiang, C.-C. Kuo, C.-J. Cho, Y.-H. Tsai, T. Satoh, C. Su, Water-resistant efficient stretchable perovskite-embedded fiber membranes for light-emitting diodes. ACS Appl. Mater. Interf. **10**(3), 2210–2215 (2018). https://doi.org/10.1021/acsami.7b15989
85. K. Ma, X.-Y. Du, Y.-W. Zhang, S. Chen, In situ fabrication of halide perovskite nanocrystals embedded in polymer composites via microfluidic spinning microreactors. J. Mater. Chem. C **5**(36), 9398–9404 (2017). https://doi.org/10.1039/C7TC02847D
86. M. Meyns, M. Perálvarez, A. Heuer-Jungemann, W. Hertog, M. Ibáñez, R. Nafria, A. Genç, J. Arbiol, M.V. Kovalenko, J. Carreras, A. Cabot, A.G. Kanaras, Polymer-enhanced stability of inorganic perovskite nanocrystals and their application in color conversion LEDs. ACS Appl. Mater. Interf. **8**(30), 19579–19586 (2016). https://doi.org/10.1021/acsami.6b02529
87. S.N. Raja, Y. Bekenstein, M.A. Koc, S. Fischer, D. Zhang, L. Lin, R.O. Ritchie, P. Yang, A.P. Alivisatos, Encapsulation of perovskite nanocrystals into macroscale polymer matrices: enhanced stability and polarization. ACS Appl. Mater. Interf. **8**(51), 35523–35533 (2016). https://doi.org/10.1021/acsami.6b09443
88. Y. Li, Y. Lv, Z. Guo, L. Dong, J. Zheng, C. Chai, N. Chen, Y. Lu, C. Chen, One-step preparation of long-term stable and flexible $CsPbBr_3$ perovskite quantum dots/ethylene vinyl acetate copolymer composite films for white light-emitting diodes. ACS Appl. Mater. Interf. **10**(18), 15888–15894 (2018). https://doi.org/10.1021/acsami.8b02857
89. Q.A. Akkerman, G. Rainò, M.V. Kovalenko, L. Manna, Genesis, challenges and opportunities for colloidal lead halide perovskite nanocrystals. Nat. Mater. **17**(5), 394–405 (2018). https://doi.org/10.1038/s41563-018-0018-4

# Index




Y. Zhou and Y. Wang (eds.), *Perovskite Quantum Dots*, Springer Series in Materials Science 303, https://doi.org/10.1007/978-981-15-6637-0

**T**



**X**

The manufacturer's authorised representative in the EU is Springer
Nature Customer Service Centre GmbH, Europaplatz 3, 69115 Heidelberg,
Germany. If you have any concerns regarding our products, please
contact ProductSafety@springernature.com

Printed and bound by CPI Group (UK) Ltd, Croydon, CR0 4YY
30/09/2026
02225235-0009